Preface to the Third Edition

In the second edition of *Probability and Statistics*, which appeared in 2000, the guiding principle was to make changes in the first edition only where necessary to bring the work in line with the emphasis on topics in contemporary texts. In addition to refinements throughout the text, a chapter on nonparametric statistics was added to extend the applicability of the text without raising its level. This theme is continued in the third edition in which the book has been reformatted and a chapter on Bayesian methods has been added. In recent years, the Bayesian paradigm has come to enjoy increased popularity and impact in such areas as economics, environmental science, medicine, and finance. Since Bayesian statistical analysis is highly computational, it is gaining even wider acceptance with advances in computer technology. We feel that an introduction to the basic principles of Bayesian data analysis is therefore in order and is consistent with Professor Murray R. Spiegel's main purpose in writing the original text—to present a modern introduction to probability and statistics using a background of calculus.

J. SCHILLER
R. A. SRINIVASAN

Preface to the Second Edition

The first edition of Schaum's *Probability and Statistics* by Murray R. Spiegel appeared in 1975, and it has gone through 21 printings since then. Its close cousin, Schaum's *Statistics* by the same author, was described as the clearest introduction to statistics in print by Gian-Carlo Rota in his book *Indiscrete Thoughts*. So it was with a degree of reverence and some caution that we undertook this revision. Our guiding principle was to make changes only where necessary to bring the text in line with the emphasis of topics in contemporary texts. The extensive treatment of sets, standard introductory material in texts of the 1960s and early 1970s, is considerably reduced. The definition of a continuous random variable is now the standard one, and more emphasis is placed on the cumulative distribution function since it is a more fundamental concept than the probability density function. Also, more emphasis is placed on the *P* values of hypotheses tests, since technology has made it possible to easily determine these values, which provide more specific information than whether or not tests meet a prespecified level of significance. Technology has also made it possible to eliminate logarithmic tables. A chapter on nonparametric statistics has been added to extend the applicability of the text without raising its level. Some problem sets have been trimmed, but mostly in cases that called for proofs of theorems for which no hints or help of any kind was given. Overall we believe that the main purpose of the first edition—to present a modern introduction to probability and statistics using a background of calculus—and the features that made the first edition such a great success have been preserved, and we hope that this edition can serve an even broader range of students.

J. SCHILLER
R. A. SRINIVASAN

Preface to the First Edition

The important and fascinating subject of probability began in the seventeenth century through efforts of such mathematicians as Fermat and Pascal to answer questions concerning games of chance. It was not until the twentieth century that a rigorous mathematical theory based on axioms, definitions, and theorems was developed. As time progressed, probability theory found its way into many applications, not only in engineering, science, and mathematics but in fields ranging from actuarial science, agriculture, and business to medicine and psychology. In many instances the applications themselves contributed to the further development of the theory.

The subject of statistics originated much earlier than probability and dealt mainly with the collection, organization, and presentation of data in tables and charts. With the advent of probability it was realized that statistics could be used in drawing valid conclusions and making reasonable decisions on the basis of analysis of data, such as in sampling theory and prediction or forecasting.

The purpose of this book is to present a modern introduction to probability and statistics using a background of calculus. For convenience the book is divided into two parts. The first deals with probability (and by itself can be used to provide an introduction to the subject), while the second deals with statistics.

The book is designed to be used either as a textbook for a formal course in probability and statistics or as a comprehensive supplement to all current standard texts. It should also be of considerable value as a book of reference for research workers or to those interested in the field for self-study. The book can be used for a one-year course, or by a judicious choice of topics, a one-semester course.

I am grateful to the Literary Executor of the late Sir Ronald A. Fisher, F.R.S., to Dr. Frank Yates, F.R.S., and to Longman Group Ltd., London, for permission to use Table III from their book *Statistical Tables for Biological, Agricultural and Medical Research* (6th edition, 1974). I also wish to take this opportunity to thank David Beckwith for his outstanding editing and Nicola Monti for his able artwork.

M. R. SPIEGEL

Contents

PART I

Probability

Basic Probability

Random Experiments

We are all familiar with the importance of experiments in science and engineering. Experimentation is useful to us because we can assume that if we perform certain experiments under very nearly identical conditions, we will arrive at results that are essentially the same. In these circumstances, we are able to control the value of the variables that affect the outcome of the experiment.

However, in some experiments, we are not able to ascertain or control the value of certain variables so that the results will vary from one performance of the experiment to the next even though most of the conditions are the same. These experiments are described as *random*. The following are some examples.

EXAMPLE 1.1 If we toss a coin, the result of the experiment is that it will either come up "tails," symbolized by T (or 0), or "heads," symbolized by H (or 1), i.e., one of the elements of the set $\{H, T\}$ (or $\{0, 1\}$).

EXAMPLE 1.2 If we toss a die, the result of the experiment is that it will come up with one of the numbers in the set $\{1, 2, 3, 4, 5, 6\}$.

EXAMPLE 1.3 If we toss a coin twice, there are four results possible, as indicated by $\{HH, HT, TH, TT\}$, i.e., both heads, heads on first and tails on second, etc.

EXAMPLE 1.4 If we are making bolts with a machine, the result of the experiment is that some may be defective. Thus when a bolt is made, it will be a member of the set {defective, nondefective}.

EXAMPLE 1.5 If an experiment consists of measuring "lifetimes" of electric light bulbs produced by a company, then the result of the experiment is a time t in hours that lies in some interval—say, $0 \leq t \leq 4000$—where we assume that no bulb lasts more than 4000 hours.

Sample Spaces

A set S that consists of all possible outcomes of a random experiment is called a *sample space*, and each outcome is called a *sample point*. Often there will be more than one sample space that can describe outcomes of an experiment, but there is usually only one that will provide the most information.

EXAMPLE 1.6 If we toss a die, one sample space, or set of all possible outcomes, is given by $\{1, 2, 3, 4, 5, 6\}$ while another is {odd, even}. It is clear, however, that the latter would not be adequate to determine, for example, whether an outcome is divisible by 3.

It is often useful to portray a sample space graphically. In such cases it is desirable to use numbers in place of letters whenever possible.

EXAMPLE 1.7 If we toss a coin twice and use 0 to represent tails and 1 to represent heads, the sample space (see Example 1.3) can be portrayed by points as in Fig. 1-1 where, for example, (0, 1) represents tails on first toss and heads on second toss, i.e., TH.

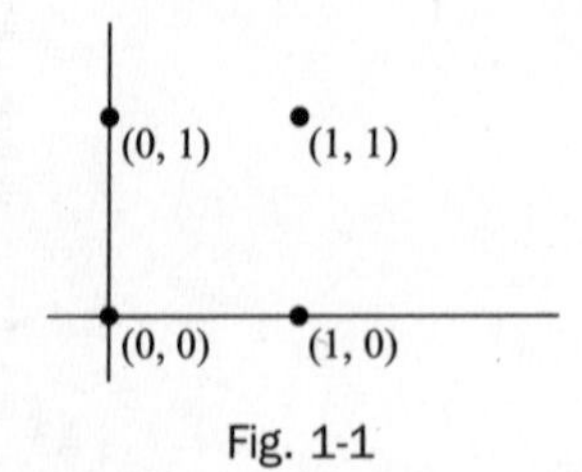

Fig. 1-1

If a sample space has a finite number of points, as in Example 1.7, it is called a *finite sample space*. If it has as many points as there are natural numbers 1, 2, 3, . . . , it is called a *countably infinite sample space*. If it has as many points as there are in some interval on the x axis, such as $0 \leq x \leq 1$, it is called a *noncountably infinite sample space*. A sample space that is finite or countably infinite is often called a *discrete sample space*, while one that is noncountably infinite is called a *nondiscrete sample space*.

Events

An *event* is a subset A of the sample space S, i.e., it is a set of possible outcomes. If the outcome of an experiment is an element of A, we say that the event A *has occurred.* An event consisting of a single point of S is often called a *simple* or *elementary event*.

EXAMPLE 1.8 If we toss a coin twice, the event that only one head comes up is the subset of the sample space that consists of points (0, 1) and (1, 0), as indicated in Fig. 1-2.

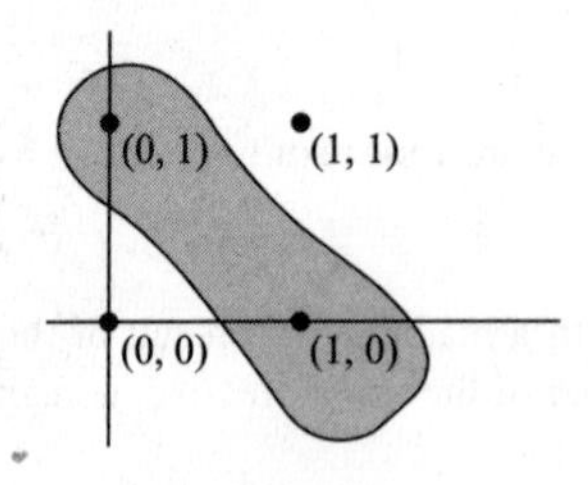

Fig. 1-2

As particular events, we have S itself, which is the *sure* or *certain event* since an element of S must occur, and the empty set $\varnothing$, which is called the *impossible event* because an element of $\varnothing$ cannot occur.

By using set operations on events in S, we can obtain other events in S. For example, if A and B are events, then

1. $A \cup B$ is the event "either A or B or both." $A \cup B$ is called the *union* of A and B.
2. $A \cap B$ is the event "both A and B." $A \cap B$ is called the *intersection* of A and B.
3. A' is the event "not A." A' is called the *complement* of A.
4. $A - B = A \cap B'$ is the event "A but not B." In particular, $A' = S - A$.

If the sets corresponding to events A and B are disjoint, i.e., $A \cap B = \varnothing$, we often say that the events are *mutually exclusive*. This means that they cannot both occur. We say that a collection of events $A_1, A_2, \ldots, A_n$ is mutually exclusive if every pair in the collection is mutually exclusive.

EXAMPLE 1.9 Referring to the experiment of tossing a coin twice, let A be the event "at least one head occurs" and B the event "the second toss results in a tail." Then $A = \{HT, TH, HH\}$, $B = \{HT, TT\}$, and so we have

$$A \cup B = \{HT, TH, HH, TT\} = S \qquad A \cap B = \{HT\}$$

$$A' = \{TT\} \qquad A - B = \{TH, HH\}$$

The Concept of Probability

In any random experiment there is always uncertainty as to whether a particular event will or will not occur. As a measure of the *chance*, or *probability*, with which we can expect the event to occur, it is convenient to assign a number between 0 and 1. If we are sure or certain that the event will occur, we say that its probability is 100% or 1, but if we are sure that the event will not occur, we say that its probability is zero. If, for example, the probability is $\frac{1}{4}$, we would say that there is a 25% chance it will occur and a 75% chance that it will not occur. Equivalently, we can say that the *odds* against its occurrence are 75% to 25%, or 3 to 1.

There are two important procedures by means of which we can estimate the probability of an event.

1. **CLASSICAL APPROACH.** If an event can occur in h different ways out of a total number of n possible ways, all of which are equally likely, then the probability of the event is h/n.

EXAMPLE 1.10 Suppose we want to know the probability that a head will turn up in a single toss of a coin. Since there are two equally likely ways in which the coin can come up—namely, heads and tails (assuming it does not roll away or stand on its edge)—and of these two ways a head can arise in only one way, we reason that the required probability is $1/2$. In arriving at this, we assume that the coin is *fair*, i.e., not *loaded* in any way.

2. **FREQUENCY APPROACH.** If after n repetitions of an experiment, where n is very large, an event is observed to occur in h of these, then the probability of the event is h/n. This is also called the *empirical probability* of the event.

EXAMPLE 1.11 If we toss a coin 1000 times and find that it comes up heads 532 times, we estimate the probability of a head coming up to be $532/1000 = 0.532$.

Both the classical and frequency approaches have serious drawbacks, the first because the words "equally likely" are vague and the second because the "large number" involved is vague. Because of these difficulties, mathematicians have been led to an *axiomatic approach* to probability.

The Axioms of Probability

Suppose we have a sample space S. If S is discrete, all subsets correspond to events and conversely, but if S is nondiscrete, only special subsets (called *measurable*) correspond to events. To each event A in the class C of events, we associate a real number $P(A)$. Then P is called a *probability function*, and $P(A)$ the *probability* of the event A, if the following axioms are satisfied.

Axiom 1 For every event A in the class C,

$$P(A) \geq 0 \tag{1}$$

Axiom 2 For the sure or certain event S in the class C,

$$P(S) = 1 \tag{2}$$

Axiom 3 For any number of mutually exclusive events $A_1, A_2, \ldots,$ in the class C,

$$P(A_1 \cup A_2 \cup \cdots) = P(A_1) + P(A_2) + \cdots \tag{3}$$

In particular, for two mutually exclusive events A_1, A_2,

$$P(A_1 \cup A_2) = P(A_1) + P(A_2) \tag{4}$$

Some Important Theorems on Probability

From the above axioms we can now prove various theorems on probability that are important in further work.

Theorem 1-1 If A_1 (A_2, then $P(A_1) \leq P(A_2)$ and $P(A_2 - A_1) = P(A_2) - P(A_1)$.

Theorem 1-2 For every event A,

$$0 \leq P(A) \leq 1, \tag{5}$$

i.e., a probability is between 0 and 1.

Theorem 1-3

$$P(\varnothing) = 0 \tag{6}$$

i.e., the impossible event has probability zero.

Theorem 1-4 If A' is the complement of A, then

$$P(A') = 1 - P(A) \tag{7}$$

Theorem 1-5 If $A = A_1 \cup A_2 \cup \cdots \cup A_n$, where $A_1, A_2, \ldots, A_n$ are mutually exclusive events, then

$$P(A) = P(A_1) + P(A_2) + \cdots + P(A_n) \tag{8}$$

In particular, if $A = S$, the sample space, then

$$P(A_1) + P(A_2) + \cdots + P(A_n) = 1 \tag{9}$$

Theorem 1-6 If A and B are any two events, then

$$P(A \cup B) = P(A) + P(B) - P(A \cap B) \tag{10}$$

More generally, if A_1, A_2, A_3 are any three events, then

$$\begin{aligned} P(A_1 \cup A_2 \cup A_3) = {} & P(A_1) + P(A_2) + P(A_3) \\ & - P(A_1 \cap A_2) - P(A_2 \cap A_3) - P(A_3 \cap A_1) \\ & + P(A_1 \cap A_2 \cap A_3) \end{aligned} \tag{11}$$

Generalizations to n events can also be made.

Theorem 1-7 For any events A and B,

$$P(A) = P(A \cap B) + P(A \cap B') \tag{12}$$

Theorem 1-8 If an event A must result in the occurrence of one of the mutually exclusive events $A_1, A_2, \ldots, A_n$, then

$$P(A) = P(A \cap A_1) + P(A \cap A_2) + \cdots + P(A \cap A_n) \tag{13}$$

Assignment of Probabilities

If a sample space S consists of a finite number of outcomes $a_1, a_2, \ldots, a_n$, then by Theorem 1-5,

$$P(A_1) + P(A_2) + \cdots + P(A_n) = 1 \tag{14}$$

where $A_1, A_2, \ldots, A_n$ are elementary events given by $A_i = \{a_i\}$.

It follows that we can arbitrarily choose any nonnegative numbers for the probabilities of these simple events as long as (14) is satisfied. In particular, if we assume *equal probabilities* for all simple events, then

$$P(A_k) = \frac{1}{n}, \quad k = 1, 2, \ldots, n \tag{15}$$

and if A is any event made up of h such simple events, we have

$$P(A) = \frac{h}{n} \tag{16}$$

This is equivalent to the classical approach to probability given on page 5. We could of course use other procedures for assigning probabilities, such as the frequency approach of page 5.

Assigning probabilities provides a *mathematical model*, the success of which must be tested by experiment in much the same manner that theories in physics or other sciences must be tested by experiment.

EXAMPLE 1.12 A single die is tossed once. Find the probability of a 2 or 5 turning up.

The sample space is $S = \{1, 2, 3, 4, 5, 6\}$. If we assign equal probabilities to the sample points, i.e., if we assume that the die is fair, then

$$P(1) = P(2) = \cdots = P(6) = \frac{1}{6}$$

The event that either 2 or 5 turns up is indicated by $2 \cup 5$. Therefore,

$$P(2 \cup 5) = P(2) + P(5) = \frac{1}{6} + \frac{1}{6} = \frac{1}{3}$$

Conditional Probability

Let A and B be two events (Fig. 1-3) such that $P(A) > 0$. Denote by $P(B \mid A)$ the probability of B *given that* A has occurred. Since A is known to have occurred, it becomes the new sample space replacing the original S. From this we are led to the definition

$$P(B \mid A) \equiv \frac{P(A \cap B)}{P(A)} \tag{17}$$

or

$$P(A \cap B) \equiv P(A)\, P(B \mid A) \tag{18}$$

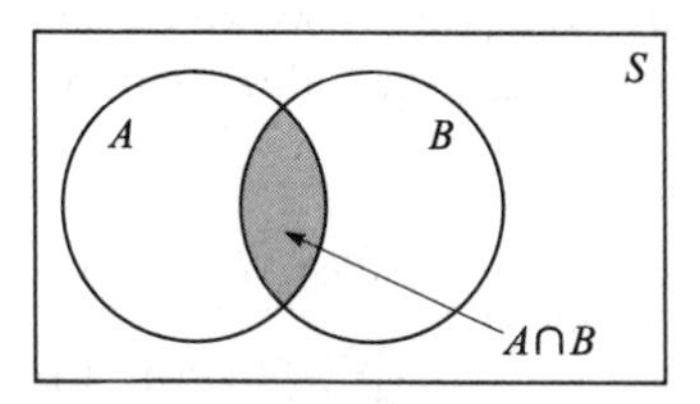

Fig. 1-3

In words, (18) says that the probability that both A and B occur is equal to the probability that A occurs times the probability that B occurs given that A has occurred. We call $P(B \mid A)$ the *conditional probability* of B given A, i.e., the probability that B will occur given that A has occurred. It is easy to show that conditional probability satisfies the axioms on page 5.

EXAMPLE 1.13 Find the probability that a single toss of a die will result in a number less than 4 if (a) no other information is given and (b) it is given that the toss resulted in an odd number.

(a) Let B denote the event {less than 4}. Since B is the union of the events 1, 2, or 3 turning up, we see by Theorem 1-5 that

$$P(B) = P(1) + P(2) + P(3) = \frac{1}{6} + \frac{1}{6} + \frac{1}{6} = \frac{1}{2}$$

assuming equal probabilities for the sample points.

(b) Letting A be the event {odd number}, we see that $P(A) = \frac{3}{6} = \frac{1}{2}$. Also $P(A \cap B) = \frac{2}{6} = \frac{1}{3}$. Then

$$P(B \mid A) = \frac{P(A \cap B)}{P(A)} = \frac{1/3}{1/2} = \frac{2}{3}$$

Hence, the added knowledge that the toss results in an odd number raises the probability from 1/2 to 2/3.

Theorems on Conditional Probability

Theorem 1-9 For any three events A_1, A_2, A_3, we have

$$P(A_1 \cap A_2 \cap A_3) = P(A_1)\, P(A_2 \mid A_1)\, P(A_3 \mid A_1 \cap A_2) \tag{19}$$

In words, the probability that A_1 and A_2 and A_3 all occur is equal to the probability that A_1 occurs times the probability that A_2 occurs given that A_1 has occurred times the probability that A_3 occurs given that both A_1 and A_2 have occurred. The result is easily generalized to n events.

Theorem 1-10 If an event A must result in one of the mutually exclusive events $A_1, A_2, \ldots, A_n$, then

$$P(A) = P(A_1)\, P(A \mid A_1) + P(A_2)\, P(A \mid A_2) + \cdots + P(A_n)\, P(A \mid A_n) \tag{20}$$

Independent Events

If $P(B \mid A) = P(B)$, i.e., the probability of B occurring is not affected by the occurrence or non-occurrence of A, then we say that A and B are *independent events*. This is equivalent to

$$P(A \cap B) = P(A)P(B) \tag{21}$$

as seen from (18). Conversely, if (21) holds, then A and B are independent.

We say that three events A_1, A_2, A_3 are *independent* if they are pairwise independent:

$$P(A_j \cap A_k) = P(A_j)P(A_k) \qquad j \neq k \quad \text{where} \quad j, k = 1, 2, 3 \tag{22}$$

and

$$P(A_1 \cap A_2 \cap A_3) = P(A_1)P(A_2)P(A_3) \tag{23}$$

Note that neither (22) nor (23) is by itself sufficient. Independence of more than three events is easily defined.

Bayes' Theorem or Rule

Suppose that $A_1, A_2, \ldots, A_n$ are mutually exclusive events whose union is the sample space S, i.e., one of the events must occur. Then if A is any event, we have the following important theorem:

Theorem 1-11 **(Bayes' Rule):**

$$P(A_k \mid A) = \frac{P(A_k)P(A \mid A_k)}{\sum_{j=1}^{n} P(A_j)P(A \mid A_j)} \tag{24}$$

This enables us to find the probabilities of the various events $A_1, A_2, \ldots, A_n$ that can *cause* A to occur. For this reason Bayes' theorem is often referred to as a *theorem on the probability of causes.*

Combinatorial Analysis

In many cases the number of sample points in a sample space is not very large, and so direct enumeration or counting of sample points needed to obtain probabilities is not difficult. However, problems arise where direct counting becomes a practical impossibility. In such cases use is made of *combinatorial analysis,* which could also be called a *sophisticated way of counting.*

Fundamental Principle of Counting: Tree Diagrams

If one thing can be accomplished in n_1 different ways and after this a second thing can be accomplished in n_2 different ways, . . . , and finally a kth thing can be accomplished in n_k different ways, then all k things can be accomplished in the specified order in $n_1 n_2 \ldots n_k$ different ways.

EXAMPLE 1.14 If a man has 2 shirts and 4 ties, then he has $2 \cdot 4 = 8$ ways of choosing a shirt and then a tie.

A diagram, called a *tree diagram* because of its appearance (Fig. 1-4), is often used in connection with the above principle.

EXAMPLE 1.15 Letting the shirts be represented by S_1, S_2 and the ties by T_1, T_2, T_3, T_4, the various ways of choosing a shirt and then a tie are indicated in the tree diagram of Fig. 1-4.

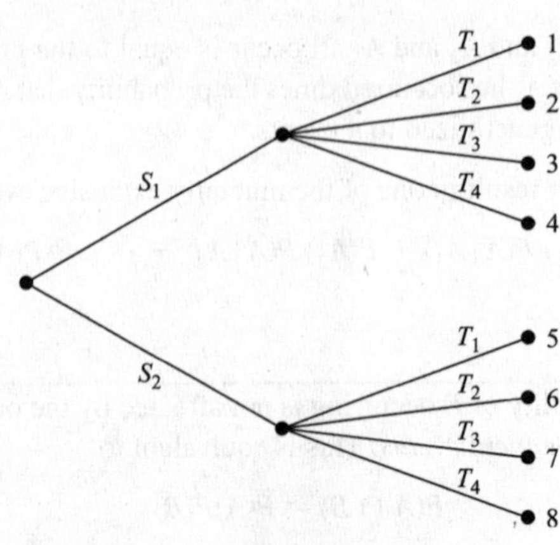

Fig. 1-4

Permutations

Suppose that we are given n distinct objects and wish to *arrange* r of these objects in a line. Since there are n ways of choosing the 1st object, and after this is done, $n - 1$ ways of choosing the 2nd object, . . . , and finally $n - r + 1$ ways of choosing the rth object, it follows by the fundamental principle of counting that the number of different *arrangements*, or *permutations* as they are often called, is given by

$$_nP_r = n(n-1)(n-2)\cdots(n-r+1) \tag{25}$$

where it is noted that the product has r factors. We call $_nP_r$ the *number of permutations of n objects taken r at a time.*

In the particular case where $r = n$, (25) becomes

$$_nP_n = n(n-1)(n-2)\cdots 1 = n! \tag{26}$$

which is called *n factorial*. We can write (25) in terms of factorials as

$$_nP_r = \frac{n!}{(n-r)!} \tag{27}$$

If $r = n$, we see that (27) and (26) agree only if we have $0! = 1$, and we shall actually take this as the definition of $0!$.

EXAMPLE 1.16 The number of different arrangements, or permutations, consisting of 3 letters each that can be formed from the 7 letters A, B, C, D, E, F, G is

$$_7P_3 = \frac{7!}{4!} = 7 \cdot 6 \cdot 5 = 210$$

Suppose that a set consists of n objects of which n_1 are of one type (i.e., indistinguishable from each other), n_2 are of a second type, . . . , n_k are of a kth type. Here, of course, $n = n_1 + n_2 + \cdots + n_k$. Then the number of different permutations of the objects is

$$_nP_{n_1, n_2, \ldots, n_k} = \frac{n!}{n_1! n_2! \cdots n_k!} \tag{28}$$

See Problem 1.25.

EXAMPLE 1.17 The number of different permutations of the 11 letters of the word *M I S S I S S I P P I*, which consists of 1 *M*, 4 *I*'s, 4 *S*'s, and 2 *P*'s, is

$$\frac{11!}{1!4!4!2!} = 34{,}650$$

Combinations

In a permutation we are interested in the order of arrangement of the objects. For example, *abc* is a different permutation from *bca*. In many problems, however, we are interested only in selecting or choosing objects without regard to order. Such selections are called *combinations*. For example, *abc* and *bca* are the same combination.

The total number of combinations of r objects selected from n (also called the *combinations of n things taken r at a time*) is denoted by $_nC_r$ or $\binom{n}{r}$. We have (see Problem 1.27)

$$\binom{n}{r} = {_nC_r} = \frac{n!}{r!(n-r)!} \tag{29}$$

It can also be written

$$\binom{n}{r} = \frac{n(n-1)\cdots(n-r+1)}{r!} = \frac{_nP_r}{r!} \tag{30}$$

It is easy to show that

$$\binom{n}{r} = \binom{n}{n-r} \quad \text{or} \quad {_nC_r} = {_nC_{n-r}} \tag{31}$$

EXAMPLE 1.18 The number of ways in which 3 cards can be chosen or selected from a total of 8 different cards is

$$_8C_3 = \binom{8}{3} = \frac{8 \cdot 7 \cdot 6}{3!} = 56$$

Binomial Coefficient

The numbers (29) are often called *binomial coefficients* because they arise in the *binomial expansion*

$$(x + y)^n = x^n + \binom{n}{1}x^{n-1}y + \binom{n}{2}x^{n-2}y^2 + \cdots + \binom{n}{n}y^n \tag{32}$$

They have many interesting properties.

EXAMPLE 1.19

$$\begin{aligned}(x + y)^4 &= x^4 + \binom{4}{1}x^3y + \binom{4}{2}x^2y^2 + \binom{4}{3}xy^3 + \binom{4}{4}y^4 \\ &= x^4 + 4x^3y + 6x^2y^2 + 4xy^3 + y^4\end{aligned}$$

Stirling's Approximation to *n*!

When n is large, a direct evaluation of $n!$ may be impractical. In such cases use can be made of the approximate formula

$$n! \sim \sqrt{2\pi n}\, n^n e^{-n} \tag{33}$$

where $e = 2.71828\ldots$, which is the base of natural logarithms. The symbol $\sim$ in (33) means that the ratio of the left side to the right side approaches 1 as $n \to \infty$.

Computing technology has largely eclipsed the value of Stirling's formula for numerical computations, but the approximation remains valuable for theoretical estimates (see Appendix A).

SOLVED PROBLEMS

Random experiments, sample spaces, and events

1.1. A card is drawn at random from an ordinary deck of 52 playing cards. Describe the sample space if consideration of suits (a) is not, (b) is, taken into account.

(a) If we do not take into account the suits, the sample space consists of ace, two, . . . , ten, jack, queen, king, and it can be indicated as $\{1, 2, \ldots, 13\}$.

(b) If we do take into account the suits, the sample space consists of ace of hearts, spades, diamonds, and clubs; . . . ; king of hearts, spades, diamonds, and clubs. Denoting hearts, spades, diamonds, and clubs, respectively, by 1, 2, 3, 4, for example, we can indicate a jack of spades by (11, 2). The sample space then consists of the 52 points shown in Fig. 1-5.

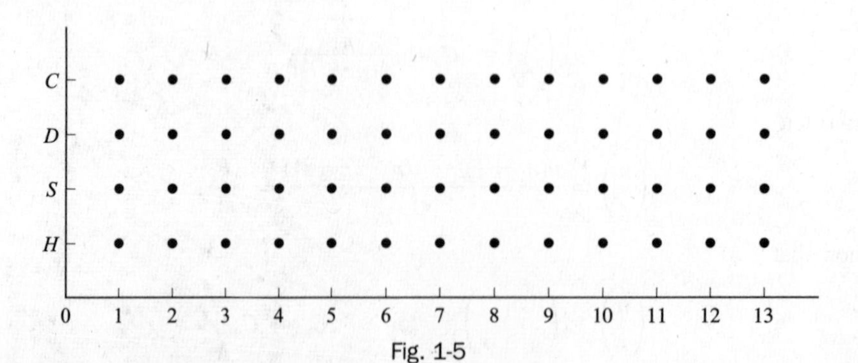

Fig. 1-5

1.2. Referring to the experiment of Problem 1.1, let A be the event {king is drawn} or simply {king} and B the event {club is drawn} or simply {club}. Describe the events (a) $A \cup B$, (b) $A \cap B$, (c) $A \cup B'$, (d) $A' \cup B'$, (e) $A - B$, (f) $A' - B'$, (g) $(A \cap B) \cup (A \cap B')$.

(a) $A \cup B$ = {either king or club (or both, i.e., king of clubs)}.

(b) $A \cap B$ = {both king and club} = {king of clubs}.

(c) Since B = {club}, B' = {not club} = {heart, diamond, spade}.

Then $A \cup B'$ = {king or heart or diamond or spade}.

(d) $A' \cup B'$ = {not king or not club} = {not king of clubs} = {any card but king of clubs}.

This can also be seen by noting that $A' \cup B' = (A \cap B)'$ and using (b).

(e) $A - B$ = {king but not club}.

This is the same as $A \cap B'$ = {king and not club}.

(f) $A' - B'$ = {not king and not "not club"} = {not king and club} = {any club except king}.

This can also be seen by noting that $A' - B' = A' \cap (B')' = A' \cap B$.

(g) $(A \cap B) \cup (A \cap B')$ = {(king and club) or (king and not club)} = {king}.

This can also be seen by noting that $(A \cap B) \cup (A \cap B') = A$.

1.3. Use Fig. 1-5 to describe the events (a) $A \cup B$, (b) $A' \cap B'$.

The required events are indicated in Fig. 1-6. In a similar manner, all the events of Problem 1.2 can also be indicated by such diagrams. It should be observed from Fig. 1-6 that $A' \cap B'$ is the complement of $A \cup B$.

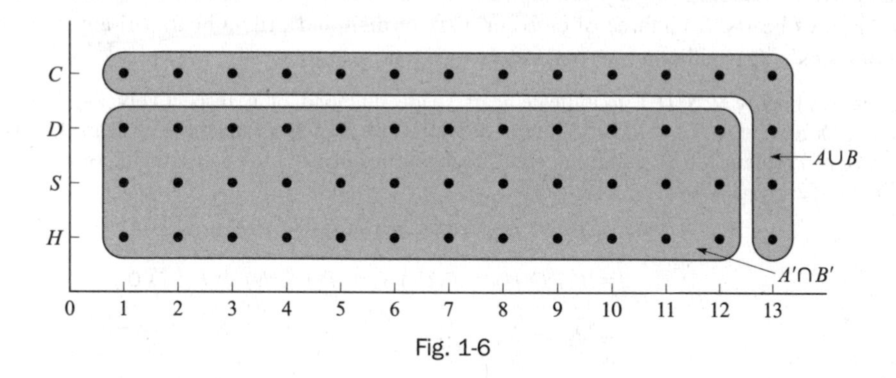

Fig. 1-6

Theorems on probability

1.4. Prove (a) Theorem 1-1, (b) Theorem 1-2, (c) Theorem 1-3, page 5.

(a) We have $A_2 = A_1 \cup (A_2 - A_1)$ where A_1 and $A_2 - A_1$ are mutually exclusive. Then by Axiom 3, page 5:

$$P(A_2) = P(A_1) + P(A_2 - A_1)$$

so that

$$P(A_2 - A_1) = P(A_2) - P(A_1)$$

Since $P(A_2 - A_1) \geq 0$ by Axiom 1, page 5, it also follows that $P(A_2) \geq P(A_1)$.

(b) We already know that $P(A) \geq 0$ by Axiom 1. To prove that $P(A) \leq 1$, we first note that A (S. Therefore, by Theorem 1-1 [part (a)] and Axiom 2,

$$P(A) \leq P(S) = 1$$

(c) We have $S = S \cup \varnothing$. Since $S \cap \varnothing = \varnothing$, it follows from Axiom 3 that

$$P(S) = P(S) + P(\varnothing) \qquad \text{or} \qquad P(\varnothing) = 0$$

1.5. Prove (a) Theorem 1-4, (b) Theorem 1-6.

(a) We have $A \cup A' = S$. Then since $A \cap A' = \varnothing$, we have

$$P(A \cup A') = P(S) \quad \text{or} \quad P(A) + P(A') = 1$$

i.e.,

$$P(A') = 1 - P(A)$$

(b) We have from the Venn diagram of Fig. 1-7,

(1) $$A \cup B = A \cup [B - (A \cap B)]$$

Then since the sets A and $B - (A \cap B)$ are mutually exclusive, we have, using Axiom 3 and Theorem 1-1,

$$P(A \cup B) = P(A) + P[B - (A \cap B)]$$
$$= P(A) + P(B) - P(A \cap B)$$

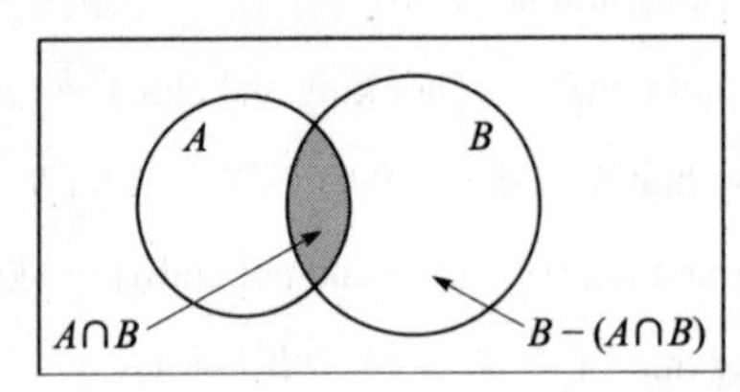

Fig. 1-7

Calculation of probabilities

1.6. A card is drawn at random from an ordinary deck of 52 playing cards. Find the probability that it is (a) an ace, (b) a jack of hearts, (c) a three of clubs or a six of diamonds, (d) a heart, (e) any suit except hearts, (f) a ten or a spade, (g) neither a four nor a club.

Let us use for brevity H, S, D, C to indicate heart, spade, diamond, club, respectively, and $1, 2, \ldots, 13$ for ace, two, . . . , king. Then $3 \cap H$ means three of hearts, while $3 \cup H$ means three or heart. Let us use the sample space of Problem 1.1(b), assigning equal probabilities of 1/52 to each sample point. For example, $P(6 \cap C) = 1/52$.

(a) $$P(1) = P(1 \cap H \text{ or } 1 \cap S \text{ or } 1 \cap D \text{ or } 1 \cap C)$$
$$= P(1 \cap H) + P(1 \cap S) + P(1 \cap D) + P(1 \cap C)$$
$$= \frac{1}{52} + \frac{1}{52} + \frac{1}{52} + \frac{1}{52} = \frac{1}{13}$$

This could also have been achieved from the sample space of Problem 1.1(a) where each sample point, in particular ace, has probability 1/13. It could also have been arrived at by simply reasoning that there are 13 numbers and so each has probability 1/13 of being drawn.

(b) $P(11 \cap H) = \frac{1}{52}$

(c) $P(3 \cap C \text{ or } 6 \cap D) = P(3 \cap C) + P(6 \cap D) = \frac{1}{52} + \frac{1}{52} = \frac{1}{26}$

(d) $P(H) = P(1 \cap H \text{ or } 2 \cap H \text{ or} \ldots 13 \cap H) = \frac{1}{52} + \frac{1}{52} + \cdots + \frac{1}{52} = \frac{13}{52} = \frac{1}{4}$

This could also have been arrived at by noting that there are four suits and each has equal probability1/2 of being drawn.

(e) $P(H') = 1 - P(H) = 1 - \frac{1}{4} = \frac{3}{4}$ using part (d) and Theorem 1-4, page 6.

(f) Since 10 and S are not mutually exclusive, we have, from Theorem 1-6,

$$P(10 \cup S) = P(10) + P(S) - P(10 \cap S) = \frac{1}{13} + \frac{1}{4} - \frac{1}{52} = \frac{4}{13}$$

(g) The probability of neither four nor club can be denoted by $P(4' \cap C')$. But $4' \cap C' = (4 \cup C)'$.

Therefore,

$$P(4' \cap C') = P[(4 \cup C)'] = 1 - P(4 \cup C)$$
$$= 1 - [P(4) + P(C) - P(4 \cap C)]$$
$$= 1 - \left[\frac{1}{13} + \frac{1}{4} - \frac{1}{52}\right] = \frac{9}{13}$$

We could also get this by noting that the diagram favorable to this event is the complement of the event shown circled in Fig. 1-8. Since this complement has $52 - 16 = 36$ sample points in it and each sample point is assigned probability $1/52$, the required probability is $36/52 = 9/13$.

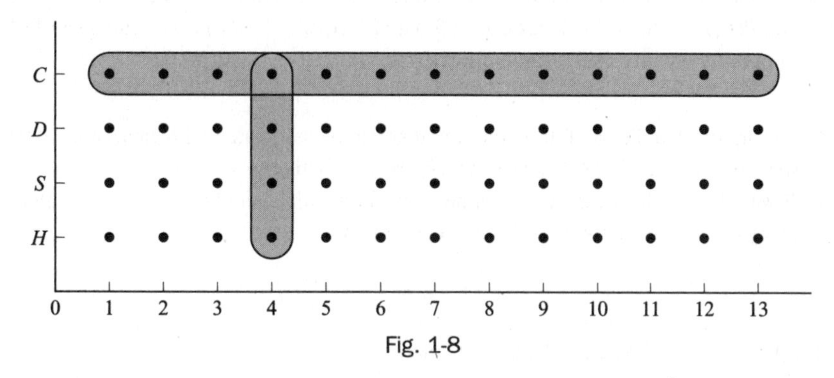

Fig. 1-8

1.7. A ball is drawn at random from a box containing 6 red balls, 4 white balls, and 5 blue balls. Determine the probability that it is (a) red, (b) white, (c) blue, (d) not red, (e) red or white.

(a) **Method 1**

Let R, W, and B denote the events of drawing a red ball, white ball, and blue ball, respectively. Then

$$P(R) = \frac{\text{ways of choosing a red ball}}{\text{total ways of choosing a ball}} = \frac{6}{6 + 4 + 5} = \frac{6}{15} = \frac{2}{5}$$

Method 2

Our sample space consists of $6 + 4 + 5 = 15$ sample points. Then if we assign equal probabilities $1/15$ to each sample point, we see that $P(R) = 6/15 = 2/5$, since there are 6 sample points corresponding to "red ball."

(b) $P(W) = \dfrac{4}{6 + 4 + 5} = \dfrac{4}{15}$

(c) $P(B) = \dfrac{5}{6 + 4 + 5} = \dfrac{5}{15} = \dfrac{1}{3}$

(d) $P(\text{not red}) = P(R') = 1 - P(R) = 1 - \dfrac{2}{5} = \dfrac{3}{5}$ by part (a).

(e) **Method 1**

$$P(\text{red or white}) = P(R \cup W) = \frac{\text{ways of choosing a red or white ball}}{\text{total ways of choosing a ball}}$$
$$= \frac{6 + 4}{6 + 4 + 5} = \frac{10}{15} = \frac{2}{3}$$

This can also be worked using the sample space as in part (a).

Method 2

$$P(R \cup W) = P(B') = 1 - P(B) = 1 - \frac{1}{3} = \frac{2}{3} \text{ by part (c).}$$

Method 3

Since events R and W are mutually exclusive, it follows from (4), page 5, that

$$P(R \cup W) = P(R) + P(W) = \frac{2}{5} + \frac{4}{15} = \frac{2}{3}$$

Conditional probability and independent events

1.8. A fair die is tossed twice. Find the probability of getting a 4, 5, or 6 on the first toss and a 1, 2, 3, or 4 on the second toss.

Let A_1 be the event "4, 5, or 6 on first toss," and A_2 be the event "1, 2, 3, or 4 on second toss." Then we are looking for $P(A_1 \cap A_2)$.

Method 1

$$P(A_1 \cap A_2) = P(A_1)P(A_2 \mid A_1) = P(A_1)P(A_2) = \left(\frac{3}{6}\right)\left(\frac{4}{6}\right) = \frac{1}{3}$$

We have used here the fact that the result of the second toss is *independent* of the first so that $P(A_2 \mid A_1) = P(A_2)$. Also we have used $P(A_1) = 3/6$ (since 4, 5, or 6 are 3 out of 6 equally likely possibilities) and $P(A_2) = 4/6$ (since 1, 2, 3, or 4 are 4 out of 6 equally likely possibilities).

Method 2

Each of the 6 ways in which a die can fall on the first toss can be associated with each of the 6 ways in which it can fall on the second toss, a total of $6 \cdot 6 = 36$ ways, all equally likely.

Each of the 3 ways in which A_1 can occur can be associated with each of the 4 ways in which A_2 can occur to give $3 \cdot 4 = 12$ ways in which both A_1 and A_2 can occur. Then

$$P(A_1 \cap A_2) = \frac{12}{36} = \frac{1}{3}$$

This shows directly that A_1 and A_2 are independent since

$$P(A_1 \cap A_2) = \frac{1}{3} = \left(\frac{3}{6}\right)\left(\frac{4}{6}\right) = P(A_1)P(A_2)$$

1.9. Find the probability of not getting a 7 or 11 total on either of two tosses of a pair of fair dice.

The sample space for each toss of the dice is shown in Fig. 1-9. For example, (5, 2) means that 5 comes up on the first die and 2 on the second. Since the dice are fair and there are 36 sample points, we assign probability 1/36 to each.

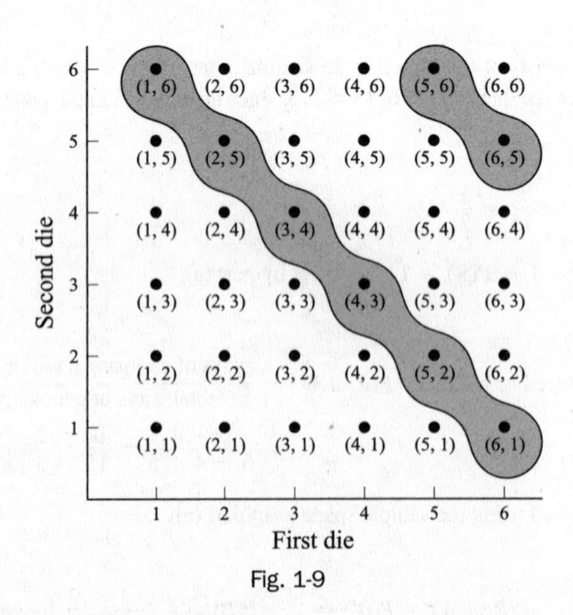

Fig. 1-9

If we let A be the event "7 or 11," then A is indicated by the circled portion in Fig. 1-9. Since 8 points are included, we have $P(A) = 8/36 = 2/9$. It follows that the probability of no 7 or 11 is given by

$$P(A') = 1 - P(A) = 1 - \frac{2}{9} = \frac{7}{9}$$

Using subscripts 1, 2 to denote 1st and 2nd tosses of the dice, we see that the probability of no 7 or 11 on either the first or second tosses is given by

$$P(A_1')P(A_2' \mid A_1') = P(A_1')P(A_2') = \left(\frac{7}{9}\right)\left(\frac{7}{9}\right) = \frac{49}{81},$$

using the fact that the tosses are independent.

1.10. Two cards are drawn from a well-shuffled ordinary deck of 52 cards. Find the probability that they are both aces if the first card is (a) replaced, (b) not replaced.

Method 1

Let A_1 = event "ace on first draw" and A_2 = event "ace on second draw." Then we are looking for $P(A_1 \cap A_2) = P(A_1)\,P(A_2 \mid A_1)$.

(a) Since for the first drawing there are 4 aces in 52 cards, $P(A_1) = 4/52$. Also, if the card is replaced for the second drawing, then $P(A_2 \mid A_1) = 4/52$, since there are also 4 aces out of 52 cards for the second drawing. Then

$$P(A_1 \cap A_2) = P(A_1)P(A_2 \mid A_1) = \left(\frac{4}{52}\right)\left(\frac{4}{52}\right) = \frac{1}{169}$$

(b) As in part (a), $P(A_1) = 4/52$. However, if an ace occurs on the first drawing, there will be only 3 aces left in the remaining 51 cards, so that $P(A_2 \mid A_1) = 3/51$. Then

$$P(A_1 \cap A_2) = P(A_1)P(A_2 \mid A_1) = \left(\frac{4}{52}\right)\left(\frac{3}{51}\right) = \frac{1}{221}$$

Method 2

(a) The first card can be drawn in any one of 52 ways, and since there is replacement, the second card can also be drawn in any one of 52 ways. Then both cards can be drawn in (52)(52) ways, all equally likely.

In such a case there are 4 ways of choosing an ace on the first draw and 4 ways of choosing an ace on the second draw so that the number of ways of choosing aces on the first and second draws is (4)(4). Then the required probability is

$$\frac{(4)(4)}{(52)(52)} = \frac{1}{169}$$

(b) The first card can be drawn in any one of 52 ways, and since there is no replacement, the second card can be drawn in any one of 51 ways. Then both cards can be drawn in (52)(51) ways, all equally likely.

In such a case there are 4 ways of choosing an ace on the first draw and 3 ways of choosing an ace on the second draw so that the number of ways of choosing aces on the first and second draws is (4)(3). Then the required probability is

$$\frac{(4)(3)}{(52)(51)} = \frac{1}{221}$$

1.11. Three balls are drawn successively from the box of Problem 1.7. Find the probability that they are drawn in the order red, white, and blue if each ball is (a) replaced, (b) not replaced.

Let R_1 = event "red on first draw," W_2 = event "white on second draw," B_3 = event "blue on third draw." We require $P(R_1 \cap W_2 \cap B_3)$.

(a) If each ball is replaced, then the events are independent and

$$\begin{aligned} P(R_1 \cap W_2 \cap B_3) &= P(R_1)P(W_2 \mid R_1)P(B_3 \mid R_2 \cap W_2) \\ &= P(R_1)P(W_2)P(B_3) \\ &= \left(\frac{6}{6+4+5}\right)\left(\frac{4}{6+4+5}\right)\left(\frac{5}{6+4+5}\right) = \frac{8}{225} \end{aligned}$$

(b) If each ball is not replaced, then the events are dependent and

$$P(R_1 \cap W_2 \cap B_3) = P(R_1)P(W_2 \mid R_1)P(B_3 \mid R_1 \cap W_2)$$

$$= \left(\frac{6}{6+4+5}\right)\left(\frac{4}{5+4+5}\right)\left(\frac{5}{5+3+5}\right) = \frac{4}{91}$$

1.12. Find the probability of a 4 turning up at least once in two tosses of a fair die.

Let A_1 = event "4 on first toss" and A_2 = event "4 on second toss." Then

$$A_1 \cup A_2 = \text{event "4 on first toss or 4 on second toss or both"}$$
$$= \text{event "at least one 4 turns up,"}$$

and we require $P(A_1 \cup A_2)$.

Method 1

Events A_1 and A_2 are not mutually exclusive, but they are independent. Hence, by (*10*) and (*21*),

$$P(A_1 \cup A_2) = P(A_1) + P(A_2) - P(A_1 \cap A_2)$$
$$= P(A_1) + P(A_2) - P(A_1)P(A_2)$$
$$= \frac{1}{6} + \frac{1}{6} - \left(\frac{1}{6}\right)\left(\frac{1}{6}\right) = \frac{11}{36}$$

Method 2

$$P(\text{at least one 4 comes up}) + P(\text{no 4 comes up}) = 1$$

Then

$$P(\text{at least one 4 comes up}) = 1 - P(\text{no 4 comes up})$$
$$= 1 - P(\text{no 4 on 1st toss and no 4 on 2nd toss})$$
$$= 1 - P(A'_1 \cap A'_2) = 1 - P(A'_1)P(A'_2)$$
$$= 1 - \left(\frac{5}{6}\right)\left(\frac{5}{6}\right) = \frac{11}{36}$$

Method 3

Total number of equally likely ways in which both dice can fall = $6 \cdot 6 = 36$.

Also Number of ways in which A_1 occurs but not $A_2 = 5$

Number of ways in which A_2 occurs but not $A_1 = 5$

Number of ways in which both A_1 and A_2 occur = 1

Then the number of ways in which at least one of the events A_1 or A_2 occurs = $5 + 5 + 1 = 11$. Therefore, $P(A_1 \cup A_2) = 11/36$.

1.13. One bag contains 4 white balls and 2 black balls; another contains 3 white balls and 5 black balls. If one ball is drawn from each bag, find the probability that (a) both are white, (b) both are black, (c) one is white and one is black.

Let W_1 = event "white ball from first bag," W_2 = event "white ball from second bag."

(a) $P(W_1 \cap W_2) = P(W_1)P(W_2 \mid W_1) = P(W_1)P(W_2) = \left(\frac{4}{4+2}\right)\left(\frac{3}{3+5}\right) = \frac{1}{4}$

(b) $P(W'_1 \cap W'_2) = P(W'_1)P(W'_2 \mid W'_1) = P(W'_1)P(W'_2) = \left(\frac{2}{4+2}\right)\left(\frac{5}{3+5}\right) = \frac{5}{24}$

(c) The required probability is

$$1 - P(W_1 \cap W_2) - P(W'_1 \cap W'_2) = 1 - \frac{1}{4} - \frac{5}{24} = \frac{13}{24}$$

1.14. Prove Theorem 1-10, page 7.

We prove the theorem for the case $n = 2$. Extensions to larger values of n are easily made. If event A must result in one of the two mutually exclusive events A_1, A_2, then

$$A = (A \cap A_1) \cup (A \cap A_2)$$

But $A \cap A_1$ and $A \cap A_2$ are mutually exclusive since A_1 and A_2 are. Therefore, by Axiom 3,

$$P(A) = P(A \cap A_1) + P(A \cap A_2)$$
$$= P(A_1)\,P(A \mid A_1) + P(A_2)\,P(A \mid A_2)$$

using (18), page 7.

1.15. Box *I* contains 3 red and 2 blue marbles while Box *II* contains 2 red and 8 blue marbles. A fair coin is tossed. If the coin turns up heads, a marble is chosen from Box *I*; if it turns up tails, a marble is chosen from Box *II*. Find the probability that a red marble is chosen.

Let R denote the event "a red marble is chosen" while I and II denote the events that Box I and Box II are chosen, respectively. Since a red marble can result by choosing either Box I or II, we can use the results of Problem 1.14 with $A = R, A_1 = I, A_2 = II$. Therefore, the probability of choosing a red marble is

$$P(R) = P(I)P(R \mid I) + P(II)P(R \mid II) = \left(\frac{1}{2}\right)\left(\frac{3}{3+2}\right) + \left(\frac{1}{2}\right)\left(\frac{2}{2+8}\right) = \frac{2}{5}$$

Bayes' theorem

1.16. Prove Bayes' theorem (Theorem 1-11, page 8).

Since A results in one of the mutually exclusive events $A_1, A_2, \ldots, A_n$, we have by Theorem 1-10 (Problem 1.14),

$$P(A) = P(A_1)P(A \mid A_1) + \cdots + P(A_n)P(A \mid A_n) = \sum_{j=1}^{n} P(A_j)P(A \mid A_j)$$

Therefore,

$$P(A_k \mid A) = \frac{P(A_k \cap A)}{P(A)} = \frac{P(A_k)P(A \mid A_k)}{\sum_{j=1}^{n} P(A_j)P(A \mid A_j)}$$

1.17. Suppose in Problem 1.15 that the one who tosses the coin does not reveal whether it has turned up heads or tails (so that the box from which a marble was chosen is not revealed) but does reveal that a red marble was chosen. What is the probability that Box *I* was chosen (i.e., the coin turned up heads)?

Let us use the same terminology as in Problem 1.15, i.e., $A = R, A_1 = I, A_2 = II$. We seek the probability that Box I was chosen given that a red marble is known to have been chosen. Using Bayes' rule with $n = 2$, this probability is given by

$$P(I \mid R) = \frac{P(I)P(R \mid I)}{P(I)P(R \mid I) + P(II)P(R \mid II)} = \frac{\left(\frac{1}{2}\right)\left(\frac{3}{3+2}\right)}{\left(\frac{1}{2}\right)\left(\frac{3}{3+2}\right) + \left(\frac{1}{2}\right)\left(\frac{2}{2+8}\right)} = \frac{3}{4}$$

Combinational analysis, counting, and tree diagrams

1.18. A committee of 3 members is to be formed consisting of one representative each from labor, management, and the public. If there are 3 possible representatives from labor, 2 from management, and 4 from the public, determine how many different committees can be formed using (a) the fundamental principle of counting and (b) a tree diagram.

(a) We can choose a labor representative in 3 different ways, and after this a management representative in 2 different ways. Then there are $3 \cdot 2 = 6$ different ways of choosing a labor and management representative. With each of these ways we can choose a public representative in 4 different ways. Therefore, the number of different committees that can be formed is $3 \cdot 2 \cdot 4 = 24$.

(b) Denote the 3 labor representatives by L_1, L_2, L_3; the management representatives by M_1, M_2; and the public representatives by P_1, P_2, P_3, P_4. Then the tree diagram of Fig. 1-10 shows that there are 24 different committees in all. From this tree diagram we can list all these different committees, e.g., $L_1M_1P_1$, $L_1M_1P_2$, etc.

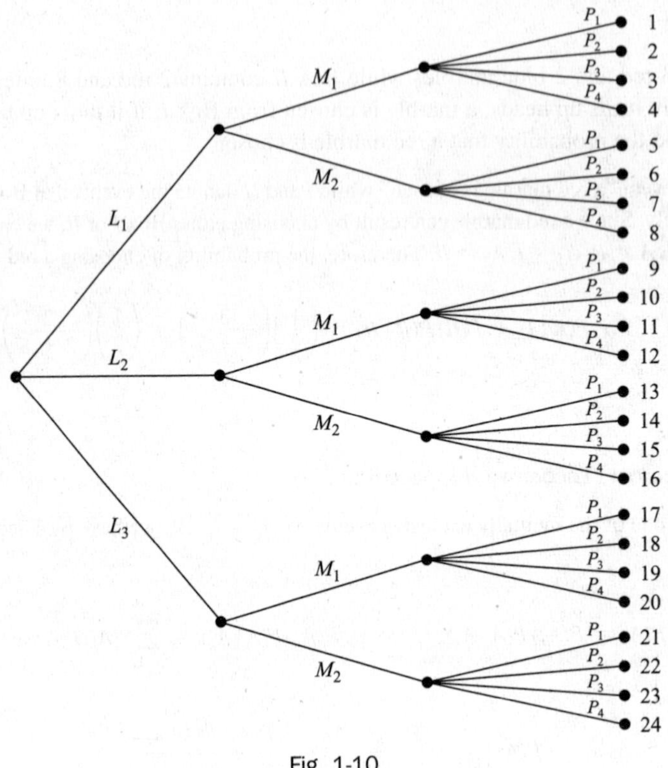

Fig. 1-10

Permutations

1.19. In how many ways can 5 differently colored marbles be arranged in a row?

We must arrange the 5 marbles in 5 positions thus: – – – – –. The first position can be occupied by any one of 5 marbles, i.e., there are 5 ways of filling the first position. When this has been done, there are 4 ways of filling the second position. Then there are 3 ways of filling the third position, 2 ways of filling the fourth position, and finally only 1 way of filling the last position. Therefore:

$$\text{Number of arrangements of 5 marbles in a row} = 5 \cdot 4 \cdot 3 \cdot 2 \cdot 1 = 5! = 120$$

In general,

$$\text{Number of arrangements of } n \text{ different objects in a row} = n(n-1)(n-2)\cdots 1 = n!$$

This is also called the *number of permutations of n different objects taken n at a time* and is denoted by ${}_nP_n$.

1.20. In how many ways can 10 people be seated on a bench if only 4 seats are available?

The first seat can be filled in any one of 10 ways, and when this has been done, there are 9 ways of filling the second seat, 8 ways of filling the third seat, and 7 ways of filling the fourth seat. Therefore:

$$\text{Number of arrangements of 10 people taken 4 at a time} = 10 \cdot 9 \cdot 8 \cdot 7 = 5040$$

In general,

$$\text{Number of arrangements of } n \text{ different objects taken } r \text{ at a time} = n(n-1)\cdots(n-r+1)$$

This is also called the *number of permutations of n different objects taken r at a time* and is denoted by ${}_nP_r$. Note that when $r = n$, ${}_nP_n = n!$ as in Problem 1.19.

1.21. Evaluate (a) ${}_8P_3$, (b) ${}_6P_4$, (c) ${}_{15}P_1$, (d) ${}_3P_3$.

(a) ${}_8P_3 = 8 \cdot 7 \cdot 6 = 336$ (b) ${}_6P_4 = 6 \cdot 5 \cdot 4 \cdot 3 = 360$ (c) ${}_{15}P_1 = 15$ (d) ${}_3P_3 = 3 \cdot 2 \cdot 1 = 6$

1.22. It is required to seat 5 men and 4 women in a row so that the women occupy the even places. How many such arrangements are possible?

The men may be seated in ${}_5P_5$ ways, and the women in ${}_4P_4$ ways. Each arrangement of the men may be associated with each arrangement of the women. Hence,

$$\text{Number of arrangements} = {}_5P_5 \cdot {}_4P_4 = 5!\,4! = (120)(24) = 2880$$

1.23. How many 4-digit numbers can be formed with the 10 digits 0, 1, 2, 3, . . . , 9 if (a) repetitions are allowed, (b) repetitions are not allowed, (c) the last digit must be zero and repetitions are not allowed?

(a) The first digit can be any one of 9 (since 0 is not allowed). The second, third, and fourth digits can be any one of 10. Then $9 \cdot 10 \cdot 10 \cdot 10 = 9000$ numbers can be formed.

(b) The first digit can be any one of 9 (any one but 0).

The second digit can be any one of 9 (any but that used for the first digit).

The third digit can be any one of 8 (any but those used for the first two digits).

The fourth digit can be any one of 7 (any but those used for the first three digits).

Then $9 \cdot 9 \cdot 8 \cdot 7 = 4536$ numbers can be formed.

Another method

The first digit can be any one of 9, and the remaining three can be chosen in ${}_9P_3$ ways. Then $9 \cdot 9P_3 = 9 \cdot 9 \cdot 8 \cdot 7 = 4536$ numbers can be formed.

(c) The first digit can be chosen in 9 ways, the second in 8 ways, and the third in 7 ways. Then $9 \cdot 8 \cdot 7 = 504$ numbers can be formed.

Another method

The first digit can be chosen in 9 ways, and the next two digits in ${}_8P_2$ ways. Then $9 \cdot {}_8P_2 = 9 \cdot 8 \cdot 7 = 504$ numbers can be formed.

1.24. Four different mathematics books, six different physics books, and two different chemistry books are to be arranged on a shelf. How many different arrangements are possible if (a) the books in each particular subject must all stand together, (b) only the mathematics books must stand together?

(a) The mathematics books can be arranged among themselves in ${}_4P_4 = 4!$ ways, the physics books in ${}_6P_6 = 6!$ ways, the chemistry books in ${}_2P_2 = 2!$ ways, and the three groups in ${}_3P_3 = 3!$ ways. Therefore,

$$\text{Number of arrangements} = 4!6!2!3! = 207{,}360.$$

(b) Consider the four mathematics books as one big book. Then we have 9 books which can be arranged in ${}_9P_9 = 9!$ ways. In all of these ways the mathematics books are together. But the mathematics books can be arranged among themselves in ${}_4P_4 = 4!$ ways. Hence,

$$\text{Number of arrangements} = 9!4! = 8{,}709{,}120$$

1.25. Five red marbles, two white marbles, and three blue marbles are arranged in a row. If all the marbles of the same color are not distinguishable from each other, how many different arrangements are possible?

Assume that there are N different arrangements. Multiplying N by the numbers of ways of arranging (a) the five red marbles among themselves, (b) the two white marbles among themselves, and (c) the three blue marbles among themselves (i.e., multiplying N by 5!2!3!), we obtain the number of ways of arranging the 10 marbles if they were all distinguishable, i.e., 10!.

Then $$(5!2!3!)N = 10! \quad \text{and} \quad N = 10!/(5!2!3!)$$

In general, the number of different arrangements of n objects of which n_1 are alike, n_2 are alike, . . . , n_k are alike is $\dfrac{n!}{n_1!n_2! \cdots n_k!}$ where $n_1 + n_2 + \cdots + n_k = n$.

1.26. In how many ways can 7 people be seated at a round table if (a) they can sit anywhere, (b) 2 particular people must not sit next to each other?

(a) Let 1 of them be seated anywhere. Then the remaining 6 people can be seated in $6! = 720$ ways, which is the total number of ways of arranging the 7 people in a circle.

(b) Consider the 2 particular people as 1 person. Then there are 6 people altogether and they can be arranged in $5!$ ways. But the 2 people considered as 1 can be arranged in $2!$ ways. Therefore, the number of ways of arranging 7 people at a round table with 2 particular people sitting together $= 5!2! = 240$.

Then using (a), the total number of ways in which 7 people can be seated at a round table so that the 2 particular people do not sit together $= 730 - 240 = 480$ ways.

Combinations

1.27. In how many ways can 10 objects be split into two groups containing 4 and 6 objects, respectively?

This is the same as the number of arrangements of 10 objects of which 4 objects are alike and 6 other objects are alike. By Problem 1.25, this is $\frac{10!}{4!6!} = \frac{10 \cdot 9 \cdot 8 \cdot 7}{4!} = 210$.

The problem is equivalent to finding the number of selections of 4 out of 10 objects (or 6 out of 10 objects), the order of selection being immaterial. In general, the number of selections of r out of n objects, called the *number of combinations of n things taken r at a time,* is denoted by ${}_nC_r$ or $\binom{n}{r}$ and is given by

$${}_nC_r = \binom{n}{r} = \frac{n!}{r!(n-r)!} = \frac{n(n-1)\cdots(n-r+1)}{r!} = \frac{{}_nP_r}{r!}$$

1.28. Evaluate (a) ${}_7C_4$, (b) ${}_6C_5$, (c) ${}_4C_4$.

(a) ${}_7C_4 = \frac{7!}{4!3!} = \frac{7 \cdot 6 \cdot 5 \cdot 4}{4!} = \frac{7 \cdot 6 \cdot 5}{3 \cdot 2 \cdot 1} = 35.$

(b) ${}_6C_5 = \frac{6!}{5!1!} = \frac{6 \cdot 5 \cdot 4 \cdot 3 \cdot 2}{5!} = 6,$ or ${}_6C_5 = {}_6C_1 = 6.$

(c) ${}_4C_4$ is the number of selections of 4 objects taken 4 at a time, and there is only one such selection. Then ${}_4C_4 = 1$. Note that formally

$${}_4C_4 = \frac{4!}{4!0!} = 1 \quad \text{if we } \textit{define } 0! = 1.$$

1.29. In how many ways can a committee of 5 people be chosen out of 9 people?

$$\binom{9}{5} = {}_9C_5 = \frac{9!}{5!4!} = \frac{9 \cdot 8 \cdot 7 \cdot 6 \cdot 5}{5!} = 126$$

1.30. Out of 5 mathematicians and 7 physicists, a committee consisting of 2 mathematicians and 3 physicists is to be formed. In how many ways can this be done if (a) any mathematician and any physicist can be included, (b) one particular physicist must be on the committee, (c) two particular mathematicians cannot be on the committee?

(a) 2 mathematicians out of 5 can be selected in ${}_5C_2$ ways.
3 physicists out of 7 can be selected in ${}_7C_3$ ways.

$$\text{Total number of possible selections} = {}_5C_2 \cdot {}_7C_3 = 10 \cdot 35 = 350$$

(b) 2 mathematicians out of 5 can be selected in ${}_5C_2$ ways.
2 physicists out of 6 can be selected in ${}_6C_2$ ways.

$$\text{Total number of possible selections} = {}_5C_2 \cdot {}_6C_2 = 10 \cdot 15 = 150$$

(c) 2 mathematicians out of 3 can be selected in ${}_3C_2$ ways.
3 physicists out of 7 can be selected in ${}_7C_3$ ways.

$$\text{Total number of possible selections} = {}_3C_2 \cdot {}_7C_3 = 3 \cdot 35 = 105$$

1.31. How many different salads can be made from lettuce, escarole, endive, watercress, and chicory?

Each green can be dealt with in 2 ways, as it can be chosen or not chosen. Since each of the 2 ways of dealing with a green is associated with 2 ways of dealing with each of the other greens, the number of ways of dealing with the 5 greens = 2^5 ways. But 2^5 ways includes the case in which no greens is chosen. Hence,

$$\text{Number of salads} = 2^5 - 1 = 31$$

Another method

One can select either 1 out of 5 greens, 2 out of 5 greens, . . . , 5 out of 5 greens. Then the required number of salads is

$$_5C_1 + {_5C_2} + {_5C_3} + {_5C_4} + {_5C_5} = 5 + 10 + 10 + 5 + 1 = 31$$

In general, for any positive integer n, $_nC_1 + {_nC_2} + {_nC_3} + \cdots + {_nC_n} = 2^n - 1$.

1.32. From 7 consonants and 5 vowels, how many words can be formed consisting of 4 different consonants and 3 different vowels? The words need not have meaning.

The 4 different consonants can be selected in $_7C_4$ ways, the 3 different vowels can be selected in $_5C_3$ ways, and the resulting 7 different letters (4 consonants, 3 vowels) can then be arranged among themselves in $_7P_7 = 7!$ ways. Then

$$\text{Number of words} = {_7C_4} \cdot {_5C_3} \cdot 7! = 35 \cdot 10 \cdot 5040 = 1{,}764{,}000$$

The Binomial Coefficients

1.33. Prove that $\binom{n}{r} = \binom{n-1}{r} + \binom{n-1}{r-1}$.

We have

$$\binom{n}{r} = \frac{n!}{r!(n-r)!} = \frac{n(n-1)!}{r!(n-r)!} = \frac{(n-r+r)(n-1)!}{r!(n-r)!}$$

$$= \frac{(n-r)(n-1)!}{r!(n-r)!} + \frac{r(n-1)!}{r!(n-r)!}$$

$$= \frac{(n-1)!}{r!(n-r-1)!} + \frac{(n-1)!}{(r-1)!(n-r)!}$$

$$= \binom{n-1}{r} + \binom{n-1}{r-1}$$

The result has the following interesting application. If we write out the coefficients in the binomial expansion of $(x + y)^n$ for $n = 0, 1, 2, \ldots$, we obtain the following arrangement, called *Pascal's triangle*:

$n = 0$							1						
$n = 1$						1		1					
$n = 2$					1		2		1				
$n = 3$				1		3		3		1			
$n = 4$			1		4		6		4		1		
$n = 5$		1		5		10		10		5		1	
$n = 6$	1		6		15		20		15		6		1
etc.													

An entry in any line can be obtained by adding the two entries in the preceding line that are to its immediate left and right. Therefore, $10 = 4 + 6$, $15 = 10 + 5$, etc.

1.34. Find the constant term in the expansion of $\left(x^2 + \frac{1}{x}\right)^{12}$.

According to the binomial theorem,

$$\left(x^2 + \frac{1}{x}\right)^{12} = \sum_{k=0}^{12}\binom{12}{k}(x^2)^k\left(\frac{1}{x}\right)^{12-k} = \sum_{k=0}^{12}\binom{12}{k}x^{3k-12}.$$

The constant term corresponds to the one for which $3k - 12 = 0$, i.e., $k = 4$, and is therefore given by

$$\binom{12}{4} = \frac{12 \cdot 11 \cdot 10 \cdot 9}{4 \cdot 3 \cdot 2 \cdot 1} = 495$$

Probability using combinational analysis

1.35. A box contains 8 red, 3 white, and 9 blue balls. If 3 balls are drawn at random without replacement, determine the probability that (a) all 3 are red, (b) all 3 are white, (c) 2 are red and 1 is white, (d) at least 1 is white, (e) 1 of each color is drawn, (f) the balls are drawn in the order red, white, blue.

(a) **Method 1**

Let R_1, R_2, R_3 denote the events, "red ball on 1st draw," "red ball on 2nd draw," "red ball on 3rd draw," respectively. Then $R_1 \cap R_2 \cap R_3$ denotes the event "all 3 balls drawn are red." We therefore have

$$P(R_1 \cap R_2 \cap R_3) = P(R_1)P(R_2 \mid R_1)P(R_3 \mid R_1 \cap R_2)$$

$$= \left(\frac{8}{20}\right)\left(\frac{7}{19}\right)\left(\frac{6}{18}\right) = \frac{14}{285}$$

Method 2

$$\text{Required probability} = \frac{\text{number of selections of 3 out of 8 red balls}}{\text{number of selections of 3 out of 20 balls}} = \frac{{}_8C_3}{{}_{20}C_3} = \frac{14}{285}$$

(b) Using the second method indicated in part (a),

$$P(\text{all 3 are white}) = \frac{{}_3C_3}{{}_{20}C_3} = \frac{1}{1140}$$

The first method indicated in part (a) can also be used.

(c) $P(\text{2 are red and 1 is white})$

$$= \frac{(\text{selections of 2 out of 8 red balls})(\text{selections of 1 out of 3 white balls})}{\text{number of selections of 3 out of 20 balls}}$$

$$= \frac{({}_8C_2)({}_3C_1)}{{}_{20}C_3} = \frac{7}{95}$$

(d) $P(\text{none is white}) = \frac{{}_{17}C_3}{{}_{20}C_3} = \frac{34}{57}$. Then

$$P(\text{at least 1 is white}) = 1 - \frac{34}{57} = \frac{23}{57}$$

(e) $P(\text{1 of each color is drawn}) = \frac{({}_8C_1)({}_3C_1)({}_9C_1)}{{}_{20}C_3} = \frac{18}{95}$

(f) $P(\text{balls drawn in order red, white, blue}) = \frac{1}{3!}P(\text{1 of each color is drawn})$

$$= \frac{1}{6}\left(\frac{18}{95}\right) = \frac{3}{95}, \text{ using (e)}$$

Another method

$$P(R_1 \cap W_2 \cap B_3) = P(R_1)P(W_2 \mid R_1)P(B_3 \mid R_1 \cap W_2)$$

$$= \left(\frac{8}{20}\right)\left(\frac{3}{19}\right)\left(\frac{9}{18}\right) = \frac{3}{95}$$

1.36. In the game of *poker* 5 cards are drawn from a pack of 52 well-shuffled cards. Find the probability that (a) 4 are aces, (b) 4 are aces and 1 is a king, (c) 3 are tens and 2 are jacks, (d) a nine, ten, jack, queen, king are obtained in any order, (e) 3 are of any one suit and 2 are of another, (f) at least 1 ace is obtained.

(a) $P(\text{4 aces}) = \dfrac{({}_4C_4)({}_{48}C_1)}{{}_{52}C_5} = \dfrac{1}{54{,}145}.$

(b) $P(\text{4 aces and 1 king}) = \dfrac{({}_4C_4)({}_4C_1)}{{}_{52}C_5} = \dfrac{1}{649{,}740}.$

(c) $P(\text{3 are tens and 2 are jacks}) = \dfrac{({}_4C_3)({}_4C_2)}{{}_{52}C_5} = \dfrac{1}{108{,}290}.$

(d) $P(\text{nine, ten, jack, queen, king in any order}) = \dfrac{({}_4C_1)({}_4C_1)({}_4C_1)({}_4C_1)({}_4C_1)}{{}_{52}C_5} = \dfrac{64}{162{,}435}.$

(e) $P(\text{3 of any one suit, 2 of another}) = \dfrac{(4 \cdot {}_{13}C_3)(3 \cdot {}_{13}C_2)}{{}_{52}C_5} = \dfrac{429}{4165},$

since there are 4 ways of choosing the first suit and 3 ways of choosing the second suit.

(f) $P(\text{no ace}) = \dfrac{{}_{48}C_5}{{}_{52}C_5} = \dfrac{35{,}673}{54{,}145}.$ Then $P(\text{at least one ace}) = 1 - \dfrac{35{,}673}{54{,}145} = \dfrac{18{,}472}{54{,}145}.$

1.37. Determine the probability of three 6s in 5 tosses of a fair die.

Let the tosses of the die be represented by the 5 spaces – – – – –. In each space we will have the events 6 or not 6 (6′). For example, three 6s and two not 6s can occur as 6 6 6′6 6′ or 6 6′6 6′6, etc.

Now the probability of the outcome 6 6 6′ 6 6′ is

$$P(6\,6\,6'\,6\,6') = P(6)\,P(6)\,P(6')\,P(6)\,P(6') = \frac{1}{6}\cdot\frac{1}{6}\cdot\frac{5}{6}\cdot\frac{1}{6}\cdot\frac{5}{6} = \left(\frac{1}{6}\right)^3\left(\frac{5}{6}\right)^2$$

since we assume independence. Similarly,

$$P = \left(\frac{1}{6}\right)^3\left(\frac{5}{6}\right)^2$$

for all other outcomes in which three 6s and two not 6s occur. But there are ${}_5C_3 = 10$ such outcomes, and these are mutually exclusive. Hence, the required probability is

$$P(6\,6\,6'6\,6' \text{ or } 6\,6'6\,6'6 \text{ or} \ldots) = {}_5C_3\left(\frac{1}{6}\right)^3\left(\frac{5}{6}\right)^2 = \frac{5!}{3!2!}\left(\frac{1}{6}\right)^3\left(\frac{5}{6}\right)^2 = \frac{125}{3888}$$

In general, if $p = P(A)$ and $q = 1 - p = P(A')$, then by using the same reasoning as given above, the probability of getting exactly x A's in n independent trials is

$${}_nC_x p^x q^{n-x} = \binom{n}{x} p^x q^{n-x}$$

1.38. A shelf has 6 mathematics books and 4 physics books. Find the probability that 3 particular mathematics books will be together.

All the books can be arranged among themselves in ${}_{10}P_{10} = 10!$ ways. Let us assume that the 3 particular mathematics books actually are replaced by 1 book. Then we have a total of 8 books that can be arranged among themselves in ${}_8P_8 = 8!$ ways. But the 3 mathematics books themselves can be arranged in ${}_3P_3 = 3!$ ways. The required probability is thus given by

$$\frac{8!\,3!}{10!} = \frac{1}{15}$$

Miscellaneous problems

1.39. A and B play 12 games of chess of which 6 are won by A, 4 are won by B, and 2 end in a draw. They agree to play a tournament consisting of 3 games. Find the probability that (a) A wins all 3 games, (b) 2 games end in a draw, (c) A and B win alternately, (d) B wins at least 1 game.

Let A_1, A_2, A_3 denote the events "A wins" in 1st, 2nd, and 3rd games, respectively, B_1, B_2, B_3 denote the events "B wins" in 1st, 2nd, and 3rd games, respectively. On the basis of their past performance (empirical probability), we shall assume that

$$P(\text{A wins any one game}) = \frac{6}{12} = \frac{1}{2}, \qquad P(\text{B wins any one game}) = \frac{4}{12} = \frac{1}{3}$$

(a) $$P(\text{A wins all 3 games}) = P(A_1 \cap A_2 \cap A_3) = P(A_1)\,P(A_2)\,P(A_3) = \left(\frac{1}{2}\right)\left(\frac{1}{2}\right)\left(\frac{1}{2}\right) = \frac{1}{8}$$

assuming that the results of each game are independent of the results of any others. (This assumption would not be justifiable if either player were *psychologically influenced* by the other one's winning or losing.)

(b) In any one game the probability of a nondraw (i.e., either A or B wins) is $q = \frac{1}{2} + \frac{1}{3} = \frac{5}{6}$ and the probability of a draw is $p = 1 - q = \frac{1}{6}$. Then the probability of 2 draws in 3 trials is (see Problem 1.37)

$$\binom{3}{2} p^2 q^{3-2} = 3\left(\frac{1}{6}\right)^2\left(\frac{5}{6}\right) = \frac{5}{72}$$

(c)
$$\begin{aligned} P(\text{A and B win alternately}) &= P(\text{A wins then B wins then A wins} \\ &\quad \textit{or} \text{ B wins then A wins then B wins}) \\ &= P(A_1 \cap B_2 \cap A_3) + P(B_1 \cap A_2 \cap B_3) \\ &= P(A_1)P(B_2)P(A_3) + P(B_1)P(A_2)P(B_3) \\ &= \left(\frac{1}{2}\right)\left(\frac{1}{3}\right)\left(\frac{1}{2}\right) + \left(\frac{1}{3}\right)\left(\frac{1}{2}\right)\left(\frac{1}{3}\right) = \frac{5}{36} \end{aligned}$$

(d)
$$\begin{aligned} P(\text{B wins at least one game}) &= 1 - P(\text{B wins no game}) \\ &= 1 - P(B'_1 \cap B'_2 \cap B'_3) \\ &= 1 - P(B'_1)P(B'_2)P(B'_3) \\ &= 1 - \left(\frac{2}{3}\right)\left(\frac{2}{3}\right)\left(\frac{2}{3}\right) = \frac{19}{27} \end{aligned}$$

1.40. A and B play a game in which they alternately toss a pair of dice. The one who is first to get a total of 7 wins the game. Find the probability that (a) the one who tosses first will win the game, (b) the one who tosses second will win the game.

(a) The probability of getting a 7 on a single toss of a pair of dice, assumed fair, is 1/6 as seen from Problem 1.9 and Fig. 1-9. If we suppose that A is the first to toss, then A will win in any of the following mutually exclusive cases with indicated associated probabilities:

(1) A wins on 1st toss. Probability $= \frac{1}{6}$.

(2) A loses on 1st toss, B then loses, A then wins. Probability $= \left(\frac{5}{6}\right)\left(\frac{5}{6}\right)\left(\frac{1}{6}\right)$.

(3) A loses on 1st toss, B loses, A loses, B loses, A wins. Probability $= \left(\frac{5}{6}\right)\left(\frac{5}{6}\right)\left(\frac{5}{6}\right)\left(\frac{5}{6}\right)\left(\frac{1}{6}\right)$.

....

Then the probability that A wins is

$$\left(\frac{1}{6}\right) + \left(\frac{5}{6}\right)\left(\frac{5}{6}\right)\left(\frac{1}{6}\right) + \left(\frac{5}{6}\right)\left(\frac{5}{6}\right)\left(\frac{5}{6}\right)\left(\frac{5}{6}\right)\left(\frac{1}{6}\right) + \cdots$$
$$= \frac{1}{6}\left[1 + \left(\frac{5}{6}\right)^2 + \left(\frac{5}{6}\right)^4 + \cdots\right] = \frac{1/6}{1 - (5/6)^2} = \frac{6}{11}$$

where we have used the result 6 of Appendix A with $x = (5/6)^2$.

(b) The probability that B wins the game is similarly

$$\left(\frac{5}{6}\right)\left(\frac{1}{6}\right) + \left(\frac{5}{6}\right)\left(\frac{5}{6}\right)\left(\frac{5}{6}\right)\left(\frac{1}{6}\right) + \cdots = \left(\frac{5}{6}\right)\left(\frac{1}{6}\right)\left[1 + \left(\frac{5}{6}\right)^2 + \left(\frac{5}{6}\right)^4 + \cdots\right]$$
$$= \frac{5/36}{1 - (5/6)^2} = \frac{5}{11}$$

Therefore, we would give 6 to 5 odds that the first one to toss will win. Note that since

$$\frac{6}{11} + \frac{5}{11} = 1$$

the probability of a tie is zero. This would not be true if the game was limited. See Problem 1.100.

1.41. A machine produces a total of 12,000 bolts a day, which are on the average 3% defective. Find the probability that out of 600 bolts chosen at random, 12 will be defective.

Of the 12,000 bolts, 3%, or 360, are defective and 11,640 are not. Then:

$$\text{Required probability} = \frac{{}_{360}C_{12}\ {}_{11,640}C_{588}}{{}_{12,000}C_{600}}$$

1.42. A box contains 5 red and 4 white marbles. Two marbles are drawn successively from the box without replacement, and it is noted that the second one is white. What is the probability that the first is also white?

Method 1

If W_1, W_2 are the events "white on 1st draw," "white on 2nd draw," respectivley, we are looking for $P(W_1 \mid W_2)$. This is given by

$$P(W_1 \mid W_2) = \frac{P(W_1 \cap W_2)}{P(W_2)} = \frac{(4/9)(3/8)}{4/9} = \frac{3}{8}$$

Method 2

Since the second is known to be white, there are only 3 ways out of the remaining 8 in which the first can be white, so that the probability is 3/8.

1.43. The probabilities that a husband and wife will be alive 20 years from now are given by 0.8 and 0.9, respectively. Find the probability that in 20 years (a) both, (b) neither, (c) at least one, will be alive.

Let H, W be the events that the husband and wife, respectively, will be alive in 20 years. Then $P(H) = 0.8$, $P(W) = 0.9$. We suppose that H and W are independent events, which may or may not be reasonable.

(a) $P(\text{both will be alive}) = P(H \cap W) = P(H)P(W) = (0.8)(0.9) = 0.72$.

(b) $P(\text{neither will be alive}) = P(H' \cap W') = P(H')\,P(W') = (0.2)(0.1) = 0.02$.

(c) $P(\text{at least one will be alive}) = 1 - P(\text{neither will be alive}) = 1 - 0.02 = 0.98$.

1.44. An inefficient secretary places n different letters into n differently addressed envelopes at random. Find the probability that at least one of the letters will arrive at the proper destination.

Let $A_1, A_2, \ldots A_n$ denote the events that the 1st, 2nd, . . . , nth letter is in the correct envelope. Then the event that at least one letter is in the correct envelope is $A_1 \cup A_2 \cup \cdots \cup A_n$, and we want to find $P(A_1 \cup A_2 \cup \cdots \cup A_n)$. From a generalization of the results (*10*) and (*11*), page 6, we have

$$(1) \qquad P(A_1 \cup A_2 \cup \cdots \cup A_n) = \sum P(A_k) - \sum P(A_j \cap A_k) + \sum P(A_i \cap A_j \cap A_k) - \cdots + (-1)^{n-1} P(A_1 \cap A_2 \cap \cdots \cap A_n)$$

where $\sum P(A_k)$ the sum of the probabilities of A_k from 1 to n, $\sum P(A_j \cap A_k)$ is the sum of the probabilities of $A_j \cap A_k$ with j and k from 1 to n and $k > j$, etc. We have, for example, the following:

$$(2) \qquad P(A_1) = \frac{1}{n} \quad \text{and similarly} \quad P(A_k) = \frac{1}{n}$$

since, of the n envelopes, only 1 will have the proper address. Also

$$(3) \qquad P(A_1 \cap A_2) = P(A_1)P(A_2 \mid A_1) = \left(\frac{1}{n}\right)\left(\frac{1}{n-1}\right)$$

since, if the 1st letter is in the proper envelope, then only 1 of the remaining $n - 1$ envelopes will be proper. In a similar way we find

$$(4) \qquad P(A_1 \cap A_2 \cap A_3) = P(A_1)P(A_2 \mid A_1)P(A_3 \mid A_1 \cap A_2) = \left(\frac{1}{n}\right)\left(\frac{1}{n-1}\right)\left(\frac{1}{n-2}\right)$$

etc., and finally

$$(5) \qquad P(A_1 \cap A_2 \cap \cdots \cap A_n) = \left(\frac{1}{n}\right)\left(\frac{1}{n-1}\right)\cdots\left(\frac{1}{1}\right) = \frac{1}{n!}$$

Now in the sum $\sum P(A_j \cap A_k)$ there are $\binom{n}{2} = {}_nC_2$ terms all having the value given by (3). Similarly in $\sum P(A_i \cap A_j \cap A_k)$, there are $\binom{n}{3} = {}_nC_3$ terms all having the value given by (4). Therefore, the required probability is

$$P(A_1 \cup A_2 \cup \cdots \cup A_n) = \binom{n}{1}\left(\frac{1}{n}\right) - \binom{n}{2}\left(\frac{1}{n}\right)\left(\frac{1}{n-1}\right) + \binom{n}{3}\left(\frac{1}{n}\right)\left(\frac{1}{n-1}\right)\left(\frac{1}{n-2}\right)$$

$$- \cdots + (-1)^{n-1}\binom{n}{n}\left(\frac{1}{n!}\right)$$

$$= 1 - \frac{1}{2!} + \frac{1}{3!} - \cdots + (-1)^{n-1}\frac{1}{n!}$$

From calculus we know that (see Appendix A)

$$e^x = 1 + x + \frac{x^2}{2!} + \frac{x^3}{3!} + \cdots$$

so that for $x = -1$

$$e^{-1} = 1 - \left(1 - \frac{1}{2!} + \frac{1}{3!} - \cdots\right)$$

or

$$1 - \frac{1}{2!} + \frac{1}{3!} - \cdots = 1 - e^{-1}$$

It follows that if n is large, the required probability is very nearly $1 - e^{-1} = 0.6321$. This means that there is a good chance of at least 1 letter arriving at the proper destination. The result is remarkable in that the probability remains practically constant for all $n > 10$. Therefore, the probability that at least 1 letter will arrive at its proper destination is practically the same whether n is 10 or 10,000.

1.45. Find the probability that n people ($n \leq 365$) selected at random will have n different birthdays.

We assume that there are only 365 days in a year and that all birthdays are equally probable, assumptions which are not quite met in reality.

The first of the n people has of course some birthday with probability $365/365 = 1$. Then, if the second is to have a different birthday, it must occur on one of the other 364 days. Therefore, the probability that the second person has a birthday different from the first is $364/365$. Similarly the probability that the third person has a birthday different from the first two is $363/365$. Finally, the probability that the nth person has a birthday different from the others is $(365 - n + 1)/365$. We therefore have

$$P(\text{all } n \text{ birthdays are different}) = \frac{365}{365} \cdot \frac{364}{365} \cdot \frac{363}{365} \cdots \frac{365-n+1}{365}$$

$$= \left(1 - \frac{1}{365}\right)\left(1 - \frac{2}{365}\right)\cdots\left(1 - \frac{n-1}{365}\right)$$

1.46. Determine how many people are required in Problem 1.45 to make the probability of distinct birthdays less than 1/2.

Denoting the given probability by p and taking natural logarithms, we find

$$(1) \qquad \ln p = \ln\left(1 - \frac{1}{365}\right) + \ln\left(1 - \frac{2}{365}\right) + \cdots + \ln\left(1 - \frac{n-1}{365}\right)$$

But we know from calculus (Appendix A, formula 7) that

$$(2) \qquad \ln(1 - x) = -x - \frac{x^2}{2} - \frac{x^3}{3} - \cdots$$

so that (1) can be written

$$\text{(3)} \qquad \ln p = -\left[\frac{1 + 2 + \cdots + (n-1)}{365}\right] - \frac{1}{2}\left[\frac{1^2 + 2^2 + \cdots + (n-1)^2}{(365)^2}\right] - \cdots$$

Using the facts that for $n = 2, 3, \ldots$ (Appendix A, formulas 1 and 2)

$$\text{(4)} \qquad 1 + 2 + \cdots + (n-1) = \frac{n(n-1)}{2}, \qquad 1^2 + 2^2 + \cdots + (n-1)^2 = \frac{n(n-1)(2n-1)}{6}$$

we obtain for (3)

$$\text{(5)} \qquad \ln p = -\frac{n(n-1)}{730} - \frac{n(n-1)(2n-1)}{12(365)^2} - \cdots$$

For n small compared to 365, say, $n < 30$, the second and higher terms on the right of (5) are negligible compared to the first term, so that a good approximation in this case is

$$\text{(6)} \qquad \ln p = \frac{n(n-1)}{730}$$

For $p = \frac{1}{2}$, $\ln p = -\ln 2 = -0.693$. Therefore, we have

$$\text{(7)} \qquad \frac{n(n-1)}{730} = 0.693 \qquad \text{or} \qquad n^2 - n - 506 = 0 \qquad \text{or} \qquad (n-23)(n+22) = 0$$

so that $n = 23$. Our conclusion therefore is that, if n is larger than 23, we can give better than even odds that at least 2 people will have the same birthday.

SUPPLEMENTARY PROBLEMS

Calculation of probabilities

1.47. Determine the probability p, or an estimate of it, for each of the following events:

(a) A king, ace, jack of clubs, or queen of diamonds appears in drawing a single card from a well-shuffled ordinary deck of cards.

(b) The sum 8 appears in a single toss of a pair of fair dice.

(c) A nondefective bolt will be found next if out of 600 bolts already examined, 12 were defective.

(d) A 7 or 11 comes up in a single toss of a pair of fair dice.

(e) At least 1 head appears in 3 tosses of a fair coin.

1.48. An experiment consists of drawing 3 cards in succession from a well-shuffled ordinary deck of cards. Let A_1 be the event "king on first draw," A_2 the event "king on second draw," and A_3 the event "king on third draw." State in words the meaning of each of the following:

(a) $P(A_1 \cap A'_2)$, (b) $P(A_1 \cup A_2)$, (c) $P(A'_1 \cup A'_2)$, (d) $P(A'_1 \cap A'_2 \cap A'_3)$, (e) $P[(A_1 \cap A_2) \cup (A'_2 \cap A_3)]$.

1.49. A marble is drawn at random from a box containing 10 red, 30 white, 20 blue, and 15 orange marbles. Find the probability that it is (a) orange or red, (b) not red or blue, (c) not blue, (d) white, (e) red, white, or blue.

1.50. Two marbles are drawn in succession from the box of Problem 1.49, replacement being made after each drawing. Find the probability that (a) both are white, (b) the first is red and the second is white, (c) neither is orange, (d) they are either red or white or both (red and white), (e) the second is not blue, (f) the first is orange, (g) at least one is blue, (h) at most one is red, (i) the first is white but the second is not, (j) only one is red.

1.51. Work Problem 1.50 with no replacement after each drawing.

Conditional probability and independent events

1.52. A box contains 2 red and 3 blue marbles. Find the probability that if two marbles are drawn at random (without replacement), (a) both are blue, (b) both are red, (c) one is red and one is blue.

1.53. Find the probability of drawing 3 aces at random from a deck of 52 ordinary cards if the cards are (a) replaced, (b) not replaced.

1.54. If at least one child in a family with 2 children is a boy, what is the probability that both children are boys?

1.55. Box *I* contains 3 red and 5 white balls, while Box *II* contains 4 red and 2 white balls. A ball is chosen at random from the first box and placed in the second box without observing its color. Then a ball is drawn from the second box. Find the probability that it is white.

Bayes' theorem or rule

1.56. A box contains 3 blue and 2 red marbles while another box contains 2 blue and 5 red marbles. A marble drawn at random from one of the boxes turns out to be blue. What is the probability that it came from the first box?

1.57. Each of three identical jewelry boxes has two drawers. In each drawer of the first box there is a gold watch. In each drawer of the second box there is a silver watch. In one drawer of the third box there is a gold watch while in the other there is a silver watch. If we select a box at random, open one of the drawers and find it to contain a silver watch, what is the probability that the other drawer has the gold watch?

1.58. Urn *I* has 2 white and 3 black balls; Urn *II*, 4 white and 1 black; and Urn *III*, 3 white and 4 black. An urn is selected at random and a ball drawn at random is found to be white. Find the probability that Urn *I* was selected.

Combinatorial analysis, counting, and tree diagrams

1.59. A coin is tossed 3 times. Use a tree diagram to determine the various possibilities that can arise.

1.60. Three cards are drawn at random (without replacement) from an ordinary deck of 52 cards. Find the number of ways in which one can draw (a) a diamond and a club and a heart in succession, (b) two hearts and then a club or a spade.

1.61. In how many ways can 3 different coins be placed in 2 different purses?

Permutations

1.62. Evaluate (a) ${}_4P_2$, (b) ${}_7P_5$, (c) ${}_{10}P_3$.

1.63. For what value of n is ${}_{n+1}P_3 = {}_nP_4$?

1.64. In how many ways can 5 people be seated on a sofa if there are only 3 seats available?

1.65. In how many ways can 7 books be arranged on a shelf if (a) any arrangement is possible, (b) 3 particular books must always stand together, (c) two particular books must occupy the ends?

1.66. How many numbers consisting of five different digits each can be made from the digits 1, 2, 3, . . . , 9 if (a) the numbers must be odd, (b) the first two digits of each number are even?

1.67. Solve Problem 1.66 if repetitions of the digits are allowed.

1.68. How many different three-digit numbers can be made with 3 fours, 4 twos, and 2 threes?

1.69. In how many ways can 3 men and 3 women be seated at a round table if (a) no restriction is imposed, (b) 2 particular women must not sit together, (c) each woman is to be between 2 men?

Combinations

1.70. Evaluate (a) ${}_5C_3$, (b) ${}_8C_4$, (c) ${}_{10}C_8$.

1.71. For what value of n is $3 \cdot {}_{n+1}C_3 = 7 \cdot {}_nC_2$?

1.72. In how many ways can 6 questions be selected out of 10?

1.73. How many different committees of 3 men and 4 women can be formed from 8 men and 6 women?

1.74. In how many ways can 2 men, 4 women, 3 boys, and 3 girls be selected from 6 men, 8 women, 4 boys and 5 girls if (a) no restrictions are imposed, (b) a particular man and woman must be selected?

1.75. In how many ways can a group of 10 people be divided into (a) two groups consisting of 7 and 3 people, (b) three groups consisting of 5, 3, and 2 people?

1.76. From 5 statisticians and 6 economists, a committee consisting of 3 statisticians and 2 economists is to be formed. How many different committees can be formed if (a) no restrictions are imposed, (b) 2 particular statisticians must be on the committee, (c) 1 particular economist cannot be on the committee?

1.77. Find the number of (a) combinations and (b) permutations of 4 letters each that can be made from the letters of the word *Tennessee*.

Binomial coefficients

1.78. Calculate (a) ${}_6C_3$, (b) $\binom{11}{4}$, (c) $({}_8C_2)({}_4C_3)/{}_{12}C_5$.

1.79. Expand (a) $(x + y)^6$, (b) $(x - y)^4$, (c) $(x - x^{-1})^5$, (d) $(x^2 + 2)^4$.

1.80. Find the coefficient of x in $\left(x + \frac{2}{x}\right)^9$.

Probability using combinatorial analysis

1.81. Find the probability of scoring a total of 7 points (a) once, (b) at least once, (c) twice, in 2 tosses of a pair of fair dice.

1.82. Two cards are drawn successively from an ordinary deck of 52 well-shuffled cards. Find the probability that (a) the first card is not a ten of clubs or an ace; (b) the first card is an ace but the second is not; (c) at least one card is a diamond; (d) the cards are not of the same suit; (e) not more than 1 card is a picture card (jack, queen, king); (f) the second card is not a picture card; (g) the second card is not a picture card given that the first was a picture card; (h) the cards are picture cards or spades or both.

1.83. A box contains 9 tickets numbered from 1 to 9, inclusive. If 3 tickets are drawn from the box 1 at a time, find the probability that they are alternately either odd, even, odd or even, odd, even.

1.84. The odds in favor of A winning a game of chess against B are 3:2. If 3 games are to be played, what are the odds (a) in favor of A winning at least 2 games out of the 3, (b) against A losing the first 2 games to B?

1.85. In the game of *bridge*, each of 4 players is dealt 13 cards from an ordinary well-shuffled deck of 52 cards. Find the probability that one of the players (say, the eldest) gets (a) 7 diamonds, 2 clubs, 3 hearts, and 1 spade; (b) a complete suit.

1.86. An urn contains 6 red and 8 blue marbles. Five marbles are drawn at random from it without replacement. Find the probability that 3 are red and 2 are blue.

1.87. (a) Find the probability of getting the sum 7 on at least 1 of 3 tosses of a pair of fair dice, (b) How many tosses are needed in order that the probability in (a) be greater than 0.95?

1.88. Three cards are drawn from an ordinary deck of 52 cards. Find the probability that (a) all cards are of one suit, (b) at least 2 aces are drawn.

1.89. Find the probability that a bridge player is given 13 cards of which 9 cards are of one suit.

Miscellaneous problems

1.90. A sample space consists of 3 sample points with associated probabilities given by $2p$, p^2, and $4p - 1$. Find the value of p.

1.91. How many words can be made from 5 letters if (a) all letters are different, (b) 2 letters are identical, (c) all letters are different but 2 particular letters cannot be adjacent?

1.92. Four integers are chosen at random between 0 and 9, inclusive. Find the probability that (a) they are all different, (b) not more than 2 are the same.

1.93. A pair of dice is tossed repeatedly. Find the probability that an 11 occurs for the first time on the 6th toss.

1.94. What is the least number of tosses needed in Problem 1.93 so that the probability of getting an 11 will be greater than (a) 0.5, (b) 0.95?

1.95. In a game of poker find the probability of getting (a) a *royal flush*, which consists of the ten, jack, queen, king, and ace of a single suit; (b) a *full house*, which consists of 3 cards of one face value and 2 of another (such as 3 tens and 2 jacks); (c) all different cards; (d) 4 aces.

1.96. The probability that a man will hit a target is $\frac{2}{3}$. If he shoots at the target until he hits it for the first time, find the probability that it will take him 5 shots to hit the target.

1.97. (a) A shelf contains 6 separate compartments. In how many ways can 4 indistinguishable marbles be placed in the compartments? (b) Work the problem if there are n compartments and r marbles. This type of problem arises in physics in connection with *Bose-Einstein statistics*.

1.98. (a) A shelf contains 6 separate compartments. In how many ways can 12 indistinguishable marbles be placed in the compartments so that no compartment is empty? (b) Work the problem if there are n compartments and r marbles where $r > n$. This type of problem arises in physics in connection with *Fermi-Dirac statistics*.

1.99. A poker player has cards 2, 3, 4, 6, 8. He wishes to discard the 8 and replace it by another card which he hopes will be a 5 (in which case he gets an "inside straight"). What is the probability that he will succeed assuming that the other three players together have (a) one 5, (b) two 5s, (c) three 5s, (d) no 5? Can the problem be worked if the number of 5s in the other players' hands is unknown? Explain.

1.100. Work Problem 1.40 if the game is limited to 3 tosses.

1.101. Find the probability that in a game of bridge (a) 2, (b) 3, (c) all 4 players have a complete suit.

ANSWERS TO SUPPLEMENTARY PROBLEMS

1.47. (a) 5/26 (b) 5/36 (c) 0.98 (d) 2/9 (e) 7/8

1.48. (a) Probability of king on first draw and no king on second draw.

(b) Probability of either a king on first draw or a king on second draw or both.

(c) No king on first draw or no king on second draw or both (no king on first and second draws).

(d) No king on first, second, and third draws.

(e) Probability of either king on first draw and king on second draw or no king on second draw and king on third draw.

1.49. (a) 1/3 (b) 3/5 (c) 11/15 (d) 2/5 (e) 4/5

1.50. (a) 4/25 (c) 16/25 (e) 11/15 (g) 104/225 (i) 6/25
(b) 4/75 (d) 64/225 (f) 1/5 (h) 221/225 (j) 52/225

1.51. (a) 29/185 (c) 118/185 (e) 11/15 (g) 86/185 (i) 9/37
(b) 2/37 (d) 52/185 (f) 1/5 (h) 182/185 (j) 26/111

1.52. (a) 3/10 (b) 1/10 (c) 3/5 **1.53.** (a) 1/2197 (b) 1/17,576

1.54. 1/3 **1.55.** 21/56 **1.56.** 21/31 **1.57.** 1/3 **1.58.** 14/57

1.59.

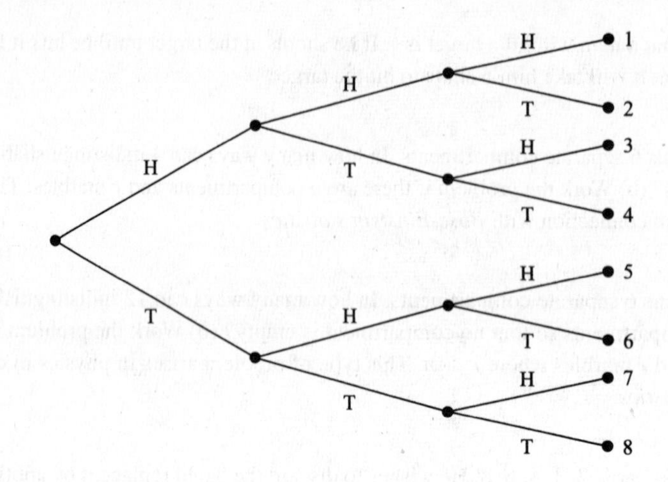

1.60. (a) $13 \times 13 \times 13$ (b) $13 \times 12 \times 26$ **1.61.** 8 **1.62.** (a) 12 (b) 2520 (c) 720

1.63. $n = 5$ **1.64.** 60 **1.65.** (a) 5040 (b) 720 (c) 240 **1.66.** (a) 8400 (b) 2520

1.67. (a) 32,805 (b) 11,664 **1.68.** 26 **1.69.** (a) 120 (b) 72 (c) 12

1.70. (a) 10 (b) 70 (c) 45 **1.71.** $n = 6$ **1.72.** 210 **1.73.** 840

1.74. (a) 42,000 (b) 7000 **1.75.** (a) 120 (b) 2520 **1.76.** (a) 150 (b) 45 (c) 100

1.77. (a) 17 (b) 163 **1.78.** (a) 20 (b) 330 (c) 14/99

1.79. (a) $x^6 + 6x^5y + 15x^4y^2 + 20x^3y^3 + 15x^2y^3 + 6xy^5 + y^6$
(b) $x^4 - 4x^3y + 6x^2y^2 - 4xy^3 + y^4$
(c) $x^5 - 5x^3 + 10x - 10x^{-1} + 5x^{-3} - x^{-5}$
(d) $x^8 + 8x^6 + 24x^4 + 32x^2 + 16$

1.80. 2016 **1.81.** (a) 5/18 (b) 11/36 (c) 1/36

1.82. (a) 47/52 (b) 16/221 (c) 15/34 (d) 13/17 (e) 210/221 (f) 10/13 (g) 40/51 (h) 77/442

1.83. 5/18 **1.84.** (a) 81 : 44 (b) 21 : 4

1.85. (a) $({}_{13}C_7)({}_{13}C_2)({}_{13}C_3)({}_{13}C_1)/{}_{52}C_{13}$ (b) $4/{}_{52}C_{13}$ **1.86.** $({}_6C_3)({}_8C_2)/{}_{14}C_5$

1.87. (a) 91/216 (b) at least 17 **1.88.** (a) $4 \cdot {}_{13}C_{3//52}C_3$ (b) $({}_4C_2 \cdot {}_{48}C_1 + {}_4C_3)/{}_{52}C_3$

1.89. $4({}_{13}C_9)({}_{39}C_4)/{}_{52}C_{13}$ **1.90.** $\sqrt{11} - 3$ **1.91.** (a) 120 (b) 60 (c) 72

1.92. (a) 63/125 (b) 963/1000 **1.93.** 1,419,857/34,012,224 **1.94.** (a) 13 (b) 53

1.95. (a) $4/{}_{52}C_5$ (b) $(13)(2)(4)(6)/{}_{52}C_5$ (c) $4^5\,({}_{13}C_5)/{}_{52}C_5$ (d) (5)(4)(3)(2)/(52)(51)(50)(49)

1.96. 2/243 **1.97.** (a) 126 (b) ${}_{n+r-1}C_{n-1}$ **1.98.** (a) 462 (b) ${}_{r-1}C_{n-1}$

1.99. (a) 3/32 (b) 1/16 (c) 1/32 (d) 1/8

1.100. prob. A wins = 61/216, prob. B wins = 5/36, prob. of tie = 125/216

1.101. (a) $12/({}_{52}C_{13})({}_{39}C_{13})$ (b) $24/({}_{52}C_{13})({}_{39}C_{13})({}_{26}C_{13})$

Random Variables and Probability Distributions

Random Variables

Suppose that to each point of a sample space we assign a number. We then have a *function* defined on the sample space. This function is called a *random variable* (or *stochastic variable*) or more precisely a *random function* (*stochastic function*). It is usually denoted by a capital letter such as X or Y. In general, a random variable has some specified physical, geometrical, or other significance.

EXAMPLE 2.1 Suppose that a coin is tossed twice so that the sample space is $S = \{HH, HT, TH, TT\}$. Let X represent the number of heads that can come up. With each sample point we can associate a number for X as shown in Table 2-1. Thus, for example, in the case of HH (i.e., 2 heads), $X = 2$ while for TH (1 head), $X = 1$. It follows that X is a random variable.

Table 2-1

Sample Point	HH	HT	TH	TT
X	2	1	1	0

It should be noted that many other random variables could also be defined on this sample space, for example, the square of the number of heads or the number of heads minus the number of tails.

A random variable that takes on a finite or countably infinite number of values (see page 4) is called a *discrete random variable* while one which takes on a noncountably infinite number of values is called a *nondiscrete random variable*.

Discrete Probability Distributions

Let X be a discrete random variable, and suppose that the possible values that it can assume are given by $x_1, x_2, x_3, \ldots$, arranged in some order. Suppose also that these values are assumed with probabilities given by

$$P(X = x_k) = f(x_k) \qquad k = 1, 2, \ldots \tag{1}$$

It is convenient to introduce the *probability function*, also referred to as *probability distribution*, given by

$$P(X = x) = f(x) \tag{2}$$

For $x = x_k$, this reduces to (1) while for other values of x, $f(x) = 0$.

In general, $f(x)$ is a probability function if

1. $f(x) \geq 0$
2. $\sum_x f(x) = 1$

where the sum in 2 is taken over all possible values of x.

EXAMPLE 2.2 Find the probability function corresponding to the random variable X of Example 2.1. Assuming that the coin is fair, we have

$$P(HH) = \frac{1}{4} \qquad P(HT) = \frac{1}{4} \qquad P(TH) = \frac{1}{4} \qquad P(TT) = \frac{1}{4}$$

Then

$$P(X = 0) = P(TT) = \frac{1}{4}$$

$$P(X = 1) = P(HT \cup TH) = P(HT) + P(TH) = \frac{1}{4} + \frac{1}{4} = \frac{1}{2}$$

$$P(X = 2) = P(HH) = \frac{1}{4}$$

The probability function is thus given by Table 2-2.

Table 2-2

x	0	1	2
$f(x)$	1/4	1/2	1/4

Distribution Functions for Random Variables

The *cumulative distribution function*, or briefly the *distribution function*, for a random variable X is defined by

$$F(x) = P(X \le x) \tag{3}$$

where x is any real number, i.e., $-\infty < x < \infty$.

The distribution function $F(x)$ has the following properties:

1. $F(x)$ is nondecreasing [i.e., $F(x) \le F(y)$ if $x \le y$].
2. $\lim_{x \to -\infty} F(x) = 0$; $\lim_{x \to \infty} F(x) = 1$.
3. $F(x)$ is continuous from the right [i.e., $\lim_{h \to 0^+} F(x + h) = F(x)$ for all x].

Distribution Functions for Discrete Random Variables

The distribution function for a discrete random variable X can be obtained from its probability function by noting that, for all x in $(-\infty, \infty)$,

$$F(x) = P(X \le x) = \sum_{u \le x} f(u) \tag{4}$$

where the sum is taken over all values u taken on by X for which $u \le x$.

If X takes on only a finite number of values $x_1, x_2, \ldots, x_n$, then the distribution function is given by

$$F(x) = \begin{cases} 0 & -\infty < x < x_1 \\ f(x_1) & x_1 \le x < x_2 \\ f(x_1) + f(x_2) & x_2 \le x < x_3 \\ \vdots & \vdots \\ f(x_1) + \cdots + f(x_n) & x_n \le x < \infty \end{cases} \tag{5}$$

EXAMPLE 2.3 (a) Find the distribution function for the random variable X of Example 2.2. (b) Obtain its graph.

(a) The distribution function is

$$F(x) = \begin{cases} 0 & -\infty < x < 0 \\ \frac{1}{4} & 0 \le x < 1 \\ \frac{3}{4} & 1 \le x < 2 \\ 1 & 2 \le x < \infty \end{cases}$$

(b) The graph of $F(x)$ is shown in Fig. 2-1.

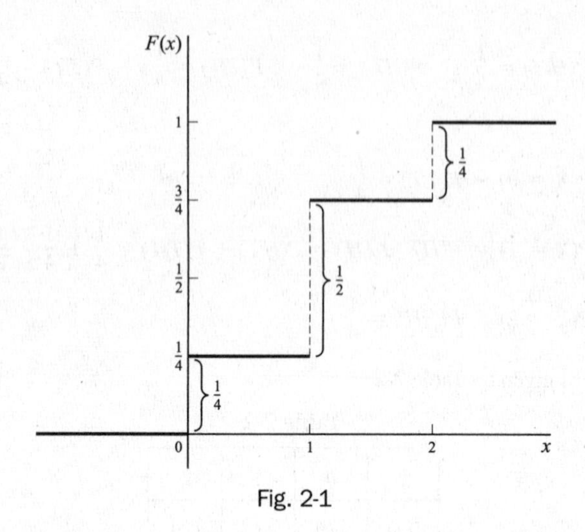

Fig. 2-1

The following things about the above distribution function, which are true in general, should be noted.

1. The magnitudes of the jumps at 0, 1, 2 are $\frac{1}{4}, \frac{1}{2}, \frac{1}{4}$ which are precisely the probabilities in Table 2-2. This fact enables one to obtain the probability function from the distribution function.
2. Because of the appearance of the graph of Fig. 2-1, it is often called a *staircase function* or *step function*. The value of the function at an integer is obtained from the higher step; thus the value at 1 is $\frac{3}{4}$ and not $\frac{1}{4}$. This is expressed mathematically by stating that the distribution function is *continuous from the right* at 0, 1, 2.
3. As we proceed from left to right (i.e. going *upstairs*), the distribution function either remains the same or increases, taking on values from 0 to 1. Because of this, it is said to be a *monotonically increasing function*.

It is clear from the above remarks and the properties of distribution functions that the probability function of a discrete random variable can be obtained from the distribution function by noting that

$$f(x) = F(x) - \lim_{u \to x^-} F(u). \tag{6}$$

Continuous Random Variables

A nondiscrete random variable X is said to be *absolutely continuous*, or simply *continuous*, if its distribution function may be represented as

$$F(x) = P(X \leq x) = \int_{-\infty}^{x} f(u)\,du \qquad (-\infty < x < \infty) \tag{7}$$

where the function $f(x)$ has the properties

1. $f(x) \geq 0$
2. $\int_{-\infty}^{\infty} f(x)dx = 1$

It follows from the above that if X is a continuous random variable, then the probability that X takes on any one particular value is zero, whereas the *interval probability* that X lies *between two different values*, say, a and b, is given by

$$P(a < X < b) = \int_{a}^{b} f(x)\,dx \tag{8}$$

EXAMPLE 2.4 If an individual is selected at random from a large group of adult males, the probability that his height X is precisely 68 inches (i.e., 68.000 . . . inches) would be zero. However, there is a probability greater than zero than X is between 67.000 . . . inches and 68.500 . . . inches, for example.

A function $f(x)$ that satisfies the above requirements is called a *probability function* or *probability distribution* for a continuous random variable, but it is more often called a *probability density function* or simply *density function*. Any function $f(x)$ satisfying Properties 1 and 2 above will automatically be a density function, and required probabilities can then be obtained from (8).

EXAMPLE 2.5 (a) Find the constant c such that the function

$$f(x) = \begin{cases} cx^2 & 0 < x < 3 \\ 0 & \text{otherwise} \end{cases}$$

is a density function, and (b) compute $P(1 < X < 2)$.

(a) Since $f(x)$ satisfies Property 1 if $c \geq 0$, it must satisfy Property 2 in order to be a density function. Now

$$\int_{-\infty}^{\infty} f(x)dx = \int_0^3 cx^2\,dx = \frac{cx^3}{3}\bigg|_0^3 = 9c$$

and since this must equal 1, we have $c = 1/9$.

(b)
$$P(1 < X < 2) = \int_1^2 \frac{1}{9}x^2\,dx = \frac{x^3}{27}\bigg|_1^2 = \frac{8}{27} - \frac{1}{27} = \frac{7}{27}$$

In case $f(x)$ is continuous, which we shall assume unless otherwise stated, the probability that X is equal to any particular value is zero. In such case we can replace either or both of the signs $<$ in (8) by $\leq$. Thus, in Example 2.5,

$$P(1 \leq X \leq 2) = P(1 \leq X < 2) = P(1 < X \leq 2) = P(1 < X < 2) = \frac{7}{27}$$

EXAMPLE 2.6 (a) Find the distribution function for the random variable of Example 2.5. (b) Use the result of (a) to find $P(1 < x \leq 2)$.

(a) We have

$$F(x) = P(X \leq x) = \int_{-\infty}^{x} f(u)\,du$$

If $x < 0$, then $F(x) = 0$. If $0 \leq x < 3$, then

$$F(x) = \int_0^x f(u)\,du = \int_0^x \frac{1}{9}u^2\,du = \frac{x^3}{27}$$

If $x \geq 3$, then

$$F(x) = \int_0^3 f(u)\,du + \int_3^x f(u)\,du = \int_0^3 \frac{1}{9}u^2\,du + \int_3^x 0\,du = 1$$

Thus the required distribution function is

$$F(x) = \begin{cases} 0 & x < 0 \\ x^3/27 & 0 \leq x < 3 \\ 1 & x \geq 3 \end{cases}$$

Note that $F(x)$ increases monotonically from 0 to 1 as is required for a distribution function. It should also be noted that $F(x)$ in this case is continuous.

(b) We have

$$P(1 < X \le 2) = P(X \le 2) - P(X \le 1) \\ = F(2) - F(1) \\ = \frac{2^3}{27} - \frac{1^3}{27} = \frac{7}{27}$$

as in Example 2.5.

The probability that X is between x and $x + \Delta x$ is given by

$$P(x \le X \le x + \Delta x) = \int_x^{x+\Delta x} f(u)\,du \tag{9}$$

so that if Δx is small, we have approximately

$$P(x \le X \le x + \Delta x) = f(x)\Delta x \tag{10}$$

We also see from (7) on differentiating both sides that

$$\frac{dF(x)}{dx} = f(x) \tag{11}$$

at all points where $f(x)$ is continuous; i.e., the derivative of the distribution function is the density function.

It should be pointed out that random variables exist that are neither discrete nor continuous. It can be shown that the random variable X with the following distribution function is an example.

$$F(x) = \begin{cases} 0 & x < 1 \\ \frac{x}{2} & 1 \le x < 2 \\ 1 & x \ge 2 \end{cases}$$

In order to obtain (11), we used the basic property

$$\frac{d}{dx}\int_a^x f(u)\,du = f(x) \tag{12}$$

which is one version of the Fundamental Theorem of Calculus.

Graphical Interpretations

If $f(x)$ is the density function for a random variable X, then we can represent $y = f(x)$ graphically by a curve as in Fig. 2-2. Since $f(x) \ge 0$, the curve cannot fall below the x axis. The entire area bounded by the curve and the x axis must be 1 because of Property 2 on page 36. Geometrically the probability that X is between a and b, i.e., $P(a < X < b)$, is then represented by the area shown shaded, in Fig. 2-2.

The distribution function $F(x) = P(X \le x)$ is a monotonically increasing function which increases from 0 to 1 and is represented by a curve as in Fig. 2-3.

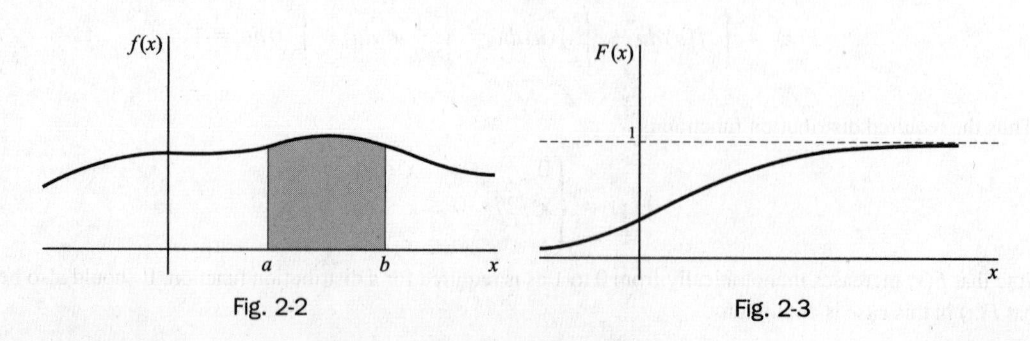

Fig. 2-2

Fig. 2-3

Joint Distributions

The above ideas are easily generalized to two or more random variables. We consider the typical case of two random variables that are either both discrete or both continuous. In cases where one variable is discrete and the other continuous, appropriate modifications are easily made. Generalizations to more than two variables can also be made.

1. DISCRETE CASE. If X and Y are two discrete random variables, we define the *joint probability function* of X and Y by

$$P(X = x, Y = y) = f(x, y) \tag{13}$$

where 1. $f(x, y) \geq 0$

2. $\sum_x \sum_y f(x, y) = 1$

i.e., the sum over all values of x and y is 1.

Suppose that X can assume any one of m values $x_1, x_2, \ldots, x_m$ and Y can assume any one of n values $y_1, y_2, \ldots, y_n$. Then the probability of the event that $X = x_j$ and $Y = y_k$ is given by

$$P(X = x_j, Y = y_k) = f(x_j, y_k) \tag{14}$$

A joint probability function for X and Y can be represented by a *joint probability table* as in Table 2-3. The probability that $X = x_j$ is obtained by adding all entries in the row corresponding to x_i and is given by

$$P(X = x_j) = f_1(x_j) = \sum_{k=1}^{n} f(x_j, y_k) \tag{15}$$

Table 2-3

$X \backslash Y$	y_1	y_2	$\cdots$	y_n	Totals ↓
x_1	$f(x_1, y_1)$	$f(x_1, y_2)$	$\cdots$	$f(x_1, y_n)$	$f_1(x_1)$
x_2	$f(x_2, y_1)$	$f(x_2, y_2)$	$\cdots$	$f(x_2, y_n)$	$f_1(x_2)$
$\vdots$	$\vdots$	$\vdots$		$\vdots$	$\vdots$
x_m	$f(x_m, y_1)$	$f(x_m, y_2)$	$\cdots$	$f(x_m, y_n)$	$f_1(x_m)$
Totals →	$f_2(y_1)$	$f_2(y_2)$	$\cdots$	$f_2(y_n)$	1 ← Grand Total

For $j = 1, 2, \ldots, m$, these are indicated by the entry totals in the extreme right-hand column or margin of Table 2-3. Similarly the probability that $Y = y_k$ is obtained by adding all entries in the column corresponding to y_k and is given by

$$P(Y = y_k) = f_2(y_k) = \sum_{j=1}^{m} f(x_j, y_k) \tag{16}$$

For $k = 1, 2, \ldots, n$, these are indicated by the entry totals in the bottom row or margin of Table 2-3.

Because the probabilities (15) and (16) are obtained from the margins of the table, we often refer to $f_1(x_j)$ and $f_2(y_k)$ [or simply $f_1(x)$ and $f_2(y)$] as the *marginal probability functions* of X and Y, respectively.

It should also be noted that

$$\sum_{j=1}^{m} f_1(x_j) = 1 \quad \sum_{k=1}^{n} f_2(y_k) = 1 \tag{17}$$

which can be written

$$\sum_{j=1}^{m} \sum_{k=1}^{n} f(x_j, y_k) = 1 \tag{18}$$

This is simply the statement that the total probability of all entries is 1. The *grand total* of 1 is indicated in the lower right-hand corner of the table.

The *joint distribution function* of X and Y is defined by

$$F(x, y) = P(X \leq x, Y \leq y) = \sum_{u \leq x} \sum_{v \leq y} f(u, v) \tag{19}$$

In Table 2-3, $F(x, y)$ is the sum of all entries for which $x_j \leq x$ and $y_k \leq y$.

2. CONTINUOUS CASE. The case where both variables are continuous is obtained easily by analogy with the discrete case on replacing sums by integrals. Thus the *joint probability function* for the random variables X and Y (or, as it is more commonly called, the *joint density function* of X and Y) is defined by

1. $f(x, y) \geq 0$
2. $\int_{-\infty}^{\infty} \int_{-\infty}^{\infty} f(x, y)\, dx\, dy = 1$

Graphically $z = f(x, y)$ represents a surface, called the *probability surface*, as indicated in Fig. 2-4. The total volume bounded by this surface and the xy plane is equal to 1 in accordance with Property 2 above. The probability that X lies between a and b while Y lies between c and d is given graphically by the shaded volume of Fig. 2-4 and mathematically by

$$P(a < X < b, c < Y < d) = \int_{x=a}^{b} \int_{y=c}^{d} f(x, y)\, dx\, dy \tag{20}$$

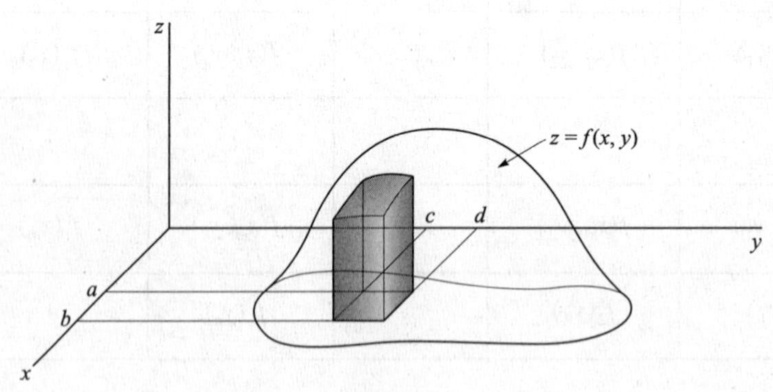

Fig. 2-4

More generally, if A represents any event, there will be a region $\mathcal{R}_A$ of the xy plane that corresponds to it. In such case we can find the probability of A by performing the integration over $\mathcal{R}_A$, i.e.,

$$P(A) = \iint_{\mathcal{R}_A} f(x, y)\, dx\, dy \tag{21}$$

The *joint distribution function* of X and Y in this case is defined by

$$F(x, y) = P(X \leq x, Y \leq y) = \int_{u=-\infty}^{x} \int_{v=-\infty}^{y} f(u, v)\, du\, dv \tag{22}$$

It follows in analogy with (11), page 38, that

$$\frac{\partial^2 F}{\partial x\,\partial y} = f(x, y) \tag{23}$$

i.e., the density function is obtained by differentiating the distribution function with respect to x and y.

From (22) we obtain

$$P(X \le x) = F_1(x) = \int_{u=-\infty}^{x}\int_{v=-\infty}^{\infty} f(u, v)\,du\,dv \tag{24}$$

$$P(Y \le y) = F_2(y) = \int_{u=-\infty}^{\infty}\int_{v=-\infty}^{y} f(u, v)\,du\,dv \tag{25}$$

We call (24) and (25) the *marginal distribution functions*, or simply the *distribution functions*, of X and Y, respectively. The derivatives of (24) and (25) with respect to x and y are then called the *marginal density functions*, or simply the *density functions*, of X and Y and are given by

$$f_1(x) = \int_{v=-\infty}^{\infty} f(x, v)\,dv \qquad f_2(y) = \int_{u=-\infty}^{\infty} f(u, y)\,du \tag{26}$$

Independent Random Variables

Suppose that X and Y are discrete random variables. If the events $X = x$ and $Y = y$ are independent events for all x and y, then we say that X and Y are *independent random variables*. In such case,

$$P(X = x, Y = y) = P(X = x)P(Y = y) \tag{27}$$

or equivalently

$$f(x, y) = f_1(x)f_2(y) \tag{28}$$

Conversely, if for all x and y the joint probability function $f(x, y)$ can be expressed as the product of a function of x alone and a function of y alone (which are then the marginal probability functions of X and Y), X and Y are independent. If, however, $f(x, y)$ cannot be so expressed, then X and Y are *dependent*.

If X and Y are continuous random variables, we say that they are *independent random variables* if the events $X \le x$ and $Y \le y$ are independent events for all x and y. In such case we can write

$$P(X \le x, Y \le y) = P(X \le x)P(Y \le y) \tag{29}$$

or equivalently

$$F(x, y) = F_1(x)F_2(y) \tag{30}$$

where $F_1(z)$ and $F_2(y)$ are the (marginal) distribution functions of X and Y, respectively. Conversely, X and Y are independent random variables if for all x and y, their joint distribution function $F(x, y)$ can be expressed as a product of a function of x alone and a function of y alone (which are the marginal distributions of X and Y, respectively). If, however, $F(x, y)$ cannot be so expressed, then X and Y are dependent.

For continuous independent random variables, it is also true that the joint density function $f(x, y)$ is the product of a function of x alone, $f_1(x)$, and a function of y alone, $f_2(y)$, and these are the (marginal) density functions of X and Y, respectively.

Change of Variables

Given the probability distributions of one or more random variables, we are often interested in finding distributions of other random variables that depend on them in some specified manner. Procedures for obtaining these distributions are presented in the following theorems for the case of discrete and continuous variables.

1. DISCRETE VARIABLES

Theorem 2-1 Let X be a discrete random variable whose probability function is $f(x)$. Suppose that a discrete random variable U is defined in terms of X by $U = \phi(X)$, where to each value of X there corresponds one and only one value of U and conversely, so that $X = \psi(U)$. Then the probability function for U is given by

$$g(u) = f[\psi(u)] \tag{31}$$

Theorem 2-2 Let X and Y be discrete random variables having joint probability function $f(x, y)$. Suppose that two discrete random variables U and V are defined in terms of X and Y by $U = \phi_1(X, Y)$, $V = \phi_2(X, Y)$, where to each pair of values of X and Y there corresponds one and only one pair of values of U and V and conversely, so that $X = \psi_1(U, V)$, $Y = \psi_2(U, V)$. Then the joint probability function of U and V is given by

$$g(u, v) = f[\psi_1(u, v), \psi_2(u, v)] \tag{32}$$

2. CONTINUOUS VARIABLES

Theorem 2-3 Let X be a continuous random variable with probability density $f(x)$. Let us define $U = \phi(X)$ where $X = \psi(U)$ as in Theorem 2-1. Then the probability density of U is given by $g(u)$ where

$$g(u)|du| = f(x)|dx| \tag{33}$$

or

$$g(u) = f(x)\left|\frac{dx}{du}\right| = f[\psi(u)]|\psi'(u)| \tag{34}$$

Theorem 2-4 Let X and Y be continuous random variables having joint density function $f(x, y)$. Let us define $U = \phi_1(X, Y)$, $V = \phi_2(X, Y)$ where $X = \psi_1(U, V)$, $Y = \psi_2(U, V)$ as in Theorem 2-2. Then the joint density function of U and V is given by $g(u, v)$ where

$$g(u, v)|du\, dv| = f(x, y)|dx\, dy| \tag{35}$$

or

$$g(u, v) = f(x, y)\left|\frac{\partial(x, y)}{\partial(u, v)}\right| = f[\psi_1(u, v), \psi_2(u, v)]|J| \tag{36}$$

In (36) the *Jacobian determinant*, or briefly *Jacobian*, is given by

$$J = \frac{\partial(x, y)}{\partial(u, v)} = \begin{vmatrix} \dfrac{\partial x}{\partial u} & \dfrac{\partial x}{\partial v} \\ \dfrac{\partial y}{\partial u} & \dfrac{\partial y}{\partial v} \end{vmatrix} \tag{37}$$

Probability Distributions of Functions of Random Variables

Theorems 2-2 and 2-4 specifically involve joint probability functions of two random variables. In practice one often needs to find the probability distribution of some specified function of several random variables. Either of the following theorems is often useful for this purpose.

Theorem 2-5 Let X and Y be continuous random variables and let $U = \phi_1(X, Y)$, $V = X$ (the second choice is arbitrary). Then the density function for U is the marginal density obtained from the joint density of U and V as found in Theorem 2-4. A similar result holds for probability functions of discrete variables.

Theorem 2-6 Let $f(x, y)$ be the joint density function of X and Y. Then the density function $g(u)$ of the random variable $U = \phi_1(X, Y)$ is found by differentiating with respect to u the distribution function given by

$$G(u) = P[\phi_1(X, Y) \le u] = \iint_{\mathcal{R}} f(x, y)\,dx\,dy \tag{38}$$

Where $\mathcal{R}$ is the region for which $\phi_1(x, y) \le u$.

Convolutions

As a particular consequence of the above theorems, we can show (see Problem 2.23) that the density function of the sum of two continuous random variables X and Y, i.e., of $U = X + Y$, having joint density function $f(x, y)$ is given by

$$g(u) = \int_{-\infty}^{\infty} f(x, u - x)\,dx \tag{39}$$

In the special case where X and Y are independent, $f(x, y) = f_1(x) f_2(y)$, and (39) reduces to

$$g(u) = \int_{-\infty}^{\infty} f_1(x)\, f_2(u - x)\,dx \tag{40}$$

which is called the *convolution* of f_1 and f_2, abbreviated, $f_1 * f_2$.

The following are some important properties of the convolution:

1. $f_1 * f_2 = f_2 * f_1$
2. $f_1 * (f_2 * f_3) = (f_1 * f_2) * f_3$
3. $f_1 * (f_2 + f_3) = f_1 * f_2 + f_1 * f_3$

These results show that f_1, f_2, f_3 obey the *commutative*, *associative*, and *distributive laws* of algebra with respect to the operation of convolution.

Conditional Distributions

We already know that if $P(A) > 0$,

$$P(B \mid A) = \frac{P(A \cap B)}{P(A)} \tag{41}$$

If X and Y are discrete random variables and we have the events (A: $X = x$), (B: $Y = y$), then (41) becomes

$$P(Y = y \mid X = x) = \frac{f(x, y)}{f_1(x)} \tag{42}$$

where $f(x, y) = P(X = x, Y = y)$ is the joint probability function and $f_1(x)$ is the marginal probability function for X. We define

$$f(y \mid x) \equiv \frac{f(x, y)}{f_1(x)} \tag{43}$$

and call it the *conditional probability function of Y given X*. Similarly, the conditional probability function of X given Y is

$$f(x \mid y) \equiv \frac{f(x, y)}{f_2(y)} \tag{44}$$

We shall sometimes denote $f(x \mid y)$ and $f(y \mid x)$ by $f_1(x \mid y)$ and $f_2(y \mid x)$, respectively.

These ideas are easily extended to the case where X, Y are continuous random variables. For example, the *conditional density function of Y given X* is

$$f(y \mid x) \equiv \frac{f(x, y)}{f_1(x)} \tag{45}$$

where $f(x, y)$ is the joint density function of X and Y, and $f_1(x)$ is the marginal density function of X. Using (45) we can, for example, find that the probability of Y being between c and d given that $x < X < x + dx$ is

$$P(c < Y < d \mid x < X < x + dx) = \int_c^d f(y \mid x)\,dy \tag{46}$$

Generalizations of these results are also available.

Applications to Geometric Probability

Various problems in probability arise from geometric considerations or have geometric interpretations. For example, suppose that we have a target in the form of a plane region of area K and a portion of it with area K_1, as in Fig. 2-5. Then it is reasonable to suppose that the probability of hitting the region of area K_1 is proportional to K_1. We thus define

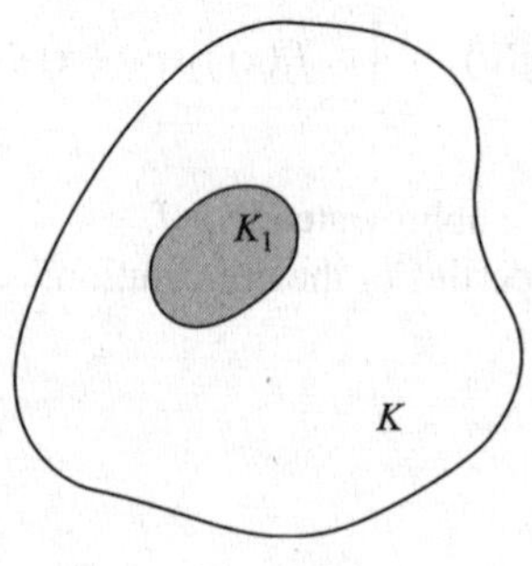

Fig. 2-5

$$P(\text{hitting region of area } K_1) = \frac{K_1}{K} \tag{47}$$

where it is assumed that the probability of hitting the target is 1. Other assumptions can of course be made. For example, there could be less probability of hitting outer areas. The type of assumption used defines the probability distribution function.

SOLVED PROBLEMS

Discrete random variables and probability distributions

2.1. Suppose that a pair of fair dice are to be tossed, and let the random variable X denote the sum of the points. Obtain the probability distribution for X.

The sample points for tosses of a pair of dice are given in Fig. 1-9, page 14. The random variable X is the sum of the coordinates for each point. Thus for (3, 2) we have $X = 5$. Using the fact that all 36 sample points are equally probable, so that each sample point has probability 1/36, we obtain Table 2-4. For example, corresponding to $X = 5$, we have the sample points (1, 4), (2, 3), (3, 2), (4, 1), so that the associated probability is 4/36.

Table 2-4

x	2	3	4	5	6	7	8	9	10	11	12
$f(x)$	1/36	2/36	3/36	4/36	5/36	6/36	5/36	4/36	3/36	2/36	1/36

2.2. Find the probability distribution of boys and girls in families with 3 children, assuming equal probabilities for boys and girls.

Problem 1.37 treated the case of n mutually independent trials, where each trial had just two possible outcomes, A and A', with respective probabilities p and $q = 1 - p$. It was found that the probability of getting exactly x A's in the n trials is ${}_nC_x p^x q^{n-x}$. This result applies to the present problem, under the assumption that successive births (the "trials") are independent as far as the sex of the child is concerned. Thus, with A being the event "a boy," $n = 3$, and $p = q = \frac{1}{2}$, we have

$$P(\text{exactly } x \text{ boys}) = P(X = x) = {}_3C_x\left(\frac{1}{2}\right)^x\left(\frac{1}{2}\right)^{3-x} = {}_3C_x\left(\frac{1}{2}\right)^3$$

where the random variable X represents the number of boys in the family. (Note that X is defined on the sample space of 3 trials.) The probability function for X,

$$f(x) = {}_3C_x\left(\frac{1}{2}\right)^3$$

is displayed in Table 2-5.

Table 2-5

x	0	1	2	3
$f(x)$	1/8	3/8	3/8	1/8

Discrete distribution functions

2.3. (a) Find the distribution function $F(x)$ for the random variable X of Problem 2.1, and (b) graph this distribution function.

(a) We have $F(x) = P(X \le x) = \sum_{u \le x} f(u)$. Then from the results of Problem 2.1, we find

$$F(x) = \begin{cases} 0 & -\infty < x < 2 \\ 1/36 & 2 \le x < 3 \\ 3/36 & 3 \le x < 4 \\ 6/36 & 4 \le x < 5 \\ \vdots & \vdots \\ 35/36 & 11 \le x < 12 \\ 1 & 12 \le x < \infty \end{cases}$$

(b) See Fig. 2-6.

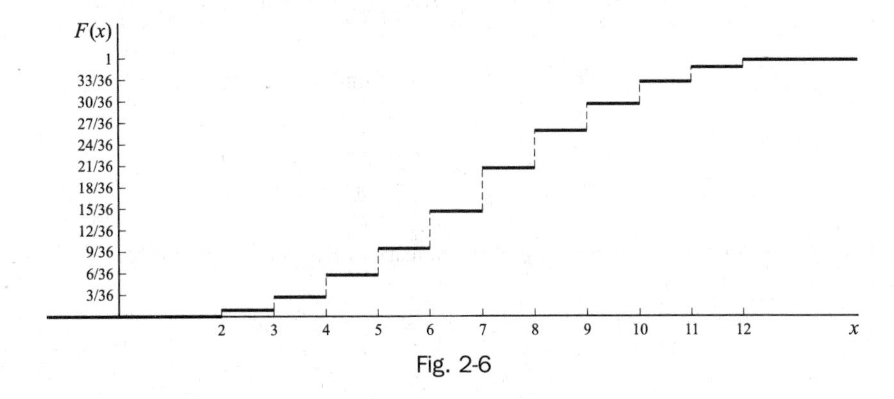

Fig. 2-6

2.4. (a) Find the distribution function $F(x)$ for the random variable X of Problem 2.2, and (b) graph this distribution function.

(a) Using Table 2-5 from Problem 2.2, we obtain

$$F(x) = \begin{cases} 0 & -\infty < x < 0 \\ 1/8 & 0 \le x < 1 \\ 1/2 & 1 \le x < 2 \\ 7/8 & 2 \le x < 3 \\ 1 & 3 \le x < \infty \end{cases}$$

(b) The graph of the distribution function of (a) is shown in Fig. 2-7.

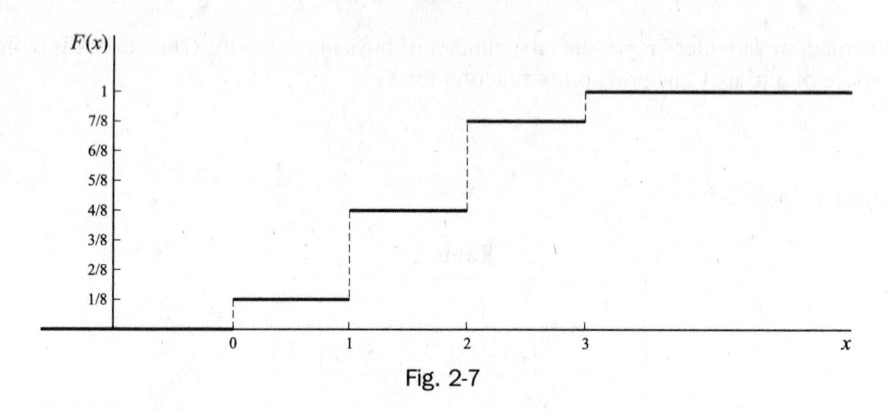

Fig. 2-7

Continuous random variables and probability distributions

2.5. A random variable X has the density function $f(x) = c/(x^2 + 1)$, where $-\infty < x < \infty$. (*a*) Find the value of the constant c. (b) Find the probability that X^2 lies between 1/3 and 1.

(a) We must have $\int_{-\infty}^{\infty} f(x)\,dx = 1$, i.e.,

$$\int_{-\infty}^{\infty} \frac{c\,dx}{x^2 + 1} = c \tan^{-1} x \Big|_{-\infty}^{\infty} = c\left[\frac{\pi}{2} - \left(-\frac{\pi}{2}\right)\right] = 1$$

so that $c = 1/\pi$.

(b) If $\frac{1}{3} \le X^2 \le 1$, then either $\frac{\sqrt{3}}{3} \le X \le 1$ or $-1 \le X \le -\frac{\sqrt{3}}{3}$. Thus the required probability is

$$\frac{1}{\pi}\int_{-1}^{-\sqrt{3}/3} \frac{dx}{x^2 + 1} + \frac{1}{\pi}\int_{\sqrt{3}/3}^{1} \frac{dx}{x^2 + 1} = \frac{2}{\pi}\int_{\sqrt{3}/3}^{1} \frac{dx}{x^2 + 1}$$

$$= \frac{2}{\pi}\left[\tan^{-1}(1) - \tan^{-1}\left(\frac{\sqrt{3}}{3}\right)\right]$$

$$= \frac{2}{\pi}\left(\frac{\pi}{4} - \frac{\pi}{6}\right) = \frac{1}{6}$$

2.6. Find the distribution function corresponding to the density function of Problem 2.5.

$$F(x) = \int_{-\infty}^{x} f(u)\,du = \frac{1}{\pi}\int_{-\infty}^{x} \frac{du}{u^2 + 1} = \frac{1}{\pi}\left[\tan^{-1} u \Big|_{-\infty}^{x}\right]$$

$$= \frac{1}{\pi}[\tan^{-1} x - \tan^{-1}(-\infty)] = \frac{1}{\pi}\left[\tan^{-1} x + \frac{\pi}{2}\right]$$

$$= \frac{1}{2} + \frac{1}{\pi}\tan^{-1} x$$

2.7. The distribution function for a random variable X is

$$F(x) = \begin{cases} 1 - e^{-2x} & x \geq 0 \\ 0 & x < 0 \end{cases}$$

Find (a) the density function, (b) the probability that $X > 2$, and (c) the probability that $-3 < X \leq 4$.

(a) $$f(x) = \frac{d}{dx}F(x) = \begin{cases} 2e^{-2x} & x > 0 \\ 0 & x < 0 \end{cases}$$

(b) $$P(X > 2) = \int_2^{\infty} 2e^{-2u}\,du = -e^{-2u}\Big|_2^{\infty} = e^{-4}$$

Another method

By definition, $P(X \leq 2) = F(2) = 1 - e^{-4}$. Hence,

$$P(X > 2) = 1 - (1 - e^{-4}) = e^{-4}$$

(c) $$P(-3 < X \leq 4) = \int_{-3}^{4} f(u)\,du = \int_{-3}^{0} 0\,du + \int_0^4 2e^{-2u}\,du$$
$$= -e^{-2u}\Big|_0^4 = 1 - e^{-8}$$

Another method

$$P(-3 < X \leq 4) = P(X \leq 4) - P(X \leq -3)$$
$$= F(4) - F(-3)$$
$$= (1 - e^{-8}) - (0) = 1 - e^{-8}$$

Joint distributions and independent variables

2.8. The joint probability function of two discrete random variables X and Y is given by $f(x, y) = c(2x + y)$, where x and y can assume all integers such that $0 \leq x \leq 2$, $0 \leq y \leq 3$, and $f(x, y) = 0$ otherwise.

(a) Find the value of the constant c.
(b) Find $P(X = 2, Y = 1)$.
(c) Find $P(X \geq 1, Y \leq 2)$.

(a) The sample points (x, y) for which probabilities are different from zero are indicated in Fig. 2-8. The probabilities associated with these points, given by $c(2x + y)$, are shown in Table 2-6. Since the grand total, $42c$, must equal 1, we have $c = 1/42$.

Table 2-6

X \ Y	0	1	2	3	Totals ↓
0	0	c	$2c$	$3c$	$6c$
1	$2c$	$3c$	$4c$	$5c$	$14c$
2	$4c$	$5c$	$6c$	$7c$	$22c$
Totals →	$6c$	$9c$	$12c$	$15c$	$42c$

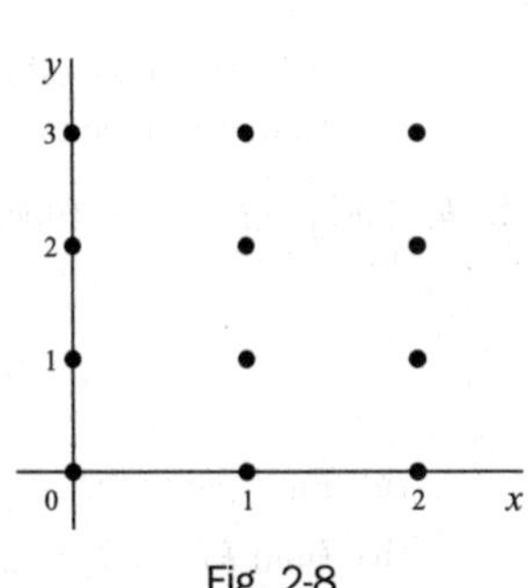

Fig. 2-8

(b) From Table 2-6 we see that

$$P(X = 2, Y = 1) = 5c + \frac{5}{42}$$

(c) From Table 2-6 we see that

$$P(X \geq 1, Y \leq 2) = \sum_{x \geq 1} \sum_{y \leq 2} f(x, y)$$
$$= (2c + 3c + 4c)(4c + 5c + 6c)$$
$$= 24c = \frac{24}{42} = \frac{4}{7}$$

as indicated by the entries shown shaded in the table.

2.9. Find the marginal probability functions (a) of X and (b) of Y for the random variables of Problem 2.8.

(a) The marginal probability function for X is given by $P(X = x) = f_1(x)$ and can be obtained from the margin totals in the right-hand column of Table 2-6. From these we see that

$$P(X = x) = f_1(x) = \begin{cases} 6c = 1/7 & x = 0 \\ 14c = 1/3 & x = 1 \\ 22c = 11/21 & x = 2 \end{cases}$$

Check: $\frac{1}{7} + \frac{1}{3} + \frac{11}{21} = 1$

(b) The marginal probability function for Y is given by $P(Y = y) = f_2(y)$ and can be obtained from the margin totals in the last row of Table 2-6. From these we see that

$$P(Y = y) = f_2(y) = \begin{cases} 6c = 1/7 & y = 0 \\ 9c = 3/14 & y = 1 \\ 12c = 2/7 & y = 2 \\ 15c = 5/14 & y = 3 \end{cases}$$

Check: $\frac{1}{7} + \frac{3}{14} + \frac{2}{7} + \frac{5}{14} = 1$

2.10. Show that the random variables X and Y of Problem 2.8 are dependent.

If the random variables X and Y are independent, then we must have, for all x and y,

$$P(X = x, Y = y) = P(X = x)P(Y = y)$$

But, as seen from Problems 2.8(b) and 2.9,

$$P(X = 2, Y = 1) = \frac{5}{42} \qquad P(X = 2) = \frac{11}{21} \qquad P(Y = 1) = \frac{3}{14}$$

so that

$$P(X = 2, Y = 1) \neq P(X = 2)P(Y = 1)$$

The result also follows from the fact that the joint probability function $(2x + y)/42$ cannot be expressed as a function of x alone times a function of y alone.

2.11. The joint density function of two continuous random variables X and Y is

$$f(x, y) = \begin{cases} cxy & 0 < x < 4, 1 < y < 5 \\ 0 & \text{otherwise} \end{cases}$$

(a) Find the value of the constant c.

(b) Find $P(1 < X < 2, 2 < Y < 3)$.

(c) Find $P(X \geq 3, Y \leq 2)$.

(a) We must have the total probability equal to 1, i.e.,

$$\int_{-\infty}^{\infty} \int_{-\infty}^{\infty} f(x, y)\, dx\, dy = 1$$

Using the definition of $f(x, y)$, the integral has the value

$$\int_{x=0}^{4}\int_{y=1}^{5} cxy\,dx\,dy = c\int_{x=0}^{4}\left[\int_{y=1}^{5} xy\,dy\right]dx$$

$$= c\int_{z=0}^{4} \frac{xy^2}{2}\bigg|_{y=1}^{5} dx = c\int_{x=0}^{4}\left(\frac{25x}{2} - \frac{x}{2}\right)dx$$

$$= c\int_{x=0}^{4} 12x\,dx = c(6x^2)\bigg|_{x=0}^{4} = 96c$$

Then $96c = 1$ and $c = 1/96$.

(b) Using the value of c found in (a), we have

$$P(1 < X < 2, 2 < Y < 3) = \int_{x=1}^{2}\int_{y=2}^{3} \frac{xy}{96}\,dx\,dy$$

$$= \frac{1}{96}\int_{x=1}^{2}\left[\int_{y=2}^{3} xy\,dy\right]dx = \frac{1}{96}\int_{x=1}^{2} \frac{xy^2}{2}\bigg|_{y=2}^{3} dx$$

$$= \frac{1}{96}\int_{x=1}^{2} \frac{5x}{2}\,dx = \frac{5}{192}\left(\frac{x^2}{2}\right)\bigg|_{1}^{2} = \frac{5}{128}$$

(c)

$$P(X \geq 3, Y \leq 2) = \int_{x=3}^{4}\int_{y=1}^{2} \frac{xy}{96}\,dx\,dy$$

$$= \frac{1}{96}\int_{x=3}^{4}\left[\int_{y=1}^{2} xy\,dy\right]dx = \frac{1}{96}\int_{x=3}^{4} \frac{xy^2}{2}\bigg|_{y=1}^{2} dx$$

$$= \frac{1}{96}\int_{x=3}^{4} \frac{3x}{2}\,dx = \frac{7}{128}$$

2.12. Find the marginal distribution functions (a) of X and (b) of Y for Problem 2.11.

(a) The marginal distribution function for X if $0 \leq x < 4$ is

$$F_1(x) = P(X \leq x) = \int_{u=-\infty}^{x}\int_{v=-\infty}^{\infty} f(u, v)\,du\,dv$$

$$= \int_{u=0}^{x}\int_{v=1}^{5} \frac{uv}{96}\,du\,dv$$

$$= \frac{1}{96}\int_{u=0}^{x}\left[\int_{v=1}^{5} uv\,dv\right]du = \frac{x^2}{16}$$

For $x \geq 4$, $F_1(x) = 1$; for $x < 0$, $F_1(x) = 0$. Thus

$$F_1(x) = \begin{cases} 0 & x < 0 \\ x^2/16 & 0 \leq x < 4 \\ 1 & x \geq 4 \end{cases}$$

As $F_1(x)$ is continuous at $x = 0$ and $x = 4$, we could replace $<$ by $\leq$ in the above expression.

(b) The marginal distribution function for Y if $1 \leq y < 5$ is

$$F_2(y) = P(Y \leq y) = \int_{u=-\infty}^{\infty} \int_{v=1}^{y} f(u, v)\, du\, dv$$

$$= \int_{u=0}^{4} \int_{v=1}^{y} \frac{uv}{96}\, du\, dv = \frac{y^2 - 1}{24}$$

For $y \geq 5$, $F_2(y) = 1$. For $y < 1$, $F_2(y) = 0$. Thus

$$F_2(y) = \begin{cases} 0 & y < 1 \\ (y^2 - 1)/24 & 1 \leq y < 5 \\ 1 & y \geq 5 \end{cases}$$

As $F_2(y)$ is continuous at $y = 1$ and $y = 5$, we could replace $<$ by $\leq$ in the above expression.

2.13. Find the joint distribution function for the random variables X, Y of Problem 2.11.

From Problem 2.11 it is seen that the joint density function for X and Y can be written as the product of a function of x alone and a function of y alone. In fact, $f(x, y) = f_1(x)f_2(y)$, where

$$f_1(x) = \begin{cases} c_1 x & 0 < x < 4 \\ 0 & \text{otherwise} \end{cases} \qquad f_2(y) = \begin{cases} c_2 y & 1 < y < 5 \\ 0 & \text{otherwise} \end{cases}$$

and $c_1 c_2 = c = 1/96$. It follows that X and Y are independent, so that their joint distribution function is given by $F(x, y) = F_1(x)F_2(y)$. The marginal distributions $F_1(x)$ and $F_2(y)$ were determined in Problem 2.12, and Fig. 2-9 shows the resulting piecewise definition of $F(x, y)$.

2.14. In Problem 2.11 find $P(X + Y < 3)$.

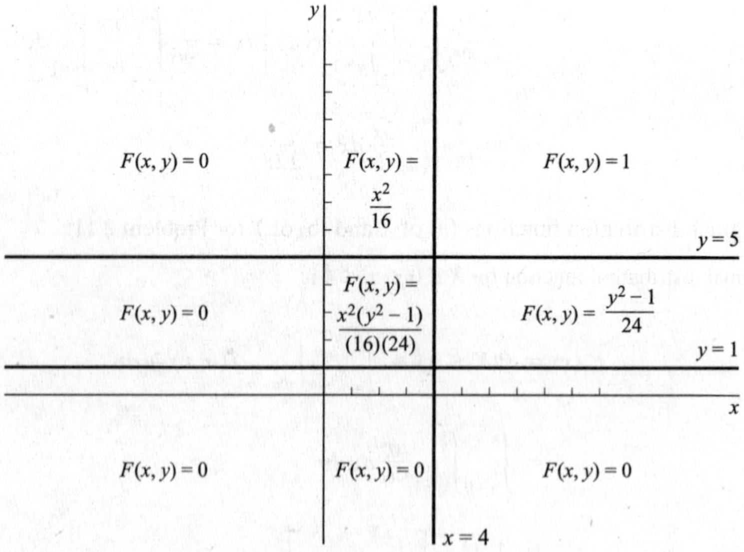

Fig. 2-9

In Fig. 2-10 we have indicated the square region $0 < x < 4$, $1 < y < 5$ within which the joint density function of X and Y is different from zero. The required probability is given by

$$P(X + Y < 3) = \iint_{\mathcal{R}} f(x, y)\, dx\, dy$$

where $\mathcal{R}$ is the part of the square over which $x + y < 3$, shown shaded in Fig. 2-10. Since $f(x, y) = xy/96$ over $\mathcal{R}$, this probability is given by

$$\int_{x=0}^{2}\int_{y=1}^{3-x} \frac{xy}{96}\,dx\,dy$$

$$= \frac{1}{96}\int_{x=0}^{2}\left[\int_{y=1}^{3-x} xy\,dy\right]dx$$

$$= \frac{1}{96}\int_{x=0}^{2} \frac{xy^2}{2}\bigg|_{y=1}^{3-x} dx = \frac{1}{192}\int_{x=0}^{2} [x(3-x)^2 - x] = \frac{1}{48}$$

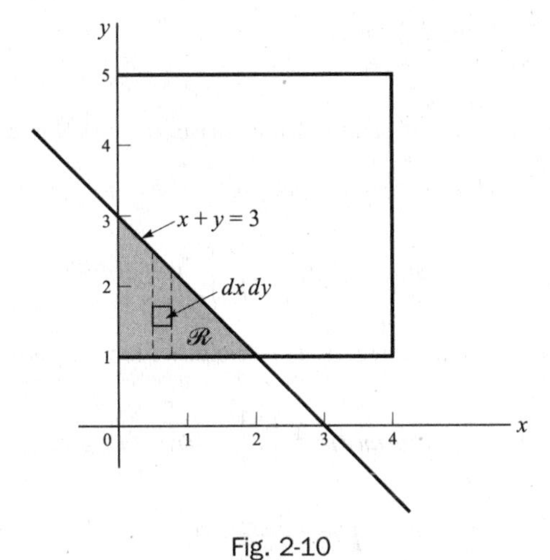

Fig. 2-10

Change of variables

2.15. Prove Theorem 2-1, page 42.

The probability function for U is given by

$$g(u) = P(U = u) = P[\phi(X) = u] = P[X = \psi(u)] = f[\psi(u)]$$

In a similar manner Theorem 2-2, page 42, can be proved.

2.16. Prove Theorem 2-3, page 42.

Consider first the case where $u = \phi(x)$ or $x = \psi(u)$ is an increasing function, i.e., u increases as x increases (Fig. 2-11). There, as is clear from the figure, we have

(1) $$P(u_1 < U < u_2) = P(x_1 < X < x_2)$$

or

(2) $$\int_{u_1}^{u_2} g(u)\,du = \int_{x_1}^{x_2} f(x)\,dx$$

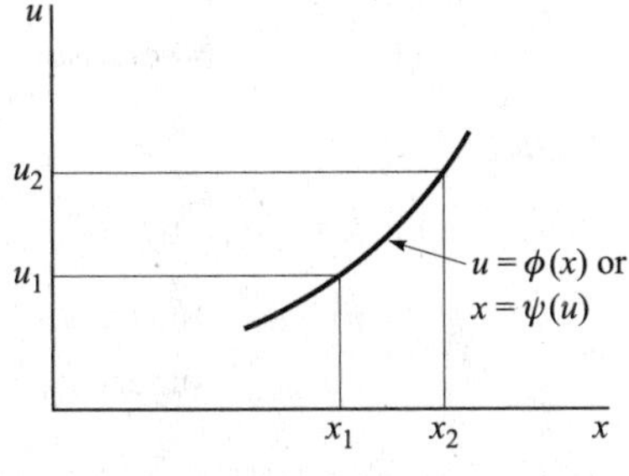

Fig. 2-11

Letting $x = \psi(u)$ in the integral on the right, (2) can be written

$$\int_{u_1}^{u_2} g(u)\,du = \int_{u_1}^{u_2} f[\psi(u)]\psi'(u)\,du$$

This can hold for all u_1 and u_2 only if the integrands are identical, i.e.,

$$g(u) = f[\psi(u)]\psi'(u)$$

This is a special case of (34), page 42, where $\psi'(u) > 0$ (i.e., the slope is positive). For the case where $\psi'(u) \le 0$, i.e., u is a decreasing function of x, we can also show that (34) holds (see Problem 2.67). The theorem can also be proved if $\psi'(u) \ge 0$ or $\psi'(u) < 0$.

2.17. Prove Theorem 2-4, page 42.

We suppose first that as x and y increase, u and v also increase. As in Problem 2.16 we can then show that

$$P(u_1 < U < u_2, v_1 < V < v_2) = P(x_1 < X < x_2, y_1 < Y < y_2)$$

or
$$\int_{v_1}^{u_2}\int_{v_1}^{v_2} g(u, v)\,du\,dv = \int_{x_1}^{x_2}\int_{y_1}^{y_2} f(x, y)\,dx\,dy$$

Letting $x = \psi_1(u, v)$, $y = \psi_2(u, v)$ in the integral on the right, we have, by a theorem of advanced calculus,

$$\int_{v_1}^{u_2}\int_{v_1}^{v_2} g(u, v)\,du\,dv = \int_{u_1}^{u_2}\int_{v_1}^{v_2} f[\psi_1(u, v), \psi_2(u, v)]J\,du\,dv$$

where
$$J = \frac{\partial(x, y)}{\partial(u, v)}$$

is the *Jacobian*. Thus

$$g(u, v) = f[\psi_1(u, v), \psi_2(u, v)]J$$

which is (36), page 42, in the case where $J > 0$. Similarly, we can prove (36) for the case where $J < 0$.

2.18. The probability function of a random variable X is

$$f(x) = \begin{cases} 2^{-x} & x = 1, 2, 3, \ldots \\ 0 & \text{otherwise} \end{cases}$$

Find the probability function for the random variable $U = X^4 + 1$.

Since $U = X^4 + 1$, the relationship between the values u and x of the random variables U and X is given by $u = x^4 + 1$ or $x = \sqrt[4]{u - 1}$, where $u = 2, 17, 82, \ldots$ and the real positive root is taken. Then the required probability function for U is given by

$$g(u) = \begin{cases} 2^{-\sqrt[4]{u-1}} & u = 2, 17, 82, \ldots \\ 0 & \text{otherwise} \end{cases}$$

using Theorem 2-1, page 42, or Problem 2.15.

2.19. The probability function of a random variable X is given by

$$f(x) = \begin{cases} x^2/81 & -3 < x < 6 \\ 0 & \text{otherwise} \end{cases}$$

Find the probability density for the random variable $U = \frac{1}{3}(12 - X)$.

We have $u = \frac{1}{3}(12 - x)$ or $x = 12 - 3u$. Thus to each value of x there is one and only one value of u and conversely. The values of u corresponding to $x = -3$ and $x = 6$ are $u = 5$ and $u = 2$, respectively. Since $\psi'(u) = dx/du = -3$, it follows by Theorem 2-3, page 42, or Problem 2.16 that the density function for U is

$$g(u) = \begin{cases} (12 - 3u)^2/27 & 2 < u < 5 \\ 0 & \text{otherwise} \end{cases}$$

Check:
$$\int_2^5 \frac{(12 - 3u)^2}{27}\,du = -\frac{(12 - 3u)^3}{243}\bigg|_2^5 = 1$$

2.20. Find the probability density of the random variable $U = X^2$ where X is the random variable of Problem 2.19.

We have $u = x^2$ or $x = \pm\sqrt{u}$. Thus to each value of x there corresponds one and only one value of u, but to each value of $u \neq 0$ there correspond *two* values of x. The values of x for which $-3 < x < 6$ correspond to values of u for which $0 \leq u < 36$ as shown in Fig. 2-12.

As seen in this figure, the interval $-3 < x \leq 3$ corresponds to $0 \leq u \leq 9$ while $3 < x < 6$ corresponds to $9 < u < 36$. In this case we cannot use Theorem 2-3 directly but can proceed as follows. The distribution function for U is

$$G(u) = P(U \leq u)$$

Now if $0 \leq u \leq 9$, we have

$$G(u) = P(U \leq u) = P(X^2 \leq u) = P(-\sqrt{u} \leq X \leq \sqrt{u})$$

$$= \int_{-\sqrt{u}}^{\sqrt{u}} f(x)\,dx$$

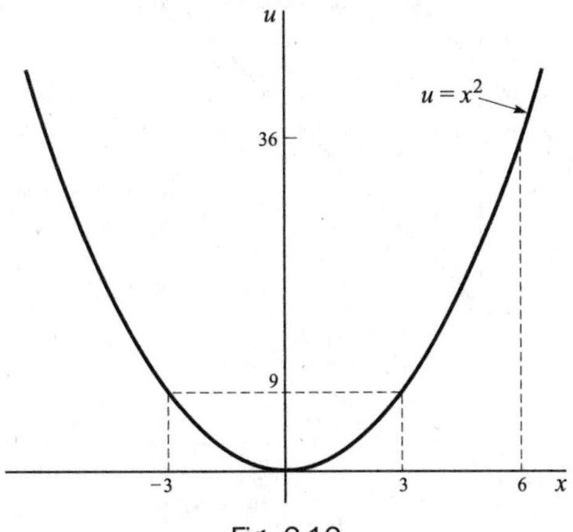

Fig. 2-12

But if $9 < u < 36$, we have

$$G(u) = P(U \leq u) = P(-3 < X < \sqrt{u}) = \int_{-3}^{\sqrt{u}} f(x)\,dx$$

Since the density function $g(u)$ is the derivative of $G(u)$, we have, using (12),

$$g(u) = \begin{cases} \dfrac{f(\sqrt{u}) + f(-\sqrt{u})}{2\sqrt{u}} & 0 \le u \le 9 \\ \dfrac{f(\sqrt{u})}{2\sqrt{u}} & 9 < u < 36 \\ 0 & \text{otherwise} \end{cases}$$

Using the given definition of $f(x)$, this becomes

$$g(u) = \begin{cases} \sqrt{u}/81 & 0 \le u \le 9 \\ \sqrt{u}/162 & 9 < u < 36 \\ 0 & \text{otherwise} \end{cases}$$

Check:

$$\int_0^9 \frac{\sqrt{u}}{81}\,du + \int_9^{36} \frac{\sqrt{u}}{162}\,du = \frac{2u^{3/2}}{243}\Bigg|_0^9 + \frac{u^{3/2}}{243}\Bigg|_9^{36} = 1$$

2.21. If the random variables X and Y have joint density function

$$f(x, y) = \begin{cases} xy/96 & 0 < x < 4,\ 1 < y < 5 \\ 0 & \text{otherwise} \end{cases}$$

(see Problem 2.11), find the density function of $U = X + 2Y$.

Method 1

Let $u = x + 2y$, $v = x$, the second relation being chosen arbitrarily. Then simultaneous solution yields $x = v$, $y = \frac{1}{2}(u - v)$. Thus the region $0 < x < 4$, $1 < y < 5$ corresponds to the region $0 < v < 4$, $2 < u - v < 10$ shown shaded in Fig. 2-13.

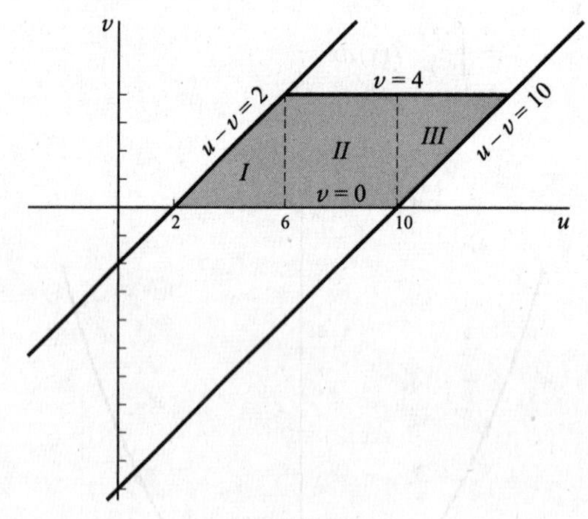

Fig. 2-13

The Jacobian is given by

$$J = \begin{vmatrix} \dfrac{\partial x}{\partial u} & \dfrac{\partial x}{\partial v} \\ \dfrac{\partial y}{\partial u} & \dfrac{\partial y}{\partial v} \end{vmatrix} = \begin{vmatrix} 0 & 1 \\ \dfrac{1}{2} & -\dfrac{1}{2} \end{vmatrix} = -\frac{1}{2}$$

Then by Theorem 2-4 the joint density function of U and V is

$$g(u, v) = \begin{cases} v(u - v)/384 & 2 < u - v < 10, 0 < v < 4 \\ 0 & \text{otherwise} \end{cases}$$

The marginal density function of U is given by

$$g_1(u) = \begin{cases} \displaystyle\int_{v=0}^{u-2} \frac{v(u - v)}{384}\,dv & 2 < u < 6 \\ \displaystyle\int_{v=0}^{4} \frac{v(u - v)}{384}\,dv & 6 < u < 10 \\ \displaystyle\int_{v=u-10}^{4} \frac{v(u - v)}{384}\,dv & 10 < u < 14 \\ 0 & \text{otherwise} \end{cases}$$

as seen by referring to the shaded regions *I*, *II*, *III* of Fig. 2-13. Carrying out the integrations, we find

$$g_1(u) = \begin{cases} (u - 2)^2(u + 4)/2304 & 2 < u < 6 \\ (3u - 8)/144 & 6 < u < 10 \\ (348u - u^3 - 2128)/2304 & 10 < u < 14 \\ 0 & \text{otherwise} \end{cases}$$

A check can be achieved by showing that the integral of $g_1(u)$ is equal to 1.

Method 2

The distribution function of the random variable $X + 2Y$ is given by

$$P(X + 2Y \le u) = \iint\limits_{x+2y \le u} f(x, y)\,dx\,dy = \iint\limits_{\substack{x+2y \le u \\ 0<x<4 \\ 1<y<5}} \frac{xy}{96}\,dx\,dy$$

For $2 < u < 6$, we see by referring to Fig. 2-14, that the last integral equals

$$\int_{x=0}^{u-2}\int_{y=1}^{(u-x)/2} \frac{xy}{96}\,dx\,dy = \int_{x=0}^{u-2}\left[\frac{x(u - x)^2}{768} - \frac{x}{192}\right]dx$$

The derivative of this with respect to u is found to be $(u - 2)^2(u + 4)/2304$. In a similar manner we can obtain the result of Method 1 for $6 < u < 10$, etc.

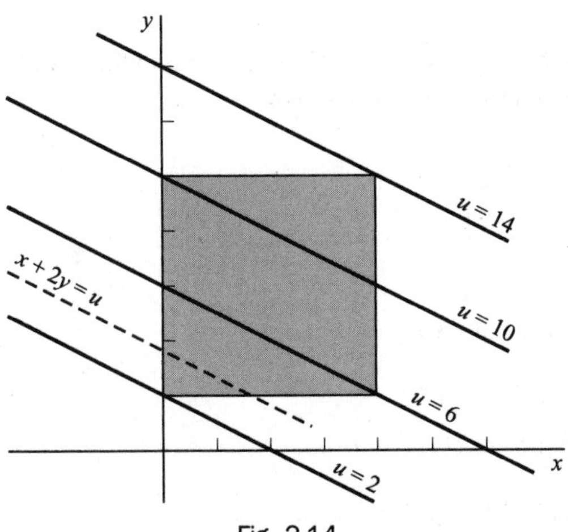

Fig. 2-14

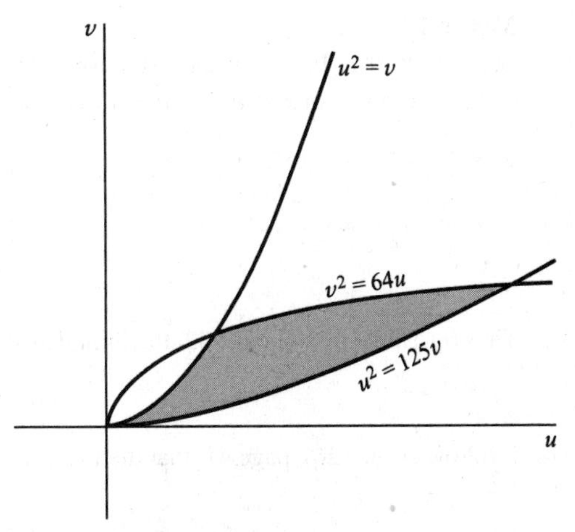

Fig. 2-15

2.22. If the random variables X and Y have joint density function

$$f(x, y) = \begin{cases} xy/96 & 0 < x < 4, 1 < y < 5 \\ 0 & \text{otherwise} \end{cases}$$

(see Problem 2.11), find the joint density function of $U = XY^2$, $V = X^2Y$.

Consider $u = xy^2$, $v = x^2y$. Dividing these equations, we obtain $y/x = u/v$ so that $y = ux/v$. This leads to the simultaneous solution $x = v^{2/3} u^{-1/3}$, $y = u^{2/3} v^{-1/3}$. The image of $0 < x < 4$, $1 < y < 5$ in the uv-plane is given by

$$0 < v^{2/3}u^{-1/3} < 4 \qquad 1 < u^{2/3}v^{-1/3} < 5$$

which are equivalent to

$$v^2 < 64u \qquad v < u^2 < 125v$$

This region is shown shaded in Fig. 2-15.

The Jacobian is given by

$$J = \begin{vmatrix} -\frac{1}{3}v^{2/3}u^{-4/3} & \frac{2}{3}v^{-1/3}u^{-1/3} \\ \frac{2}{3}u^{-1/3}v^{-1/3} & -\frac{1}{3}u^{2/3}v^{-4/3} \end{vmatrix} = -\frac{1}{3}u^{-2/3}v^{-2/3}$$

Thus the joint density function of U and V is, by Theorem 2-4,

$$g(u, v) = \begin{cases} \dfrac{(v^{2/3}u^{-1/3})(u^{2/3}v^{-1/3})}{96}\left(\frac{1}{3}u^{-2/3}v^{-2/3}\right) & v^2 < 64u, v < u^2 < 125v \\ 0 & \text{otherwise} \end{cases}$$

or

$$g(u, v) = \begin{cases} u^{-1/3}v^{-1/3}/288 & v^2 < 64u, \quad v < u^2 < 125v \\ 0 & \text{otherwise} \end{cases}$$

Convolutions

2.23. Let X and Y be random variables having joint density function $f(x, y)$. Prove that the density function of $U = X + Y$ is

$$g(u) = \int_{-\infty}^{\infty} f(v, u - v)dv$$

Method 1

Let $U = X + Y$, $V = X$, where we have arbitrarily added the second equation. Corresponding to these we have $u = x + y$, $v = x$ or $x = v$, $y = u - v$. The Jacobian of the transformation is given by

$$J = \begin{vmatrix} \dfrac{\partial x}{\partial u} & \dfrac{\partial x}{\partial v} \\ \dfrac{\partial y}{\partial u} & \dfrac{\partial y}{\partial v} \end{vmatrix} = \begin{vmatrix} 0 & 1 \\ 1 & -1 \end{vmatrix} = -1$$

Thus by Theorem 2-4, page 42, the joint density function of U and V is

$$g(u, v) = f(v, u - v)$$

It follows from (26), page 41, that the marginal density function of U is

$$g(u) = \int_{-\infty}^{\infty} f(v, u - v)dv$$

Method 2

The distribution function of $U = X + Y$ is equal to the double integral of $f(x, y)$ taken over the region defined by $x + y \leq u$, i.e.,

$$G(u) = \iint_{x+y\leq u} f(x, y)\,dx\,dy$$

Since the region is below the line $x + y = u$, as indicated by the shading in Fig. 2-16, we see that

$$G(u) = \int_{x=-\infty}^{\infty}\left[\int_{y=-\infty}^{u-x} f(x, y)\,dy\right]dx$$

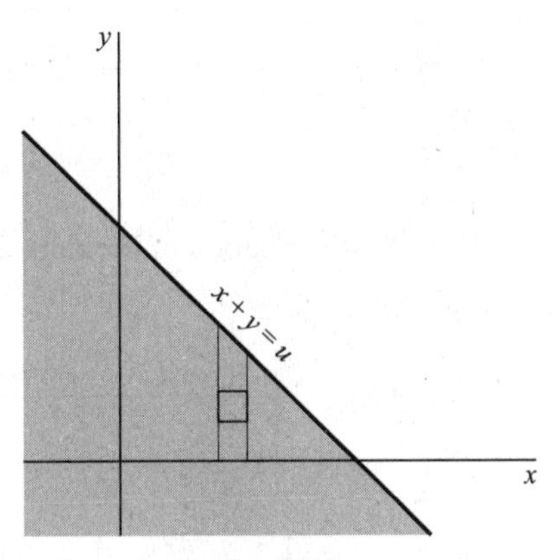

Fig. 2-16

The density function of U is the derivative of $G(u)$ with respect to u and is given by

$$g(u) = \int_{-\infty}^{\infty} f(x, u - x)\,dx$$

using (12) first on the x integral and then on the y integral.

2.24. Work Problem 2.23 if X and Y are independent random variables having density functions $f_1(x)$, $f_2(y)$, respectively.

In this case the joint density function is $f(x, y) = f_1(x) f_2(y)$, so that by Problem 2.23 the density function of $U = X + Y$ is

$$g(u) = \int_{-\infty}^{\infty} f_1(v) f_2(u - v)\,dv = f_1 * f_2$$

which is the *convolution* of f_1 and f_2.

2.25. If X and Y are independent random variables having density functions

$$f_1(x) = \begin{cases} 2e^{-2x} & x \geq 0 \\ 0 & x < 0 \end{cases} \qquad f_2(y) = \begin{cases} 3e^{-3y} & y \geq 0 \\ 0 & y < 0 \end{cases}$$

find the density function of their sum, $U = X + Y$.

By Problem 2.24 the required density function is the convolution of f_1 and f_2 and is given by

$$g(u) = f_1 * f_2 = \int_{-\infty}^{\infty} f_1(v) f_2(u - v)\,dv$$

In the integrand f_1 vanishes when $v < 0$ and f_2 vanishes when $v > u$. Hence

$$g(u) = \int_0^u (2e^{-2v})(3e^{-3(u-v)})\,dv$$

$$= 6e^{-3u}\int_0^u e^v\,dv = 6e^{-3u}(e^u - 1) = 6(e^{-2u} - e^{3u})$$

if $u \geq 0$ and $g(u) = 0$ if $u < 0$.

Check:
$$\int_{-\infty}^{\infty} g(u)\,du = 6\int_{0}^{\infty}(e^{-2u} - e^{-3u})\,du = 6\left(\frac{1}{2} - \frac{1}{3}\right) = 1$$

2.26. Prove that $f_1 * f_2 = f_2 * f_1$ (Property 1, page 43).

We have

$$f_1 * f_2 = \int_{v=-\infty}^{\infty} f_1(v)f_2(u - v)\,dv$$

Letting $w = u - v$ so that $v = u - w$, $dv = -dw$, we obtain

$$f_1 * f_2 = \int_{w=\infty}^{-\infty} f_1(u - w)f_2(w)(-dw) = \int_{w=-\infty}^{\infty} f_2(w)f_1(u - w)\,dw = f_2 * f_1$$

Conditional distributions

2.27. Find (a) $f(y\,|\,2)$, (b) $P(Y = 1\,|\,X = 2)$ for the distribution of Problem 2.8.

(a) Using the results in Problems 2.8 and 2.9, we have

$$f(y\,|\,x) = \frac{f(x, y)}{f_1(x)} = \frac{(2x + y)/42}{f_1(x)}$$

so that with $x = 2$

$$f(y\,|\,2) = \frac{(4 + y)/42}{11/21} = \frac{4 + y}{22}$$

(b)
$$P(Y = 1\,|\,X = 2) = f(1\,|\,2) = \frac{5}{22}$$

2.28. If X and Y have the joint density function

$$f(x, y) = \begin{cases} \frac{3}{4} + xy & 0 < x < 1,\ 0 < y < 1 \\ 0 & \text{otherwise} \end{cases}$$

find (a) $f(y\,|\,x)$, (b) $P(Y > \frac{1}{2}\,|\,\frac{1}{2} < X < \frac{1}{2} + dx)$.

(a) For $0 < x < 1$,

$$f_1(x) = \int_0^1 \left(\frac{3}{4} + xy\right)dy = \frac{3}{4} + \frac{x}{2}$$

and
$$f(y\,|\,x) = \frac{f(x, y)}{f_1(x)} = \begin{cases} \dfrac{3 + 4xy}{3 + 2x} & 0 < y < 1 \\ 0 & \text{other } y \end{cases}$$

For other values of x, $f(y\,|\,x)$ is not defined.

(b)
$$P(Y > \tfrac{1}{2}\,|\,\tfrac{1}{2} < X < \tfrac{1}{2} + dx) = \int_{1/2}^{\infty} f(y\,|\,\tfrac{1}{2})\,dy = \int_{1/2}^{1} \frac{3 + 2y}{4}\,dy = \frac{9}{16}$$

2.29. The joint density function of the random variables X and Y is given by

$$f(x, y) = \begin{cases} 8xy & 0 \leq x \leq 1,\ 0 \leq y \leq x \\ 0 & \text{otherwise} \end{cases}$$

Find (a) the marginal density of X, (b) the marginal density of Y, (c) the conditional density of X, (d) the conditional density of Y.

The region over which $f(x, y)$ is different from zero is shown shaded in Fig. 2-17.

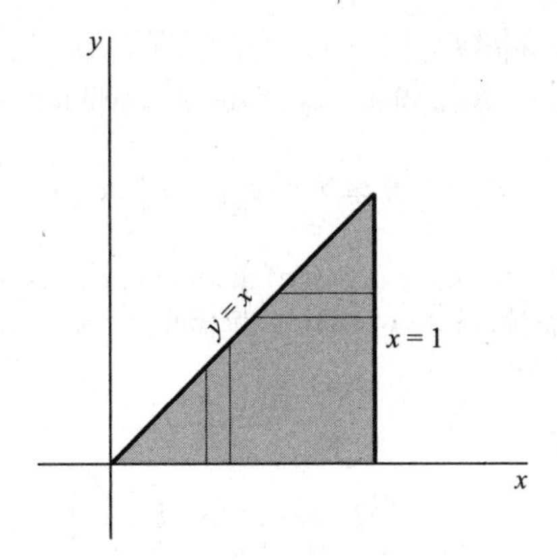

Fig. 2-17

(a) To obtain the marginal density of X, we fix x and integrate with respect to y from 0 to x as indicated by the vertical strip in Fig. 2-17. The result is

$$f_1(x) = \int_{y=0}^{x} 8xy\,dy = 4x^3$$

for $0 < x < 1$. For all other values of x, $f_1(x) = 0$.

(b) Similarly, the marginal density of Y is obtained by fixing y and integrating with respect to x from $x = y$ to $x = 1$, as indicated by the horizontal strip in Fig. 2-17. The result is, for $0 < y < 1$,

$$f_2(y) = \int_{x=y}^{1} 8xy\,dx = 4y(1 - y^2)$$

For all other values of y, $f_2(y) = 0$.

(c) The conditional density function of X is, for $0 < y < 1$,

$$f_1(x \mid y) = \frac{f(x, y)}{f_2(y)} = \begin{cases} 2x/(1 - y^2) & y \le x \le 1 \\ 0 & \text{other } x \end{cases}$$

The conditional density function is not defined when $f_2(y) = 0$.

(d) The conditional density function of Y is, for $0 < x < 1$,

$$f_2(y \mid x) = \frac{f(x, y)}{f_1(x)} = \begin{cases} 2y/x^2 & 0 \le y \le x \\ 0 & \text{other } y \end{cases}$$

The conditional density function is not defined when $f_1(x) = 0$.

Check:

$$\int_0^1 f_1(x)\,dx = \int_0^1 4x^3\,dx = 1, \quad \int_0^1 f_2(y)\,dy = \int_0^1 4y(1 - y^2)\,dy = 1$$

$$\int_y^1 f_1(x \mid y)\,dx = \int_y^1 \frac{2x}{1 - y^2}\,dx = 1$$

$$\int_0^x f_2(y \mid x)\,dy = \int_0^x \frac{2y}{x^2}\,dy = 1$$

2.30. Determine whether the random variables of Problem 2.29 are independent.

In the shaded region of Fig. 2-17, $f(x, y) = 8xy$, $f_1(x) = 4x^3$, $f_2(y) = 4y(1 - y^2)$. Hence $f(x, y) \neq f_1(x) f_2(y)$, and thus X and Y are dependent.

It should be noted that it does not follow from $f(x, y) = 8xy$ that $f(x, y)$ can be expressed as a function of x alone times a function of y alone. This is because the restriction $0 \le y \le x$ occurs. If this were replaced by some restriction on y not depending on x (as in Problem 2.21), such a conclusion would be valid.

Applications to geometric probability

2.31. A person playing darts finds that the probability of the dart striking between r and $r + dr$ is

$$P(r \leq R \leq r + dr) = c\left[1 - \left(\frac{r}{a}\right)^2\right]dr$$

Here, R is the distance of the hit from the center of the target, c is a constant, and a is the radius of the target (see Fig. 2-18). Find the probability of hitting the bull's-eye, which is assumed to have radius b. Assume that the target is always hit.

The density function is given by

$$f(r) = c\left[1 - \left(\frac{r}{a}\right)^2\right]$$

Since the target is always hit, we have

$$c\int_0^a \left[1 - \left(\frac{r}{a}\right)^2\right]dr = 1$$

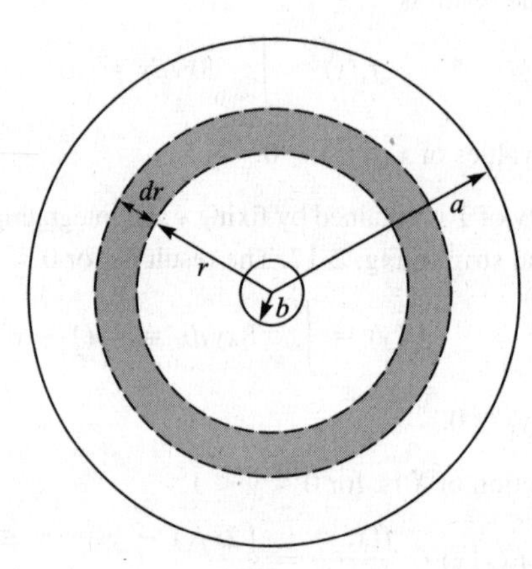

Fig. 2-18

from which $c = 3/2a$. Then the probability of hitting the bull's-eye is

$$\int_0^b f(r)\,dr = \frac{3}{2a}\int_0^b \left[1 - \left(\frac{r}{a}\right)^2\right]dr = \frac{b(3a^2 - b^2)}{2a^3}$$

2.32. Two points are selected at random in the interval $0 \leq x \leq 1$. Determine the probability that the sum of their squares is less than 1.

Let X and Y denote the random variables associated with the given points. Since equal intervals are assumed to have equal probabilities, the density functions of X and Y are given, respectively, by

(1) $$f_1(x) = \begin{cases} 1 & 0 \leq x \leq 1 \\ 0 & \text{otherwise} \end{cases} \qquad f_2(y) = \begin{cases} 1 & 0 \leq y \leq 1 \\ 0 & \text{otherwise} \end{cases}$$

Then since X and Y are independent, the joint density function is given by

(2) $$f(x, y) = f_1(x)f_2(y) = \begin{cases} 1 & 0 \leq x \leq 1, 0 \leq y \leq 1 \\ 0 & \text{otherwise} \end{cases}$$

It follows that the required probability is given by

(3) $$P(X^2 + Y^2 \leq 1) = \iint\limits_{\mathfrak{R}} dx\,dy$$

where $\mathfrak{R}$ is the region defined by $x^2 + y^2 \leq 1, x \geq 0, y \geq 0$, which is a quarter of a circle of radius 1 (Fig. 2-19). Now since (3) represents the area of $\mathfrak{R}$, we see that the required probability is $\pi/4$.

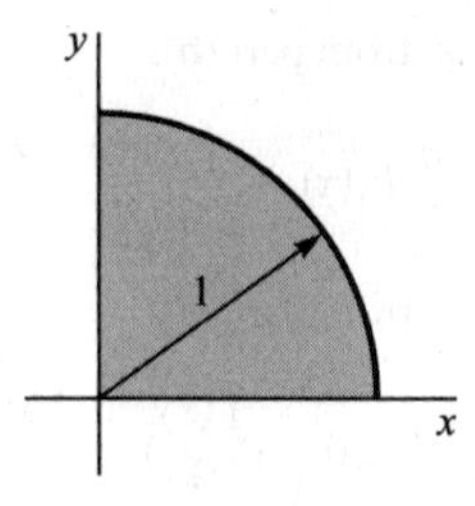

Fig. 2-19

Miscellaneous problems

2.33. Suppose that the random variables X and Y have a joint density function given by

$$f(x, y) = \begin{cases} c(2x + y) & 2 < x < 6, 0 < y < 5 \\ 0 & \text{otherwise} \end{cases}$$

Find (a) the constant c, (b) the marginal distribution functions for X and Y, (c) the marginal density functions for X and Y, (d) $P(3 < X < 4, Y > 2)$, (e) $P(X > 3)$, (f) $P(X + Y > 4)$, (g) the joint distribution function, (h) whether X and Y are independent.

(a) The total probability is given by

$$\int_{x=2}^{6}\int_{y=0}^{5} c(2x + y)\,dx\,dy = \int_{x=2}^{6} c\left(2xy + \frac{y^2}{2}\right)\Bigg|_0^5 dx$$

$$= \int_{x=2}^{6} c\left(10x + \frac{25}{2}\right)dx = 210c$$

For this to equal 1, we must have $c = 1/210$.

(b) The marginal distribution function for X is

$$F_1(x) = P(X \le x) = \int_{u=-\infty}^{x}\int_{v=-\infty}^{\infty} f(u, v)\,du\,dv$$

$$= \begin{cases} \displaystyle\int_{u=-\infty}^{x}\int_{v=-\infty}^{\infty} 0\,du\,dv = 0 & x < 2 \\ \displaystyle\int_{u=2}^{x}\int_{v=0}^{5} \frac{2u + v}{210}\,du\,dv = \frac{2x^2 + 5x - 18}{84} & 2 \le x < 6 \\ \displaystyle\int_{u=2}^{6}\int_{v=0}^{5} \frac{2u + v}{210}\,du\,dv = 1 & x \ge 6 \end{cases}$$

The marginal distribution function for Y is

$$F_2(y) = P(Y \le y) = \int_{u=-\infty}^{\infty}\int_{v=-\infty}^{y} f(u, v)\,du\,dv$$

$$= \begin{cases} \displaystyle\int_{u=-\infty}^{\infty}\int_{v=-8}^{y} 0\,du\,dv = 0 & y < 0 \\ \displaystyle\int_{u=0}^{6}\int_{v=0}^{y} \frac{2u + v}{210}\,du\,dv = \frac{y^2 + 16y}{105} & 0 \le y < 5 \\ \displaystyle\int_{u=2}^{6}\int_{v=0}^{5} \frac{2u + v}{210}\,du\,dv = 1 & y \ge 5 \end{cases}$$

(c) The marginal density function for X is, from part (b),

$$f_1(x) = \frac{d}{dx}F_1(x) = \begin{cases} (4x + 5)/84 & 2 < x < 6 \\ 0 & \text{otherwise} \end{cases}$$

The marginal density function for Y is, from part (b),

$$f_2(y) = \frac{d}{dy}F_2(y) = \begin{cases} (2y + 16)/105 & 0 < y < 5 \\ 0 & \text{otherwise} \end{cases}$$

(d)
$$P(3 < X < 4, Y > 2) = \frac{1}{210}\int_{x=3}^{4}\int_{y=2}^{5}(2x + y)\,dx\,dy = \frac{3}{20}$$

(e)
$$P(X > 3) = \frac{1}{210}\int_{x=3}^{6}\int_{y=0}^{5}(2x + y)\,dx\,dy = \frac{23}{28}$$

(f)
$$P(X + Y > 4) = \iint_{\mathcal{R}} f(x, y)\,dx\,dy$$

where $\mathcal{R}$ is the shaded region of Fig. 2-20. Although this can be found, it is easier to use the fact that

$$P(X + Y > 4) = 1 - P(X + Y \leq 4) = 1 - \iint_{\mathcal{R}} f(x, y)\,dx\,dy$$

where $\mathcal{R}'$ is the cross-hatched region of Fig. 2-20. We have

$$P(X + Y \leq 4) = \frac{1}{210}\int_{x=2}^{4}\int_{y=0}^{4-x}(2x + y)\,dx\,dy = \frac{2}{35}$$

Thus $P(X + Y > 4) = 33/35$.

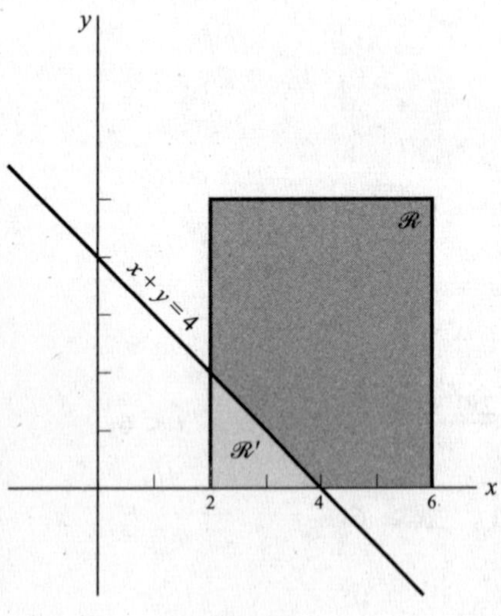

Fig. 2-20

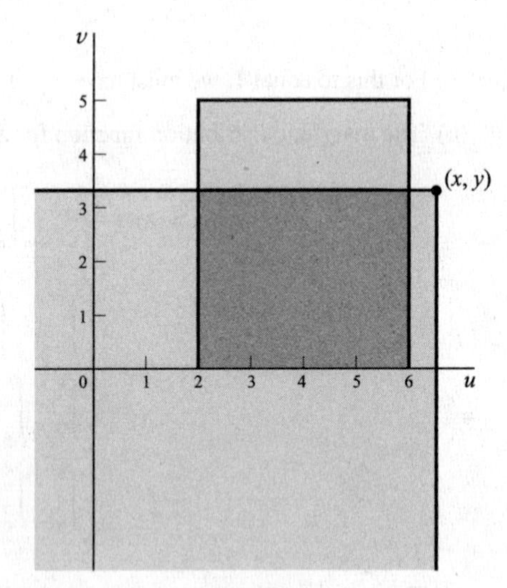

Fig. 2-21

(g) The joint distribution function is

$$F(x, y) = P(X \leq x, Y \leq y) = \int_{u=-\infty}^{x}\int_{v=-\infty}^{y} f(u, v)\,du\,dv$$

In the uv plane (Fig. 2-21) the region of integration is the intersection of the quarter plane $u \leq x$, $v \leq y$ and the rectangle $2 < u < 6$, $0 < v < 5$ [over which $f(u, v)$ is nonzero]. For (x, y) located as in the figure, we have

$$F(x, y) = \int_{u=2}^{6}\int_{v=0}^{y}\frac{2u + v}{210}\,du\,dv = \frac{16y + y^2}{105}$$

When (x, y) lies inside the rectangle, we obtain another expression, etc. The complete results are shown in Fig. 2-22.

(h) The random variables are dependent since

$$f(x, y) \neq f_1(x)f_2(y)$$

or equivalently, $F(x, y) \neq F_1(x)F_2(y)$.

2.34. Let X have the density function

$$f(x) = \begin{cases} 6x(1 - x) & 0 < x < 1 \\ 0 & \text{otherwise} \end{cases}$$

Find a function $Y = h(X)$ which has the density function

$$g(y) = \begin{cases} 12y^3(1 - y^2) & 0 < y < 1 \\ 0 & \text{otherwise} \end{cases}$$

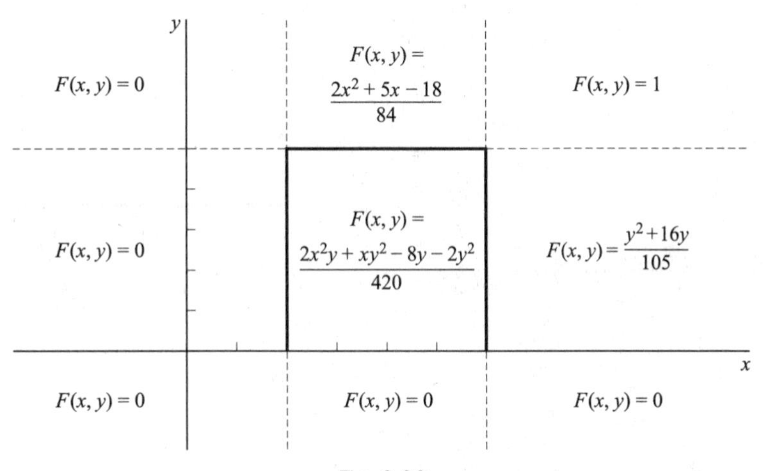

Fig. 2-22

We assume that the unknown function h is such that the intervals $X \leq x$ and $Y \leq y + h(x)$ correspond in a one-one, continuous fashion. Then $P(X \leq x) = P(Y \leq y)$, i.e., the distribution functions of X and Y must be equal. Thus, for $0 < x, y < 1$,

$$\int_0^x 6u(1 - u)\,du = \int_0^y 12v^3(1 - v^2)\,dv$$

or

$$3x^2 - 2x^3 = 3y^4 - 2y^6$$

By inspection, $x = y^2$ or $y = h(x) = +\sqrt{x}$ is a solution, and this solution has the desired properties. Thus $Y = +\sqrt{X}$.

2.35. Find the density function of $U = XY$ if the joint density function of X and Y is $f(x, y)$.

Method 1

Let $U = XY$ and $V = X$, corresponding to which $u = xy$, $v = x$ or $x = v$, $y = u/v$. Then the Jacobian is given by

$$J = \begin{vmatrix} \dfrac{\partial x}{\partial u} & \dfrac{\partial x}{\partial v} \\ \dfrac{\partial y}{\partial u} & \dfrac{\partial y}{\partial v} \end{vmatrix} = \begin{vmatrix} 0 & 1 \\ v^{-1} & -uv^{-2} \end{vmatrix} = -v^{-1}$$

Thus the joint density function of U and V is

$$g(u, v) = \frac{1}{|v|} f\left(v, \frac{u}{v}\right)$$

from which the marginal density function of U is obtained as

$$g(u) = \int_{-\infty}^{\infty} g(u, v)\,dv = \int_{-\infty}^{\infty} \frac{1}{|v|} f\left(v, \frac{u}{v}\right) dv$$

Method 2

The distribution function of U is

$$G(u) = \iint_{xy \le u} f(x, y)\,dx\,dy$$

For $u \ge 0$, the region of integration is shown shaded in Fig. 2-23. We see that

$$G(u) = \int_{-\infty}^{0} \left[\int_{u/x}^{\infty} f(x, y)\,dy\right] dx + \int_{0}^{\infty} \left[\int_{-\infty}^{u/x} f(x, y)\,dy\right] dx$$

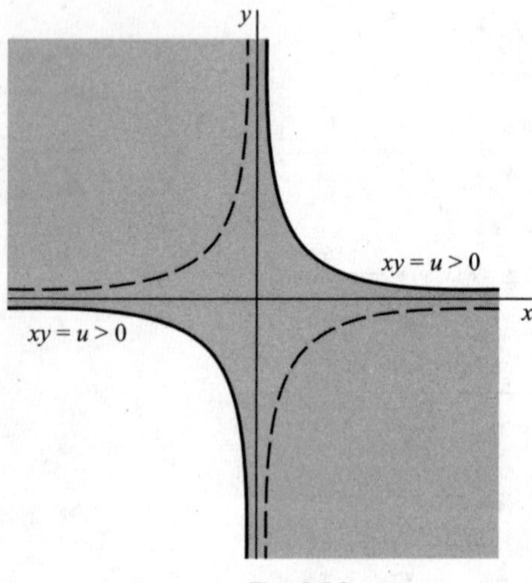

Fig. 2-23

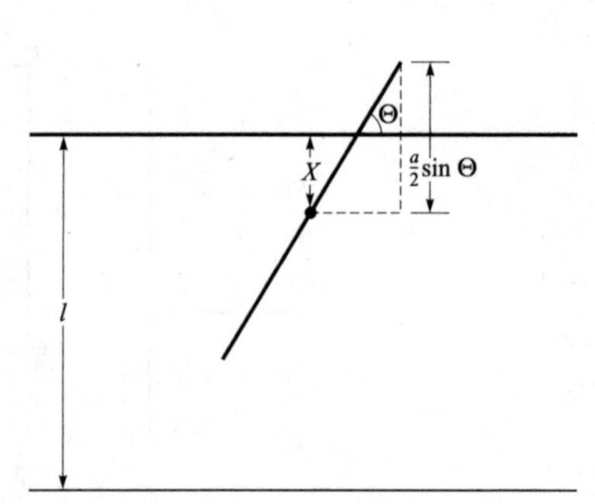

Fig. 2-24

Differentiating with respect to u, we obtain

$$g(u) = \int_{-\infty}^{0} \left(\frac{-1}{x}\right) f\left(x, \frac{u}{x}\right) dx + \int_{0}^{\infty} \frac{1}{x} f\left(x, \frac{u}{x}\right) dx = \int_{-\infty}^{\infty} \frac{1}{|x|} f\left(x, \frac{u}{x}\right) dx$$

The same result is obtained for $u < 0$, when the region of integration is bounded by the dashed hyperbola in Fig. 2-24.

2.36. A floor has parallel lines on it at equal distances l from each other. A needle of length $a < l$ is dropped at random onto the floor. Find the probability that the needle will intersect a line. (This problem is known as *Buffon's needle problem.*)

Let X be a random variable that gives the distance of the midpoint of the needle to the nearest line (Fig. 2-24). Let Θ be a random variable that gives the acute angle between the needle (or its extension) and the line. We denote by x and θ any particular values of X and Θ. It is seen that X can take on any value between 0 and $l/2$, so that $0 \le x \le l/2$. Also Θ can take on any value between 0 and $\pi/2$. It follows that

$$P(x < X \le x + dx) = \frac{2}{l}\,dx \qquad P(\theta \le \Theta + d\theta) = \frac{2}{\pi}\,d\theta$$

i.e., the density functions of X and Θ are given by $f_1(x) = 2/l$, $f_2(\theta) = 2/\pi$. As a check, we note that

$$\int_0^{l/2} \frac{2}{l}\,dx = 1 \qquad \int_0^{\pi/2} \frac{2}{\pi}\,d\theta = 1$$

Since X and Θ are independent the joint density function is

$$f(x, \theta) = \frac{2}{l} \cdot \frac{2}{\pi} = \frac{4}{l\pi}$$

From Fig. 2-24 it is seen that the needle actually hits a line when $X \leq (a/2) \sin \Theta$. The probability of this event is given by

$$\frac{4}{l\pi} \int_{\theta=0}^{\pi/2} \int_{x=0}^{(a/2)\sin\theta} dx\, d\theta = \frac{2a}{l\pi}$$

When the above expression is equated to the frequency of hits observed in actual experiments, accurate values of π are obtained. This indicates that the probability model described above is appropriate.

2.37. Two people agree to meet between 2:00 P.M. and 3:00 P.M., with the understanding that each will wait no longer than 15 minutes for the other. What is the probability that they will meet?

Let X and Y be random variables representing the times of arrival, measured in fractions of an hour after 2:00 P.M., of the two people. Assuming that equal intervals of time have equal probabilities of arrival, the density functions of X and Y are given respectively by

$$f_1(x) = \begin{cases} 1 & 0 \leq x \leq 1 \\ 0 & \text{otherwise} \end{cases}$$

$$f_2(y) = \begin{cases} 1 & 0 \leq y \leq 1 \\ 0 & \text{otherwise} \end{cases}$$

Then, since X and Y are independent, the joint density function is

(1) $$f(x, y) = f_1(x) f_2(y) = \begin{cases} 1 & 0 \leq x \leq 1, 0 \leq y \leq 1 \\ 0 & \text{otherwise} \end{cases}$$

Since 15 minutes $= \frac{1}{4}$ hour, the required probability is

(2) $$P\left(|X - Y| \leq \frac{1}{4}\right) = \iint_{\mathcal{R}} dx\, dy$$

where 5 is the region shown shaded in Fig. 2-25. The right side of (2) is the area of this region, which is equal to $1 - (\frac{3}{4})(\frac{3}{4}) = \frac{7}{16}$, since the square has area 1, while the two corner triangles have areas $\frac{1}{2}(\frac{3}{4})(\frac{3}{4})$ each. Thus the required probability is $7/16$.

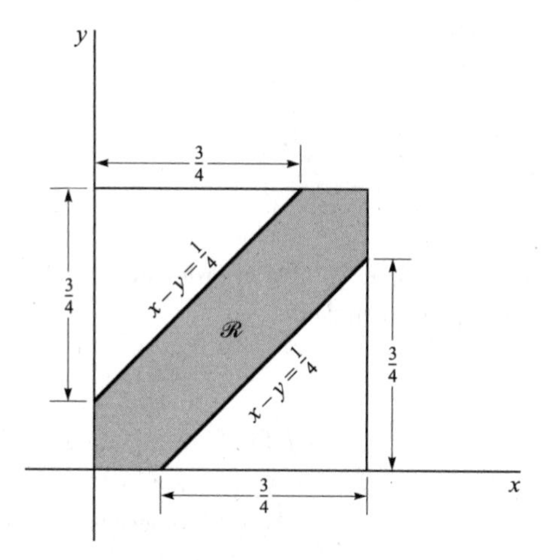

Fig. 2-25

SUPPLEMENTARY PROBLEMS

Discrete random variables and probability distributions

2.38. A coin is tossed three times. If X is a random variable giving the number of heads that arise, construct a table showing the probability distribution of X.

2.39. An urn holds 5 white and 3 black marbles. If 2 marbles are to be drawn at random without replacement and X denotes the number of white marbles, find the probability distribution for X.

2.40. Work Problem 2.39 if the marbles are to be drawn with replacement.

2.41. Let Z be a random variable giving the number of heads minus the number of tails in 2 tosses of a fair coin. Find the probability distribution of Z. Compare with the results of Examples 2.1 and 2.2.

2.42. Let X be a random variable giving the number of aces in a random draw of 4 cards from an ordinary deck of 52 cards. Construct a table showing the probability distribution of X.

Discrete distribution functions

2.43. The probability function of a random variable X is shown in Table 2-7. Construct a table giving the distribution function of X.

Table 2-7

x	1	2	3
$f(x)$	1/2	1/3	1/6

Table 2-8

x	1	2	3	4
$F(x)$	1/8	3/8	3/4	1

2.44. Obtain the distribution function for (a) Problem 2.38, (b) Problem 2.39, (c) Problem 2.40.

2.45. Obtain the distribution function for (a) Problem 2.41, (b) Problem 2.42.

2.46. Table 2-8 shows the distribution function of a random variable X. Determine (a) the probability function, (b) $P(1 \le X \le 3)$, (c) $P(X \ge 2)$, (d) $P(X < 3)$, (e) $P(X > 1.4)$.

Continuous random variables and probability distributions

2.47. A random variable X has density function

$$f(x) = \begin{cases} ce^{-3x} & x > 0 \\ 0 & x \le 0 \end{cases}$$

Find (a) the constant c, (b) $P(1 < X < 2)$, (c) $P(X \ge 3)$, (d) $P(X < 1)$.

2.48. Find the distribution function for the random variable of Problem 2.47. Graph the density and distribution functions, describing the relationship between them.

2.49. A random variable X has density function

$$f(x) = \begin{cases} cx^2 & 1 \le x \le 2 \\ cx & 2 < x < 3 \\ 0 & \text{otherwise} \end{cases}$$

Find (a) the constant c, (b) $P(X > 2)$, (c) $P(1/2 < X < 3/2)$.

2.50. Find the distribution function for the random variable X of Problem 2.49.

2.51. The distribution function of a random variable X is given by

$$F(x) = \begin{cases} cx^3 & 0 \le x < 3 \\ 1 & x \ge 3 \\ 0 & x < 0 \end{cases}$$

If $P(X = 3) = 0$, find (a) the constant c, (b) the density function, (c) $P(X > 1)$, (d) $P(1 < X < 2)$.

2.52. Can the function

$$F(x) = \begin{cases} c(1 - x^2) & 0 \le x \le 1 \\ 0 & \text{otherwise} \end{cases}$$

be a distribution function? Explain.

2.53. Let X be a random variable having density function

$$f(x) = \begin{cases} cx & 0 \le x \le 2 \\ 0 & \text{otherwise} \end{cases}$$

Find (a) the value of the constant c, (b) $P(\frac{1}{2} < X < \frac{3}{2})$, (c) $P(X > 1)$, (d) the distribution function.

Joint distributions and independent variables

2.54. The joint probability function of two discrete random variables X and Y is given by $f(x, y) = cxy$ for $x = 1, 2, 3$ and $y = 1, 2, 3$, and equals zero otherwise. Find (a) the constant c, (b) $P(X = 2, Y = 3)$, (c) $P(1 \le X \le 2, Y \le 2)$, (d) $P(X \ge 2)$, (e) $P(Y < 2)$, (f) $P(X = 1)$, (g) $P(Y = 3)$.

2.55. Find the marginal probability functions of (a) X and (b) Y for the random variables of Problem 2.54. (c) Determine whether X and Y are independent.

2.56. Let X and Y be continuous random variables having joint density function

$$f(x, y) = \begin{cases} c(x^2 + y^2) & 0 \le x \le 1, 0 \le y \le 1 \\ 0 & \text{otherwise} \end{cases}$$

Determine (a) the constant c, (b) $P(X < \frac{1}{2}, Y > \frac{1}{2})$, (c) $P(\frac{1}{4} < X < \frac{3}{4})$, (d) $P(Y < \frac{1}{2})$, (e) whether X and Y are independent.

2.57. Find the marginal distribution functions (a) of X and (b) of Y for the density function of Problem 2.56.

Conditional distributions and density functions

2.58. Find the conditional probability function (a) of X given Y, (b) of Y given X, for the distribution of Problem 2.54.

2.59. Let

$$f(x, y) = \begin{cases} x + y & 0 \le x \le 1, 0 \le y \le 1 \\ 0 & \text{otherwise} \end{cases}$$

Find the conditional density function of (a) X given Y, (b) Y given X.

2.60. Find the conditional density of (a) X given Y, (b) Y given X, for the distribution of Problem 2.56.

2.61. Let

$$f(x, y) = \begin{cases} e^{-(x+y)} & x \ge 0, y \ge 0 \\ 0 & \text{otherwise} \end{cases}$$

be the joint density function of X and Y. Find the conditional density function of (a) X given Y, (b) Y given X.

Change of variables

2.62. Let X have density function

$$f(x) = \begin{cases} e^{-x} & x > 0 \\ 0 & x \le 0 \end{cases}$$

Find the density function of $Y = X^2$.

2.63. (a) If the density function of X is $f(x)$ find the density function of X^3. (b) Illustrate the result in part (a) by choosing

$$f(x) = \begin{cases} 2e^{-2x} & x \ge 0 \\ 0 & x < 0 \end{cases}$$

and check the answer.

2.64. If X has density function $f(x) = 2(\pi)^{-1/2}e^{-x^2/2}$, $-\infty < x < \infty$, find the density function of $Y = X^2$.

2.65. Verify that the integral of $g_1(u)$ in Method 1 of Problem 2.21 is equal to 1.

2.66. If the density of X is $f(x) = 1/\pi(x^2 + 1)$, $-\infty < x < \infty$, find the density of $Y = \tan^{-1} X$.

2.67. Complete the work needed to find $g_1(u)$ in Method 2 of Problem 2.21 and check your answer.

2.68. Let the density of X be

$$f(x) = \begin{cases} 1/2 & -1 < x < 1 \\ 0 & \text{otherwise} \end{cases}$$

Find the density of (a) $3X - 2$, (b) $X^3 + 1$.

2.69. Check by direct integration the joint density function found in Problem 2.22.

2.70. Let X and Y have joint density function

$$f(x, y) = \begin{cases} e^{-(x+y)} & x \ge 0, y \ge 0 \\ 0 & \text{otherwise} \end{cases}$$

If $U = X/Y$, $V = X + Y$, find the joint density function of U and V.

2.71. Use Problem 2.22 to find the density function of (a) $U = XY^2$, (b) $V = X^2Y$.

2.72. Let X and Y be random variables having joint density function $f(x, y) = (2\pi)^{-1} e^{-(x^2+y^2)}$, $-\infty < x < \infty$, $-\infty < y < \infty$. If R and Θ are new random variables such that $X = R \cos \Theta$, $Y = R \sin \Theta$, show that the density function of R is

$$g(r) = \begin{cases} re^{-r^2/2} & r \ge 0 \\ 0 & r < 0 \end{cases}$$

2.73. Let
$$f(x, y) = \begin{cases} 1 & 0 \le x \le 1, 0 \le y \le 1 \\ 0 & \text{otherwise} \end{cases}$$
be the joint density function of X and Y. Find the density function of $Z = XY$.

Convolutions

2.74. Let X and Y be identically distributed independent random variables with density function
$$f(t) = \begin{cases} 1 & 0 \le t \le 1 \\ 0 & \text{otherwise} \end{cases}$$
Find the density function of $X + Y$ and check your answer.

2.75. Let X and Y be identically distributed independent random variables with density function
$$f(t) = \begin{cases} e^{-t} & t \ge 0 \\ 0 & \text{otherwise} \end{cases}$$
Find the density function of $X + Y$ and check your answer.

2.76. Work Problem 2.21 by first making the transformation $2Y = Z$ and then using convolutions to find the density function of $U = X + Z$.

2.77. If the independent random variables X_1 and X_2 are identically distributed with density function
$$f(t) = \begin{cases} te^{-t} & t \ge 0 \\ 0 & t < 0 \end{cases}$$
find the density function of $X_1 + X_2$.

Applications to geometric probability

2.78. Two points are to be chosen at random on a line segment whose length is $a > 0$. Find the probability that the three line segments thus formed will be the sides of a triangle.

2.79. It is known that a bus will arrive at random at a certain location sometime between 3:00 P.M. and 3:30 P.M. A man decides that he will go at random to this location between these two times and will wait at most 5 minutes for the bus. If he misses it, he will take the subway. What is the probability that he will take the subway?

2.80. Two line segments, AB and CD, have lengths 8 and 6 units, respectively. Two points P and Q are to be chosen at random on AB and CD, respectively. Show that the probability that the area of a triangle will have height AP and that the base CQ will be greater than 12 square units is equal to $(1 - \ln 2)/2$.

Miscellaneous problems

2.81. Suppose that $f(x) = c/3^x$, $x = 1, 2, \ldots$, is the probability function for a random variable X. (a) Determine c. (b) Find the distribution function. (c) Graph the probability function and the distribution function. (d) Find $P(2 \le X < 5)$. (e) Find $P(X \ge 3)$.

2.82. Suppose that
$$f(x) = \begin{cases} cxe^{-2x} & x \ge 0 \\ 0 & \text{otherwise} \end{cases}$$
is the density function for a random variable X. (a) Determine c. (b) Find the distribution function. (c) Graph the density function and the distribution function. (d) Find $P(X \ge 1)$. (e) Find $P(2 \le X < 3)$.

2.83. The probability function of a random variable X is given by

$$f(x) = \begin{cases} 2p & x = 1 \\ p & x = 2 \\ 4p & x = 3 \\ 0 & \text{otherwise} \end{cases}$$

where p is a constant. Find (a) $P(0 \le X < 3)$, (b) $P(X > 1)$.

2.84. (a) Prove that for a suitable constant c,

$$F(x) = \begin{cases} 0 & x \le 0 \\ c(1 - e^{-x})^2 & x > 0 \end{cases}$$

is the distribution function for a random variable X, and find this c. (b) Determine $P(1 < X < 2)$.

2.85. A random variable X has density function

$$f(x) = \begin{cases} \frac{3}{2}(1 - x^2) & 0 \le x \le 1 \\ 0 & \text{otherwise} \end{cases}$$

Find the density function of the random variable $Y = X^2$ and check your answer.

2.86. Two independent random variables, X and Y, have respective density functions

$$f(x) = \begin{cases} c_1 e^{-2x} & x > 0 \\ 0 & x \le 0 \end{cases} \qquad g(y) = \begin{cases} c_2 y e^{-3y} & y > 0 \\ 0 & y \le 0 \end{cases}$$

Find (a) c_1 and c_2, (b) $P(X + Y > 1)$, (c) $P(1 < X < 2, Y \ge 1)$, (d) $P(1 < X < 2)$, (e) $P(Y \ge 1)$.

2.87. In Problem 2.86 what is the relationship between the answers to (c), (d), and (e)? Justify your answer.

2.88. Let X and Y be random variables having joint density function

$$f(x, y) = \begin{cases} c(2x + y) & 0 < x < 1, 0 < y < 2 \\ 0 & \text{otherwise} \end{cases}$$

Find (a) the constant c, (b) $P(X > \frac{1}{2}, Y < \frac{3}{2})$, (c) the (marginal) density function of X, (d) the (marginal) density function of Y.

2.89. In Problem 2.88 is $P(X > \frac{1}{2}, Y < \frac{3}{2}) = P(X > \frac{1}{2})P(Y < \frac{3}{2})$? Why?

2.90. In Problem 2.86 find the density function (a) of X^2, (b) of $X + Y$.

2.91. Let X and Y have joint density function

$$f(x, y) = \begin{cases} 1/y & 0 < x < y, 0 < y < 1 \\ 0 & \text{otherwise} \end{cases}$$

(a) Determine whether X and Y are independent, (b) Find $P(X > \frac{1}{2})$. (c) Find $P(X < \frac{1}{2}, Y > \frac{1}{3})$. (d) Find $P(X + Y > \frac{1}{2})$.

2.92. Generalize (a) Problem 2.74 and (b) Problem 2.75 to three or more variables.

2.93. Let X and Y be identically distributed independent random variables having density function $f(u) = (2\pi)^{-1/2}e^{-u^2/2}$, $-\infty < u < \infty$. Find the density function of $Z = X^2 + Y^2$.

2.94. The joint probability function for the random variables X and Y is given in Table 2-9. (a) Find the marginal probability functions of X and Y. (b) Find $P(1 \le X < 3, Y \ge 1)$. (c) Determine whether X and Y are independent.

Table 2-9

X \ Y	0	1	2
0	1/18	1/9	1/6
1	1/9	1/18	1/9
2	1/6	1/6	1/18

2.95. Suppose that the joint probability function of random variables X and Y is given by

$$f(x, y) = \begin{cases} cxy & 0 \le x \le 2, 0 \le y \le x \\ 0 & \text{otherwise} \end{cases}$$

(a) Determine whether X and Y are independent. (b) Find $P(\frac{1}{2} < X < 1)$. (c) Find $P(Y \ge 1)$. (d) Find $P(\frac{1}{2} < X < 1, Y \ge 1)$.

2.96. Let X and Y be independent random variables each having density function

$$f(u) = \frac{\lambda^u e^{-\lambda}}{u} \qquad u = 0, 1, 2, \ldots$$

where $\lambda > 0$. Prove that the density function of $X + Y$ is

$$g(u) = \frac{(2\lambda)^u e^{-2\lambda}}{u!} \qquad u = 0, 1, 2, \ldots$$

2.97. A stick of length L is to be broken into two parts. What is the probability that one part will have a length of more than double the other? State clearly what assumptions would you have made. Discuss whether you believe these assumptions are realistic and how you might improve them if they are not.

2.98. A floor is made up of squares of side l. A needle of length $a < l$ is to be tossed onto the floor. Prove that the probability of the needle intersecting at least one side is equal to $a(4l - a)/\pi l^2$.

2.99. For a needle of given length, what should be the side of a square in Problem 2.98 so that the probability of intersection is a maximum? Explain your answer.

2.100. Let

$$f(x, y, z) = \begin{cases} 24xy^2z^3 & 0 < x < 1, 0 < y < 1, 0 < z < 1 \\ 0 & \text{otherwise} \end{cases}$$

be the joint density function of three random variables X, Y, and Z. Find (a) $P(X > \frac{1}{2}, Y < \frac{1}{2}, Z > \frac{1}{2})$, (b) $P(Z < X + Y)$.

2.101. A cylindrical stream of particles, of radius a, is directed toward a hemispherical target ABC with center at O as indicated in Fig. 2-26. Assume that the distribution of particles is given by

$$f(r) = \begin{cases} 1/a & 0 < r < a \\ 0 & \text{otherwise} \end{cases}$$

where r is the distance from the axis OB. Show that the distribution of particles along the target is given by

$$g(\theta) = \begin{cases} \cos\theta & 0 < \theta < \pi/2 \\ 0 & \text{otherwise} \end{cases}$$

where θ is the angle that line OP (from O to any point P on the target) makes with the axis.

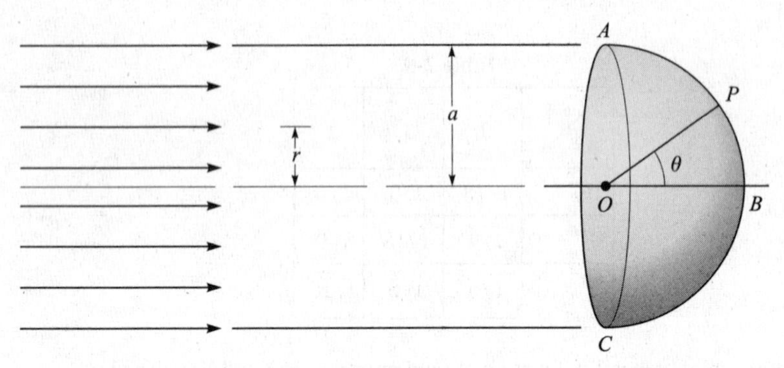

Fig. 2-26

2.102. In Problem 2.101 find the probability that a particle will hit the target between $\theta = 0$ and $\theta = \pi/4$.

2.103. Suppose that random variables X, Y, and Z have joint density function

$$f(x, y, z) = \begin{cases} 1 - \cos\pi x \cos\pi y \cos\pi z & 0 < x < 1, 0 < y < 1, 0 < z < 1 \\ 0 & \text{otherwise} \end{cases}$$

Show that although any two of these random variables are independent, i.e., their marginal density function factors, all three are not independent.

ANSWERS TO SUPPLEMENTARY PROBLEMS

2.38.

x	0	1	2	3
$f(x)$	1/8	3/8	3/8	1/8

2.39.

x	0	1	2
$f(x)$	3/28	15/28	5/14

2.40.

x	0	1	2
$f(x)$	9/64	15/32	25/64

2.42.

x	0	1	2	3	4
$f(x)$	$\frac{194,580}{270,725}$	$\frac{69,184}{270,725}$	$\frac{6768}{270,725}$	$\frac{192}{270,725}$	$\frac{1}{270,725}$

2.43.

x	0	1	2	3
$f(x)$	1/8	1/2	7/8	1

2.46. (a)

x	1	2	3	4
$f(x)$	1/8	1/4	3/8	1/4

(b) 3/4 (c) 7/8 (d) 3/8 (e) 7/8

2.47. (a) 3 (b) $e^{-3} - e^{-6}$ (c) e^{-9} (d) $1 - e^{-3}$ **2.48.** $F(x) = \begin{cases} 1 - e^{-3x} & x \geq 0 \\ 0 & x \leq 0 \end{cases}$

2.49. (a) 6/29 (b) 15/29 (c) 19/116 **2.50.** $F(x) = \begin{cases} 0 & x \leq 1 \\ (2x^3 - 2)/29 & 1 \leq x \leq 2 \\ (3x^2 + 2)/29 & 2 \leq x \leq 3 \\ 1 & x \geq 3 \end{cases}$

2.51. (a) 1/27 (b) $f(x) = \begin{cases} x^2/9 & 0 \leq x < 3 \\ 0 & \text{otherwise} \end{cases}$ (c) 26/27 (d) 7/27

2.53. (a) 1/2 (b) 1/2 (c) 3/4 (d) $F(x) = \begin{cases} 0 & x \leq 0 \\ x^2/4 & 0 \leq x \leq 2 \\ 1 & x \geq 2 \end{cases}$

2.54. (a) 1/36 (b) 1/6 (c) 1/4 (d) 5/6 (e) 1/6 (f) 1/6 (g) 1/2

2.55. (a) $f_1(x) = \begin{cases} x/6 & x = 1, 2, 3 \\ 0 & \text{other } x \end{cases}$ (b) $f_2(y) = \begin{cases} y/6 & y = 1, 2, 3 \\ 0 & \text{other } y \end{cases}$

2.56. (a) 3/2 (b) 1/4 (c) 29/64 (d) 5/16

2.57. (a) $F_1(x) = \begin{cases} 0 & x \leq 0 \\ \frac{1}{2}(x^3 + x) & 0 \leq x \leq 1 \\ 1 & x \geq 1 \end{cases}$ (b) $F_2(y) = \begin{cases} 0 & y \leq 0 \\ \frac{1}{2}(y^3 + y) & 0 \leq y \leq 1 \\ 1 & y \geq 1 \end{cases}$

2.58. (a) $f(x \mid y) = f_1(x)$ for $y = 1, 2, 3$ (see Problem 2.55)

(b) $f(y \mid x) = f_2(y)$ for $x = 1, 2, 3$ (see Problem 2.55)

2.59. (a) $f(x \mid y) = \begin{cases} (x + y)/(y + \frac{1}{2}) & 0 \leq x \leq 1, 0 \leq y \leq 1 \\ 0 & \text{other } x, 0 \leq y \leq 1 \end{cases}$

(b) $f(y \mid x) = \begin{cases} (x + y)/(x + \frac{1}{2}) & 0 \leq x \leq 1, 0 \leq y \leq 1 \\ 0 & 0 \leq x \leq 1, \text{other } y \end{cases}$

2.60. (a) $f(x \mid y) = \begin{cases} (x^2 + y^2)/(y^2 + \frac{1}{3}) & 0 \leq x \leq 1, 0 \leq y \leq 1 \\ 0 & \text{other } x, 0 \leq y \leq 1 \end{cases}$

(b) $f(y \mid x) = \begin{cases} (x^2 + y^2)/(x^2 + \frac{1}{3}) & 0 \leq x \leq 1, 0 \leq y \leq 1 \\ 0 & 0 \leq x \leq 1, \text{other } y \end{cases}$

2.61. (a) $f(x \mid y) = \begin{cases} e^{-x} & x \geq 0, y \geq 0 \\ 0 & x < 0, y \geq 0 \end{cases}$ (b) $f(y \mid x) = \begin{cases} e^{-y} & x \geq 0, y \geq 0 \\ 0 & x \geq 0, y < 0 \end{cases}$

2.62. $e^{-\sqrt{y}}/2\sqrt{y}$ for $y > 0$; 0 otherwise **2.64.** $(2\pi)^{-1/2}y^{-1/2}e^{-y/2}$ for $y > 0$; 0 otherwise

2.66. $1/\pi$ for $-\pi/2 < y < \pi/2$; 0 otherwise

2.68. (a) $g(y) = \begin{cases} \frac{1}{6} & -5 < y < 1 \\ 0 & \text{otherwise} \end{cases}$ (b) $g(y) = \begin{cases} \frac{1}{6}(1 - y)^{-2/3} & 0 < y < 1 \\ \frac{1}{6}(y - 1)^{-2/3} & 1 < y < 2 \\ 0 & \text{otherwise} \end{cases}$

2.70. $ve^{-v}/(1 + u)^2$ for $u \geq 0$, $v \geq 0$; 0 otherwise

2.73. $g(z) = \begin{cases} -\ln z & 0 < z < 1 \\ 0 & \text{otherwise} \end{cases}$ **2.77.** $g(x) = \begin{cases} x^3 e^{-x}/6 & x \geq 0 \\ 0 & x < 0 \end{cases}$

2.74. $g(u) = \begin{cases} u & 0 \leq u \leq 1 \\ 2 - u & 1 \leq u \leq 2 \\ 0 & \text{otherwise} \end{cases}$ **2.78.** 1/4

2.75. $g(u) = \begin{cases} ue^{-u} & u \geq 0 \\ 0 & u < 0 \end{cases}$ **2.79.** 61/72

2.81. (a) 2 (b) $F(x) = \begin{cases} 0 & x < 1 \\ 1 - 3^{-y} & y \leq x < y + 1; y = 1, 2, 3, \ldots \end{cases}$ (d) 26/81 (e) 1/9

2.82. (a) 4 (b) $F(x) = \begin{cases} 1 - e^{-2x}(2x + 1) & x \geq 0 \\ 0 & x < 0 \end{cases}$ (d) $3e^{-2}$ (e) $5e^{-4} - 7e^{-6}$

2.83. (a) 3/7 (b) 5/7 **2.84.** (a) $c = 1$ (b) $e^{-4} - 3e^{-2} + 2e^{-1}$

2.86. (a) $c_1 = 2, c_2 = 9$ (b) $9e^{-2} - 14e^{-3}$ (c) $4e^{-5} - 4e^{-7}$ (d) $e^{-2} - e^{-4}$ (e) $4e^{-3}$

2.88. (a) 1/4 (b) 27/64 (c) $f_1(x) = \begin{cases} x + \frac{1}{2} & 0 < x < 1 \\ 0 & \text{otherwise} \end{cases}$ (d) $f_2(y) = \begin{cases} \frac{1}{4}(y + 1) & 0 < y < 2 \\ 0 & \text{otherwise} \end{cases}$

2.90. (a) $\begin{cases} e^{-2y}/\sqrt{y} & y > 0 \\ 0 & \text{otherwise} \end{cases}$ (b) $\begin{cases} 18e^{-2u} & u > 0 \\ 0 & \text{otherwise} \end{cases}$

2.91. (b) $\frac{1}{2}(1 - \ln 2)$ (c) $\frac{1}{6} + \frac{1}{2}\ln 2$ (d) $\frac{1}{2}\ln 2$ **2.95.** (b) 15/256 (c) 9/16 (d) 0

2.93. $g(z) = \begin{cases} \frac{1}{2}e^{-z/2} & z \geq 0 \\ 0 & z < 0 \end{cases}$ **2.100.** (a) 45/512 (b) 1/14

2.94. (b) 7/18 **2.102.** $\sqrt{2}/2$

CHAPTER 3

Mathematical Expectation

Definition of Mathematical Expectation

A very important concept in probability and statistics is that of the *mathematical expectation, expected value*, or briefly the *expectation*, of a random variable. For a discrete random variable X having the possible values $x_1, \ldots, x_n$, the expectation of X is defined as

$$E(X) = x_1 P(X = x_1) + \cdots + x_n P(X = x_n) = \sum_{j=1}^{n} x_j P(X = x_j) \tag{1}$$

or equivalently, if $P(X = x_j) = f(x_j)$,

$$E(X) = x_1 f(x_1) + \cdots + x_n f(x_n) = \sum_{j=1}^{n} x_j f(x_j) = \sum x f(x) \tag{2}$$

where the last summation is taken over all appropriate values of x. As a special case of (2), where the probabilities are all equal, we have

$$E(X) = \frac{x_1 + x_2 + \cdots + x_n}{n} \tag{3}$$

which is called the *arithmetic mean*, or simply the *mean*, of $x_1, x_2, \ldots, x_n$.

If X takes on an infinite number of values $x_1, x_2, \ldots$, then $E(X) = \sum_{j=1}^{\infty} x_j f(x_j)$ provided that the infinite series converges absolutely.

For a continuous random variable X having density function $f(x)$, the expectation of X is defined as

$$E(X) = \int_{-\infty}^{\infty} x f(x)\, dx \tag{4}$$

provided that the integral converges absolutely.

The expectation of X is very often called the *mean* of X and is denoted by μ_X, or simply μ, when the particular random variable is understood.

The mean, or expectation, of X gives a single value that acts as a representative or average of the values of X, and for this reason it is often called a *measure of central tendency*. Other measures are considered on page 83.

EXAMPLE 3.1 Suppose that a game is to be played with a single die assumed fair. In this game a player wins \$20 if a 2 turns up, \$40 if a 4 turns up; loses \$30 if a 6 turns up; while the player neither wins nor loses if any other face turns up. Find the expected sum of money to be won.

Let X be the random variable giving the amount of money won on any toss. The possible amounts won when the die turns up $1, 2, \ldots, 6$ are $x_1, x_2, \ldots, x_6$, respectively, while the probabilities of these are $f(x_1), f(x_2), \ldots, f(x_6)$. The probability function for X is displayed in Table 3-1. Therefore, the expected value or expectation is

$$E(X) = (0)\left(\frac{1}{6}\right) + (20)\left(\frac{1}{6}\right) + (0)\left(\frac{1}{6}\right) + (40)\left(\frac{1}{6}\right) + (0)\left(\frac{1}{6}\right) + (-30)\left(\frac{1}{6}\right) = 5$$

Table 3-1

x_j	0	+20	0	+40	0	−30
$f(x_j)$	1/6	1/6	1/6	1/6	1/6	1/6

It follows that the player can expect to win $5. In a fair game, therefore, the player should be expected to pay $5 in order to play the game.

EXAMPLE 3.2 The density function of a random variable X is given by

$$f(x) = \begin{cases} \frac{1}{2}x & 0 < x < 2 \\ 0 & \text{otherwise} \end{cases}$$

The expected value of X is then

$$E(X) = \int_{-\infty}^{\infty} xf(x)\,dx = \int_0^2 x\left(\frac{1}{2}x\right)dx = \int_0^2 \frac{x^2}{2}\,dx = \frac{x^3}{6}\bigg|_0^2 = \frac{4}{3}$$

Functions of Random Variables

Let X be a discrete random variable with probability function $f(x)$. Then $Y = g(X)$ is also a discrete random variable, and the probability function of Y is

$$h(y) = P(Y = y) = \sum_{\{x|g(x)=y\}} P(X = x) = \sum_{\{x|g(x)=y\}} f(x)$$

If X takes on the values $x_1, x_2, \ldots, x_n$, and Y the values $y_1, y_2, \ldots, y_m$ $(m \leq n)$, then $y_1h(y_1) + y_2h(y_2) + \cdots + y_mh(y_m) = g(x_1)f(x_1) + g(x_2)f(x_2) + \cdots + g(x_n)f(x_n)$. Therefore,

$$\begin{aligned} E[g(X)] &= g(x_1)f(x_1) + g(x_2)f(x_2) + \cdots + g(x_n)f(x_n) \\ &= \sum_{j=1}^{n} g(x_j)f(x_j) = \sum g(x)f(x) \end{aligned} \tag{5}$$

Similarly, if X is a continuous random variable having probability density $f(x)$, then it can be shown that

$$E[g(X)] = \int_{-\infty}^{\infty} g(x)f(x)\,dx \tag{6}$$

Note that (5) and (6) do not involve, respectively, the probability function and the probability density function of $Y = g(X)$.

Generalizations are easily made to functions of two or more random variables. For example, if X and Y are two continuous random variables having joint density function $f(x, y)$, then the expectation of $g(X, Y)$ is given by

$$E[g(X, Y)] = \int_{-\infty}^{\infty}\int_{-\infty}^{\infty} g(x, y)f(x, y)\,dx\,dy \tag{7}$$

EXAMPLE 3.3 If X is the random variable of Example 3.2,

$$E(3X^2 - 2X) = \int_{-\infty}^{\infty} (3x^2 - 2x)f(x)\,dx = \int_0^2 (3x^2 - 2x)\left(\frac{1}{2}x\right)dx = \frac{10}{3}$$

Some Theorems on Expectation

Theorem 3-1 If c is any constant, then

$$E(cX) = cE(X) \tag{8}$$

Theorem 3-2 If X and Y are any random variables, then

$$E(X + Y) = E(X) + E(Y) \tag{9}$$

Theorem 3-3 If X and Y are independent random variables, then

$$E(XY) = E(X)E(Y) \tag{10}$$

Generalizations of these theorems are easily made.

The Variance and Standard Deviation

We have already noted on page 75 that the expectation of a random variable X is often called the *mean* and is denoted by μ. Another quantity of great importance in probability and statistics is called the *variance* and is defined by

$$\text{Var}(X) = E[(X - \mu)^2] \tag{11}$$

The variance is a nonnegative number. The positive square root of the variance is called the *standard deviation* and is given by

$$\sigma_X = \sqrt{\text{Var}(X)} = \sqrt{E[(X - \mu)^2]} \tag{12}$$

Where no confusion can result, the standard deviation is often denoted by σ instead of σ_X, and the variance in such case is σ^2.

If X is a discrete random variable taking the values $x_1, x_2, \ldots, x_n$ and having probability function $f(x)$, then the variance is given by

$$\sigma_X^2 = E[(X - \mu)^2] = \sum_{j=1}^{n}(x_j - \mu)^2 f(x_j) = \sum (x - \mu)^2 f(x) \tag{13}$$

In the special case of (13) where the probabilities are all equal, we have

$$\sigma^2 = [(x_1 - \mu)^2 + (x_2 - \mu)^2 + \cdots + (x_n - \mu)^2]/n \tag{14}$$

which is the variance for a set of n numbers $x_1, \ldots, x_n$.

If X takes on an infinite number of values $x_1, x_2, \ldots$, then $\sigma_X^2 = \sum_{j=1}^{\infty}(x_j - \mu)^2 f(x_j)$, provided that the series converges.

If X is a continuous random variable having density function $f(x)$, then the variance is given by

$$\sigma_X^2 = E[(X - \mu)^2] = \int_{-\infty}^{\infty} (x - \mu)^2 f(x)\, dx \tag{15}$$

provided that the integral converges.

The variance (or the standard deviation) is a measure of the *dispersion*, or *scatter*, of the values of the random variable about the mean μ. If the values tend to be concentrated near the mean, the variance is small; while if the values tend to be distributed far from the mean, the variance is large. The situation is indicated graphically in Fig. 3-1 for the case of two continuous distributions having the same mean μ.

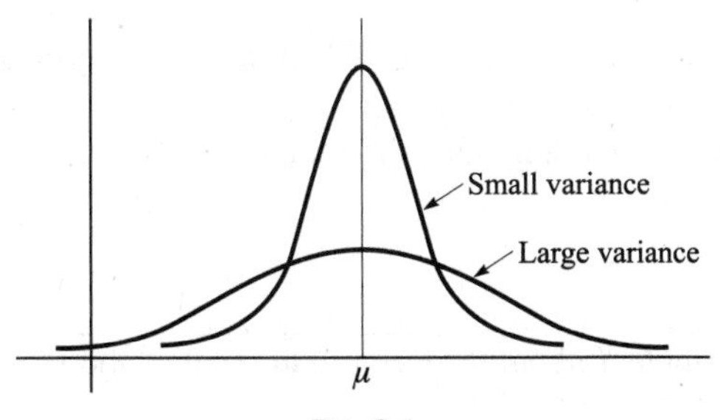

Fig. 3-1

EXAMPLE 3.4 Find the variance and standard deviation of the random variable of Example 3.2. As found in Example 3.2, the mean is $\mu = E(X) = 4/3$. Then the variance is given by

$$\sigma^2 = E\left[\left(X - \frac{4}{3}\right)^2\right] = \int_{-\infty}^{\infty}\left(x - \frac{4}{3}\right)^2 f(x)\,dx = \int_0^2\left(x - \frac{4}{3}\right)^2\left(\frac{1}{2}x\right)dx = \frac{2}{9}$$

and so the standard deviation is $\sigma = \sqrt{\frac{2}{9}} = \frac{\sqrt{2}}{3}$

Note that if X has certain *dimensions* or *units*, such as *centimeters* (cm), then the variance of X has units cm^2 while the standard deviation has the same unit as X, i.e., cm. It is for this reason that the standard deviation is often used.

Some Theorems on Variance

Theorem 3-4

$$\sigma^2 = E[(X - \mu)^2] = E(X^2) - \mu^2 = E(X^2) - [E(X)]^2 \tag{16}$$

where $\mu = E(X)$.

Theorem 3-5 If c is any constant,

$$\text{Var}(cX) = c^2\,\text{Var}(X) \tag{17}$$

Theorem 3-6 The quantity $E[(X - a)^2]$ is a minimum when $a = \mu = E(X)$.

Theorem 3-7 If X and Y are independent random variables,

$$\text{Var}(X + Y) = \text{Var}(X) + \text{Var}(Y) \quad \text{or} \quad \sigma^2_{X+Y} = \sigma^2_X + \sigma^2_Y \tag{18}$$

$$\text{Var}(X - Y) = \text{Var}(X) + \text{Var}(Y) \quad \text{or} \quad \sigma^2_{X-Y} = \sigma^2_X + \sigma^2_Y \tag{19}$$

Generalizations of Theorem 3-7 to more than two independent variables are easily made. In words, the variance of a sum of independent variables equals the sum of their variances.

Standardized Random Variables

Let X be a random variable with mean μ and standard deviation σ ($\sigma > 0$). Then we can define an associated *standardized random variable* given by

$$X^* = \frac{X - \mu}{\sigma} \tag{20}$$

An important property of X^* is that it has a mean of zero and a variance of 1, which accounts for the name *standardized,* i.e.,

$$E(X^*) = 0, \qquad \text{Var}(X^*) = 1 \tag{21}$$

The values of a standardized variable are sometimes called *standard scores,* and X is then said to be expressed in *standard units* (i.e., σ is taken as the unit in measuring $X - \mu$).

Standardized variables are useful for comparing different distributions.

Moments

The *rth moment of a random variable X about the mean* μ, also called the *rth central moment*, is defined as

$$\mu_r = E[(X - \mu)^r] \tag{22}$$

where $r = 0, 1, 2, \ldots$. It follows that $\mu_0 = 1$, $\mu_1 = 0$, and $\mu_2 = \sigma^2$, i.e., the second central moment or second moment about the mean is the variance. We have, assuming absolute convergence,

$$\mu_r = \sum (x - \mu)^r f(x) \qquad \text{(discrete variable)} \tag{23}$$

$$\mu_r = \int_{-\infty}^{\infty} (x - \mu)^r f(x)\, dx \qquad \text{(continuous variable)} \tag{24}$$

The *rth moment of X about the origin*, also called the *rth raw moment*, is defined as

$$\mu_r' = E(X^r) \tag{25}$$

where $r = 0, 1, 2, \ldots$, and in this case there are formulas analogous to (23) and (24) in which $\mu = 0$.

The relationship between these moments is given by

$$\mu_r = \mu_r' - \binom{r}{1}\mu_{r-1}'\mu + \cdots + (-1)^j\binom{r}{j}\mu_{r-j}'\mu^j + \cdots + (-1)^r\mu_0'\mu^r \tag{26}$$

As special cases we have, using $\mu_1' = \mu$ and $\mu_0' = 1$,

$$\begin{aligned} \mu_2 &= \mu_2' - \mu^2 \\ \mu_3 &= \mu_3' - 3\mu_2'\mu + 2\mu^3 \\ \mu_4 &= \mu_4' - 4\mu_3'\mu + 6\mu_2'\mu^2 - 3\mu^4 \end{aligned} \tag{27}$$

Moment Generating Functions

The *moment generating function* of X is defined by

$$M_X(t) = E(e^{tX}) \tag{28}$$

that is, assuming convergence,

$$M_X(t) = \sum e^{tx} f(x) \qquad \text{(discrete variable)} \tag{29}$$

$$M_X(t) = \int_{-\infty}^{\infty} e^{tx} f(x)\, dx \qquad \text{(continuous variable)} \tag{30}$$

We can show that the Taylor series expansion is [Problem 3.15(a)]

$$M_X(t) = 1 + \mu t + \mu_2'\frac{t^2}{2!} + \cdots + \mu_r'\frac{t^r}{r!} + \cdots \tag{31}$$

Since the coefficients in this expansion enable us to find the moments, the reason for the name *moment generating function* is apparent. From the expansion we can show that [Problem 3.15(b)]

$$\mu_r' = \left.\frac{d^r}{dt^r} M_X(t)\right|_{t=0} \tag{32}$$

i.e., μ_r' is the rth derivative of $M_X(t)$ evaluated at $t = 0$. Where no confusion can result, we often write $M(t)$ instead of $M_X(t)$.

Some Theorems on Moment Generating Functions

Theorem 3-8 If $M_X(t)$ is the moment generating function of the random variable X and a and b ($b \neq 0$) are constants, then the moment generating function of $(X + a)/b$ is

$$M_{(X+a)/b}(t) = e^{at/b} M_X\left(\frac{t}{b}\right) \tag{33}$$

Theorem 3-9 If X and Y are independent random variables having moment generating functions $M_X(t)$ and $M_Y(t)$, respectively, then

$$M_{X+Y}(t) = M_X(t)\, M_Y(t) \tag{34}$$

Generalizations of Theorem 3-9 to more than two independent random variables are easily made. In words, the moment generating function of a sum of independent random variables is equal to the product of their moment generating functions.

Theorem 3-10 **(Uniqueness Theorem)** Suppose that X and Y are random variables having moment generating functions $M_X(t)$ and $M_Y(t)$, respectively. Then X and Y have the same probability distribution if and only if $M_X(t) = M_Y(t)$ identically.

Characteristic Functions

If we let $t = i\omega$, where i is the imaginary unit, in the moment generating function we obtain an important function called the *characteristic function.* We denote this by

$$\phi_X(\omega) = M_X(i\omega) = E(e^{i\omega X}) \tag{35}$$

It follows that

$$\phi_X(\omega) = \sum e^{i\omega x} f(x) \qquad \text{(discrete variable)} \tag{36}$$

$$\phi_X(\omega) = \int_{-\infty}^{\infty} e^{i\omega x} f(x)\, dx \qquad \text{(continuous variable)} \tag{37}$$

Since $|e^{i\omega x}| = 1$, the series and the integral always converge absolutely.

The corresponding results (31) and (32) become

$$\phi_X(\omega) = 1 + i\mu\omega - \mu_2' \frac{\omega^2}{2!} + \cdots + i^r \mu_r' \frac{\omega^r}{r!} + \cdots \tag{38}$$

where

$$\mu_r' = (-1)^r i^r \frac{d^r}{d\omega^r} \phi_X(\omega) \bigg|_{\omega=0} \tag{39}$$

When no confusion can result, we often write $\phi(\omega)$ instead of $\phi_X(\omega)$.

Theorems for characteristic functions corresponding to Theorems 3-8, 3-9, and 3-10 are as follows.

Theorem 3-11 If $\phi_X(\omega)$ is the characteristic function of the random variable X and a and b ($b \neq 0$) are constants, then the characteristic function of $(X + a)/b$ is

$$\phi_{(X+a)/b}(\omega) = e^{ai\omega/b} \phi_X\left(\frac{\omega}{b}\right) \tag{40}$$

Theorem 3-12 If X and Y are independent random variables having characteristic functions $\phi_X(\omega)$ and $\phi_Y(\omega)$, respectively, then

$$\phi_{X+Y}(\omega) = \phi_X(\omega)\,\phi_Y(\omega) \tag{41}$$

More generally, the characteristic function of a sum of independent random variables is equal to the product of their characteristic functions.

Theorem 3-13 **(Uniqueness Theorem)** Suppose that X and Y are random variables having characteristic functions $\phi_X(\omega)$ and $\phi_Y(\omega)$, respectively. Then X and Y have the same probability distribution if and only if $\phi_X(\omega) = \phi_Y(\omega)$ identically.

An important reason for introducing the characteristic function is that (37) represents the *Fourier transform* of the density function $f(x)$. From the theory of Fourier transforms, we can easily determine the density function from the characteristic function. In fact,

$$f(x) = \frac{1}{2\pi}\int_{-\infty}^{\infty} e^{-i\omega x}\phi_X(\omega)\,d\omega \tag{42}$$

which is often called an *inversion formula*, or *inverse Fourier transform*. In a similar manner we can show in the discrete case that the probability function $f(x)$ can be obtained from (36) by use of *Fourier series*, which is the analog of the Fourier integral for the discrete case. See Problem 3.39.

Another reason for using the characteristic function is that it always exists whereas the moment generating function may not exist.

Variance for Joint Distributions. Covariance

The results given above for one variable can be extended to two or more variables. For example, if X and Y are two continuous random variables having joint density function $f(x, y)$, the means, or expectations, of X and Y are

$$\mu_X = E(X) = \int_{-\infty}^{\infty}\int_{-\infty}^{\infty} xf(x, y)\,dx\,dy, \qquad \mu_Y = E(Y) = \int_{-\infty}^{\infty}\int_{-\infty}^{\infty} yf(x, y)\,dx\,dy \tag{43}$$

and the variances are

$$\begin{aligned} \sigma_X^2 &= E[(X - \mu_X)^2] = \int_{-\infty}^{\infty}\int_{-\infty}^{\infty} (x - \mu_X)^2 f(x, y)\,dx\,dy \\ \sigma_Y^2 &= E[(Y - \mu_Y)^2] = \int_{-\infty}^{\infty}\int_{-\infty}^{\infty} (y - \mu_Y)^2 f(x, y)\,dx\,dy \end{aligned} \tag{44}$$

Note that the marginal density functions of X and Y are not directly involved in (43) and (44).

Another quantity that arises in the case of two variables X and Y is the *covariance* defined by

$$\sigma_{XY} = \text{Cov}(X, Y) = E[(X - \mu_X)(Y - \mu_Y)] \tag{45}$$

In terms of the joint density function $f(x, y)$, we have

$$\sigma_{XY} = \int_{-\infty}^{\infty}\int_{-\infty}^{\infty} (x - \mu_X)(y - \mu_Y)f(x, y)\,dx\,dy \tag{46}$$

Similar remarks can be made for two discrete random variables. In such cases (43) and (46) are replaced by

$$\mu_X = \sum_x\sum_y xf(x, y) \qquad \mu_Y = \sum_x\sum_y yf(x, y) \tag{47}$$

$$\sigma_{XY} = \sum_x\sum_y (x - \mu_X)(y - \mu_Y)f(x, y) \tag{48}$$

where the sums are taken over all the discrete values of X and Y.

The following are some important theorems on covariance.

Theorem 3-14

$$\sigma_{XY} = E(XY) - E(X)E(Y) = E(XY) - \mu_X\mu_Y \tag{49}$$

Theorem 3-15 If X and Y are independent random variables, then

$$\sigma_{XY} = \text{Cov}(X, Y) = 0 \tag{50}$$

Theorem 3-16

$$\text{Var}(X \pm Y) = \text{Var}(X) + \text{Var}(Y) \pm 2\,\text{Cov}(X, Y) \tag{51}$$

or

$$\sigma_{X\pm Y}^2 = \sigma_X^2 + \sigma_Y^2 \pm 2\sigma_{XY} \tag{52}$$

Theorem 3-17

$$|\sigma_{XY}| \le \sigma_X\sigma_Y \tag{53}$$

The converse of Theorem 3-15 is not necessarily true. If X and Y are independent, Theorem 3-16 reduces to Theorem 3-7.

Correlation Coefficient

If X and Y are independent, then $\text{Cov}(X, Y) = \sigma_{XY} = 0$. On the other hand, if X and Y are completely dependent, for example, when $X = Y$, then $\text{Cov}(X, Y) = \sigma_{XY} = \sigma_X \sigma_Y$. From this we are led to a *measure of the dependence* of the variables X and Y given by

$$\rho = \frac{\sigma_{XY}}{\sigma_X \sigma_Y} \tag{54}$$

We call ρ the *correlation coefficient*, or *coefficient of correlation*. From Theorem 3-17 we see that $-1 \leq \rho \leq 1$. In the case where $\rho = 0$ (i.e., the covariance is zero), we call the variables X and Y *uncorrelated*. In such cases, however, the variables may or may not be independent. Further discussion of correlation cases will be given in Chapter 8.

Conditional Expectation, Variance, and Moments

If X and Y have joint density function $f(x, y)$, then as we have seen in Chapter 2, the conditional density function of Y given X is $f(y \mid x) = f(x, y)/f_1(x)$ where $f_1(x)$ is the marginal density function of X. We can define the *conditional expectation*, or *conditional mean*, of Y given X by

$$E(Y \mid X = x) = \int_{-\infty}^{\infty} y f(y \mid x)\, dy \tag{55}$$

where "$X = x$" is to be interpreted as $x < X \leq x + dx$ in the continuous case. Theorems 3-1 and 3-2 also hold for conditional expectation.

We note the following properties:

1. $E(Y \mid X = x) = E(Y)$ when X and Y are independent.
2. $E(Y) = \int_{-\infty}^{\infty} E(Y \mid X = x) f_1(x)\, dx$.

It is often convenient to calculate expectations by use of Property 2, rather than directly.

EXAMPLE 3.5 The average travel time to a distant city is c hours by car or b hours by bus. A woman cannot decide whether to drive or take the bus, so she tosses a coin. What is her expected travel time?

Here we are dealing with the joint distribution of the outcome of the toss, X, and the travel time, Y, where $Y = Y_{\text{car}}$ if $X = 0$ and $Y = Y_{\text{bus}}$ if $X = 1$. Presumably, both Y_{car} and Y_{bus} are independent of X, so that by Property 1 above

$$E(Y \mid X = 0) = E(Y_{\text{car}} \mid X = 0) = E(Y_{\text{car}}) = c$$

and

$$E(Y \mid X = 1) = E(Y_{\text{bus}} \mid X = 1) = E(Y_{\text{bus}}) = b$$

Then Property 2 (with the integral replaced by a sum) gives, for a fair coin,

$$E(Y) = E(Y \mid X = 0)P(X = 0) + E(Y \mid X = 1)P(X = 1) = \frac{c + b}{2}$$

In a similar manner we can define the *conditional variance* of Y given X as

$$E[(Y - \mu_2)^2 \mid X = x] = \int_{-\infty}^{\infty} (y - \mu_2)^2 f(y \mid x)\, dy \tag{56}$$

where $\mu_2 = E(Y \mid X = x)$. Also we can define the rth *conditional moment* of Y about any value a given X as

$$E[(Y - a)^r \mid X = x] = \int_{-\infty}^{\infty} (y - a)^r f(y \mid x)\, dy \tag{57}$$

The usual theorems for variance and moments extend to conditional variance and moments.

Chebyshev's Inequality

An important theorem in probability and statistics that reveals a general property of discrete or continuous random variables having finite mean and variance is known under the name of *Chebyshev's inequality.*

Theorem 3-18 **(Chebyshev's Inequality)** Suppose that X is a random variable (discrete or continuous) having mean μ and variance σ^2, which are finite. Then if ϵ is any positive number,

$$P(|X - \mu| \geq \epsilon) \leq \frac{\sigma^2}{\epsilon^2} \tag{58}$$

or, with $\epsilon = k\sigma$,

$$P(|X - \mu| \geq k\sigma) \leq \frac{1}{k^2} \tag{59}$$

EXAMPLE 3.6 Letting $k = 2$ in Chebyshev's inequality (59), we see that

$$P(|X - \mu| \geq 2\sigma) \leq 0.25 \qquad \text{or} \qquad P(|X - \mu| < 2\sigma) \geq 0.75$$

In words, the probability of X differing from its mean by more than 2 standard deviations is less than or equal to 0.25; equivalently, the probability that X will lie within 2 standard deviations of its mean is greater than or equal to 0.75. This is quite remarkable in view of the fact that we have not even specified the probability distribution of X.

Law of Large Numbers

The following theorem, called the *law of large numbers*, is an interesting consequence of Chebyshev's inequality.

Theorem 3-19 **(Law of Large Numbers):** Let $X_1, X_2, \ldots, X_n$ be mutually independent random variables (discrete or continuous), each having finite mean μ and variance σ^2. Then if $S_n = X_1 + X_2 + \cdots + X_n (n = 1, 2, \ldots)$,

$$\lim_{n \to \infty} P\left(\left|\frac{S_n}{n} - \mu\right| \geq \epsilon\right) = 0 \tag{60}$$

Since S_n/n is the arithmetic mean of $X_1, \ldots, X_n$, this theorem states that the probability of the arithmetic mean S_n/n differing from its expected value μ by more than ϵ approaches zero as $n \to \infty$. A stronger result, which we might expect to be true, is that $\lim_{n \to \infty} S_n/n = \mu$, but this is actually false. However, we can prove that $\lim_{n \to \infty} S_n/n = \mu$ *with probability one.* This result is often called the *strong law of large numbers*, and, by contrast, that of Theorem 3-19 is called the *weak law of large numbers.* When the "law of large numbers" is referred to without qualification, the weak law is implied.

Other Measures of Central Tendency

As we have already seen, the mean, or expectation, of a random variable X provides a measure of central tendency for the values of a distribution. Although the mean is used most, two other measures of central tendency are also employed. These are the *mode* and the *median.*

1. **MODE.** The *mode* of a discrete random variable is that value which occurs most often or, in other words, has the greatest probability of occurring. Sometimes we have two, three, or more values that have relatively large probabilities of occurrence. In such cases, we say that the distribution is *bimodal, trimodal,* or *multimodal*, respectively. The mode of a continuous random variable X is the value (or values) of X where the probability density function has a relative maximum.

2. **MEDIAN.** The *median* is that value x for which $P(X < x) \leq \frac{1}{2}$ and $P(X > x) \leq \frac{1}{2}$. In the case of a continuous distribution we have $P(X < x) = \frac{1}{2} = P(X > x)$, and the median separates the density curve into two parts having equal areas of $1/2$ each. In the case of a discrete distribution a unique median may not exist (see Problem 3.34).

Percentiles

It is often convenient to subdivide the area under a density curve by use of ordinates so that the area to the left of the ordinate is some percentage of the total unit area. The values corresponding to such areas are called *percentile values*, or briefly *percentiles*. Thus, for example, the area to the left of the ordinate at x_α in Fig. 3-2 is α. For instance, the area to the left of $x_{0.10}$ would be 0.10, or 10%, and $x_{0.10}$ would be called the *10th percentile* (also called the *first decile*). The median would be the *50th percentile* (or *fifth decile*).

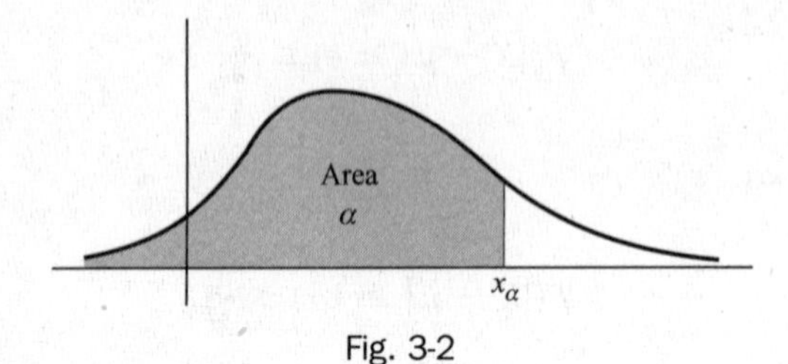

Fig. 3-2

Other Measures of Dispersion

Just as there are various measures of central tendency besides the mean, there are various measures of dispersion or scatter of a random variable besides the variance or standard deviation. Some of the most common are the following.

1. **SEMI-INTERQUARTILE RANGE.** If $x_{0.25}$ and $x_{0.75}$ represent the 25th and 75th percentile values, the difference $x_{0.75} - x_{0.25}$ is called the *interquartile range* and $\frac{1}{2}(x_{0.75} - x_{0.25})$ is the *semi-interquartile range.*
2. **MEAN DEVIATION.** The *mean deviation* (M.D.) of a random variable X is defined as the expectation of $|X - \mu|$, i.e., assuming convergence,

$$\text{M.D.}(X) = E[|X - \mu|] = \sum |x - \mu| f(x) \qquad \text{(discrete variable)} \tag{61}$$

$$\text{M.D.}(X) = E[|X - \mu|] = \int_{-\infty}^{\infty} |x - \mu| f(x)\, dx \qquad \text{(continuous variable)} \tag{62}$$

Skewness and Kurtosis

1. **SKEWNESS.** Often a distribution is not symmetric about any value but instead has one of its tails longer than the other. If the longer tail occurs to the right, as in Fig. 3-3, the distribution is said to be *skewed to the right*, while if the longer tail occurs to the left, as in Fig. 3-4, it is said to be *skewed to the left*. Measures describing this asymmetry are called *coefficients of skewness*, or briefly *skewness*. One such measure is given by

$$\alpha_3 = \frac{E[(X - \mu)^3]}{\sigma^3} = \frac{\mu_3}{\sigma^3} \tag{63}$$

The measure σ_3 will be positive or negative according to whether the distribution is skewed to the right or left, respectively. For a symmetric distribution, $\sigma_3 = 0$.

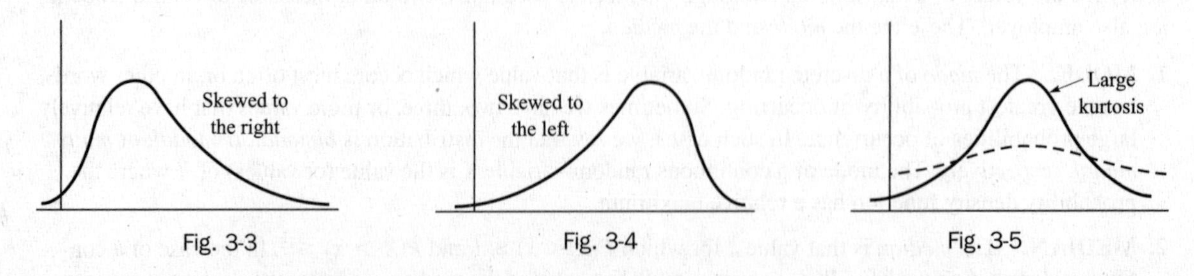

Fig. 3-3 Fig. 3-4 Fig. 3-5

2. **KURTOSIS.** In some cases a distribution may have its values concentrated near the mean so that the distribution has a large peak as indicated by the solid curve of Fig. 3-5. In other cases the distribution may be

relatively flat as in the dashed curve of Fig. 3-5. Measures of the degree of peakedness of a distribution are called *coefficients of kurtosis*, or briefly *kurtosis*. A measure often used is given by

$$\alpha_4 = \frac{E[(X-\mu)^4]}{\sigma^4} = \frac{\mu_4}{\sigma^4} \tag{64}$$

This is usually compared with the normal curve (see Chapter 4), which has a coefficient of kurtosis equal to 3. See also Problem 3.41.

SOLVED PROBLEMS

Expectation of random variables

3.1. In a lottery there are 200 prizes of \$5, 20 prizes of \$25, and 5 prizes of \$100. Assuming that 10,000 tickets are to be issued and sold, what is a fair price to pay for a ticket?

Let X be a random variable denoting the amount of money to be won on a ticket. The various values of X together with their probabilities are shown in Table 3-2. For example, the probability of getting one of the 20 tickets giving a \$25 prize is $20/10{,}000 = 0.002$. The expectation of X in dollars is thus

$$E(X) = (5)(0.02) + (25)(0.002) + (100)(0.0005) + (0)(0.9775) = 0.2$$

or 20 cents. Thus the fair price to pay for a ticket is 20 cents. However, since a lottery is usually designed to raise money, the price per ticket would be higher.

Table 3-2

x (dollars)	5	25	100	0
$P(X = x)$	0.02	0.002	0.0005	0.9775

3.2. Find the expectation of the sum of points in tossing a pair of fair dice.

Let X and Y be the points showing on the two dice. We have

$$E(X) = E(Y) = 1\left(\frac{1}{6}\right) + 2\left(\frac{1}{6}\right) + \cdots + 6\left(\frac{1}{6}\right) = \frac{7}{2}$$

Then, by Theorem 3-2,

$$E(X+Y) = E(X) + E(Y) = 7$$

3.3. Find the expectation of a discrete random variable X whose probability function is given by

$$f(x) = \left(\frac{1}{2}\right)^x \qquad (x = 1, 2, 3, \ldots)$$

We have

$$E(X) = \sum_{x=1}^{\infty} x\left(\frac{1}{2}\right)^x = \frac{1}{2} + 2\left(\frac{1}{4}\right) + 3\left(\frac{1}{8}\right) + \cdots$$

To find this sum, let

$$S = \frac{1}{2} + 2\left(\frac{1}{4}\right) + 3\left(\frac{1}{8}\right) + 4\left(\frac{1}{16}\right) + \cdots$$

Then

$$\frac{1}{2}S = \frac{1}{4} + 2\left(\frac{1}{8}\right) + 3\left(\frac{1}{16}\right) + \cdots$$

Subtracting,

$$\frac{1}{2}S = \frac{1}{2} + \frac{1}{4} + \frac{1}{8} + \frac{1}{16} + \cdots = 1$$

Therefore, $S = 2$.

3.4. A continuous random variable X has probability density given by

$$f(x) = \begin{cases} 2e^{-2x} & x > 0 \\ 0 & x \leq 0 \end{cases}$$

Find (a) $E(X)$, (b) $E(X^2)$.

(a)
$$E(X) = \int_{-\infty}^{\infty} xf(x)\,dx = \int_0^{\infty} x(2e^{-2x})\,dx = 2\int_0^{\infty} xe^{-2x}\,dx$$

$$= 2\left[(x)\left(\frac{e^{-2x}}{-2}\right) - (1)\left(\frac{e^{-2x}}{4}\right)\right]\Bigg|_0^{\infty} = \frac{1}{2}$$

(b)
$$E(X^2) = \int_{-\infty}^{\infty} x^2 f(x)\,dx = 2\int_0^{\infty} x^2 e^{-2x}\,dx$$

$$= 2\left[(x^2)\left(\frac{e^{-2x}}{-2}\right) - (2x)\left(\frac{e^{-2x}}{4}\right) + (2)\left(\frac{e^{-2x}}{-8}\right)\right]\Bigg|_0^{\infty} = \frac{1}{2}$$

3.5. The joint density function of two random variables X and Y is given by

$$f(x, y) = \begin{cases} xy/96 & 0 < x < 4, 1 < y < 5 \\ 0 & \text{otherwise} \end{cases}$$

Find (a) $E(X)$, (b) $E(Y)$, (c) $E(XY)$, (d) $E(2X + 3Y)$.

(a)
$$E(X) = \int_{-\infty}^{\infty}\int_{-\infty}^{\infty} xf(x, y)\,dx\,dy = \int_{x=0}^{4}\int_{y=1}^{5} x\left(\frac{xy}{96}\right)dx\,dy = \frac{8}{3}$$

(b)
$$E(Y) = \int_{-\infty}^{\infty}\int_{-\infty}^{\infty} yf(x, y)\,dx\,dy = \int_{x=0}^{4}\int_{y=1}^{5} y\left(\frac{xy}{96}\right)dx\,dy = \frac{31}{9}$$

(c)
$$E(XY) = \int_{-\infty}^{\infty}\int_{-\infty}^{\infty} (xy)f(x, y)\,dx\,dy = \int_{x=0}^{4}\int_{y=1}^{5} (xy)\left(\frac{xy}{96}\right)dx\,dy = \frac{248}{27}$$

(d)
$$E(2X + 3Y) = \int_{-\infty}^{\infty}\int_{-\infty}^{\infty} (2x + 3y)f(x, y)\,dx\,dy = \int_{x=0}^{4}\int_{y=1}^{5} (2x + 3y)\left(\frac{xy}{96}\right)dx\,dy = \frac{47}{3}$$

Another method

(c) Since X and Y are independent, we have, using parts (a) and (b),

$$E(XY) = E(X)E(Y) = \left(\frac{8}{3}\right)\left(\frac{31}{9}\right) = \frac{248}{27}$$

(d) By Theorems 3-1 and 3-2, pages 76–77, together with (a) and (b),

$$E(2X + 3Y) = 2E(X) + 3E(Y) = 2\left(\frac{8}{3}\right) + 3\left(\frac{31}{9}\right) = \frac{47}{3}$$

3.6. Prove Theorem 3-2, page 77.

Let $f(x, y)$ be the joint probability function of X and Y, assumed discrete. Then

$$E(X + Y) = \sum_x\sum_y (x + y)f(x, y)$$

$$= \sum_x\sum_y xf(x, y) + \sum_x\sum_y yf(x, y)$$

$$= E(X) + E(Y)$$

If either variable is continuous, the proof goes through as before, with the appropriate summations replaced by integrations. Note that the theorem is true whether or not X and Y are independent.

3.7. Prove Theorem 3-3, page 77.

Let $f(x, y)$ be the joint probability function of X and Y, assumed discrete. If the variables X and Y are independent, we have $f(x, y) = f_1(x)f_2(y)$. Therefore,

$$E(XY) = \sum_x \sum_y xyf(x, y) = \sum_x \sum_y xyf_1(x)f_2(y)$$
$$= \sum_x \left[xf_1(x) \sum_y yf_2(y) \right]$$
$$= \sum_x [(xf_1(x)E(y)]$$
$$= E(X)E(Y)$$

If either variable is continuous, the proof goes through as before, with the appropriate summations replaced by integrations. Note that the validity of this theorem hinges on whether $f(x, y)$ can be expressed as a function of x multiplied by a function of y, for all x and y, i.e., on whether X and Y are independent. For dependent variables it is not true in general.

Variance and standard deviation

3.8. Find (a) the variance, (b) the standard deviation of the sum obtained in tossing a pair of fair dice.

(a) Referring to Problem 3.2, we have $E(X) = E(Y) = 1/2$. Moreover,

$$E(X^2) = E(Y^2) = 1^2\left(\frac{1}{6}\right) + 2^2\left(\frac{1}{6}\right) + \cdots + 6^2\left(\frac{1}{6}\right) = \frac{91}{6}$$

Then, by Theorem 3-4,

$$\text{Var}(X) = \text{Var}(Y) = \frac{91}{6} - \left(\frac{7}{2}\right)^2 = \frac{35}{12}$$

and, since X and Y are independent, Theorem 3-7 gives

$$\text{Var}(X + Y) = \text{Var}(X) + \text{Var}(Y) = \frac{35}{6}$$

(b)
$$\sigma_{X+Y} = \sqrt{\text{Var}(X + Y)} = \sqrt{\frac{35}{6}}$$

3.9. Find (a) the variance, (b) the standard deviation for the random variable of Problem 3.4.

(a) As in Problem 3.4, the mean of X is $\mu = E(X) = \frac{1}{2}$. Then the variance is

$$\text{Var}(X) = E[(X - \mu)^2] = E\left[\left(X - \frac{1}{2}\right)^2\right] = \int_{-\infty}^{\infty}\left(x - \frac{1}{2}\right)^2 f(x)\,dx$$
$$= \int_0^{\infty}\left(x - \frac{1}{2}\right)^2 (2e^{-2x})\,dx = \frac{1}{4}$$

Another method
By Theorem 3-4,

$$\text{Var}(X) = E[(X - \mu)^2] = E(X^2) - [E(X)]^2 = \frac{1}{2} - \left(\frac{1}{2}\right)^2 = \frac{1}{4}$$

(b)
$$\sigma = \sqrt{\text{Var}(X)} = \sqrt{\frac{1}{4}} = \frac{1}{2}$$

3.10. Prove Theorem 3-4, page 78.

We have

$$\begin{aligned} E[(X-\mu)^2] &= E(X^2 - 2\mu X + \mu^2) = E(X^2) - 2\mu E(X) + \mu^2 \\ &= E(X^2) - 2\mu^2 + \mu^2 = E(X^2) - \mu^2 \\ &= E(X^2) - [E(X)]^2 \end{aligned}$$

3.11. Prove Theorem 3-6, page 78.

$$\begin{aligned} E[(X-a)^2] &= E[\{(X-\mu) + (\mu - a)\}^2] \\ &= E[(X-\mu)^2 + 2(X-\mu)(\mu-a) + (\mu-a)^2] \\ &= E[(X-\mu)^2] + 2(\mu-a)E(X-\mu) + (\mu-a)^2 \\ &= E[(X-\mu)^2] + (\mu-a)^2 \end{aligned}$$

since $E(X-\mu) = E(X) - \mu = 0$. From this we see that the minimum value of $E[(X-a)^2]$ occurs when $(\mu-a)^2 = 0$, i.e., when $a = \mu$.

3.12. If $X^* = (X-\mu)/\sigma$ is a standardized random variable, prove that (a) $E(X^*) = 0$, (b) $\text{Var}(X^*) = 1$.

(a)
$$E(X^*) = E\left(\frac{X-\mu}{\sigma}\right) = \frac{1}{\sigma}[E(X-\mu)] = \frac{1}{\sigma}[E(X) - \mu] = 0$$

since $E(X) = \mu$.

(b)
$$\text{Var}(X^*) = \text{Var}\left(\frac{X-\mu}{\sigma}\right) = \frac{1}{\sigma^2}E[(X-\mu)^2] = 1$$

using Theorem 3-5, page 78, and the fact that $E[(X-\mu)^2] = \sigma^2$.

3.13. Prove Theorem 3-7, page 78.

$$\begin{aligned} \text{Var}(X+Y) &= E[\{(X+Y) - (\mu_X + \mu_Y)\}^2] \\ &= E[\{(X-\mu_X) + (Y-\mu_Y)\}^2] \\ &= E[(X-\mu_X)^2 + 2(X-\mu_X)(Y-\mu_Y) + (Y-\mu_Y)^2] \\ &= E[(X-\mu_X)^2] + 2E[(X-\mu_X)(Y-\mu_Y)] + E[(Y-\mu_Y)^2] \\ &= \text{Var}(X) + \text{Var}(Y) \end{aligned}$$

using the fact that

$$E[(X-\mu_X)(Y-\mu_Y)] = E(X-\mu_X)E(Y-\mu_Y) = 0$$

since X and Y, and therefore $X-\mu_X$ and $Y-\mu_Y$, are independent. The proof of (*19*), page 78, follows on replacing Y by $-Y$ and using Theorem 3-5.

Moments and moment generating functions

3.14. Prove the result (26), page 79.

$$\begin{aligned} \mu_r &= E[(X-\mu)^r] \\ &= E\left[X^r - \binom{r}{1}X^{r-1}\mu + \cdots + (-1)^j\binom{r}{j}X^{r-j}\mu^j \right. \\ &\qquad \left. + \cdots + (-1)^{r-1}\binom{r}{r-1}X\mu^{r-1} + (-1)^r\mu^r\right] \end{aligned}$$

$$= E(X^r) - \binom{r}{1}E(X^{r-1})\mu + \cdots + (-1)^j\binom{r}{j}E(X^{r-j})\mu^j$$

$$+ \cdots + (-1)^{r-1}\binom{r}{r-1}E(X)\mu^{r-1} + (-1)^r\mu^r$$

$$= \mu'_r - \binom{r}{1}\mu'_{r-1}\mu + \cdots + (-1)^j\binom{r}{j}\mu'_{r-j}\mu^j$$

$$+ \cdots + (-1)^{r-1}r\mu^r + (-1)^{-r}\mu^r$$

where the last two terms can be combined to give $(-1)^{r-1}(r-1)\mu^r$.

3.15. Prove (a) result (31), (b) result (32), page 79.

(a) Using the power series expansion for e^u (3., Appendix A), we have

$$M_X(t) = E(e^{tX}) = E\left(1 + tX + \frac{t^2X^2}{2!} + \frac{t^3X^3}{3!} + \cdots\right)$$

$$= 1 + tE(X) + \frac{t^2}{2!}E(X^2) + \frac{t^3}{3!}E(X^3) + \cdots$$

$$= 1 + \mu t + \mu'_2\frac{t^2}{2!} + \mu'_3\frac{t^3}{3!} + \cdots$$

(b) This follows immediately from the fact known from calculus that if the Taylor series of $f(t)$ about $t = a$ is

$$f(t) = \sum_{n=0}^{\infty} c_n(t-a)^n$$

then
$$c_n = \frac{1}{n!}\frac{d^n}{dt^n}f(t)\bigg|_{t=a}$$

3.16. Prove Theorem 3-9, page 80.

Since X and Y are independent, any function of X and any function of Y are independent. Hence,

$$M_{X+Y}(t) = E[e^{t(X+Y)}] = E(e^{tX}e^{tY}) = E(e^{tX})E(e^{tY}) = M_X(t)M_Y(t)$$

3.17. The random variable X can assume the values 1 and -1 with probability $\frac{1}{2}$ each. Find (a) the moment generating function, (b) the first four moments about the origin.

(a)
$$E(e^{tX}) = e^{t(1)}\left(\frac{1}{2}\right) + e^{t(-1)}\left(\frac{1}{2}\right) = \frac{1}{2}(e^t + e^{-t})$$

(b) We have
$$e^t = 1 + t + \frac{t^2}{2!} + \frac{t^3}{3!} + \frac{t^4}{4!} + \cdots$$

$$e^{-t} = 1 - t + \frac{t^2}{2!} - \frac{t^3}{3!} + \frac{t^4}{4!} - \cdots$$

Then (1)
$$\frac{1}{2}(e^t + e^{-t}) = 1 + \frac{t^2}{2!} + \frac{t^4}{4!} + \cdots$$

But (2)
$$M_X(t) = 1 + \mu t + \mu'_2\frac{t^2}{2!} + \mu'_3\frac{t^3}{3!} + \mu'_4\frac{t^4}{4!} + \cdots$$

Then, comparing (1) and (2), we have

$$\mu = 0, \quad \mu'_2 = 1, \quad \mu'_3 = 0, \quad \mu'_4 = 1, \ldots$$

The odd moments are all zero, and the even moments are all one.

3.18. A random variable X has density function given by

$$f(x) = \begin{cases} 2e^{-2x} & x \geq 0 \\ 0 & x < 0 \end{cases}$$

Find (a) the moment generating function, (b) the first four moments about the origin.

(a)
$$M(t) = E(e^{tX}) = \int_{-\infty}^{\infty} e^{tx} f(x)\,dx$$

$$= \int_0^{\infty} e^{tx}(2e^{-2x})\,dx = 2\int_0^{\infty} e^{(t-2)x}\,dx$$

$$= \frac{2e^{(t-2)x}}{t-2}\bigg|_0^{\infty} = \frac{2}{2-t}, \quad \text{assuming } t < 2$$

(b) If $|t| < 2$ we have

$$\frac{2}{2-t} = \frac{1}{1-t/2} = 1 + \frac{t}{2} + \frac{t^2}{4} + \frac{t^3}{8} + \frac{t^4}{16} + \cdots$$

But
$$M(t) = 1 + \mu t + \mu_2' \frac{t^2}{2!} + \mu_3' \frac{t^3}{3!} + \mu_4' \frac{t^4}{4!} + \cdots$$

Therefore, on comparing terms, $\mu = \frac{1}{2}, \mu_2' = \frac{1}{2}, \mu_3' = \frac{3}{4}, \mu_4' = \frac{3}{2}$.

3.19. Find the first four moments (a) about the origin, (b) about the mean, for a random variable X having density function

$$f(x) = \begin{cases} 4x(9-x^2)/81 & 0 \leq x \leq 3 \\ 0 & \text{otherwise} \end{cases}$$

(a)
$$\mu_1' = E(X) = \frac{4}{81}\int_0^3 x^2(9-x^2)\,dx = \frac{8}{5} = \mu$$

$$\mu_2' = E(X^2) = \frac{4}{81}\int_0^3 x^3(9-x^2)\,dx = 3$$

$$\mu_3' = E(X^3) = \frac{4}{81}\int_0^3 x^4(9-x^2)\,dx = \frac{216}{35}$$

$$\mu_4' = E(X^4) = \frac{4}{81}\int_0^3 x^5(9-x^2)\,dx = \frac{27}{2}$$

(b) Using the result (27), page 79, we have

$$\mu_1 = 0$$

$$\mu_2 = 3 - \left(\frac{8}{5}\right)^2 = \frac{11}{25} = \sigma^2$$

$$\mu_3 = \frac{216}{35} - 3(3)\left(\frac{8}{5}\right) + 2\left(\frac{8}{5}\right)^3 = -\frac{32}{875}$$

$$\mu_4 = \frac{27}{2} - 4\left(\frac{216}{35}\right)\left(\frac{8}{5}\right) + 6(3)\left(\frac{8}{5}\right)^2 - 3\left(\frac{8}{5}\right)^4 = \frac{3693}{8750}$$

Characteristic functions

3.20. Find the characteristic function of the random variable X of Problem 3.17.

The characteristic function is given by

$$E(e^{i\omega X}) = e^{i\omega(1)}\left(\frac{1}{2}\right) + e^{i\omega(-1)}\left(\frac{1}{2}\right) = \frac{1}{2}(e^{i\omega} + e^{-i\omega}) = \cos\omega$$

using Euler's formulas,

$$e^{i\theta} = \cos\theta + i\sin\theta \qquad e^{-i\theta} = \cos\theta - i\sin\theta$$

with $\theta = \omega$. The result can also be obtained from Problem 3.17(a) on putting $t = i\omega$.

3.21. Find the characteristic function of the random variable X having density function given by

$$f(x) = \begin{cases} 1/2a & |x| < a \\ 0 & \text{otherwise} \end{cases}$$

The characteristic function is given by

$$E(e^{i\omega X}) = \int_{-\infty}^{\infty} e^{i\omega x} f(x)\,dx = \frac{1}{2a}\int_{-a}^{a} e^{i\omega x}\,dx$$

$$= \frac{1}{2a}\frac{e^{i\omega x}}{i\omega}\bigg|_{-a}^{a} = \frac{e^{ia\omega} - e^{-ia\omega}}{2ia\omega} = \frac{\sin a\omega}{a\omega}$$

using Euler's formulas (see Problem 3.20) with $\theta = a\omega$.

3.22. Find the characteristic function of the random variable X having density function $f(x) = ce^{-a|x|}$, $-\infty < x < \infty$, where $a > 0$, and c is a suitable constant.

Since $f(x)$ is a density function, we must have

$$\int_{-\infty}^{\infty} f(x)\,dx = 1$$

so that

$$c\int_{-\infty}^{\infty} e^{-a|x|}\,dx = c\left[\int_{-\infty}^{0} e^{-a(-x)}\,dx + \int_{0}^{\infty} e^{-a(x)}\,dx\right]$$

$$= c\frac{e^{ax}}{a}\bigg|_{-\infty}^{0} + c\frac{e^{-ax}}{-a}\bigg|_{0}^{\infty} = \frac{2c}{a} = 1$$

Then $c = a/2$. The characteristic function is therefore given by

$$E(e^{i\omega X}) = \int_{-\infty}^{\infty} e^{i\omega x} f(x)\,dx$$

$$= \frac{a}{2}\left[\int_{-\infty}^{0} e^{i\omega x}e^{-a(-x)}\,dx + \int_{0}^{\infty} e^{i\omega x}e^{-a(x)}\,dx\right]$$

$$= \frac{a}{2}\left[\int_{-\infty}^{0} e^{(a+i\omega)x}\,dx + \int_{0}^{\infty} e^{-(a-i\omega)x}\,dx\right]$$

$$= \frac{a}{2}\frac{e^{(a+i\omega)x}}{a+i\omega}\bigg|_{-\infty}^{0} + a\frac{e^{-(a-i\omega)x}}{-(a-i\omega)}\bigg|_{0}^{\infty}$$

$$= \frac{a}{2(a+i\omega)} + \frac{a}{2(a-i\omega)} = \frac{a^2}{a^2+\omega^2}$$

Covariance and correlation coefficient

3.23. Prove Theorem 3-14, page 81.

By definition the covariance of X and Y is

$$\begin{aligned}
\sigma_{XY} &= \text{Cov}(X, Y) = E[(X - \mu_X)(Y - \mu_Y)] \\
&= E[XY - \mu_X Y - \mu_Y X + \mu_X\mu_Y] \\
&= E(XY) - \mu_X E(Y) - \mu_Y E(X) + E(\mu_X\mu_Y) \\
&= E(XY) - \mu_X\mu_Y - \mu_Y\mu_X + \mu_X\mu_Y \\
&= E(XY) - \mu_X\mu_Y \\
&= E(XY) - E(X)E(Y)
\end{aligned}$$

3.24. Prove Theorem 3-15, page 81.

If X and Y are independent, then $E(XY) = E(X)E(Y)$. Therefore, by Problem 3.23,

$$\sigma_{XY} = \text{Cov}(X, Y) = E(XY) - E(X)E(Y) = 0$$

3.25. Find (a) $E(X)$, (b) $E(Y)$, (c) $E(XY)$, (d) $E(X^2)$, (e) $E(Y^2)$, (f) Var (X), (g) Var (Y), (h) Cov (X, Y), (i) ρ, if the random variables X and Y are defined as in Problem 2.8, pages 47–48.

(a)
$$E(X) = \sum_x \sum_y x f(x, y) = \sum_x x\left[\sum_y f(x, y)\right]$$
$$= (0)(6c) + (1)(14c) + (2)(22c) = 58c = \frac{58}{42} = \frac{29}{21}$$

(b)
$$E(Y) = \sum_x \sum_y y f(x, y) = \sum_y y\left[\sum_x f(x, y)\right]$$
$$= (0)(6c) + (1)(9c) + (2)(12c) + (3)(15c) = 78c = \frac{78}{42} = \frac{13}{7}$$

(c)
$$E(XY) = \sum_x \sum_y xy f(x, y)$$
$$= (0)(0)(0) + (0)(1)(c) + (0)(2)(2c) + (0)(3)(3c)$$
$$+ (1)(0)(2c) + (1)(1)(3c) + (1)(2)(4c) + (1)(3)(5c)$$
$$+ (2)(0)(4c) + (2)(1)(5c) + (2)(2)(6c) + (2)(3)(7c)$$
$$= 102c = \frac{102}{42} = \frac{17}{7}$$

(d)
$$E(X^2) = \sum_x \sum_y x^2 f(x, y) = \sum_x x^2\left[\sum_y f(x, y)\right]$$
$$= (0)^2(6c) + (1)^2(14c) + (2)^2(22c) = 102c = \frac{102}{42} = \frac{17}{7}$$

(e)
$$E(Y^2) = \sum_x \sum_y y^2 f(x, y) = \sum_y y^2\left[\sum_x f(x, y)\right]$$
$$= (0)^2(6c) + (1)^2(9c) + (2)^2(12c) + (3)^2(15c) = 192c = \frac{192}{42} = \frac{32}{7}$$

(f)
$$\sigma_X^2 = \text{Var}(X) = E(X^2) - [E(X)]^2 = \frac{17}{7} - \left(\frac{29}{21}\right)^2 = \frac{230}{441}$$

(g)
$$\sigma_Y^2 = \text{Var}(Y) = E(Y^2) - [E(Y)]^2 = \frac{32}{7} - \left(\frac{13}{7}\right)^2 = \frac{55}{49}$$

(h)
$$\sigma_{XY} = \text{Cov}(X, Y) = E(XY) - E(X)E(Y) = \frac{17}{7} - \left(\frac{29}{21}\right)\left(\frac{13}{7}\right) = -\frac{20}{147}$$

(i)
$$\rho = \frac{\sigma_{XY}}{\sigma_X \sigma_Y} = \frac{-20/147}{\sqrt{230/441}\sqrt{55/49}} = \frac{-20}{\sqrt{230}\sqrt{55}} = -0.2103 \text{ approx.}$$

3.26. Work Problem 3.25 if the random variables X and Y are defined as in Problem 2.33, pages 61–63.

Using $c = 1/210$, we have:

(a)
$$E(X) = \frac{1}{210}\int_{x=2}^{6}\int_{y=0}^{5} (x)(2x + y)\,dx\,dy = \frac{268}{63}$$

(b)
$$E(Y) = \frac{1}{210}\int_{x=2}^{6}\int_{y=0}^{5} (y)(2x + y)\,dx\,dy = \frac{170}{63}$$

(c)
$$E(XY) = \frac{1}{210}\int_{x=2}^{6}\int_{y=0}^{5} (xy)(2x + y)\,dx\,dy = \frac{80}{7}$$

(d) $$E(X^2) = \frac{1}{210}\int_{x=2}^{6}\int_{y=0}^{5}(x^2)(2x + y)\,dx\,dy = \frac{1220}{63}$$

(e) $$E(Y^2) = \frac{1}{210}\int_{x=2}^{6}\int_{y=0}^{5}(y^2)(2x + y)\,dx\,dy = \frac{1175}{126}$$

(f) $$\sigma_X^2 = \text{Var}(X) = E(X^2) - [E(X)]^2 = \frac{1220}{63} - \left(\frac{268}{63}\right)^2 = \frac{5036}{3969}$$

(g) $$\sigma_Y^2 = \text{Var}(Y) = E(Y^2) - [E(Y)]^2 = \frac{1175}{126} - \left(\frac{170}{63}\right)^2 = \frac{16{,}225}{7938}$$

(h) $$\sigma_{XY} = \text{Cov}(X, Y) = E(XY) - E(X)E(Y) = \frac{80}{7} - \left(\frac{268}{63}\right)\left(\frac{170}{63}\right) = -\frac{200}{3969}$$

(i) $$\rho = \frac{\sigma_{XY}}{\sigma_X\sigma_Y} = \frac{-200/3969}{\sqrt{5036/3969}\sqrt{16{,}225/7938}} = \frac{-200}{\sqrt{2518}\sqrt{16{,}225}} = -0.03129 \text{ approx.}$$

Conditional expectation, variance, and moments

3.27. Find the conditional expectation of Y given $X = 2$ in Problem 2.8, pages 47–48.

As in Problem 2.27, page 58, the conditional probability function of Y given $X = 2$ is

$$f(y\,|\,2) = \frac{4 + y}{22}$$

Then the conditional expectation of Y given $X = 2$ is

$$E(Y\,|\,X = 2) = \sum_y y\left(\frac{4 + y}{22}\right)$$

where the sum is taken over all y corresponding to $X = 2$. This is given by

$$E(Y\,|\,X = 2) = (0)\left(\frac{4}{22}\right) + 1\left(\frac{5}{22}\right) + 2\left(\frac{6}{22}\right) + 3\left(\frac{7}{22}\right) = \frac{19}{11}$$

3.28. Find the conditional expectation of (a) Y given X, (b) X given Y in Problem 2.29, pages 58–59.

(a) $$E(Y\,|\,X = x)\int_{-\infty}^{\infty} yf_2(y\,|\,x)\,dy = \int_0^x y\left(\frac{2y}{x^2}\right)dy = \frac{2x}{3}$$

(b) $$E(X\,|\,Y = y) = \int_{-\infty}^{\infty} xf_1(x\,|\,y)\,dx = \int_y^1 x\left(\frac{2x}{1 - y^2}\right)dx$$

$$= \frac{2(1 - y^3)}{3(1 - y^2)} = \frac{2(1 + y + y^2)}{3(1 + y)}$$

3.29. Find the conditional variance of Y given X for Problem 2.29, pages 58–59.

The required variance (second moment about the mean) is given by

$$E[(Y - \mu_2)^2\,|\,X = x] = \int_{-\infty}^{\infty}(y - \mu_2)^2 f_2(y\,|\,x)\,dy = \int_0^x\left(y - \frac{2x}{3}\right)^2\left(\frac{2y}{x^2}\right)dy = \frac{x^2}{18}$$

where we have used the fact that $\mu_2 = E(Y\,|\,X = x) = 2x/3$ from Problem 3.28(a).

Chebyshev's inequality

3.30. Prove Chebyshev's inequality.

We shall present the proof for continuous random variables. A proof for discrete variables is similar if integrals are replaced by sums. If $f(x)$ is the density function of X, then

$$\sigma^2 = E[(X - \mu)^2] = \int_{-\infty}^{\infty}(x - \mu)^2 f(x)\,dx$$

Since the integrand is nonnegative, the value of the integral can only decrease when the range of integration is diminished. Therefore,

$$\sigma^2 \geq \int_{|x-\mu| \geq \epsilon} (x-\mu)^2 f(x)\,dx \geq \int_{|x-\mu| \geq \epsilon} \epsilon^2 f(x)\,dx = \epsilon^2 \int_{|x-\mu| \geq \epsilon} f(x)\,dx$$

But the last integral is equal to $P(|X - \mu| \geq \epsilon)$. Hence,

$$P(|X - \mu| \geq \epsilon) \leq \frac{\sigma^2}{\epsilon^2}$$

3.31. For the random variable of Problem 3.18, (a) find $P(|X - \mu| > 1)$. (b) Use Chebyshev's inequality to obtain an upper bound on $P(|X - \mu| > 1)$ and compare with the result in (a).

(a) From Problem 3.18, $\mu = 1/2$. Then

$$P(|X - \mu| < 1) = P\left(\left|X - \frac{1}{2}\right| < 1\right) = P\left(-\frac{1}{2} < X < \frac{3}{2}\right)$$

$$= \int_0^{3/2} 2e^{-2x}\,dx = 1 - e^{-3}$$

Therefore $$P\left(\left|X - \frac{1}{2}\right| \geq 1\right) = 1 - (1 - e^{-3}) = e^{-3} = 0.04979$$

(b) From Problem 3.18, $\sigma^2 = \mu_2' - \mu^2 = 1/4$. Chebyshev's inequality with $\epsilon = 1$ then gives

$$P(|X - \mu| \geq 1) \leq \sigma^2 = 0.25$$

Comparing with (a), we see that the bound furnished by Chebyshev's inequality is here quite crude. In practice, Chebyshev's inequality is used to provide estimates when it is inconvenient or impossible to obtain exact values.

Law of large numbers

3.32. Prove the law of large numbers stated in Theorem 3-19, page 83.

We have $$E(X_1) = E(X_2) = \cdots = E(X_n) = \mu$$

$$\text{Var}(X_1) = \text{Var}(X_2) = \cdots = \text{Var}(X_n) = \sigma^2$$

Then $$E\left(\frac{S_n}{n}\right) = E\left(\frac{X_1 + \cdots + X_n}{n}\right) = \frac{1}{n}[E(X_1) + \cdots + E(X_n)] = \frac{1}{n}(n\mu) = \mu$$

$$\text{Var}(S_n) = \text{Var}(X_1 + \cdots + X_n) = \text{Var}(X_1) + \cdots + \text{Var}(X_n) = n\sigma^2$$

so that $$\text{Var}\left(\frac{S_n}{n}\right) = \frac{1}{n^2}\text{Var}(S_n) = \frac{\sigma^2}{n}$$

where we have used Theorem 3-5 and an extension of Theorem 3-7.

Therefore, by Chebyshev's inequality with $X = S_n/n$, we have

$$P\left(\left|\frac{S_n}{n} - \mu\right| \geq \epsilon\right) \leq \frac{\sigma^2}{n\epsilon^2}$$

Taking the limit as $n \to \infty$, this becomes, as required,

$$\lim_{n\to\infty} P\left(\left|\frac{S_n}{n} - \mu\right| \geq \epsilon\right) = 0$$

Other measures of central tendency

3.33. The density function of a continuous random variable X is

$$f(x) = \begin{cases} 4x(9 - x^2)/81 & 0 \leq x \leq 3 \\ 0 & \text{otherwise} \end{cases}$$

(a) Find the mode. (b) Find the median. (c) Compare mode, median, and mean.

(a) The mode is obtained by finding where the density $f(x)$ has a relative maximum. The relative maxima of $f(x)$ occur where the derivative is zero, i.e.,

$$\frac{d}{dx}\left[\frac{4x(9-x^2)}{81}\right] = \frac{36-12x^2}{81} = 0$$

Then $x = \sqrt{3} = 1.73$ approx., which is the required mode. Note that this does give the maximum since the second derivative, $-24x/81$, is negative for $x = \sqrt{3}$.

(b) The median is that value a for which $P(X \le a) = 1/2$. Now, for $0 < a < 3$,

$$P(X \le a) = \frac{4}{81}\int_0^a x(9-x^2)\,dx = \frac{4}{81}\left(\frac{9a^2}{2} - \frac{a^4}{4}\right)$$

Setting this equal to $1/2$, we find that

$$2a^4 - 36a^2 + 81 = 0$$

from which

$$a^2 = \frac{36 \pm \sqrt{(36)^2 - 4(2)(81)}}{2(2)} = \frac{36 \pm \sqrt{648}}{4} = 9 \pm \frac{9}{2}\sqrt{2}$$

Therefore, the required median, which must lie between 0 and 3, is given by

$$a^2 = 9 - \frac{9}{2}\sqrt{2}$$

from which $a = 1.62$ approx.

(c)
$$E(X) = \frac{4}{81}\int_0^3 x^2(9-x^2)\,dx = \frac{4}{81}\left(3x^3 - \frac{x^5}{5}\right)\Bigg|_0^3 = 1.60$$

which is practically equal to the median. The mode, median, and mean are shown in Fig. 3-6.

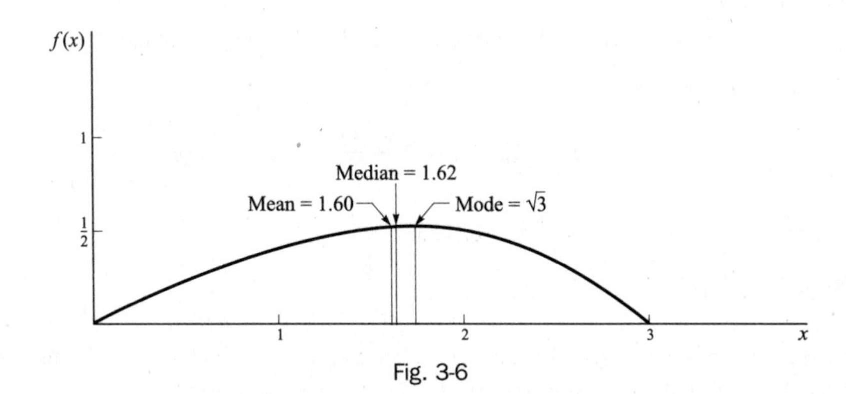

Fig. 3-6

3.34. A discrete random variable has probability function $f(x) = 1/2^x$ where $x = 1, 2, \ldots$. Find (a) the mode, (b) the median, and (c) compare them with the mean.

(a) The mode is the value x having largest associated probability. In this case it is $x = 1$, for which the probability is $1/2$.

(b) If x is any value between 1 and 2, $P(X < x) = \frac{1}{2}$ and $P(X > x) = \frac{1}{2}$. Therefore, *any number* between 1 and 2 could represent the median. For convenience, we choose the midpoint of the interval, i.e., $3/2$.

(c) As found in Problem 3.3, $\mu = 2$. Therefore, the ordering of the three measures is just the reverse of that in Problem 3.33.

Percentiles

3.35. Determine the (a) 10th, (b) 25th, (c) 75th percentile values for the distribution of Problem 3.33.

From Problem 3.33(b) we have

$$P(X \leq a) = \frac{4}{81}\left(\frac{9a^2}{2} - \frac{a^4}{4}\right) = \frac{18a^2 - a^4}{81}$$

(a) The 10th percentile is the value of a for which $P(X \leq a) = 0.10$, i.e., the solution of $(18a^2 - a^4)/81 = 0.10$. Using the method of Problem 3.33, we find $a = 0.68$ approx.

(b) The 25th percentile is the value of a such that $(18a^2 - a^4)/81 = 0.25$, and we find $a = 1.098$ approx.

(c) The 75th percentile is the value of a such that $(18a^2 - a^4)/81 = 0.75$, and we find $a = 2.121$ approx.

Other measures of dispersion

3.36. Determine, (a) the semi-interquartile range, (b) the mean deviation for the distribution of Problem 3.33.

(a) By Problem 3.35 the 25th and 75th percentile values are 1.098 and 2.121, respectively. Therefore,

$$\text{Semi-interquartile range} = \frac{2.121 - 1.098}{2} = 0.51 \text{ approx.}$$

(b) From Problem 3.33 the mean is $\mu = 1.60 = 8/5$. Then

$$\begin{aligned}
\text{Mean deviation} = \text{M.D.} = E(|X - \mu|) &= \int_{-\infty}^{\infty} |x - \mu| f(x)\,dx \\
&= \int_0^3 \left|x - \frac{8}{5}\right| \left[\frac{4x}{81}(9 - x^2)\right] dx \\
&= \int_0^{8/5} \left(\frac{8}{5} - x\right)\left[\frac{4x}{81}(9 - x^2)\right] dx + \int_{8/5}^{3} \left(x - \frac{8}{5}\right)\left[\frac{4x}{81}(9 - x^2)\right] dx \\
&= 0.555 \text{ approx.}
\end{aligned}$$

Skewness and kurtosis

3.37. Find the coefficient of (a) skewness, (b) kurtosis for the distribution of Problem 3.19.

From Problem 3.19(b) we have

$$\sigma^2 = \frac{11}{25} \qquad \mu_3 = -\frac{32}{875} \qquad \mu_4 = \frac{3693}{8750}$$

(a) Coefficient of skewness $= \alpha_3 = \dfrac{\mu_3}{\sigma^3} = -0.1253$

(b) Coefficient of kurtosis $= \alpha_4 = \dfrac{\mu_4}{\sigma^4} = 2.172$

It follows that there is a moderate skewness to the left, as is indicated in Fig. 3-6. Also the distribution is somewhat less peaked than the normal distribution, which has a kurtosis of 3.

Miscellaneous problems

3.38. If $M(t)$ is the moment generating function for a random variable X, prove that the mean is $\mu = M'(0)$ and the variance is $\sigma^2 = M''(0) - [M'(0)]^2$.

From (32), page 79, we have on letting $r = 1$ and $r = 2$,

$$\mu'_1 = M'(0) \qquad \mu'_2 = M''(0)$$

Then from (27)

$$\mu = M'(0) \qquad \mu_2 = \sigma^2 = M''(0) - [M'(0)]^2$$

3.39. Let X be a random variable that takes on the values $x_k = k$ with probabilities p_k where $k = \pm 1, \ldots, \pm n$. (a) Find the characteristic function $\phi(\omega)$ of X, (b) obtain p_k in terms of $\phi(\omega)$.

(a) The characteristic function is

$$\phi(\omega) = E(e^{i\omega X}) = \sum_{k=-n}^{n} e^{i\omega x_k} p_k = \sum_{k=-n}^{n} p_k e^{ik\omega}$$

(b) Multiply both sides of the expression in (a) by $e^{-ij\omega}$ and integrate with respect to ω from 0 to 2π. Then

$$\int_{\omega=0}^{2\pi} e^{-ij\omega}\phi(\omega)\,d\omega = \sum_{k=-n}^{n} p_k \int_{\omega=0}^{2\pi} e^{i(k-j)\omega}\,d\omega = 2\pi p_j$$

since

$$\int_{\omega=0}^{2\pi} e^{i(k-j)\omega}\,d\omega = \begin{cases} \left.\dfrac{e^{i(k-j)\omega}}{i(k-j)}\right|_0^{2\pi} = 0 & k \neq j \\ 2\pi & k = j \end{cases}$$

Therefore,

$$p_j = \frac{1}{2\pi}\int_{\omega=0}^{2\pi} e^{-ij\omega}\phi(\omega)\,d\omega$$

or, replacing j by k,

$$p_k = \frac{1}{2\pi}\int_{\omega=0}^{2\pi} e^{-ik\omega}\phi(\omega)\,d\omega$$

We often call $\sum_{k=-n}^{n} p_k e^{ik\omega}$ (where n can theoretically be infinite) the *Fourier series* of $\phi(\omega)$ and p_k the *Fourier coefficients*. For a continuous random variable, the Fourier series is replaced by the Fourier integral (see page 81).

3.40. Use Problem 3.39 to obtain the probability distribution of a random variable X whose characteristic function is $\phi(\omega) = \cos\omega$.

From Problem 3.39

$$\begin{aligned} p_k &= \frac{1}{2\pi}\int_{\omega=0}^{2\pi} e^{-ik\omega}\cos\omega\,d\omega \\ &= \frac{1}{2\pi}\int_{\omega=0}^{2\pi} e^{-ik\omega}\left[\frac{e^{i\omega}+e^{-i\omega}}{2}\right]d\omega \\ &= \frac{1}{4\pi}\int_{\omega=0}^{2\pi} e^{i(1-k)\omega}\,d\omega + \frac{1}{4\pi}\int_{\omega=0}^{2\pi} e^{-i(1+k)\omega}\,d\omega \end{aligned}$$

If $k = 1$, we find $p_1 = \frac{1}{2}$; if $k = -1$, we find $p_{-1} = \frac{1}{2}$. For all other values of k, we have $p_k = 0$. Therefore, the random variable is given by

$$X = \begin{cases} 1 & \text{probability } 1/2 \\ -1 & \text{probability } 1/2 \end{cases}$$

As a check, see Problem 3.20.

3.41. Find the coefficient of (a) skewness, (b) kurtosis of the distribution defined by the *normal curve*, having density

$$f(x) = \frac{1}{\sqrt{2\pi}}e^{-x^2/2} \qquad -\infty < x < \infty$$

(a) The distribution has the appearance of Fig. 3-7. By symmetry, $\mu_1' = \mu = 0$ and $\mu_3' = 0$. Therefore the coefficient of skewness is zero.

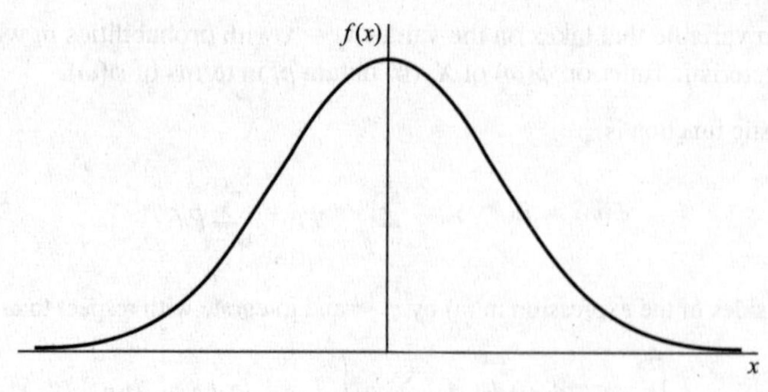

Fig. 3-7

(b) We have

$$\mu_2' = E(X^2) = \frac{1}{\sqrt{2\pi}}\int_{-\infty}^{\infty} x^2 e^{-x^2/2}\,dx = \frac{2}{\sqrt{2\pi}}\int_0^{\infty} x^2 e^{-x^2/2}\,dx$$

$$= \frac{2}{\sqrt{\pi}}\int_0^{\infty} v^{1/2} e^{-v}\,dv$$

$$= \frac{2}{\sqrt{\pi}}\Gamma\left(\frac{3}{2}\right) = \frac{2}{\sqrt{\pi}}\cdot\frac{1}{2}\Gamma\left(\frac{1}{2}\right) = 1$$

where we have made the transformation $x^2/2 = v$ and used properties of the gamma function given in (2) and (5) of Appendix A. Similarly we obtain

$$\mu_4' = E(X^4) = \frac{1}{\sqrt{2\pi}}\int_{-\infty}^{\infty} x^4 e^{-x^2/2}\,dx = \frac{2}{\sqrt{2\pi}}\int_0^{\infty} x^4 e^{-x^2/2}\,dx$$

$$= \frac{4}{\sqrt{\pi}}\int_0^{\infty} v^{3/2} e^{-v}\,dv$$

$$= \frac{4}{\sqrt{\pi}}\Gamma\left(\frac{5}{2}\right) = \frac{4}{\sqrt{\pi}}\cdot\frac{3}{2}\cdot\frac{1}{2}\Gamma\left(\frac{1}{2}\right) = 3$$

Now

$$\sigma^2 = E[(X-\mu)^2] = E(X)^2 = \mu_2' = 1$$

$$\mu_4 = E[(X-\mu)^4] = E(X^4) = \mu_4' = 3$$

Thus the coefficient of kurtosis is

$$\frac{\mu_4}{\sigma^4} = 3$$

3.42. Prove that $-1 \leq \rho \leq 1$ (see page 82).

For any real constant c, we have

$$E[\{Y - \mu_Y - c(X-\mu)\}^2] \geq 0$$

Now the left side can be written

$$E[(Y-\mu_Y)^2] + c^2E[(X-\mu_X)^2] - 2cE[(X-\mu_X)(Y-\mu_Y)] = \sigma_Y^2 + c^2\sigma_X^2 - 2c\sigma_{XY}$$

$$= \sigma_Y^2 + \sigma_X^2\left(c^2 - \frac{2c\sigma_{XY}}{\sigma_X^2}\right)$$

$$= \sigma_Y^2 + \sigma_X^2\left(c^2 - \frac{\sigma_{XY}}{\sigma_X^2}\right)^2 - \frac{\sigma_{XY}^2}{\sigma_X^2}$$

$$= \frac{\sigma_X^2\sigma_Y^2 - \sigma_{XY}^2}{\sigma_X^2} + \sigma_X^2\left(c - \frac{\sigma_{XY}}{\sigma_X^2}\right)^2$$

In order for this last quantity to be greater than or equal to zero for every value of c, we must have

$$\sigma_X^2\sigma_Y^2 - \sigma_{XY}^2 \geq 0 \quad \text{or} \quad \frac{\sigma_{XY}^2}{\sigma_X^2\sigma_Y^2} \leq 1$$

which is equivalent to $\rho^2 \leq 1$ or $-1 \leq \rho \leq 1$.

SUPPLEMENTARY PROBLEMS

Expectation of random variables

3.43. A random variable X is defined by $X = \begin{cases} -2 & \text{prob. } 1/3 \\ 3 & \text{prob. } 1/2. \\ 1 & \text{prob. } 1/6 \end{cases}$ Find (a) $E(X)$, (b) $E(2X + 5)$, (c) $E(X^2)$.

3.44. Let X be a random variable defined by the density function $f(x) = \begin{cases} 3x^2 & 0 \leq x \leq 1 \\ 0 & \text{otherwise} \end{cases}$.

Find (a) $E(X)$, (b) $E(3X - 2)$, (c) $E(X^2)$.

3.45. The density function of a random variable X is $f(x) = \begin{cases} e^{-x} & x \geq 0 \\ 0 & \text{otherwise} \end{cases}$.

Find (a) $E(X)$, (b) $E(X^2)$, (c) $E[(X - 1)^2]$.

3.46. What is the expected number of points that will come up in 3 successive tosses of a fair die? Does your answer seem reasonable? Explain.

3.47. A random variable X has the density function $f(x) = \begin{cases} e^{-x} & x \geq 0 \\ 0 & x < 0 \end{cases}$. Find $E(e^{2X/3})$.

3.48. Let X and Y be independent random variables each having density function

$$f(u) = \begin{cases} 2e^{-2u} & u \geq 0 \\ 0 & \text{otherwise} \end{cases}$$

Find (a) $E(X + Y)$, (b) $E(X^2 + Y^2)$, (c) $E(XY)$.

3.49. Does (a) $E(X + Y) = E(X) + E(Y)$, (b) $E(XY) = E(X)E(Y)$, in Problem 3.48? Explain.

3.50. Let X and Y be random variables having joint density function

$$f(x, y) = \begin{cases} \frac{3}{5}x(x + y) & 0 \leq x \leq 1, 0 \leq y \leq 2 \\ 0 & \text{otherwise} \end{cases}$$

Find (a) $E(X)$, (b) $E(Y)$, (c) $E(X + Y)$, (d) $E(XY)$.

3.51. Does (a) $E(X + Y) = E(X) + E(Y)$, (b) $E(XY) = E(X)E(Y)$, in Problem 3.50? Explain.

3.52. Let X and Y be random variables having joint density

$$f(x, y) = \begin{cases} 4xy & 0 \leq x \leq 1, 0 \leq y \leq 1 \\ 0 & \text{otherwise} \end{cases}$$

Find (a) $E(X)$, (b) $E(Y)$, (c) $E(X + Y)$, (d) $E(XY)$.

3.53. Does (a) $E(X + Y) = E(X) + E(Y)$, (b) $E(XY) = E(X)E(Y)$, in Problem 3.52? Explain.

3.54. Let $f(x, y) = \begin{cases} \frac{1}{4}(2x + y) & 0 \le x \le 1, 0 \le y \le 2 \\ 0 & \text{otherwise} \end{cases}$. Find (a) $E(X)$, (b) $E(Y)$, (c) $E(X^2)$, (d) $E(Y^2)$, (e) $E(X + Y)$, (f) $E(XY)$.

3.55. Let X and Y be independent random variables such that

$$X = \begin{cases} 1 & \text{prob. } 1/3 \\ 0 & \text{prob. } 2/3 \end{cases} \qquad Y = \begin{cases} 2 & \text{prob. } 3/4 \\ -3 & \text{prob. } 1/4 \end{cases}$$

Find (a) $E(3X + 2Y)$, (b) $E(2X^2 - Y^2)$, (c) $E(XY)$, (d) $E(X^2Y)$.

3.56. Let $X_1, X_2, \ldots, X_n$ be n random variables which are identically distributed such that

$$X_k = \begin{cases} 1 & \text{prob. } 1/2 \\ 2 & \text{prob. } 1/3 \\ -1 & \text{prob. } 1/6 \end{cases}$$

Find (a) $E(X_1 + X_2 + \cdots + X_n)$, (b) $E(X_1^2 + X_2^2 + \cdots + X_n^2)$.

Variance and standard deviation

3.57. Find (a) the variance, (b) the standard deviation of the number of points that will come up on a single toss of a fair die.

3.58. Let X be a random variable having density function

$$f(x) = \begin{cases} 1/4 & -2 \le x \le 2 \\ 0 & \text{otherwise} \end{cases}$$

Find (a) $\text{Var}(X)$, (b) σ_X.

3.59. Let X be a random variable having density function

$$f(x) = \begin{cases} e^{-x} & x \ge 0 \\ 0 & \text{otherwise} \end{cases}$$

Find (a) $\text{Var}(X)$, (b) σ_X.

3.60. Find the variance and standard deviation for the random variable X of (a) Problem 3.43, (b) Problem 3.44.

3.61. A random variable X has $E(X) = 2$, $E(X^2) = 8$. Find (a) $\text{Var}(X)$, (b) σ_X.

3.62. If a random variable X is such that $E[(X - 1)^2] = 10$, $E[(X - 2)^2] = 6$ find (a) $E(X)$, (b) $\text{Var}(X)$, (c) σ_X.

Moments and moment generating functions

3.63. Find (a) the moment generating function of the random variable

$$X = \begin{cases} 1/2 & \text{prob. } 1/2 \\ -1/2 & \text{prob. } 1/2 \end{cases}$$

and (b) the first four moments about the origin.

3.64. (a) Find the moment generating function of a random variable X having density function

$$f(x) = \begin{cases} x/2 & 0 \le x \le 2 \\ 0 & \text{otherwise} \end{cases}$$

(b) Use the generating function of (a) to find the first four moments about the origin.

3.65. Find the first four moments about the mean in (a) Problem 3.43, (b) Problem 3.44.

3.66. (a) Find the moment generating function of a random variable having density function

$$f(x) = \begin{cases} e^{-x} & x \ge 0 \\ 0 & \text{otherwise} \end{cases}$$

and (b) determine the first four moments about the origin.

3.67. In Problem 3.66 find the first four moments about the mean.

3.68. Let X have density function $f(x) = \begin{cases} 1/(b-a) & a \le x \le b \\ 0 & \text{otherwise} \end{cases}$. Find the kth moment about (a) the origin, (b) the mean.

3.69. If $M(t)$ is the moment generating function of the random variable X, prove that the 3rd and 4th moments about the mean are given by

$$\mu_3 = M'''(0) - 3M''(0)M'(0) + 2[M'(0)]^3$$
$$\mu_4 = M^{(\text{iv})}(0) - 4M'''(0)M'(0) + 6M''(0)[M'(0)]^2 - 3[M'(0)]^4$$

Characteristic functions

3.70. Find the characteristic function of the random variable $X = \begin{cases} a & \text{prob. } p \\ b & \text{prob. } q = 1 - p \end{cases}$.

3.71. Find the characteristic function of a random variable X that has density function

$$f(x) = \begin{cases} 1/2a & |x| \le a \\ 0 & \text{otherwise} \end{cases}$$

3.72. Find the characteristic function of a random variable with density function

$$f(x) = \begin{cases} x/2 & 0 \le x \le 2 \\ 0 & \text{otherwise} \end{cases}$$

3.73. Let $X_k = \begin{cases} 1 & \text{prob. } 1/2 \\ -1 & \text{prob. } 1/2 \end{cases}$ be independent random variables ($k = 1, 2, \ldots, n$). Prove that the characteristic function of the random variable

$$\frac{X_1 + X_2 + \cdots + X_n}{\sqrt{n}}$$

is $[\cos(\omega/\sqrt{n})]^n$.

3.74. Prove that as $n \to \infty$ the characteristic function of Problem 3.73 approaches $e^{-\omega^2/2}$. (*Hint*: Take the logarithm of the characteristic function and use L'Hospital's rule.)

Covariance and correlation coefficient

3.75. Let X and Y be random variables having joint density function

$$f(x, y) = \begin{cases} x + y & 0 \le x \le 1, 0 \le y \le 1 \\ 0 & \text{otherwise} \end{cases}$$

Find (a) $\text{Var}(X)$, (b) $\text{Var}(Y)$, (c) σ_X, (d) σ_Y, (e) σ_{XY}, (f) ρ.

3.76. Work Problem 3.75 if the joint density function is $f(x, y) = \begin{cases} e^{-(x+y)} & x \ge 0, y \ge 0 \\ 0 & \text{otherwise} \end{cases}$

3.77. Find (a) $\text{Var}(X)$, (b) $\text{Var}(Y)$, (c) σ_X, (d) σ_Y, (e) σ_{XY}, (f) ρ, for the random variables of Problem 2.56.

3.78. Work Problem 3.77 for the random variables of Problem 2.94.

3.79. Find (a) the covariance, (b) the correlation coefficient of two random variables X and Y if $E(X) = 2$, $E(Y) = 3$, $E(XY) = 10$, $E(X^2) = 9$, $E(Y^2) = 16$.

3.80. The correlation coefficient of two random variables X and Y is $-\frac{1}{4}$ while their variances are 3 and 5. Find the covariance.

Conditional expectation, variance, and moments

3.81. Let X and Y have joint density function

$$f(x, y) = \begin{cases} x + y & 0 \le x \le 1, 0 \le y \le 1 \\ 0 & \text{otherwise} \end{cases}$$

Find the conditional expectation of (a) Y given X, (b) X given Y.

3.82. Work Problem 3.81 if $f(x, y) = \begin{cases} 2e^{-(x+2y)} & x \ge 0, y \ge 0 \\ 0 & \text{otherwise} \end{cases}$

3.83. Let X and Y have the joint probability function given in Table 2-9, page 71. Find the conditional expectation of (a) Y given X, (b) X given Y.

3.84. Find the conditional variance of (a) Y given X, (b) X given Y for the distribution of Problem 3.81.

3.85. Work Problem 3.84 for the distribution of Problem 3.82.

3.86. Work Problem 3.84 for the distribution of Problem 2.94.

Chebyshev's inequality

3.87. A random variable X has mean 3 and variance 2. Use Chebyshev's inequality to obtain an upper bound for (a) $P(|X - 3| \ge 2)$, (b) $P(|X - 3| \ge 1)$.

3.88. Prove Chebyshev's inequality for a discrete variable X. (*Hint*: See Problem 3.30.)

3.89. A random variable X has the density function $f(x) = \frac{1}{2}e^{-|x|}$, $-\infty < x < \infty$. (a) Find $P(|X - \mu| > 2)$. (b) Use Chebyshev's inequality to obtain an upper bound on $P(|X - \mu| > 2)$ and compare with the result in (a).

Law of large numbers

3.90. Show that the (weak) law of large numbers can be stated as

$$\lim_{n\to\infty} P\left(\left|\frac{S_n}{n} - \mu\right| < \epsilon\right) = 1$$

and interpret.

3.91. Let X_k $(k = 1, \ldots, n)$ be n independent random variables such that

$$X_k = \begin{cases} 1 & \text{prob. } p \\ 0 & \text{prob. } q = 1 - p \end{cases}$$

(a) If we interpret X_k to be the number of heads on the kth toss of a coin, what interpretation can be given to $S_n = X_1 + \cdots + X_n$?

(b) Show that the law of large numbers in this case reduces to

$$\lim_{n\to\infty} P\left(\left|\frac{S_n}{n} - p\right| \geq \epsilon\right) = 0$$

and interpret this result.

Other measures of central tendency

3.92. Find (a) the mode, (b) the median of a random variable X having density function

$$f(x) = \begin{cases} e^{-x} & x \geq 0 \\ 0 & \text{otherwise} \end{cases}$$

and (c) compare with the mean.

3.93. Work Problem 3.100 if the density function is

$$f(x) = \begin{cases} 4x(1 - x^2) & 0 \leq x \leq 1 \\ 0 & \text{otherwise} \end{cases}$$

3.94. Find (a) the median, (b) the mode for a random variable X defined by

$$X = \begin{cases} 2 & \text{prob. } 1/3 \\ -1 & \text{prob. } 2/3 \end{cases}$$

and (c) compare with the mean.

3.95. Find (a) the median, (b) the mode of the set of numbers 1, 3, 2, 1, 5, 6, 3, 3, and (c) compare with the mean.

Percentiles

3.96. Find the (a) 25th, (b) 75th percentile values for the random variable having density function

$$f(x) = \begin{cases} 2(1 - x) & 0 \leq x \leq 1 \\ 0 & \text{otherwise} \end{cases}$$

3.97. Find the (a) 10th, (b) 25th, (c) 75th, (d) 90th percentile values for the random variable having density function

$$f(x) = \begin{cases} c(x - x^3) & 0 < x < 1 \\ 0 & \text{otherwise} \end{cases}$$

where c is an appropriate constant.

Other measures of dispersion

3.98. Find (a) the semi-interquartile range, (b) the mean deviation for the random variable of Problem 3.96.

3.99. Work Problem 3.98 for the random variable of Problem 3.97.

3.100. Find the mean deviation of the random variable X in each of the following cases.

(a) $f(x) = \begin{cases} e^{-x} & x \geq 0 \\ 0 & \text{otherwise} \end{cases}$ (b) $f(x) = \dfrac{1}{\pi(1 + x^2)}, \quad -\infty < x < \infty.$

3.101. Obtain the probability that the random variable X differs from its mean by more than the semi-interquartile range in the case of (a) Problem 3.96, (b) Problem 3.100(a).

Skewness and kurtosis

3.102. Find the coefficient of (a) skewness, (b) kurtosis for the distribution of Problem 3.100(a).

3.103. If

$$f(x) = \begin{cases} c\left(1 - \dfrac{|x|}{a}\right) & |x| \leq a \\ 0 & |x| > a \end{cases}$$

where c is an appropriate constant, is the density function of X, find the coefficient of (a) skewness, (b) kurtosis.

3.104. Find the coefficient of (a) skewness, (b) kurtosis, for the distribution with density function

$$f(x) = \begin{cases} \lambda e^{-\lambda x} & x \geq 0 \\ 0 & x < 0 \end{cases}$$

Miscellaneous problems

3.105. Let X be a random variable that can take on the values 2, 1, and 3 with respective probabilities 1/3, 1/6, and 1/2. Find (a) the mean, (b) the variance, (c) the moment generating function, (d) the characteristic function, (e) the third moment about the mean.

3.106. Work Problem 3.105 if X has density function

$$f(x) = \begin{cases} c(1 - x) & 0 < x < 1 \\ 0 & \text{otherwise} \end{cases}$$

where c is an appropriate constant.

3.107. Three dice, assumed fair, are tossed successively. Find (a) the mean, (b) the variance of the sum.

3.108. Let X be a random variable having density function

$$f(x) = \begin{cases} cx & 0 \leq x \leq 2 \\ 0 & \text{otherwise} \end{cases}$$

where c is an appropriate constant. Find (a) the mean, (b) the variance, (c) the moment generating function, (d) the characteristic function, (e) the coefficient of skewness, (f) the coefficient of kurtosis.

3.109. Let X and Y have joint density function

$$f(x, y) = \begin{cases} cxy & 0 < x < 1, 0 < y < 1 \\ 0 & \text{otherwise} \end{cases}$$

Find (a) $E(X^2 + Y^2)$, (b) $E(\sqrt{X^2 + Y^2})$.

3.110. Work Problem 3.109 if X and Y are independent identically distributed random variables having density function $f(u) = (2\pi)^{-1/2}e^{-u^2/2}, -\infty < u < \infty$.

3.111. Let X be a random variable having density function

$$f(x) = \begin{cases} \frac{1}{2} & -1 < x < 1 \\ 0 & \text{otherwise} \end{cases}$$

and let $Y = X^2$. Find (a) $E(X)$, (b) $E(Y)$, (c) $E(XY)$.

ANSWERS TO SUPPLEMENTARY PROBLEMS

3.43. (a) 1 (b) 7 (c) 6 **3.44.** (a) 3/4 (b) 1/4 (c) 3/5

3.45. (a) 1 (b) 2 (c) 1 **3.46.** 10.5 **3.47.** 3

3.48. (a) 1 (b) 1 (c) 1/4

3.50. (a) 7/10 (b) 6/5 (c) 19/10 (d) 5/6

3.52. (a) 2/3 (b) 2/3 (c) 4/3 (d) 4/9

3.54. (a) 7/12 (b) 7/6 (c) 5/12 (d) 5/3 (e) 7/4 (f) 2/3

3.55. (a) 5/2 (b) –55/12 (c) 1/4 (d) 1/4

3.56. (a) n (b) $2n$ **3.57.** (a) 35/12 (b) $\sqrt{35/12}$

3.58. (a) 4/3 (b) $\sqrt{4/3}$ **3.59.** (a) 1 (b) 1

3.60. (a) $\text{Var}(X) = 5, \sigma_X = \sqrt{5}$ (b) $\text{Var}(X) = 3/80, \sigma_X = \sqrt{15}/20$

3.61. (a) 4 (b) 2 **3.62.** (a) 7/2 (b) 15/4 (c) $\sqrt{15}/2$

3.63. (a) $\frac{1}{2}(e^{t/2} + e^{-t/2}) = \cosh(t/2)$ (b) $\mu = 0, \mu_2' = 1, \mu_3' = 0, \mu_4' = 1$

3.64. (a) $(1 + 2te^{2t} - e^{2t})/2t^2$ (b) $\mu = 4/3, \mu_2' = 2, \mu_3' = 16/5, \mu_4' = 16/3$

3.65. (a) $\mu_1 = 0, \mu_2 = 5, \mu_3 = -5, \mu_4 = 35$ (b) $\mu_1 = 0, \mu_2 = 3/80, \mu_3 = -121/160, \mu_4 = 2307/8960$

3.66. (a) $1/(1 - t), |t| < 1$ (b) $\mu = 1, \mu_2' = 2, \mu_3' = 6, \mu_4' = 24$

3.67. $\mu_1 = 0, \mu_2 = 1, \mu_3 = 2, \mu_4 = 33$

3.68. (a) $(b^{k+1} - a^{k+1})/(k + 1)(b - a)$ (b) $[1 + (-1)^k](b - a)^k/2^{k+1}(k + 1)$

3.70. $pe^{i\omega a} + qe^{i\omega b}$ **3.71.** $(\sin a\omega)/a\omega$ **3.72.** $(e^{2i\omega} - 2i\omega e^{2i\omega} - 1)/2\omega^2$

3.75. (a) $11/144$ (b) $11/144$ (c) $\sqrt{11}/12$ (d) $\sqrt{11}/12$ (e) $-1/144$ (f) $-1/11$

3.76. (a) 1 (b) 1 (c) 1 (d) 1 (e) 0 (f) 0

3.77. (a) $73/960$ (b) $73/960$ (c) $\sqrt{73/960}$ (d) $\sqrt{73/960}$ (e) $-1/64$ (f) $-15/73$

3.78. (a) $233/324$ (b) $233/324$ (c) $\sqrt{233}/18$ (d) $\sqrt{233}/18$ (e) $-91/324$ (f) $-91/233$

3.79. (a) 4 (b) $4/\sqrt{35}$ **3.80.** $-\sqrt{15}/4$

3.81. (a) $(3x + 2)/(6x + 3)$ for $0 \le x \le 1$ (b) $(3y + 2)/(6y + 3)$ for $0 \le y \le 1$

3.82. (a) $1/2$ for $x \ge 0$ (b) 1 for $y \ge 0$

3.83. (a)

X	0	1	2
$E(Y \mid X)$	4/3	1	5/7

(b)

Y	0	1	2
$E(X \mid Y)$	4/3	7/6	1/2

3.84. (a) $\dfrac{6x^2 + 6x + 1}{18(2x + 1)^2}$ for $0 \le x \le 1$ (b) $\dfrac{6y^2 + 6y + 1}{18(2y + 1)^2}$ for $0 \le y \le 1$

3.85. (a) $1/9$ (b) 1

3.86. (a)

X	0	1	2
$\mathrm{Var}(Y \mid X)$	5/9	4/5	24/49

(b)

Y	0	1	2
$\mathrm{Var}(X \mid Y)$	5/9	29/36	7/12

3.87. (a) $1/2$ (b) 2 (useless) **3.89.** (a) e^{-2} (b) 0.5

3.92. (a) $+0$ (b) $\ln 2$ (c) 1 **3.93.** (a) $1/\sqrt{3}$ (b) $\sqrt{1 - (1/\sqrt{2})}$ (c) $8/15$

3.94. (a) does not exist (b) -1 (c) 0 **3.95.** (a) 3 (b) 3 (c) 3

3.96. (a) $1 - \frac{1}{2}\sqrt{3}$ (b) $1/2$

3.97. (a) $\sqrt{1 - (3/\sqrt{10})}$ (b) $\sqrt{1 - (\sqrt{3}/2)}$ (c) $\sqrt{1/2}$ (d) $\sqrt{1 - (1/\sqrt{10})}$

3.98. (a) 1 (b) $(\sqrt{3} - 1)/4$ (c) $16/81$

3.99. (a) 1 (b) 0.17 (c) 0.051 **3.100.** (a) $1 - 2e^{-1}$ (b) does not exist

3.101. (a) $(5 - 2\sqrt{3})/3$ (b) $(3 - 2e^{-1}\sqrt{3})/3$

3.102. (a) 2 (b) 9 **3.103.** (a) 0 (b) $24/5a$ **3.104.** (a) 2 (b) 9

3.105. (a) $7/3$ (b) $5/9$ (c) $(e^t + 2e^{2t} + 3e^{3t})/6$ (d) $(e^{i\omega} + 2e^{2i\omega} + 3e^{3i\omega})/6$ (e) $-7/27$

3.106. (a) $1/3$ (b) $1/18$ (c) $2(e^t - 1 - t)/t^2$ (d) $-2(e^{i\omega} - 1 - i\omega)/\omega^2$ (e) $1/135$

3.107. (a) $21/2$ (b) $35/4$

3.108. (a) $4/3$ (b) $2/9$ (c) $(1 + 2te^{2t} - e^{2t})/2t^2$ (d) $-(1 + 2i\omega e^{2i\omega} - e^{2i\omega})/2\omega^2$
(e) $-2\sqrt{18}/15$ (f) $12/5$

3.109. (a) 1 (b) $8(2\sqrt{2} - 1)/15$

3.110. (a) 2 (b) $\sqrt{2\pi}/2$

3.111. (a) 0 (b) $1/3$ (c) 0

Special Probability Distributions

The Binomial Distribution

Suppose that we have an experiment such as tossing a coin or die repeatedly or choosing a marble from an urn repeatedly. Each toss or selection is called a *trial*. In any single trial there will be a probability associated with a particular event such as head on the coin, 4 on the die, or selection of a red marble. In some cases this probability will not change from one trial to the next (as in tossing a coin or die). Such trials are then said to be *independent* and are often called *Bernoulli trials* after James Bernoulli who investigated them at the end of the seventeenth century.

Let p be the probability that an event will happen in any single Bernoulli trial (called the *probability of success*). Then $q = 1 - p$ is the probability that the event will fail to happen in any single trial (called the *probability of failure*). The probability that the event will happen exactly x times in n trials (i.e., successes and $n - x$ failures will occur) is given by the probability function

$$f(x) = P(X = x) = \binom{n}{x} p^x q^{n-x} = \frac{n!}{x!(n-x)!} p^x q^{n-x} \tag{1}$$

where the random variable X denotes the number of successes in n trials and $x = 0, 1, \ldots, n$.

EXAMPLE 4.1 The probability of getting exactly 2 heads in 6 tosses of a fair coin is

$$P(X = 2) = \binom{6}{2}\left(\frac{1}{2}\right)^2\left(\frac{1}{2}\right)^{6-2} = \frac{6!}{2!4!}\left(\frac{1}{2}\right)^2\left(\frac{1}{2}\right)^{6-2} = \frac{15}{64}$$

The discrete probability function (1) is often called the *binomial distribution* since for $x = 0, 1, 2, \ldots, n$, it corresponds to successive terms in the *binomial expansion*

$$(q + p)^n = q^n + \binom{n}{1} q^{n-1}p + \binom{n}{2} q^{n-2}p^2 + \cdots + p^n = \sum_{x=0}^{n} \binom{n}{x} p^x q^{n-x} \tag{2}$$

The special case of a binomial distribution with $n = 1$ is also called the *Bernoulli distribution*.

Some Properties of the Binomial Distribution

Some of the important properties of the binomial distribution are listed in Table 4-1.

Table 4-1

Mean	$\mu = np$
Variance	$\sigma^2 = npq$
Standard deviation	$\sigma = \sqrt{npq}$
Coefficient of skewness	$\alpha_3 = \dfrac{q-p}{\sqrt{npq}}$
Coefficient of kurtosis	$\alpha_4 = 3 + \dfrac{1-6pq}{npq}$
Moment generating function	$M(t) = (q + pe^t)^n$
Characteristic function	$\phi(\omega) = (q + pe^{i\omega})^n$

EXAMPLE 4.2 In 100 tosses of a fair coin, the expected or mean number of heads is $\mu = (100)(\frac{1}{2}) = 50$ while the standard deviation is $\sigma = \sqrt{(100)(\frac{1}{2})(\frac{1}{2})} = 5$.

The Law of Large Numbers for Bernoulli Trials

The law of large numbers, page 83, has an interesting interpretation in the case of Bernoulli trials and is presented in the following theorem.

Theorem 4-1 **(Law of Large Numbers for Bernoulli Trials):** Let X be the random variable giving the number of successes in n Bernoulli trials, so that X/n is the proportion of successes. Then if p is the probability of success and ϵ is any positive number,

$$\lim_{n\to\infty} P\left(\left|\frac{X}{n} - p\right| \geq \epsilon\right) = 0 \tag{3}$$

In other words, in the long run it becomes extremely likely that the proportion of successes, X/n, will be as close as you like to the probability of success in a single trial, p. This law in a sense justifies use of the empirical definition of probability on page 5. A stronger result is provided by the *strong law* of large numbers (page 83), which states that with probability one, $\lim_{n\to\infty} X/n = p$, i.e., X/n actually *converges to* p except in a negligible number of cases.

The Normal Distribution

One of the most important examples of a continuous probability distribution is the *normal distribution*, sometimes called the *Gaussian distribution*. The density function for this distribution is given by

$$f(x) = \frac{1}{\sigma\sqrt{2\pi}} e^{-(x-\mu)^2/2\sigma^2} \qquad -\infty < x < \infty \tag{4}$$

where μ and σ are the mean and standard deviation, respectively. The corresponding distribution function is given by

$$F(x) = P(X \leq x) = \frac{1}{\sigma\sqrt{2\pi}} \int_{-\infty}^{x} e^{-(v-\mu)^2/2\sigma^2}\, dv \tag{5}$$

If X has the distribution function given by (5), we say that the random variable X is *normally distributed* with mean μ and variance σ^2.

If we let Z be the standardized variable corresponding to X, i.e., if we let

$$Z = \frac{X - \mu}{\sigma} \tag{6}$$

then the mean or expected value of Z is 0 and the variance is 1. In such cases the density function for Z can be obtained from (4) by formally placing $\mu = 0$ and $\sigma = 1$, yielding

$$f(z) = \frac{1}{\sqrt{2\pi}} e^{-z^2/2} \tag{7}$$

This is often referred to as the *standard normal density function.* The corresponding distribution function is given by

$$F(z) = P(Z \le z) = \frac{1}{\sqrt{2\pi}} \int_{-\infty}^{z} e^{-u^2/2}\, du = \frac{1}{2} + \frac{1}{\sqrt{2\pi}} \int_{0}^{z} e^{-u^2/2}\, du \tag{8}$$

We sometimes call the value z of the standardized variable Z the *standard score.* The function $F(z)$ is related to the extensively tabulated *error function,* erf(z). We have

$$\text{erf}(z) = \frac{2}{\sqrt{\pi}} \int_{0}^{z} e^{-u^2}\, du \quad \text{and} \quad F(z) = \frac{1}{2}\left[1 + \text{erf}\left(\frac{z}{\sqrt{2}}\right)\right] \tag{9}$$

A graph of the density function (7), sometimes called the *standard normal curve,* is shown in Fig. 4-1. In this graph we have indicated the areas within 1, 2, and 3 standard deviations of the mean (i.e., between $z = -1$ and $+1$, $z = -2$ and $+2$, $z = -3$ and $+3$) as equal, respectively, to 68.27%, 95.45% and 99.73% of the total area, which is one. This means that

$$P(-1 \le Z \le 1) = 0.6827, \qquad P(-2 \le Z \le 2) = 0.9545, \qquad P(-3 \le Z \le 3) = 0.9973 \tag{10}$$

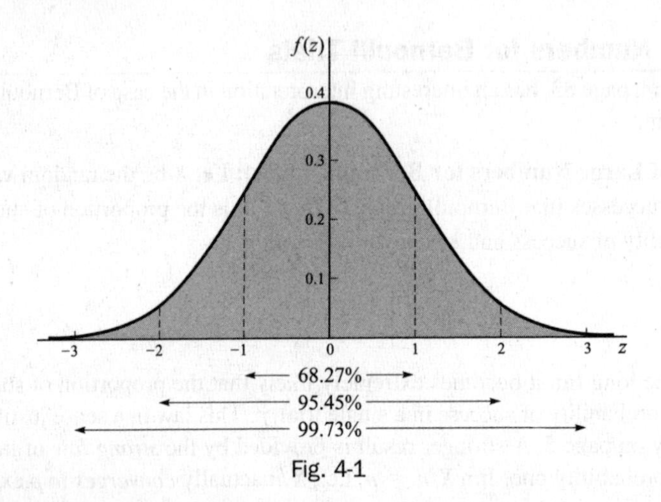

Fig. 4-1

A table giving the areas under this curve bounded by the ordinates at $z = 0$ and any positive value of z is given in Appendix C. From this table the areas between any two ordinates can be found by using the symmetry of the curve about $z = 0$.

Some Properties of the Normal Distribution

In Table 4-2 we list some important properties of the general normal distribution.

Table 4-2

Mean	μ
Variance	σ^2
Standard deviation	σ
Coefficient of skewness	$\alpha_3 = 0$
Coefficient of kurtosis	$\alpha_4 = 3$
Moment generating function	$M(t) = e^{ut+(\sigma^2 t^2/2)}$
Characteristic function	$\phi(\omega) = e^{i\mu\omega-(\sigma^2\omega^2/2)}$

Relation Between Binomial and Normal Distributions

If n is large and if neither p nor q is too close to zero, the binomial distribution can be closely approximated by a normal distribution with standardized random variable given by

$$Z = \frac{X - np}{\sqrt{npq}} \tag{11}$$

Here X is the random variable giving the number of successes in n Bernoulli trials and p is the probability of success. The approximation becomes better with increasing n and is exact in the limiting case. (See Problem 4.17.) In practice, the approximation is very good if both np and nq are greater than 5. The fact that the binomial distribution approaches the normal distribution can be described by writing

$$\lim_{n\to\infty} P\left(a \le \frac{X - np}{\sqrt{npq}} \le b\right) = \frac{1}{\sqrt{2\pi}}\int_a^b e^{-u^2/2}\,du \tag{12}$$

In words, we say that the standardized random variable $(X - np)/\sqrt{npq}$ is *asymptotically normal.*

The Poisson Distribution

Let X be a discrete random variable that can take on the values 0, 1, 2, . . . such that the probability function of X is given by

$$f(x) = P(X = x) = \frac{\lambda^x e^{-\lambda}}{x!} \qquad x = 0, 1, 2, \ldots \tag{13}$$

where λ is a given positive constant. This distribution is called the *Poisson distribution* (after S. D. Poisson, who discovered it in the early part of the nineteenth century), and a random variable having this distribution is said to be *Poisson distributed.*

The values of $f(x)$ in (13) can be obtained by using Appendix G, which gives values of $e^{-\lambda}$ for various values of λ.

Some Properties of the Poisson Distribution

Some important properties of the Poisson distribution are listed in Table 4-3.

Table 4-3

Mean	$\mu = \lambda$
Variance	$\sigma^2 = \lambda$
Standard deviation	$\sigma = \sqrt{\lambda}$
Coefficient of skewness	$\alpha_3 = 1/\sqrt{\lambda}$
Coefficient of kurtosis	$\alpha_4 = 3 + (1/\lambda)$
Moment generating function	$M(t) = e^{\lambda(e^t - 1)}$
Characteristic function	$\phi(\omega) = e^{\lambda(e^{i\omega} - 1)}$

Relation Between the Binomial and Poisson Distributions

In the binomial distribution (1), if n is large while the probability p of occurrence of an event is close to zero, so that $q = 1 - p$ is close to 1, the event is called a *rare event*. In practice we shall consider an event as rare if the number of trials is at least 50 ($n \ge 50$) while np is less than 5. For such cases the binomial distribution is very closely approximated by the Poisson distribution (13) with $\lambda = np$. This is to be expected on comparing Tables 4-1 and 4-3, since by placing $\lambda = np$, $q \approx 1$, and $p \approx 0$ in Table 4-1, we get the results in Table 4-3.

Relation Between the Poisson and Normal Distributions

Since there is a relation between the binomial and normal distributions and between the binomial and Poisson distributions, we would expect that there should also be a relation between the Poisson and normal distributions. This is in fact the case. We can show that if X is the Poisson random variable of (13) and $(X - \lambda)/\sqrt{\lambda}$ is the corresponding standardized random variable, then

$$\lim_{\lambda \to \infty} P\left(a \le \frac{X - \lambda}{\sqrt{\lambda}} \le b\right) = \frac{1}{\sqrt{2\pi}} \int_a^b e^{-u^2/2}\, du \tag{14}$$

i.e., the Poisson distribution approaches the normal distribution as $\lambda \to \infty$ or $(X - \lambda)/\sqrt{\lambda}$ is *asymptotically normal.*

The Central Limit Theorem

The similarity between (12) and (14) naturally leads us to ask whether there are any other distributions besides the binomial and Poisson that have the normal distribution as the limiting case. The following remarkable theorem reveals that actually a large class of distributions have this property.

Theorem 4-2 **(Central Limit Theorem)** Let $X_1, X_2, \ldots, X_n$ be independent random variables that are identically distributed (i.e., all have the *same* probability function in the discrete case or density function in the continuous case) and have finite mean μ and variance σ^2. Then if $S_n = X_1 + X_2 + \cdots + X_n$ $(n = 1, 2 \ldots)$,

$$\lim_{n \to \infty} P\left(a \le \frac{S_n - n\mu}{\sigma\sqrt{n}} \le b\right) = \frac{1}{\sqrt{2\pi}} \int_a^b e^{-u^2/2}\, du \tag{15}$$

that is, the random variable $(S_n - n\mu)/\sigma\sqrt{n}$, which is the standardized variable corresponding to S_n, is asymptotically normal.

The theorem is also true under more general conditions; for example, it holds when $X_1, X_2, \ldots, X_n$ are independent random variables with the same mean and the same variance but not necessarily identically distributed.

The Multinomial Distribution

Suppose that events $A_1, A_2, \ldots, A_k$ are mutually exclusive, and can occur with respective probabilities $p_1, p_2, \ldots, p_k$ where $p_1 + p_2 + \cdots + p_k = 1$. If $X_1, X_2, \ldots, X_k$ are the random variables respectively giving the number of times that $A_1, A_2, \ldots, A_k$ occur in a total of n trials, so that $X_1 + X_2 + \cdots + X_k = n$, then

$$P(X_1 = n_1, X_2 = n_2, \ldots, X_k = n_k) = \frac{n}{n_1!n_2! \cdots n_k!} p_1^{n_1} p_2^{n_2} \cdots p_k^{n_k} \tag{16}$$

where $n_1 + n_2 + \cdots + n_k = n$, is the joint probability function for the random variables $X_1, \ldots, X_k$.

This distribution, which is a generalization of the binomial distribution, is called the *multinomial distribution* since (16) is the general term in the multinomial expansion of $(p_1 + p_2 + \cdots + p_k)^n$.

EXAMPLE 4.3 If a fair die is to be tossed 12 times, the probability of getting 1, 2, 3, 4, 5 and 6 points exactly twice each is

$$P(X_1 = 2, X_2 = 2, \ldots, X_6 = 2) = \frac{12!}{2!2!2!2!2!2!}\left(\frac{1}{6}\right)^2\left(\frac{1}{6}\right)^2\left(\frac{1}{6}\right)^2\left(\frac{1}{6}\right)^2\left(\frac{1}{6}\right)^2\left(\frac{1}{6}\right)^2 = \frac{1925}{559{,}872} = 0.00344$$

The expected number of times that $A_1, A_2, \ldots, A_k$ will occur in n trials are $np_1, np_2, \ldots, np_k$ respectively, i.e.,

$$E(X_1) = np_1, \quad E(X_2) = np_2, \quad \ldots, \quad E(X_k) = np_k \tag{17}$$

The Hypergeometric Distribution

Suppose that a box contains b blue marbles and r red marbles. Let us perform n trials of an experiment in which a marble is chosen at random, its color is observed, and then the marble is put back in the box. This type of experiment is often referred to as *sampling with replacement.* In such a case, if X is the random variable denoting the number of blue marbles chosen (successes) in n trials, then using the binomial distribution (1) we see that the probability of exactly x successes is

$$P(X = x) = \binom{n}{x} \frac{b^x r^{n-x}}{(b + r)^n}, \qquad x = 0, 1, \ldots, n \tag{18}$$

since $p = b/(b + r)$, $q = 1 - p = r/(b + r)$.

If we modify the above so that *sampling is without replacement*, i.e., the marbles are not replaced after being chosen, then

$$P(X = x) = \frac{\binom{b}{x}\binom{r}{n-x}}{\binom{b+r}{n}}, \qquad x = \max(0, n - r), \ldots, \min(n, b) \tag{19}$$

This is the *hypergeometric distribution*. The mean and variance for this distribution are

$$\mu = \frac{nb}{b + r}, \qquad \sigma^2 = \frac{nbr(b + r - n)}{(b + r)^2(b + r - 1)} \tag{20}$$

If we let the total number of blue and red marbles be N, while the proportions of blue and red marbles are p and $q = 1 - p$, respectively, then

$$p = \frac{b}{b + r} = \frac{b}{N}, \qquad q = \frac{r}{b + r} = \frac{r}{N} \qquad \text{or} \qquad b = Np, \qquad r = Nq \tag{21}$$

so that (19) and (20) become, respectively,

$$P(X = x) = \frac{\binom{Np}{x}\binom{Nq}{n-x}}{\binom{N}{n}} \tag{22}$$

$$\mu = np, \qquad \sigma^2 = \frac{npq(N - n)}{N - 1} \tag{23}$$

Note that as $N \to \infty$ (or N is large compared with n), (22) reduces to (18), which can be written

$$P(X = x) = \binom{n}{x} p^x q^{n-x} \tag{24}$$

and (23) reduces to

$$\mu = np, \qquad \sigma^2 = npq \tag{25}$$

in agreement with the first two entries in Table 4-1, page 109. The results are just what we would expect, since for large N, sampling without replacement is practically identical to sampling with replacement.

The Uniform Distribution

A random variable X is said to be *uniformly distributed* in $a \le x \le b$ if its density function is

$$f(x) = \begin{cases} 1/(b - a) & a \le x \le b \\ 0 & \text{otherwise} \end{cases} \tag{26}$$

and the distribution is called a *uniform distribution*.

The distribution function is given by

$$F(x) = P(X \le x) = \begin{cases} 0 & x < a \\ (x - a)/(b - a) & a \le x < b \\ 1 & x \ge b \end{cases} \tag{27}$$

The mean and variance are, respectively,

$$\mu = \frac{1}{2}(a + b), \qquad \sigma^2 = \frac{1}{12}(b - a)^2 \tag{28}$$

The Cauchy Distribution

A random variable X is said to be *Cauchy distributed*, or to have the *Cauchy distribution*, if the density function of X is

$$f(x) = \frac{a}{\pi(x^2 + a^2)} \qquad a > 0, -\infty < x < \infty \tag{29}$$

This density function is symmetrical about $x = 0$ so that its median is zero. However, the mean, variance, and higher moments do not exist. Similarly, the moment generating function does not exist. However, the characteristic function does exist and is given by

$$\phi(\omega) = e^{-a\omega} \tag{30}$$

The Gamma Distribution

A random variable X is said to have the *gamma distribution*, or to be *gamma distributed*, if the density function is

$$f(x) = \begin{cases} \dfrac{x^{\alpha-1}e^{-x/\beta}}{\beta^\alpha \Gamma(\alpha)} & x > 0 \\ 0 & x \le 0 \end{cases} \qquad (\alpha, \beta > 0) \tag{31}$$

where $\Gamma(\alpha)$ is the *gamma function* (see Appendix A). The mean and variance are given by

$$\mu = \alpha\beta, \qquad \sigma^2 = \alpha\beta^2 \tag{32}$$

The moment generating function and characteristic function are given, respectively, by

$$M(t) = (1 - \beta t)^{-\alpha}, \qquad \phi(\omega) = (1 - \beta i\omega)^{-\alpha} \tag{33}$$

The Beta Distribution

A random variable is said to have the *beta distribution*, or to be *beta distributed*, if the density function is

$$f(x) = \begin{cases} \dfrac{x^{\alpha-1}(1 - x)^{\beta-1}}{B(\alpha, \beta)} & 0 < x < 1 \\ 0 & \text{otherwise} \end{cases} \qquad (\alpha, \beta > 0) \tag{34}$$

where $B(\alpha, \beta)$ is the *beta function* (see Appendix A). In view of the relation (9), Appendix A, between the beta and gamma functions, the beta distribution can also be defined by the density function

$$f(x) = \begin{cases} \dfrac{\Gamma(\alpha + \beta)}{\Gamma(\alpha)\Gamma(\beta)} x^{\alpha-1}(1 - x)^{\beta-1} & 0 < x < 1 \\ 0 & \text{otherwise} \end{cases} \tag{35}$$

where α, β are positive. The mean and variance are

$$\mu = \frac{\alpha}{\alpha + \beta}, \qquad \sigma^2 = \frac{\alpha\beta}{(\alpha + \beta)^2(\alpha + \beta + 1)} \tag{36}$$

For $\alpha > 1, \beta > 1$ there is a unique mode at the value

$$x_{\text{mode}} = \frac{\alpha - 1}{\alpha + \beta - 2} \tag{37}$$

The Chi-Square Distribution

Let $X_1, X_2, \ldots, X_v$ be v independent normally distributed random variables with mean zero and variance 1. Consider the random variable

$$\chi^2 = X_1^2 + X_2^2 + \cdots + X_v^2 \tag{38}$$

where χ^2 is called *chi square.* Then we can show that for $x \geq 0$,

$$P(\chi^2 \leq x) = \frac{1}{2^{v/2}\Gamma(v/2)}\int_0^x u^{(v/2)-1}e^{-u/2}\,du \tag{39}$$

and $P(\chi^2 \leq x) = 0$ for $x < 0$.

The distribution defined by (39) is called the *chi-square distribution*, and v is called the *number of degrees of freedom.* The distribution defined by (39) has corresponding density function given by

$$f(x) = \begin{cases} \dfrac{1}{2^{v/2}\Gamma(v/2)}x^{(v/2)-1}e^{-x/2} & x > 0 \\ 0 & x \leq 0 \end{cases} \tag{40}$$

It is seen that the chi-square distribution is a special case of the gamma distribution with $\alpha = v/2$ $\beta = 2$. Therefore,

$$\mu = v, \qquad \sigma^2 = 2v, \qquad M(t) = (1 - 2t)^{-v/2}, \qquad \phi(\omega) = (1 - 2i\omega)^{-v/2} \tag{41}$$

For large $v (v \geq 30)$, we can show that $\sqrt{2\chi^2} - \sqrt{2v - 1}$ is very nearly normally distributed with mean 0 and variance 1.

Three theorems that will be useful in later work are as follows:

Theorem 4-3 Let $X_1, X_2, \ldots, X_v$ be independent normally distributed random variables with mean 0 and variance 1. Then $\chi^2 = X_1^2 + X_2^2 + \cdots + X_v^2$ is chi-square distributed with v degrees of freedom.

Theorem 4-4 Let $U_1, U_2, \ldots, U_k$ be independent random variables that are chi-square distributed with $v_1, v_2, \ldots, v_k$ degrees of freedom, respectively. Then their sum $W = U_1 + U_2 + \cdots + U_k$ is chi-square distributed with $v_1 + v_2 + \cdots + v_k$ degrees of freedom.

Theorem 4-5 Let V_1 and V_2 be independent random variables. Suppose that V_1 is chi-square distributed with v_1 degrees of freedom while $V = V_1 + V_2$ is chi-square distributed with v degrees of freedom, where $v > v_1$. Then V_2 is chi-square distributed with $v - v_1$ degrees of freedom.

In connection with the chi-square distribution, the t distribution (below), the F distribution (page 116), and others, it is common in statistical work to use the *same symbol* for both the random variable and a value of that random variable. Therefore, percentile values of the chi-square distribution for v degrees of freedom are denoted by $\chi^2_{p,v}$, or briefly χ^2_p if v is understood, and not by $x_{p,v}$ or x_p. (See Appendix E.) This is an ambiguous notation, and the reader should use care with it, especially when changing variables in density functions.

Student's *t* Distribution

If a random variable has the density function

$$f(t) = \frac{\Gamma\left(\frac{v+1}{2}\right)}{\sqrt{v\pi}\,\Gamma\left(\frac{v}{2}\right)}\left(1 + \frac{t^2}{v}\right)^{-(v+1)/2} \qquad -\infty < t < \infty \tag{42}$$

it is said to have *Student's t distribution,* briefly the *t distribution*, with v degrees of freedom. If v is large ($v \geq 30$), the graph of $f(t)$ closely approximates the standard normal curve as indicated in Fig. 4-2. Percentile

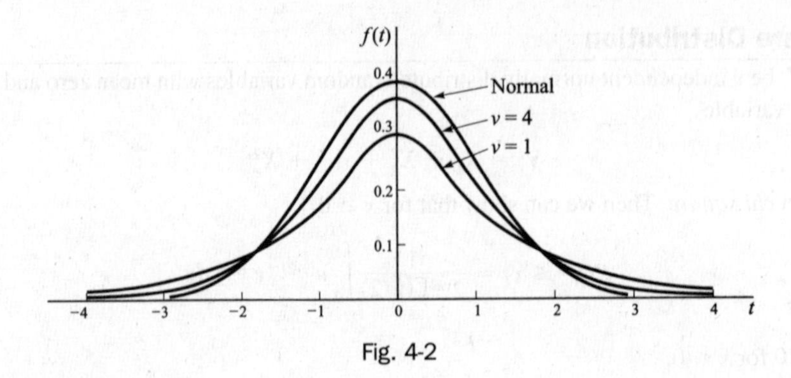

Fig. 4-2

values of the t distribution for ν degrees of freedom are denoted by $t_{p,\nu}$ or briefly t_p if ν is understood. For a table giving such values, see Appendix D. Since the t distribution is symmetrical, $t_{1-p} = -t_p$; for example, $t_{0.5} = -t_{0.95}$.

For the t distribution we have

$$\mu = 0 \quad \text{and} \quad \sigma^2 = \frac{\nu}{\nu - 2} \quad (\nu > 2). \tag{43}$$

The following theorem is important in later work.

Theorem 4-6 Let Y and Z be independent random variables, where Y is normally distributed with mean 0 and variance 1 while Z is chi-square distributed with ν degrees of freedom. Then the random variable

$$T = \frac{Y}{\sqrt{Z/\nu}} \tag{44}$$

has the t distribution with ν degrees of freedom.

The *F* Distribution

A random variable is said to have the *F distribution* (named after R. A. Fisher) *with* ν_1 *and* ν_2 *degrees of freedom* if its density function is given by

$$f(u) = \begin{cases} \dfrac{\Gamma\left(\dfrac{\nu_1 + \nu_2}{2}\right)}{\Gamma\left(\dfrac{\nu_1}{2}\right)\Gamma\left(\dfrac{\nu_2}{2}\right)} \nu_1^{\nu_1/2} \nu_2^{\nu_2/2} u^{(\nu_1/2)-1}(\nu_2 + \nu_1 u)^{-(\nu_1+\nu_2)/2} & u > 0 \\ 0 & u \le 0 \end{cases} \tag{45}$$

Percentile values of the F distribution for ν_1, ν_2 degrees of freedom are denoted by F_{p,ν_1,ν_2}, or briefly F_p if ν_1, ν_2 are understood. For a table giving such values in the case where $p = 0.95$ and $p = 0.99$, see Appendix F.

The mean and variance are given, respectively, by

$$\mu = \frac{\nu_2}{\nu_2 - 2} \; (\nu_2 > 2) \quad \text{and} \quad \sigma^2 = \frac{2\nu_2^2(\nu_1 + \nu_2 - 2)}{\nu_1(\nu_2 - 4)(\nu_2 - 2)^2} \quad (\nu_2 > 4) \tag{46}$$

The distribution has a unique mode at the value

$$u_{\text{mode}} = \left(\frac{\nu_1 - 2}{\nu_1}\right)\left(\frac{\nu_2}{\nu_2 + 2}\right) \quad (\nu_1 > 2) \tag{47}$$

The following theorems are important in later work.

Theorem 4-7 Let V_1 and V_2 be independent random variables that are chi-square distributed with v_1 and v_2 degrees of freedom, respectively. Then the random variable

$$V = \frac{V_1/v_1}{V_2/v_2} \tag{48}$$

has the F distribution with v_1 and v_2 degrees of freedom.

Theorem 4-8

$$F_{1-p,v_2,v_1} = \frac{1}{F_{p,v_1,v_2}}$$

Relationships Among Chi-Square, *t*, and *F* Distributions

Theorem 4-9 $$F_{1-p,1,v} = t^2_{1-(p/2),v}$$

Theorem 4-10 $$F_{p,v,\infty} = \frac{\chi^2_{p,v}}{v}$$

The Bivariate Normal Distribution

A generalization of the normal distribution to two continuous random variables X and Y is given by the joint density function

$$f(x, y) = \frac{1}{2\pi\sigma_1\sigma_2\sqrt{1-\rho^2}} \exp\left\{-\left[\left(\frac{x-\mu_1}{\sigma_1}\right)^2 - 2\rho\left(\frac{x-\mu_1}{\sigma_1}\right)\left(\frac{y-\mu_2}{\sigma_2}\right) + \left(\frac{y-\mu_2}{\sigma_2}\right)^2\right]\Big/2(1-\rho^2)\right\} \tag{49}$$

where $-\infty < x < \infty$, $-\infty < y < \infty$; μ_1, μ_2 are the means of X and Y; σ_1, σ_2 are the standard deviations of X and Y; and ρ is the correlation coefficient between X and Y. We often refer to (49) as the *bivariate normal distribution.*

For any joint distribution the condition $\rho = 0$ is necessary for independence of the random variables (see Theorem 3-15). In the case of (49) this condition is also sufficient (see Problem 4.51).

Miscellaneous Distributions

In the distributions listed below, the constants $\alpha, \beta, a, b, \ldots$ are taken as positive unless otherwise stated. The characteristic function $\phi(\omega)$ is obtained from the moment generating function, where given, by letting $t = i\omega$.

1. GEOMETRIC DISTRIBUTION.

$$f(x) = P(X = x) = pq^{x-1} \qquad x = 1, 2, \ldots$$

$$\mu = \frac{1}{p} \qquad \sigma^2 = \frac{q}{p^2} \qquad M(t) = \frac{pe^t}{1 - qe^t}$$

The random variable X represents the number of Bernoulli trials up to and including that in which the first success occurs. Here p is the probability of success in a single trial.

2. PASCAL'S OR NEGATIVE BINOMIAL DISTRIBUTION.

$$f(x) = P(X = x) = \binom{x-1}{r-1} p^r q^{x-r} \qquad x = r, r+1, \ldots$$

$$\mu = \frac{r}{P} \qquad \sigma^2 = \frac{rq}{p^2} \qquad M(t) = \left(\frac{pe^t}{1 - qe^t}\right)^r$$

The random variable X represents the number of Bernoulli trials up to and including that in which the rth success occurs. The special case $r = 1$ gives the geometric distribution.

3. EXPONENTIAL DISTRIBUTION.

$$f(x) = \begin{cases} \alpha e^{-\alpha x} & x > 0 \\ 0 & x \le 0 \end{cases}$$

$$\mu = \frac{1}{\alpha} \qquad \sigma^2 = \frac{1}{\alpha^2} \qquad M(t) = \frac{\alpha}{\alpha - t}$$

4. WEIBULL DISTRIBUTION.

$$f(x) = \begin{cases} abx^{b-1}e^{-ax^b} & x > 0 \\ 0 & x \le 0 \end{cases}$$

$$\mu = a^{-1/b}\Gamma\left(1 + \frac{1}{b}\right) \qquad \sigma^2 = a^{-2/b}\left[\Gamma\left(1 + \frac{2}{b}\right) - \Gamma^2\left(1 + \frac{1}{b}\right)\right]$$

5. MAXWELL DISTRIBUTION.

$$f(x) = \begin{cases} \sqrt{2/\pi}\,\alpha^{3/2}x^2e^{-\alpha x^2/2} & x > 0 \\ 0 & x \le 0 \end{cases}$$

$$\mu = 2\sqrt{\frac{2}{\pi\alpha}} \qquad \sigma^2 = \left(3 - \frac{8}{\pi}\right)\alpha^{-1}$$

SOLVED PROBLEMS

The binomial distribution

4.1. Find the probability that in tossing a fair coin three times, there will appear (a) 3 heads, (b) 2 tails and 1 head, (c) at least 1 head, (d) not more than 1 tail.

Method 1

Let H denote heads and T denote tails, and suppose that we designate HTH, for example, to mean head on first toss, tail on second toss, and then head on third toss.

Since 2 possibilities (head or tail) can occur on each toss, there are a total of (2)(2)(2) = 8 possible outcomes, i.e., sample points, in the sample space. These are

$$HHH, HHT, HTH, HTT, TTH, THH, THT, TTT$$

For a fair coin these are assigned equal probabilities of 1/8 each. Therefore,

(a) $P(3 \text{ heads}) = P(HHH) = \frac{1}{8}$

(b) $P(2 \text{ tails and } 1 \text{ head}) = P(HTT \cup TTH \cup THT)$

$$= P(HTT) + P(TTH) + P(THT) = \frac{1}{8} + \frac{1}{8} + \frac{1}{8} = \frac{3}{8}$$

(c) $P(\text{at least 1 head})$

$= P(1, 2, \text{or } 3 \text{ heads})$

$= P(1 \text{ head}) + P(2 \text{ heads}) + P(3 \text{ heads})$

$= P(HTT \cup THT \cup TTH) + P(HHT \cup HTH \cup THH) + P(HHH)$

$= P(HTT) + P(THT) + P(TTH) + P(HHT) + P(HTH) + P(THH) + P(HHH) = \frac{7}{8}$

Alternatively,

$$P(\text{at least 1 head}) = 1 - P(\text{no head}) = 1 - P(TTT) = 1 - \frac{1}{8} = \frac{7}{8}$$

(d) $P(\text{not more than 1 tail}) = P(0 \text{ tails or } 1 \text{ tail})$

$= P(0 \text{ tails}) + P(1 \text{ tail})$

$= P(HHH) + P(HHT \cup HTH \cup THH)$

$= P(HHH) + P(HHT) + P(HTH) + P(THH)$

$= \frac{4}{8} = \frac{1}{2}$

Method 2 (using formula)

(a) $P(3 \text{ heads}) = \binom{3}{3}\left(\frac{1}{2}\right)^3\left(\frac{1}{2}\right)^0 = \frac{1}{8}$

(b) $P(2 \text{ tails and } 1 \text{ head}) = \binom{3}{2}\left(\frac{1}{2}\right)^2\left(\frac{1}{2}\right)^1 = \frac{3}{8}$

(c)
$$\begin{aligned} P(\text{at least 1 head}) &= P(1, 2, \text{ or } 3 \text{ heads}) \\ &= P(1 \text{ head}) + P(2 \text{ heads}) + P(3 \text{ heads}) \\ &= \binom{3}{1}\left(\frac{1}{2}\right)^1\left(\frac{1}{2}\right)^2 + \binom{3}{2}\left(\frac{1}{2}\right)^2\left(\frac{1}{2}\right)^1 + \binom{3}{3}\left(\frac{1}{2}\right)^3\left(\frac{1}{2}\right)^0 = \frac{7}{8} \end{aligned}$$

Alternatively,

$$\begin{aligned} P(\text{at least 1 head}) &= 1 - P(\text{no head}) \\ &= 1 - \binom{3}{0}\left(\frac{1}{2}\right)^0\left(\frac{1}{2}\right)^3 = \frac{7}{8} \end{aligned}$$

(d)
$$\begin{aligned} P(\text{not more than 1 tail}) &= P(0 \text{ tails or } 1 \text{ tail}) \\ &= P(0 \text{ tails}) + P(1 \text{ tail}) \\ &= \binom{3}{3}\left(\frac{1}{2}\right)^3\left(\frac{1}{2}\right)^0 + \binom{3}{2}\left(\frac{1}{2}\right)^2\left(\frac{1}{2}\right) = \frac{1}{2} \end{aligned}$$

It should be mentioned that the notation of random variables can also be used. For example, if we let X be the random variable denoting the number of heads in 3 tosses, (c) can be written

$$P(\text{at least 1 head}) = P(X \geq 1) = P(X = 1) + P(X = 2) + P(X = 3) = \frac{7}{8}$$

We shall use both approaches interchangeably.

4.2. Find the probability that in five tosses of a fair die, a 3 will appear (a) twice, (b) at most once, (c) at least two times.

Let the random variable X be the number of times a 3 appears in five tosses of a fair die. We have

$$\text{Probability of 3 in a single toss} = p = \frac{1}{6}$$

$$\text{Probability of no 3 in a single toss} = q = 1 - p = \frac{5}{6}$$

(a) $P(3 \text{ occurs twice}) = P(X = 2) = \binom{5}{2}\left(\frac{1}{6}\right)^2\left(\frac{5}{6}\right)^3 = \frac{625}{3888}$

(b)
$$\begin{aligned} P(3 \text{ occurs at most once}) &= P(X \leq 1) = P(X = 0) + P(X = 1) \\ &= \binom{5}{0}\left(\frac{1}{6}\right)^0\left(\frac{5}{6}\right)^5 + \binom{5}{1}\left(\frac{1}{6}\right)^1\left(\frac{5}{6}\right)^4 \\ &= \frac{3125}{7776} + \frac{3125}{7776} = \frac{3125}{3888} \end{aligned}$$

(c) $P(3 \text{ occurs at least 2 times})$

$$\begin{aligned} &= P(X \geq 2) \\ &= P(X = 2) + P(X = 3) + P(X = 4) + P(X = 5) \\ &= \binom{5}{2}\left(\frac{1}{6}\right)^2\left(\frac{5}{6}\right)^3 + \binom{5}{3}\left(\frac{1}{6}\right)^3\left(\frac{5}{6}\right)^2 + \binom{5}{4}\left(\frac{1}{6}\right)^4\left(\frac{5}{6}\right)^1 + \binom{5}{5}\left(\frac{1}{6}\right)^5\left(\frac{5}{6}\right)^0 \\ &= \frac{625}{3888} + \frac{125}{3888} + \frac{25}{7776} + \frac{1}{7776} = \frac{763}{3888} \end{aligned}$$

4.3. Find the probability that in a family of 4 children there will be (a) at least 1 boy, (b) at least 1 boy and at least 1 girl. Assume that the probability of a male birth is 1/2.

(a) $P(1\text{ boy}) = \binom{4}{1}\left(\frac{1}{2}\right)^1\left(\frac{1}{2}\right)^3 = \frac{1}{4}, \quad P(2\text{ boys}) = \binom{4}{2}\left(\frac{1}{2}\right)^2\left(\frac{1}{2}\right)^2 = \frac{3}{8}$

$P(3\text{ boys}) = \binom{4}{3}\left(\frac{1}{2}\right)^3\left(\frac{1}{2}\right)^1 = \frac{1}{4}, \quad P(4\text{ boys}) = \binom{4}{4}\left(\frac{1}{2}\right)^4\left(\frac{1}{2}\right)^0 = \frac{1}{16}$

Then

$$P(\text{at least 1 boy}) = P(1\text{ boy}) + P(2\text{ boys}) + P(3\text{ boys}) + P(4\text{ boys})$$
$$= \frac{1}{4} + \frac{3}{8} + \frac{1}{4} + \frac{1}{16} = \frac{15}{16}$$

Another method

$$P(\text{at least 1 boy}) = 1 - P(\text{no boy}) = 1 - \left(\frac{1}{2}\right)^4 = 1 - \frac{1}{16} = \frac{15}{16}$$

(b) $P(\text{at least 1 boy and at least 1 girl}) = 1 - P(\text{no boy}) - P(\text{no girl})$

$$= 1 - \frac{1}{16} - \frac{1}{16} = \frac{7}{8}$$

We could also have solved this problem by letting X be a random variable denoting the number of boys in families with 4 children. Then, for example, (a) becomes

$$P(X \geq 1) = P(X = 1) + P(X = 2) + P(X = 3) + P(X = 4) = \frac{15}{16}$$

4.4. Out of 2000 families with 4 children each, how many would you expect to have (a) at least 1 boy, (b) 2 boys, (c) 1 or 2 girls, (d) no girls?

Referring to Problem 4.3, we see that

(a) Expected number of families with at least 1 boy $= 2000\left(\frac{15}{16}\right) = 1875$

(b) Expected number of families with 2 boys $= 2000 \cdot P(2\text{ boys}) = 2000\left(\frac{3}{8}\right) = 750$

(c) $P(1\text{ or }2\text{ girls}) = P(1\text{ girl}) + P(2\text{ girls})$

$$= P(1\text{ boy}) + P(2\text{ boys}) = \frac{1}{4} + \frac{3}{8} = \frac{5}{8}$$

Expected number of families with 1 or 2 girls $= (2000)\left(\frac{5}{8}\right) = 1250$

(d) Expected number of families with no girls $= (2000)\left(\frac{1}{16}\right) = 125$

4.5. If 20% of the bolts produced by a machine are defective, determine the probability that out of 4 bolts chosen at random, (a) 1, (b) 0, (c) less than 2, bolts will be defective.

The probability of a defective bolt is $p = 0.2$, of a nondefective bolt is $q = 1 - p = 0.8$. Let the random variable X be the number of defective bolts. Then

(a) $P(X = 1) = \binom{4}{1}(0.2)^1(0.8)^3 = 0.4096$

(b) $P(X = 0) = \binom{4}{0}(0.2)^0(0.8)^4 = 0.4096$

(c) $P(X < 2) = P(X = 0) + P(X = 1)$

$= 0.4096 + 0.4096 = 0.8192$

4.6. Find the probability of getting a total of 7 at least once in three tosses of a pair of fair dice.

In a single toss of a pair of fair dice the probability of a 7 is $p = 1/6$ (see Problem 2.1, page 44), so that the probability of no 7 in a single toss is $q = 1 - p = 5/6$. Then

$$P(\text{no 7 in three tosses}) = \binom{3}{0}\left(\frac{1}{6}\right)^0\left(\frac{5}{6}\right)^3 = \frac{125}{216}$$

and
$$P(\text{at least one 7 in three tosses}) = 1 - \frac{125}{216} = \frac{91}{216}$$

4.7. Find the moment generating function of a random variable X that is binomially distributed.

Method 1

If X is binomially distributed,

$$f(x) = P(X = x) = \binom{n}{x}p^x q^{n-x}$$

Then the moment generating function is given by

$$\begin{aligned} M(t) = E(e^{tx}) &= \sum e^{tx} f(x) \\ &= \sum_{x=0}^{n} e^{tx}\binom{n}{x} p^x q^{n-x} \\ &= \sum_{x=0}^{n}\binom{n}{x}(pe^t)^x q^{n-x} \\ &= (q + pe^t)^n \end{aligned}$$

Method 2

For a sequence of n Bernoulli trials, define

$$X_j = \begin{cases} 0 & \text{if failure in } j\text{th trial} \\ 1 & \text{if success in } j\text{th trial} \end{cases} \qquad (j = 1, 2, \ldots, n)$$

Then the X_j are independent and $X = X_1 + X_2 + \cdots + X_n$. For the moment generating function of X_j, we have

$$M_j(t) = e^{t0}q + e^{t1}p = q + pe^t \qquad (j = 1, 2, \ldots, n)$$

Then by Theorem 3-9, page 80,

$$M(t) = M_1(t)M_2(t)\ldots M_n(t) = (q + pe^t)^n$$

4.8. Prove that the mean and variance of a binomially distributed random variable are, respectively, $\mu = np$ and $\sigma^2 = npq$.

Proceeding as in Method 2 of Problem 4.7, we have for $j = 1, 2, \ldots, n$,

$$E(X_j) = 0q + 1p = p$$

$$\begin{aligned} \text{Var}(X_j) &= E[(X_j - p)^2] = (0 - p)^2 q + (1 - p)^2 p \\ &= p^2 q + q^2 p = pq(p + q) = pq \end{aligned}$$

Then
$$\mu = E(X) = E(X_1) + E(X_2) + \cdots + E(X_n) = np$$

$$\sigma^2 = \text{Var}(X) = \text{Var}(X_1) + \text{Var}(X_2) + \cdots + \text{Var}(X_n) = npq$$

where we have used Theorem 3-7 for σ^2.

The above results can also be obtained (but with more difficulty) by differentiating the moment generating function (see Problem 3.38) or directly from the probability function.

4.9. If the probability of a defective bolt is 0.1, find (a) the mean, (b) the standard deviation, for the number of defective bolts in a total of 400 bolts.

(a) Mean $\mu = np = (400)(0.1) = 40$, i.e., we can *expect* 40 bolts to be defective.

(b) Variance $\sigma^2 = npq = (400)(0.1)(0.9) = 36$. Hence, the standard deviation $\sigma = \sqrt{36} = 6$.

The law of large numbers for Bernoulli trials

4.10. Prove Theorem 4-1, the (weak) law of large numbers for Bernoulli trials.

By Chebyshev's inequality, page 83, if X is any random variable with finite mean μ and variance σ^2, then

$$P(|X - \mu| \geq k\sigma) \leq \frac{1}{k^2} \tag{1}$$

In particular, if X is binomially or Bernoulli distributed, then $\mu = np$, $\sigma = \sqrt{npq}$ and (1) becomes

$$P(|X - np| \geq k\sqrt{npq}) \leq \frac{1}{k^2} \tag{2}$$

or

$$P\left(\left|\frac{X}{n} - p\right| \geq k\sqrt{\frac{pq}{n}}\right) \leq \frac{1}{k^2} \tag{3}$$

If we let $\epsilon = k\sqrt{\frac{pq}{n}}$, (3) becomes

$$P\left(\left|\frac{X}{n} - p\right| \geq \epsilon\right) \leq \frac{pq}{n\epsilon^2}$$

and taking the limit as $n \to \infty$ we have, as required,

$$\lim_{n\to\infty} P\left(\left|\frac{X}{n} - p\right| \geq \epsilon\right) = 0$$

The result also follows directly from Theorem 3-19, page 83, with $S_n = X$, $\mu = np$, $\sigma = \sqrt{npq}$.

4.11. Give an interpretation of the (weak) law of large numbers for the appearances of a 3 in successive tosses of a fair die.

The law of large numbers states in this case that the probability of the proportion of 3s in n tosses differing from 1/6 by more than any value $\epsilon > 0$ approaches zero as $n \to \infty$.

The normal distribution

4.12. Find the area under the standard normal curve shown in Fig. 4-3 (a) between $z = 0$ and $z = 1.2$, (b) between $z = -0.68$ and $z = 0$, (c) between $z = -0.46$ and $z = 2.21$, (d) between $z = 0.81$ and $z = 1.94$, (e) to the right of $z = -1.28$.

(a) Using the table in Appendix C, proceed down the column marked z until entry 1.2 is reached. Then proceed right to column marked 0. The result, 0.3849, is the required area and represents the probability that Z is between 0 and 1.2 (Fig. 4-3). Therefore,

$$P(0 \leq Z \leq 1.2) = \frac{1}{\sqrt{2\pi}}\int_0^{1.2} e^{-u^2/2}\,du = 0.3849$$

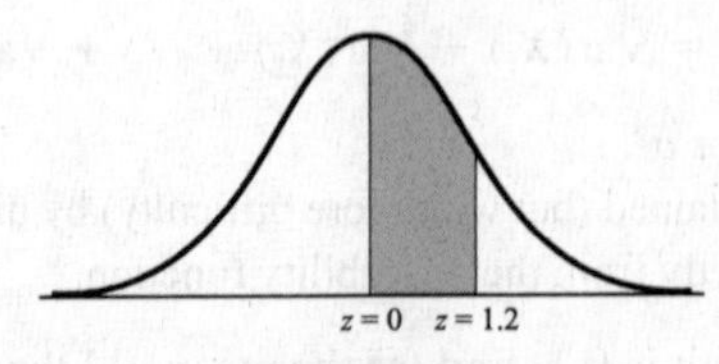

Fig. 4-3

(b) Required area = area between $z = 0$ and $z = +0.68$ (by symmetry). Therefore, proceed downward under column marked z until entry 0.6 is reached. Then proceed right to column marked 8.

The result, 0.2517, is the required area and represents the probability that Z is between -0.68 and 0 (Fig. 4-4). Therefore,

$$P(-0.68 \leq Z \leq 0) = \frac{1}{\sqrt{2\pi}}\int_{-0.68}^{0} e^{-u^2/2}\,du$$
$$= \frac{1}{\sqrt{2\pi}}\int_{0}^{0.68} e^{-u^2/2}\,du = 0.2517$$

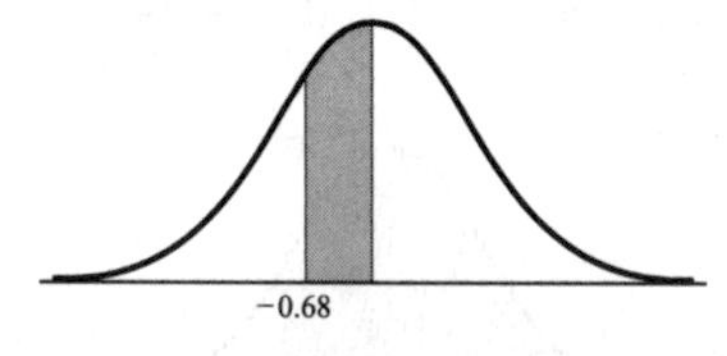

Fig. 4-4

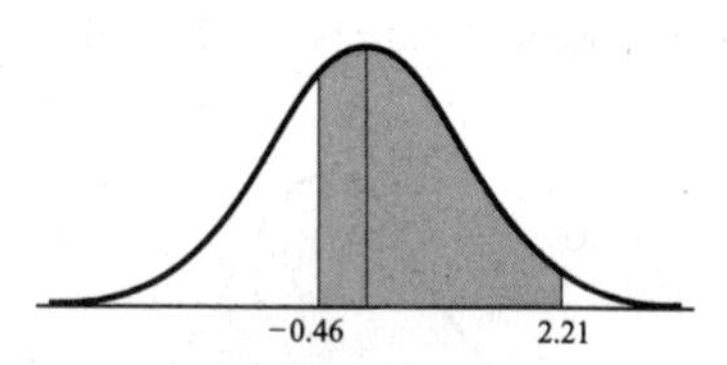

Fig. 4-5

(c) Required area = (area between $z = -0.46$ and $z = 0$)
+ (area between $z = 0$ and $z = 2.21$)
= (area between $z = 0$ and $z = 0.46$)
+ (area between $z = 0$ and $z = 2.21$)
= 0.1772 + 0.4864 = 0.6636

The area, 0.6636, represents the probability that Z is between -0.46 and 2.21 (Fig. 4-5). Therefore,

$$P(-0.46 \leq Z \leq 2.21) = \frac{1}{\sqrt{2\pi}}\int_{-0.46}^{2.21} e^{-u^2/2}\,du$$
$$= \frac{1}{\sqrt{2\pi}}\int_{-0.46}^{0} e^{-u^2/2}\,du + \frac{1}{\sqrt{2\pi}}\int_{0}^{2.21} e^{-u^2/2}\,du$$
$$= \frac{1}{\sqrt{2\pi}}\int_{0}^{0.46} e^{-u^2/2}\,du + \frac{1}{\sqrt{2\pi}}\int_{0}^{2.21} e^{-u^2/2}\,du = 0.1772 + 0.4864$$
$$= 0.6636$$

(d) Required area (Fig. 4-6) = (area between $z = 0$ and $z = 1.94$)
− (area between $z = 0$ and $z = 0.81$)
= 0.4738 − 0.2910 = 0.1828

This is the same as $P(0.81 \leq Z \leq 1.94)$.

(e) Required area (Fig. 4-7) = (area between $z = -1.28$ and $z = 0$)
+ (area to right of $z = 0$)
= 0.3997 + 0.5 = 0.8997

This is the same as $P(Z \geq -1.28)$.

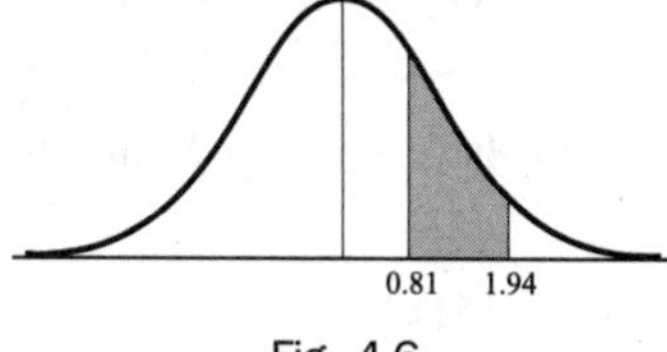

Fig. 4-6

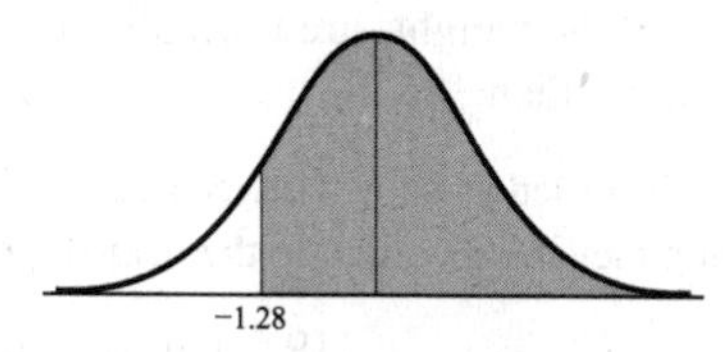

Fig. 4-7

4.13. If "area" refers to that under the standard normal curve, find the value or values of z such that (a) area between 0 and z is 0.3770, (b) area to left of z is 0.8621, (c) area between -1.5 and z is 0.0217.

(a) In the table in Appendix C the entry 0.3770 is located to the right of the row marked 1.1 and under the column marked 6. Then the required $z = 1.16$.

By symmetry, $z = -1.16$ is another value of z. Therefore, $z = \pm 1.16$ (Fig. 4-8). The problem is equivalent to solving for z the equation

$$\frac{1}{\sqrt{2\pi}}\int_0^z e^{-u^2/2}\,du = 0.3770$$

(b) Since the area is greater than 0.5, z must be positive.

Area between 0 and z is $0.8621 - 0.5 = 0.3621$, from which $z = 1.09$ (Fig. 4-9).

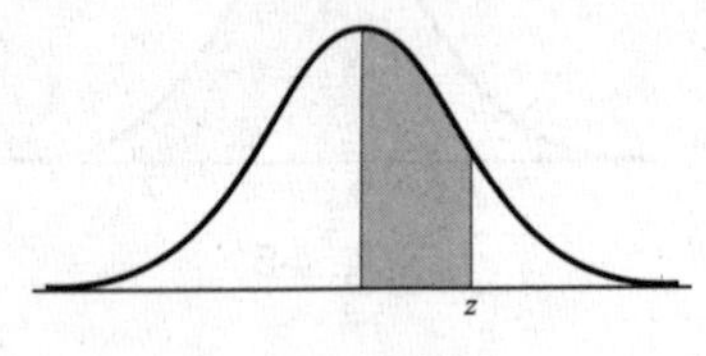

Fig. 4-8

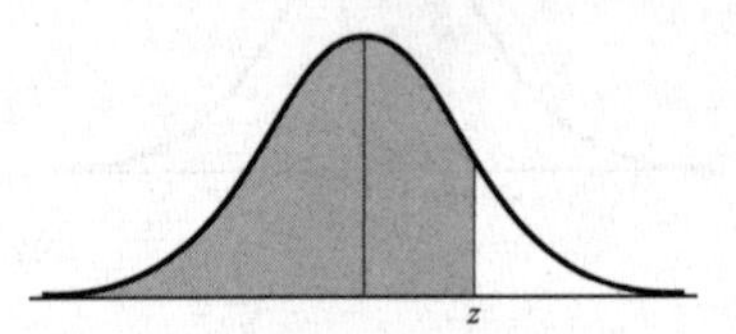

Fig. 4-9

(c) If z were positive, the area would be greater than the area between -1.5 and 0, which is 0.4332; hence z must be negative.

Case 1 z is negative but to the right of -1.5 (Fig. 4-10).

$$\text{Area between } -1.5 \text{ and } z = (\text{area between } -1.5 \text{ and } 0) - (\text{area between } 0 \text{ and } z)$$

$$0.0217 = 0.4332 - (\text{area between } 0 \text{ and } z)$$

Then the area between 0 and z is $0.4332 - 0.0217 = 0.4115$ from which $z = -1.35$.

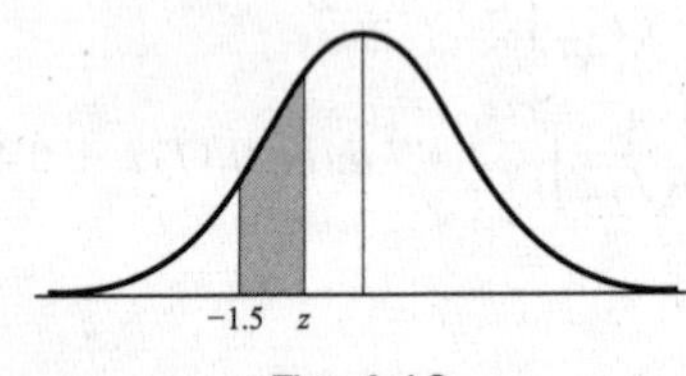

Fig. 4.10

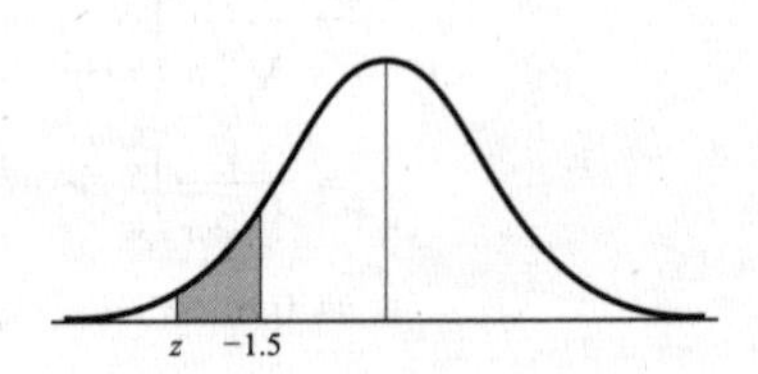

Fig. 4.11

Case 2 z is negative but to the left of -1.5 (Fig. 4-11).

$$\text{Area between } z \text{ and } -1.5 = (\text{area between } z \text{ and } 0) - (\text{area between } -1.5 \text{ and } 0)$$

$$0.0217 = (\text{area between } 0 \text{ and } z) - 0.4332$$

Then the area between 0 and z is $0.0217 + 0.4332 = 0.4549$ and $z = -1.694$ by using linear interpolation; or, with slightly less precision, $z = -1.69$.

4.14. The mean weight of 500 male students at a certain college is 151 lb and the standard deviation is 15 lb. Assuming that the weights are normally distributed, find how many students weigh (a) between 120 and 155 lb, (b) more than 185 lb.

(a) Weights recorded as being between 120 and 155 lb can actually have any value from 119.5 to 155.5 lb, assuming they are recorded to the nearest pound (Fig. 4-12).

$$119.5 \text{ lb in standard units} = (119.5 - 151)/15 = -2.10$$

$$155.5 \text{ lb in standard units} = (155.5 - 151)/15 = 0.30$$

$$\begin{aligned}\text{Required proportion of students} &= (\text{area between } z = -2.10 \text{ and } z = 0.30) \\ &= (\text{area between } z = -2.10 \text{ and } z = 0) \\ &\quad + (\text{area between } z = 0 \text{ and } z = 0.30) \\ &= 0.4821 + 0.1179 = 0.6000\end{aligned}$$

Then the number of students weighing between 120 and 155 lb is $500(0.6000) = 300$

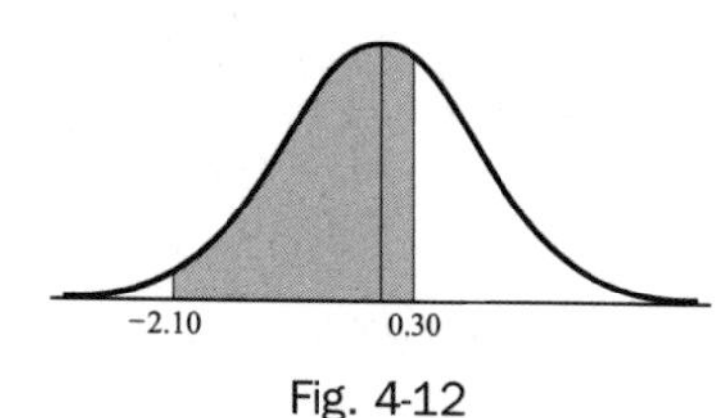

Fig. 4-12

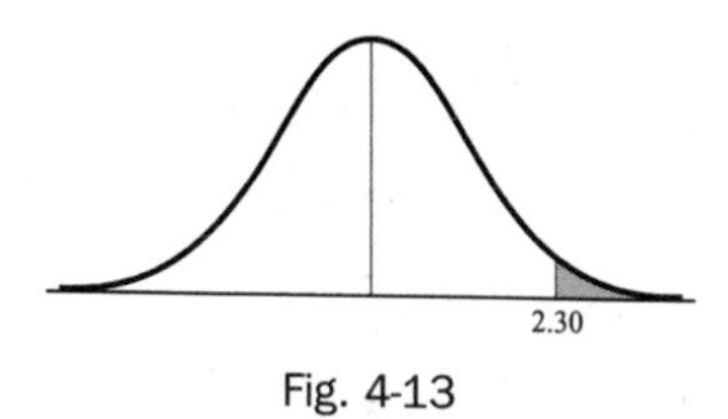

Fig. 4-13

(b) Students weighing more than 185 lb must weigh at least 185.5 lb (Fig. 4-13).

$$185.5 \text{ lb in standard units} = (185.5 - 151)/15 = 2.30$$

$$\begin{aligned}&\text{Required proportion of students} \\ &= (\text{area to right of } z = 2.30) \\ &= (\text{area to right of } z = 0) \\ &\quad - (\text{area between } z = 0 \text{ and } z = 2.30) \\ &= 0.5 - 0.4893 = 0.0107\end{aligned}$$

Then the number of students weighing more than 185 lb is $500(0.0107) = 5$.

If W denotes the weight of a student chosen at random, we can summarize the above results in terms of probability by writing

$$P(119.5 \le W \le 155.5) = 0.6000 \qquad P(W \ge 185.5) = 0.0107$$

4.15. The mean inside diameter of a sample of 200 washers produced by a machine is 0.502 inches and the standard deviation is 0.005 inches. The purpose for which these washers are intended allows a maximum tolerance in the diameter of 0.496 to 0.508 inches, otherwise the washers are considered defective. Determine the percentage of defective washers produced by the machine, assuming the diameters are normally distributed.

$$0.496 \text{ in standard units} = (0.496 - 0.502)/0.005 = -1.2$$

$$0.508 \text{ in standard units} = (0.508 - 0.502)/0.005 = 1.2$$

Proportion of nondefective washers

$$\begin{aligned}&= (\text{area under normal curve between } z = -1.2 \text{ and } z = 1.2) \\ &= (\text{twice the area between } z = 0 \text{ and } z = 1.2) \\ &= 2(0.3849) = 0.7698, \text{ or } 77\%\end{aligned}$$

Therefore, the percentage of defective washers is $100\% - 77\% = 23\%$ (Fig. 4-14).

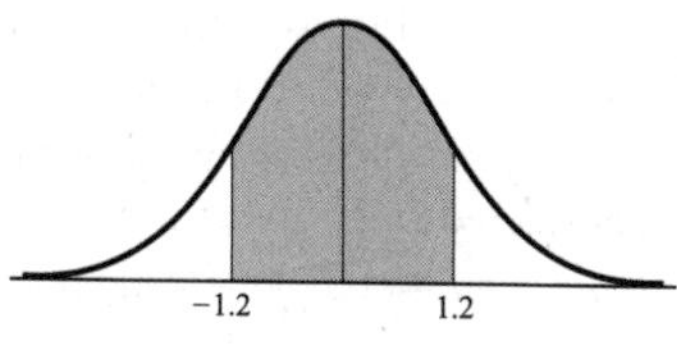

Fig. 4-14

Note that if we think of the interval 0.496 to 0.508 inches as actually representing diameters of from 0.4955 to 0.5085 inches, the above result is modified slightly. To two significant figures, however, the results are the same.

4.16. Find the moment generating function for the general normal distribution.

We have

$$M(t) = E(e^{tX}) = \frac{1}{\sigma\sqrt{2\pi}}\int_{-\infty}^{\infty} e^{tx}e^{-(x-\mu)^2/2\sigma^2}\,dx$$

Letting $(x - \mu)/\sigma = v$ in the integral so that $x = \mu + \sigma v$, $dx = \sigma\,dv$, we have

$$M(t) = \frac{1}{\sqrt{2\pi}}\int_{-\infty}^{\infty} e^{\mu t+\sigma vt-(v^2/2)}\,dv = \frac{e^{\mu t+(\sigma^2t^2/2)}}{\sqrt{2\pi}}\int_{-\infty}^{\infty} e^{-(v-\sigma t)^2/2}\,dv$$

Now letting $v - \sigma t = w$, we find that

$$M(t) = e^{\mu t+(\sigma^2t^2/2)}\left(\frac{1}{\sqrt{2\pi}}\int_{-\infty}^{\infty} e^{-w^2/2}\,dw\right) = e^{\mu t+(\sigma^2t^2/2)}$$

Normal approximation to binomial distribution

4.17. Find the probability of getting between 3 and 6 heads inclusive in 10 tosses of a fair coin by using (a) the binomial distribution, (b) the normal approximation to the binomial distribution.

(a) Let X be the random variable giving the number of heads that will turn up in 10 tosses (Fig. 4-15). Then

$$P(X = 3) = \binom{10}{3}\left(\frac{1}{2}\right)^3\left(\frac{1}{2}\right)^7 = \frac{15}{128} \qquad P(X = 4) = \binom{10}{4}\left(\frac{1}{2}\right)^4\left(\frac{1}{2}\right)^6 = \frac{105}{512}$$

$$P(X = 5) = \binom{10}{5}\left(\frac{1}{2}\right)^5\left(\frac{1}{2}\right)^5 = \frac{63}{256} \qquad P(X = 6) = \binom{10}{6}\left(\frac{1}{2}\right)^6\left(\frac{1}{2}\right)^4 = \frac{105}{512}$$

Then the required probability is

$$P(3 \le X \le 6) = \frac{15}{128} + \frac{105}{512} + \frac{63}{256} + \frac{105}{512} = \frac{99}{128} = 0.7734$$

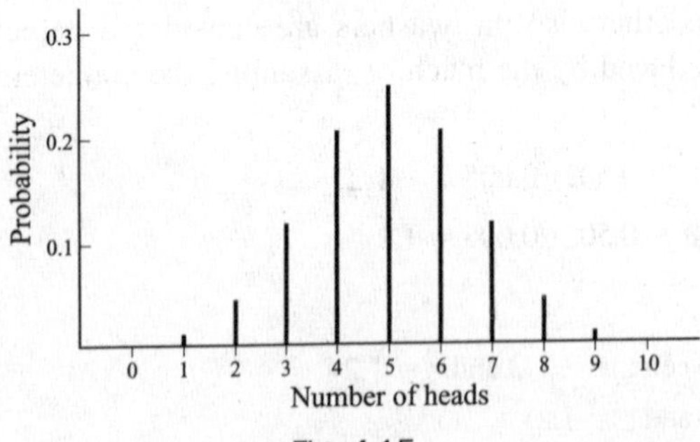

Fig. 4-15

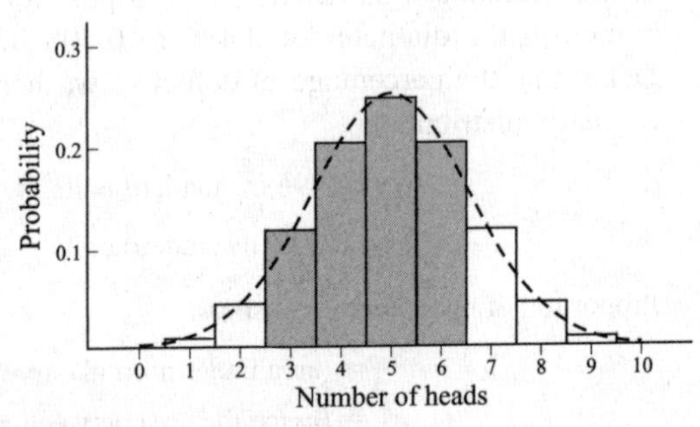

Fig. 4-16

(b) The probability distribution for the number of heads that will turn up in 10 tosses of the coin is shown graphically in Figures 4-15 and 4-16, where Fig. 4-16 treats the data as if they were continuous. The required probability is the sum of the areas of the shaded rectangles in Fig. 4-16 and can be approximated by the area under the corresponding normal curve, shown dashed. Treating the data as continuous, it follows that 3 to 6 heads can be considered as 2.5 to 6.5 heads. Also, the mean and variance for the binomial distribution are given by $\mu = np = 10(\frac{1}{2}) = 5$ and $\sigma = \sqrt{npq} = \sqrt{(10)(\frac{1}{2})(\frac{1}{2})} = 1.58$.

Now

$$2.5 \text{ in standard units} = \frac{2.5 - 5}{1.58} = -1.58$$

$$6.5 \text{ in standard units} = \frac{6.5 - 5}{1.58} = 0.95$$

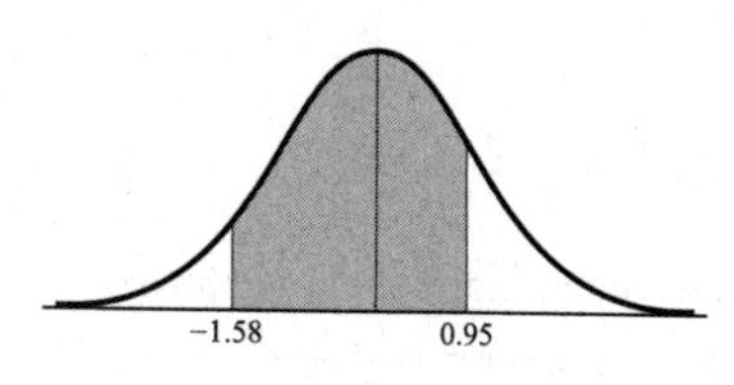

Fig. 4-17

$$\begin{aligned}\text{Required probability (Fig. 4-17)} &= (\text{area between } z = -1.58 \text{ and } z = 0.95)\\ &= (\text{area between } z = -1.58 \text{ and } z = 0)\\ &\quad + (\text{area between } z = 0 \text{ and } z = 0.95)\\ &= 0.4429 + 0.3289 = 0.7718\end{aligned}$$

which compares very well with the true value 0.7734 obtained in part (a). The accuracy is even better for larger values of n.

4.18. A fair coin is tossed 500 times. Find the probability that the number of heads will not differ from 250 by (a) more than 10, (b) more than 30.

$$\mu = np = (500)\left(\frac{1}{2}\right) = 250 \qquad \sigma = \sqrt{npq} = \sqrt{(500)\left(\frac{1}{2}\right)\left(\frac{1}{2}\right)} = 11.18$$

(a) We require the probability that the number of heads will lie between 240 and 260, or considering the data as continuous, between 239.5 and 260.5.

$$239.5 \text{ in standard units } = \frac{239.5 - 250}{11.18} = -0.94 \qquad 260.5 \text{ in standard units} = 0.94$$

$$\begin{aligned}\text{Required probability} &= (\text{area under normal curve between } z = -0.94 \text{ and } z = 0.94)\\ &= (\text{twice area between } z = 0 \text{ and } z = 0.94) = 2(0.3264) = 0.6528\end{aligned}$$

(b) We require the probability that the number of heads will lie between 220 and 280 or, considering the data as continuous, between 219.5 and 280.5.

$$219.5 \text{ in standard units } = \frac{219.5 - 250}{11.18} = -2.73 \qquad 280.5 \text{ in standard units} = 2.73$$

$$\begin{aligned}\text{Required probability} &= (\text{twice area under normal curve between } z = 0 \text{ and } z = 2.73)\\ &= 2(0.4968) = 0.9936\end{aligned}$$

It follows that we can be very confident that the number of heads will not differ from that expected (250) by more than 30. Therefore, if it turned out that the *actual* number of heads was 280, we would strongly believe that the coin was not fair, i.e., it was *loaded.*

4.19. A die is tossed 120 times. Find the probability that the face 4 will turn up (a) 18 times or less, (b) 14 times or less, assuming the die is fair.

The face 4 has probability $p = \frac{1}{6}$ of turning up and probability $q = \frac{5}{6}$ of not turning up.

(*a*) We want the probability of the number of 4s being between 0 and 18. This is given exactly by

$$\binom{120}{18}\left(\frac{1}{2}\right)^{18}\left(\frac{5}{6}\right)^{102} + \binom{120}{17}\left(\frac{1}{6}\right)^{17}\left(\frac{5}{6}\right)^{103} + \cdots + \binom{120}{0}\left(\frac{1}{6}\right)^{0}\left(\frac{5}{6}\right)^{120}$$

but since the labor involved in the computation is overwhelming, we use the normal approximation.

Considering the data as continuous, it follows that 0 to 18 4s can be treated as −0.5 to 18.5 4s. Also,

$$\mu = np = 120\left(\frac{1}{6}\right) = 20 \qquad \text{and} \qquad \sigma = \sqrt{npq} = \sqrt{(120)\left(\frac{1}{6}\right)\left(\frac{5}{6}\right)} = 4.08$$

Then

$$-0.5 \text{ in standard units} = \frac{-0.5 - 20}{4.08} = -5.02. \qquad 18.5 \text{ in standard units} = -0.37$$

$$\begin{aligned}\text{Required probability} &= (\text{area under normal curve between } z = -5.02 \text{ and } z = -0.37) \\ &= (\text{area between } z = 0 \text{ and } z = -5.02) \\ &\quad - (\text{area between } z = 0 \text{ and } z = -0.37) \\ &= 0.5 - 0.1443 = 0.3557\end{aligned}$$

(b) We proceed as in part (a), replacing 18 by 14. Then

$$-0.5 \text{ in standard units} = -5.02 \qquad 14.5 \text{ in standard units} = \frac{14.5 - 20}{4.08} = -1.35$$

$$\begin{aligned}\text{Required probability} &= (\text{area under normal curve between } z = -5.02 \text{ and } z = -1.35) \\ &= (\text{area between } z = 0 \text{ and } z = -5.02) \\ &\quad - (\text{area between } z = 0 \text{ and } z = -1.35) \\ &= 0.5 - 0.4115 = 0.0885\end{aligned}$$

It follows that if we were to take repeated samples of 120 tosses of a die, a 4 should turn up 14 times or less in about one-tenth of these samples.

The Poisson distribution

4.20. Establish the validity of the Poisson approximation to the binomial distribution.

If X is binomially distributed, then

$$(1) \qquad P(X = x) = \binom{n}{x} p^x q^{n-x}$$

where $E(X) = np$. Let $\lambda = np$ so that $p = \lambda/n$. Then (1) becomes

$$\begin{aligned}P(X = x) &= \binom{n}{x}\left(\frac{\lambda}{n}\right)^x\left(1 - \frac{\lambda}{n}\right)^{n-x} \\ &= \frac{n(n-1)(n-2)\cdots(n-x+1)}{x!n^x}\lambda^x\left(1 - \frac{\lambda}{n}\right)^{n-x} \\ &= \frac{\left(1 - \frac{1}{n}\right)\left(1 - \frac{2}{n}\right)\cdots\left(1 - \frac{x-1}{n}\right)}{x!}\lambda^x\left(1 - \frac{\lambda}{n}\right)^{n-x}\end{aligned}$$

Now as $n \to \infty$,

$$\left(1 - \frac{1}{n}\right)\left(1 - \frac{2}{n}\right)\cdots\left(1 - \frac{x-1}{n}\right) \to 1$$

while

$$\left(1 - \frac{\lambda}{n}\right)^{n-x} = \left(1 - \frac{\lambda}{n}\right)^n\left(1 - \frac{\lambda}{n}\right)^{-x} \to (e^{-\lambda})(1) = e^{-\lambda}$$

using the well-known result from calculus that

$$\lim_{n\to\infty}\left(1 + \frac{u}{n}\right)^n = e^u$$

It follows that when $n \to \infty$ but λ stays fixed (i.e., $p \to 0$),

$$(2) \qquad P(X = x) \to \frac{\lambda^x e^{-\lambda}}{x!}$$

which is the Poisson distribution.

Another method

The moment generating function for the binomial distribution is

(3) $$(q + pe^t)^n = (1 - p + pe^t)^n = [1 + p(e^t - 1)]^n$$

If $\lambda = np$ so that $p = \lambda/n$, this becomes

(4) $$\left[1 + \frac{\lambda(e^t - 1)}{n}\right]^n$$

As $n \to \infty$ this approaches

(5) $$e^{\lambda(e^t - 1)}$$

which is the moment generating function of the Poisson distribution. The required result then follows on using Theorem 3-10, page 77.

4.21. Verify that the limiting function (2) of Problem 4.20 is actually a probability function.

First, we see that $P(X = x) > 0$ for $x = 0, 1, \ldots$, given that $\lambda > 0$. Second, we have

$$\sum_{x=0}^{\infty} P(X = x) = \sum_{x=0}^{\infty} \frac{\lambda^x e^{-\lambda}}{x!} = e^{-\lambda} \sum_{x=0}^{\infty} \frac{\lambda^x}{x!} = e^{-\lambda} \cdot e^{\lambda} = 1$$

and the verification is complete.

4.22. Ten percent of the tools produced in a certain manufacturing process turn out to be defective. Find the probability that in a sample of 10 tools chosen at random, exactly 2 will be defective, by using (a) the binomial distribution, (b) the Poisson approximation to the binomial distribution.

(a) The probability of a defective tool is $p = 0.1$. Let X denote the number of defective tools out of 10 chosen. Then, according to the binomial distribution,

$$P(X = 2) = \binom{10}{2}(0.1)^2(0.9)^8 = 0.1937 \text{ or } 0.19$$

(b) We have $\lambda = np = (10)(0.1) = 1$. Then, according to the Poisson distribution,

$$P(X = x) = \frac{\lambda^x e^{-\lambda}}{x!} \quad \text{or} \quad P(X = 2) = \frac{(1)^2 e^{-1}}{2!} = 0.1839 \text{ or } 0.18$$

In general, the approximation is good if $p \le 0.1$ and $\lambda = np \le 5$.

4.23. If the probability that an individual will suffer a bad reaction from injection of a given serum is 0.001, determine the probability that out of 2000 individuals, (a) exactly 3, (b) more than 2, individuals will suffer a bad reaction.

Let X denote the number of individuals suffering a bad reaction. X is Bernoulli distributed, but since bad reactions are assumed to be rare events, we can suppose that X is Poisson distributed, i.e.,

$$P(X = x) = \frac{\lambda^x e^{-\lambda}}{x!} \quad \text{where } \lambda = np = (2000)(0.001) = 2$$

(a) $$P(X = 3) = \frac{2^3 e^{-2}}{3!} = 0.180$$

(b) $$\begin{aligned} P(X > 2) &= 1 - [P(X = 0) + P(X = 1) + P(X = 2)] \\ &= 1 - \left[\frac{2^0 e^{-2}}{0!} + \frac{2^1 e^{-2}}{1!} + \frac{2^2 e^{-2}}{2!}\right] \\ &= 1 - 5e^{-2} = 0.323 \end{aligned}$$

An exact evaluation of the probabilities using the binomial distribution would require much more labor.

The central limit theorem

4.24. Verify the central limit theorem for a random variable X that is binomially distributed, and thereby establish the validity of the normal approximation to the binomial distribution.

The standardized variable for X is $X^* = (X - np)/\sqrt{npq}$, and the moment generating function for X^* is

$$\begin{aligned}
E(e^{tX^*}) &= E(e^{t(X-np)/\sqrt{npq}}) \\
&= e^{-tnp/\sqrt{npq}}\,E(e^{tX/\sqrt{npq}}) \\
&= e^{-tnp/\sqrt{npq}}\sum_{x=0}^{n} e^{tx/\sqrt{npq}}\binom{n}{x}p^x q^{n-x} \\
&= e^{-tnp/\sqrt{npq}}\sum_{x=0}^{n}\binom{n}{x}(pe^{t/\sqrt{npq}})^x q^{n-x} \\
&= e^{-tnp/\sqrt{npq}}(q + pe^{t/\sqrt{npq}})^n \\
&= [e^{-tp/\sqrt{npq}}(q + pe^{t/\sqrt{npq}})]^n \\
&= (qe^{-tp/\sqrt{npq}} + pe^{tq/\sqrt{npq}})^n
\end{aligned}$$

Using the expansion

$$e^u = 1 + u + \frac{u^2}{2!} + \frac{u^3}{3!} + \cdots$$

we find

$$\begin{aligned}
qe^{-tp/\sqrt{npq}} + pe^{tq/\sqrt{npq}} &= q\left(1 - \frac{tp}{\sqrt{npq}} + \frac{t^2p^2}{2npq} + \cdots\right) \\
&\quad + p\left(1 + \frac{tq}{\sqrt{npq}} + \frac{t^2q^2}{2npq} + \cdots\right) \\
&= q + p + \frac{pq(p+q)t^2}{2npq} + \cdots \\
&= 1 + \frac{t^2}{2n} + \cdots
\end{aligned}$$

Therefore,

$$E(e^{tX^*}) = \left(1 + \frac{t^2}{2n} + \cdots\right)^n$$

But as $n \to \infty$, the right-hand side approaches $e^{t^2/2}$, which is the moment generating function for the standard normal distribution. Therefore, the required result follows by Theorem 3-10, page 77.

4.25. Prove the central limit theorem (Theorem 4-2, page 112).

For $n = 1, 2, \ldots$, we have $S_n = X_1 + X_2 + \cdots + X_n$. Now $X_1, X_2, \ldots, X_n$ each have mean μ and variance σ^2. Thus

$$E(S_n) = E(X_1) + E(X_2) + \cdots + E(X_n) = n\mu$$

and, because the X_k are independent,

$$\text{Var}(S_n) = \text{Var}(X_1) + \text{Var}(X_2) + \cdots + \text{Var}(X_n) = n\sigma^2$$

It follows that the standardized random variable corresponding to S_n is

$$S_n^* = \frac{S_n - n\mu}{\sigma\sqrt{n}}$$

The moment generating function for S_n^* is

$$\begin{aligned}
E(e^{tS_n^*}) &= E[e^{t(S_n - n\mu)/\sigma\sqrt{n}}] \\
&= E[e^{t(X_1-\mu)/\sigma\sqrt{n}}e^{t(X_2-\mu)/\sigma\sqrt{n}} \cdots e^{t(X_n-\mu)/\sigma\sqrt{n}}] \\
&= E[e^{t(X_1-\mu)/\sigma\sqrt{n}}] \cdot E[e^{t(X_2-\mu)/\sigma\sqrt{n}}] \cdots E[e^{t(X_n-\mu)/\sigma\sqrt{n}}] \\
&= \{E[e^{t(X_1-\mu)/\sigma\sqrt{n}}]\}^n
\end{aligned}$$

where, in the last two steps, we have respectively used the facts that the X_k are independent and are identically distributed. Now, by a Taylor series expansion,

$$E[e^{t(X_1-\mu)/\sigma\sqrt{n}}] = E\left[1 + \frac{t(X_1 - \mu)}{\sigma\sqrt{n}} + \frac{t^2(X_1 - \mu)^2}{2\sigma^2 n} + \cdots\right]$$

$$= E(1) + \frac{t}{\sigma\sqrt{n}} E(X_1 - \mu) + \frac{t^2}{2\sigma^2 n} E[(X_1 - \mu)^2] + \cdots$$

$$= 1 + \frac{t}{\sigma\sqrt{n}}(0) + \frac{t^2}{2\sigma^2 n}(\sigma^2) + \cdots = 1 + \frac{t^2}{2n} + \cdots$$

so that

$$E(e^{tS_n^*}) = \left(1 + \frac{t^2}{2n} + \cdots\right)^n$$

But the limit of this as $n \to \infty$ is $e^{t^2/2}$, which is the moment generating function of the standardized normal distribution. Hence, by Theorem 3-10, page 80, the required result follows.

Multinomial distribution

4.26. A box contains 5 red balls, 4 white balls, and 3 blue balls. A ball is selected at random from the box, its color is noted, and then the ball is replaced. Find the probability that out of 6 balls selected in this manner, 3 are red, 2 are white, and 1 is blue.

Method 1 (by formula)

$$P(\text{red at any drawing}) = \frac{5}{12} \qquad P(\text{white at any drawing}) = \frac{4}{12}$$

$$P(\text{blue at any drawing}) = \frac{3}{12}$$

Then

$$P(3\text{ red, 2 white, 1 blue}) = \frac{6!}{3!2!1!}\left(\frac{5}{12}\right)^3\left(\frac{4}{12}\right)^2\left(\frac{3}{12}\right)^1 = \frac{625}{5184}$$

Method 2

The probability of choosing any red ball is 5/12. Then the probability of choosing 3 red balls is $(5/12)^3$. Similarly, the probability of choosing 2 white balls is $(4/12)^2$, and of choosing 1 blue ball, $(3/12)^1$. Therefore, the probability of choosing 3 red, 2 white, and 1 blue in that order is

$$\left(\frac{5}{12}\right)^3\left(\frac{4}{12}\right)^2\left(\frac{3}{12}\right)^1$$

But the same selection can be achieved in various other orders, and the number of these different ways is

$$\frac{6!}{3!2!1!}$$

as shown in Chapter 1. Then the required probability is

$$\frac{6!}{3!2!1!}\left(\frac{5}{12}\right)^3\left(\frac{4}{12}\right)^2\left(\frac{3}{12}\right)^1$$

Method 3

The required probability is the term $p_r^3 p_w^2 p_b$ in the multinomial expansion of $(p_r + p_w + p_b)^6$ where $p_r = 5/12$, $p_w = 4/12$, $p_b = 3/12$. By actual expansion, the above result is obtained.

The hypergeometric distribution

4.27. A box contains 6 blue marbles and 4 red marbles. An experiment is performed in which a marble is chosen at random and its color observed, but the marble is not replaced. Find the probability that after 5 trials of the experiment, 3 blue marbles will have been chosen.

Method 1

The number of different ways of selecting 3 blue marbles out of 6 blue marbles is $\binom{6}{3}$. The number of different ways of selecting the remaining 2 marbles out of the 4 red marbles is $\binom{4}{2}$. Therefore, the number of different samples containing 3 blue marbles and 2 red marbles is $\binom{6}{3}\binom{4}{2}$.

Now the total number of different ways of selecting 5 marbles out of the 10 marbles (6 + 4) in the box is $\binom{10}{5}$. Therefore, the required probability is given by

$$\frac{\binom{6}{3}\binom{4}{2}}{\binom{10}{5}} = \frac{10}{21}$$

Method 2 (using formula)

We have $b = 6, r = 4, n = 5, x = 3$. Then by (*19*), page 113, the required probability is

$$P(X = 3) = \frac{\binom{6}{3}\binom{4}{2}}{\binom{10}{2}}$$

The uniform distribution

4.28. Show that the mean and variance of the uniform distribution (page 113) are given respectively by (a) $\mu = \frac{1}{2}(a + b)$, (b) $\sigma^2 = \frac{1}{12}(b - a)^2$.

(a)
$$\mu = E(X) = \int_a^b \frac{x\,dx}{b - a} = \frac{x^2}{2(b - a)}\bigg|_a^b = \frac{b^2 - a^2}{2(b - a)} = \frac{a + b}{2}$$

(b) We have

$$E(X^2) = \int_a^b \frac{x^2 dx}{b - a} = \frac{x^3}{3(b - a)}\bigg|_a^b = \frac{b^3 - a^3}{3(b - a)} = \frac{b^2 + ab + a^2}{3}$$

Then the variance is given by

$$\sigma^2 = E[(X - \mu)^2] = E(X^2) - \mu^2$$

$$= \frac{b^2 + ab + a^2}{3} - \left(\frac{a + b}{2}\right)^2 = \frac{1}{12}(b - a)^2$$

The Cauchy distribution

4.29. Show that (a) the moment generating function for a Cauchy distributed random variable X does not exist but that (b) the characteristic function does exist.

(a) The moment generating function of X is

$$E(e^{tX}) = \frac{a}{\pi}\int_{-\infty}^{\infty} \frac{e^{tx}}{x^2 + a^2}\,dx$$

which does not exist if t is real. This can be seen by noting, for example, that if $x \geq 0, t > 0$,

$$e^{tx} = 1 + tx + \frac{t^2x^2}{2!} + \cdots > \frac{t^2x^2}{2}$$

so that

$$\frac{a}{\pi}\int_{-\infty}^{\infty} \frac{e^{tx}}{x^2 + a^2}dx \geq \frac{at^2}{2\pi}\int_0^{\infty} \frac{x^2}{x^2 + a^2}dx$$

and the integral on the right diverges.

(b) The characteristic function of X is

$$E(e^{i\omega X}) = \frac{a}{\pi}\int_{-\infty}^{\infty} \frac{e^{i\omega x}}{x^2 + a^2}\,dx$$

$$= \frac{a}{\pi}\int_{-\infty}^{\infty} \frac{\cos \omega x}{x^2 + a^2}\,dx + \frac{ai}{\pi}\int_{-\infty}^{\infty} \frac{\sin \omega x}{x^2 + a^2}\,dx$$

$$= \frac{2a}{\pi}\int_0^{\infty} \frac{\cos \omega x}{x^2 + a^2}\,dx$$

where we have used the fact that the integrands in the next to last line are even and odd functions, respectively. The last integral can be shown to exist and to equal $e^{-a\omega}$.

4.30. Let Θ be a uniformly distributed random variable in the interval $-\frac{\pi}{2} \le \theta \le \frac{\pi}{2}$. Prove that $X = a \tan \Theta$, $a > 0$, is Cauchy distributed in $-\infty < x < \infty$.

The density function of Θ is

$$f(\theta) = \frac{1}{\pi} \qquad -\frac{\pi}{2} \le \theta \le \frac{\pi}{2}$$

Considering the transformation $x = a \tan \theta$, we have

$$\theta = \tan^{-1}\frac{x}{a} \qquad \text{and} \qquad \frac{d\theta}{dx} = \frac{a}{x^2 + a^2} > 0$$

Then by Theorem 2-3, page 42, the density function of X is given by

$$g(x) = f(\theta)\left|\frac{d\theta}{dx}\right| = \frac{1}{\pi}\frac{a}{x^2 + a^2}$$

which is the Cauchy distribution.

The gamma distribution

4.31. Show that the mean and variance of the gamma distribution are given by (a) $\mu = \alpha\beta$, (b) $\sigma^2 = \alpha\beta^2$.

(a)
$$\mu = \int_0^\infty x\left[\frac{x^{\alpha-1}e^{-x/\beta}}{\beta^\alpha\Gamma(\alpha)}\right]dx = \int_0^\infty \frac{x^\alpha e^{-x/\beta}}{\beta^\alpha\Gamma(\alpha)}dx$$

Letting $x/\beta = t$, we have

$$\mu = \frac{\beta^\alpha\beta}{\beta^\alpha\Gamma(\alpha)}\int_0^\infty t^\alpha e^{-t}\,dt = \frac{\beta}{\Gamma(\alpha)}\Gamma(\alpha + 1) = \alpha\beta$$

(b)
$$E(X^2) = \int_0^\infty x^2\left[\frac{x^{\alpha-1}e^{-x/\beta}}{\beta^\alpha\Gamma(\alpha)}\right]dx = \int_0^\infty \frac{x^{\alpha+1}e^{-x/\beta}}{\beta^\alpha\Gamma(\alpha)}dx$$

Letting $x/\beta = t$, we have

$$E(X^2) = \frac{\beta^{\alpha+1}\beta}{\beta^\alpha\Gamma(\alpha)}\int_0^\infty t^{\alpha+1}e^{-t}\,dt$$
$$= \frac{\beta^2}{\Gamma(\alpha)}\Gamma(\alpha + 2) = \beta^2(\alpha + 1)\alpha$$

since $\Gamma(\alpha + 2) = (\alpha + 1)\Gamma(\alpha + 1) = (\alpha + 1)\alpha\Gamma(\alpha)$. Therefore,

$$\sigma^2 = E(X^2) - \mu^2 = \beta^2(\alpha + 1)\alpha - (\alpha\beta)^2 = \alpha\beta^2$$

The beta distribution

4.32. Find the mean of the beta distribution.

$$\mu = E(X) = \frac{\Gamma(\alpha + \beta)}{\Gamma(\alpha)\Gamma(\beta)}\int_0^1 x[x^{\alpha-1}(1 - x)^{\beta-1}]\,dx$$
$$= \frac{\Gamma(\alpha + \beta)}{\Gamma(\alpha)\Gamma(\beta)}\int_0^1 x^\alpha(1 - x)^{\beta-1}\,dx$$
$$= \frac{\Gamma(\alpha + \beta)}{\Gamma(\alpha)\Gamma(\beta)}\frac{\Gamma(\alpha + 1)\Gamma(\beta)}{\Gamma(\alpha + 1 + \beta)}$$
$$= \frac{\Gamma(\alpha + \beta)}{\Gamma(\alpha)\Gamma(\beta)}\frac{\alpha\Gamma(\alpha)\Gamma(\beta)}{(\alpha + \beta)\Gamma(\alpha + \beta)} = \frac{\alpha}{\alpha + \beta}$$

4.33. Find the variance of the beta distribution.

The second moment about the origin is

$$
\begin{aligned}
E(X^2) &= \frac{\Gamma(\alpha+\beta)}{\Gamma(\alpha)\Gamma(\beta)}\int_0^1 x^2[x^{\alpha-1}(1-x)^{\beta-1}]\,dx \\
&= \frac{\Gamma(\alpha+\beta)}{\Gamma(\alpha)\Gamma(\beta)}\int_0^1 x^{\alpha+1}(1-x)^{\beta-1}\,dx \\
&= \frac{\Gamma(\alpha+\beta)}{\Gamma(\alpha)\Gamma(\beta)}\,\frac{\Gamma(\alpha+2)\Gamma(\beta)}{\Gamma(\alpha+2+\beta)} \\
&= \frac{\Gamma(\alpha+\beta)}{\Gamma(\alpha)\Gamma(\beta)}\,\frac{(\alpha+1)\alpha\Gamma(\alpha)\Gamma(\beta)}{(\alpha+\beta+1)(\alpha+\beta)\Gamma(\alpha+\beta)} \\
&= \frac{\alpha(\alpha+1)}{(\alpha+\beta)(\alpha+\beta+1)}
\end{aligned}
$$

Then using Problem 4.32, the variance is

$$
\sigma^2 = E(X^2) - [E(X)]^2 = \frac{\alpha(\alpha+1)}{(\alpha+\beta)(\alpha+\beta+1)} - \left(\frac{\alpha}{\alpha+\beta}\right)^2 = \frac{\alpha\beta}{(\alpha+\beta)^2(\alpha+\beta+1)}
$$

The chi-square distribution

4.34. Show that the moment generating function of a random variable X, which is chi-square distributed with v degrees of freedom, is $M(t) = (1-2t)^{-v/2}$.

$$
\begin{aligned}
M(t) = E(e^{tX}) &= \frac{1}{2^{v/2}\Gamma(v/2)}\int_0^\infty e^{tx}x^{(v-2)/2}e^{-x/2}\,dx \\
&= \frac{1}{2^{v/2}\Gamma(v/2)}\int_0^\infty x^{(v-2)/2}e^{-(1-2t)x/2}\,dx
\end{aligned}
$$

Letting $(1-2t)x/2 = u$ in the last integral, we find

$$
\begin{aligned}
M(t) &= \frac{1}{2^{v/2}\Gamma(v/2)}\int_0^\infty \left(\frac{2u}{1-2t}\right)^{(v-2)/2} e^{-u}\,\frac{2\,du}{1-2t} \\
&= \frac{(1-2t)^{-v/2}}{\Gamma(v/2)}\int_0^\infty u^{(v/2)-1}e^{-u}\,du = (1-2t)^{-v/2}
\end{aligned}
$$

4.35. Let X_1 and X_2 be independent random variables that are chi-square distributed with v_1 and v_2 degrees of freedom, respectively, (a) Show that the moment generating function of $Z = X_1 + X_2$ is $(1-2t)^{-(v_1+v_2)/2}$, thereby (b) show that Z is chi-square distributed with $v_1 + v_2$ degrees of freedom.

(a) The moment generating function of $Z = X_1 + X_2$ is

$$
M(t) = E[e^{t(X_1+X_2)}] = E(e^{tX_1})E(e^{tX_2}) = (1-2t)^{-v_1/2}(1-2t)^{-v_2/2} = (1-2t)^{-(v_1+v_2)/2}
$$

using Problem 4.34.

(b) It is seen from Problem 4.34 that a distribution whose moment generating function is $(1-2t)^{-(v_1+v_2)/2}$ is the chi-square distribution with $v_1 + v_2$ degrees of freedom. This must be the distribution of Z, by Theorem 3-10, page 77.

By generalizing the above results, we obtain a proof of Theorem 4-4, page 115.

4.36. Let X be a normally distributed random variable having mean 0 and variance 1. Show that X^2 is chi-square distributed with 1 degree of freedom.

We want to find the distribution of $Y = X^2$ given a standard normal distribution for X. Since the correspondence between X and Y is not one-one, we cannot apply Theorem 2-3 as it stands but must proceed as follows.

For $y < 0$, it is clear that $P(Y \le y) = 0$. For $y \ge 0$, we have

$$P(Y \le y) = P(X^2 \le y) = P(-\sqrt{y} \le X \le +\sqrt{y})$$

$$= \frac{1}{\sqrt{2\pi}} \int_{-\sqrt{y}}^{+\sqrt{y}} e^{-x^2/2}\,dx = \frac{2}{\sqrt{2\pi}} \int_{0}^{+\sqrt{y}} e^{-x^2/2}\,dx$$

where the last step uses the fact that the standard normal density function is even. Making the change of variable $x = +\sqrt{t}$ in the final integral, we obtain

$$P(Y \le y) = \frac{1}{\sqrt{2\pi}} \int_0^y t^{-1/2} e^{-t/2}\,dt$$

But this is a chi-square distribution with 1 degree of freedom, as is seen by putting $v = 1$ in (39), page 115, and using the fact that $\Gamma\left(\frac{1}{2}\right) = \sqrt{\pi}$.

4.37. Prove Theorem 4-3, page 115, for $v = 2$.

By Problem 4.36 we see that if X_1 and X_2 are normally distributed with mean 0 and variance 1, then X_1^2 and X_2^2 are chi square distributed with 1 degree of freedom each. Then, from Problem 4.35(b), we see that $Z = X_1^2 + X_2^2$ is chi square distributed with $1 + 1 = 2$ degrees of freedom if X_1 and X_2 are independent. The general result for all positive integers v follows in the same manner.

4.38. The graph of the chi-square distribution with 5 degrees of freedom is shown in Fig. 4-18. (See the remarks on notation on page 115.) Find the values χ_1^2, χ_2^2 for which

(a) the shaded area on the right = 0.05,

(b) the total shaded area = 0.05,

(c) the shaded area on the left = 0.10,

(d) the shaded area on the right = 0.01.

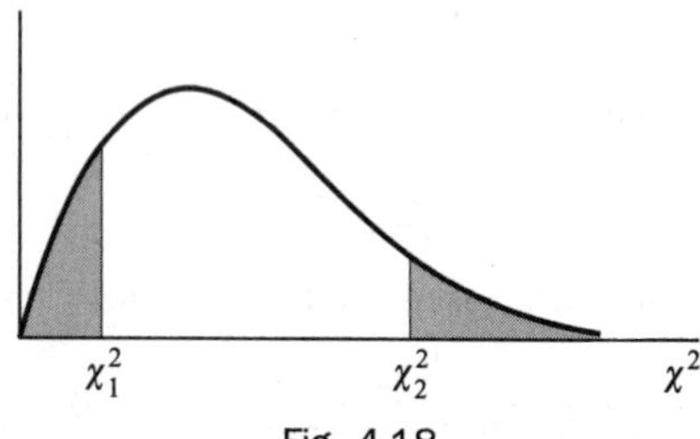

Fig. 4-18

(a) If the shaded area on the right is 0.05, then the area to the left of χ_2^2 is $(1 - 0.05) = 0.95$, and χ_2^2 represents the 95th percentile, $\chi_{0.95}^2$.

Referring to the table in Appendix E, proceed downward under column headed v until entry 5 is reached. Then proceed right to the column headed $\chi_{0.95}^2$. The result, 11.1, is the required value of χ^2.

(b) Since the distribution is not symmetric, there are many values for which the total shaded area = 0.05. For example, the right-hand shaded area could be 0.04 while the left-hand shaded area is 0.01. It is customary, however, unless otherwise specified, to choose the two areas equal. In this case, then, each area = 0.025.

If the shaded area on the right is 0.025, the area to the left of χ_2^2 is $1 - 0.025 = 0.975$ and χ_2^2 represents the 97.5th percentile, $\chi_{0.975}^2$, which from Appendix E is 12.8.

Similarly, if the shaded area on the left is 0.025, the area to the left of χ_1^2 is 0.025 and χ_1^2 represents the 2.5th percentile, $\chi_{0.025}^2$, which equals 0.831.

Therefore, the values are 0.831 and 12.8.

(c) If the shaded area on the left is 0.10, χ_1^2 represents the 10th percentile, $\chi_{0.10}^2$, which equals 1.61.

(d) If the shaded area on the right is 0.01, the area to the left of χ_2^2 is 0.99, and χ_2^2 represents the 99th percentile, $\chi_{0.99}^2$, which equals 15.1.

4.39. Find the values of χ^2 for which the area of the right-hand tail of the χ^2 distribution is 0.05, if the number of degrees of freedom v is equal to (a) 15, (b) 21, (c) 50.

Using the table in Appendix E, we find in the column headed $\chi^2_{0.95}$ the values: (a) 25.0 corresponding to $v = 15$; (b) 32.7 corresponding to $v = 21$; (c) 67.5 corresponding to $v = 50$.

4.40. Find the median value of χ^2 corresponding to (a) 9, (b) 28, (c) 40 degrees of freedom.

Using the table in Appendix E, we find in the column headed $\chi^2_{0.50}$ (since the median is the 50th percentile) the values: (a) 8.34 corresponding to $v = 9$; (b) 27.3 corresponding to $v = 28$; (c) 39.3 corresponding to $v = 40$.

It is of interest to note that the median values are very nearly equal to the number of degrees of freedom. In fact, for $v > 10$ the median values are equal to $v - 0.7$, as can be seen from the table.

4.41. Find $\chi^2_{0.95}$ for (a) $v = 50$, (b) $v = 100$ degrees of freedom.

For v greater than 30, we can use the fact that $(\sqrt{2\chi^2} - \sqrt{2v - 1})$ is very closely normally distributed with mean zero and variance one. Then if z_p is the (100p)th percentile of the standardized normal distribution, we can write, to a high degree of approximation,

$$\sqrt{2\chi^2_p} - \sqrt{2v - 1} = z_p \quad \text{or} \quad \sqrt{2\chi^2_p} = z_p + \sqrt{2v - 1}$$

from which

$$\chi^2_p = \tfrac{1}{2}(z_p + \sqrt{2v - 1})^2$$

(a) If $v = 50$, $\chi^2_{0.95} = \frac{1}{2}(z_{0.95} + \sqrt{2(50) - 1})^2 = \frac{1}{2}(1.64 + \sqrt{99})^2 = 69.2$, which agrees very well with the value 67.5 given in Appendix E.

(b) If $v = 100$, $\chi^2_{0.95} = \frac{1}{2}(z_{0.95} + \sqrt{2(100) - 1})^2 = \frac{1}{2}(1.64 + \sqrt{199})^2 = 124.0$ (actual value = 124.3).

Student's *t* distribution

4.42. Prove Theorem 4-6, page 116.

Since Y is normally distributed with mean 0 and variance 1, its density function is

$$\text{(1)} \qquad \frac{1}{\sqrt{2\pi}} e^{-y^2/2}$$

Since Z is chi-square distributed with v degrees of freedom, its density function is

$$\text{(2)} \qquad \frac{1}{2^{v/2}\Gamma(v/2)} z^{(v/2)-1} e^{-z/2} \qquad z > 0$$

Because Y and Z are independent, their joint density function is the product of (1) and (2), i.e.,

$$\frac{1}{\sqrt{2\pi}\, 2^{v/2}\Gamma(v/2)} z^{(v/2)-1} e^{-(y^2+z)/2}$$

for $-\infty < y < +\infty$, $z > 0$.

The distribution function of $T = Y/\sqrt{Z/v}$ is

$$F(x) = P(T \le x) = P(Y \le x\sqrt{Z/v})$$

$$= \frac{1}{\sqrt{2\pi} 2^{v/2}\Gamma(v/2)} \iint_{\mathcal{R}} z^{(v/2)-1} e^{-(y^2+z)/2}\, dy\, dz$$

where the integral is taken over the region $\mathcal{R}$ of the yz plane for which $y \le x\sqrt{z/v}$. We first fix z and integrate with respect to y from $-\infty$ to $x\sqrt{z/v}$. Then we integrate with respect to z from 0 to ∞. We therefore have

$$F(x) = \frac{1}{\sqrt{2\pi} 2^{v/2}\Gamma(v/2)} \int_{z=0}^{\infty} z^{(v/2)-1} e^{-z/2} \left[\int_{y=-\infty}^{x\sqrt{z/v}} e^{-y^2/2}\, dy \right] dz$$

Letting $y = u\sqrt{z/v}$ in the bracketed integral, we find

$$F(x) = \frac{1}{\sqrt{2\pi}\,2^{v/2}\Gamma(v/2)}\int_{z=0}^{\infty}\int_{u=-\infty}^{\infty} z^{(v/2)-1}e^{-z/2}\sqrt{z/v}\,e^{-u^2z/2v}\,du\,dz$$

$$= \frac{1}{\sqrt{2\pi v}\,2^{v/2}\Gamma(v/2)}\int_{u=-\infty}^{x}\left[\int_{z=0}^{\infty} z^{(v-1)/2}\,e^{-(z/2)[1+(u^2/v)]}\,dz\right]du$$

Letting $w = \frac{z}{2}\left(1 + \frac{u^2}{v}\right)$, this can then be written

$$F(x) = \frac{1}{\sqrt{2\pi v}\,2^{v/2}\Gamma(v/2)}\cdot 2^{(v+1)/2}\int_{u=-\infty}^{x}\left[\int_{w=0}^{\infty}\frac{w^{(v-1)/2}e^{-w}}{(1 + u^2/v)^{(v+1)/2}}\,dw\right]du$$

$$= \frac{\Gamma\left(\frac{v+1}{2}\right)}{\sqrt{\pi v}\,\Gamma\left(\frac{v}{2}\right)}\int_{u=-\infty}^{x}\frac{du}{(1 + u^2/v)^{(v+1)/2}}$$

as required.

4.43. The graph of Student's t distribution with 9 degrees of freedom is shown in Fig. 4-19. Find the value of t_1 for which

(a) the shaded area on the right = 0.05,

(b) the total shaded area = 0.05,

(c) the total unshaded area = 0.99,

(d) the shaded area on the left = 0.01,

(e) the area to the left of t_1 is 0.90.

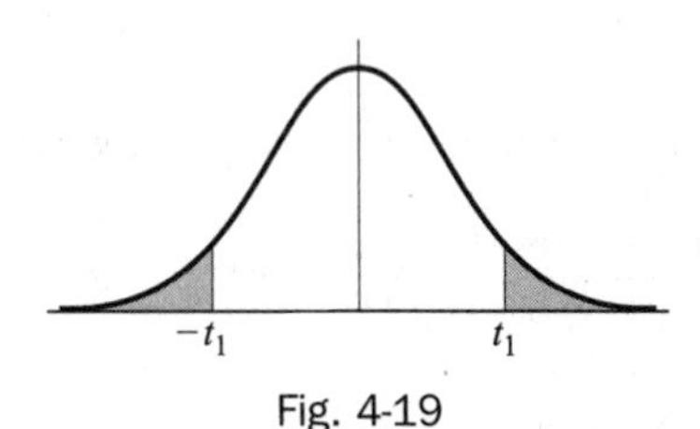

Fig. 4-19

(a) If the shaded area on the right is 0.05, then the area to the left of t_1 is $(1 - 0.05) = 0.95$, and t_1 represents the 95th percentile, $t_{0.95}$.

Referring to the table in Appendix D, proceed downward under the column headed v until entry 9 is reached. Then proceed right to the column headed $t_{0.95}$. The result 1.83 is the required value of t.

(b) If the total shaded area is 0.05, then the shaded area on the right is 0.025 by symmetry. Therefore, the area to the left of t_1 is $(1 - 0.025) = 0.975$, and t_1 represents the 97.5th percentile, $t_{0.975}$. From Appendix D, we find 2.26 as the required value of t.

(c) If the total unshaded area is 0.99, then the total shaded area is $(1 - 0.99) = 0.01$, and the shaded area to the right is $0.01/2 = 0.005$. From the table we find $t_{0.995} = 3.25$.

(d) If the shaded area on the left is 0.01, then by symmetry the shaded area on the right is 0.01. From the table, $t_{0.99} = 2.82$. Therefore, the value of t for which the shaded area on the left is 0.01 is -2.82.

(e) If the area to the left of t_1 is 0.90, then t_1 corresponds to the 90th percentile, $t_{0.90}$, which from the table equals 1.38.

4.44. Find the values of t for which the area of the right-hand tail of the t distribution is 0.05 if the number of degrees of freedom ν is equal to (a) 16, (b) 27, (c) 200.

Referring to Appendix D, we find in the column headed $t_{0.95}$ the values: (a) 1.75 corresponding to $\nu = 16$; (b) 1.70 corresponding to $\nu = 27$; (c) 1.645 corresponding to $\nu = 200$. (The latter is the value that would be obtained by using the normal curve. In Appendix D this value corresponds to the entry in the last row marked ∞.)

The *F* distribution

4.45. Prove Theorem 4-7.

The joint density function of V_1 and V_2 is given by

$$f(v_1, v_2) = \left(\frac{1}{2^{\nu_1/2}\Gamma(\nu_1/2)} v_1^{(\nu_1/2)-1} e^{-v_1/2}\right)\left(\frac{1}{2^{\nu_2/2}\Gamma(\nu_2/2)} v_2^{(\nu_2/2)-1} e^{-v_2/2}\right)$$

$$= \frac{1}{2^{(\nu_1+\nu_2)/2}\Gamma(\nu_1/2)\Gamma(\nu_2/2)} v_1^{(\nu_1/2)-1} v_2^{(\nu_2/2)-1} e^{-(v_1+v_2)/2}$$

if $v_1 > 0$, $v_2 > 0$ and 0 otherwise. Make the transformation

$$u = \frac{v_1/\nu_1}{v_2/\nu_2} = \frac{\nu_2 v_1}{\nu_1 v_2}, \qquad w = v_2 \qquad \text{or} \qquad v_1 = \frac{\nu_1 u w}{\nu_2} \qquad v_2 = w$$

Then the Jacobian is

$$\frac{\partial(v_1, v_2)}{\partial(u, w)} = \begin{vmatrix} \partial v_1/\partial u & \partial v_1/\partial w \\ \partial v_2/\partial u & \partial v_2/\partial w \end{vmatrix} = \begin{vmatrix} \nu_1 w/\nu_2 & \nu_1 u/\nu_2 \\ 0 & 1 \end{vmatrix} = \frac{\nu_1 w}{\nu_2}$$

Denoting the density as a function of u and w by $g(u, w)$, we thus have

$$g(u, w) = \frac{1}{2^{(\nu_1+\nu_2)/2}\Gamma(\nu_1/2)\Gamma(\nu_2/2)} \left(\frac{\nu_1 u w}{\nu_2}\right)^{(\nu_1/2)-1} w^{(\nu_2/2)-1} e^{-[1+(\nu_1 u/\nu_2)](w/2)} \frac{\nu_1 w}{\nu_2}$$

if $u > 0$, $w > 0$ and 0 otherwise.

The (marginal) density function of U can now be found by integrating with respect to w from 0 to ∞, i.e.,

$$h(u) = \frac{(\nu_1/\nu_2)^{\nu_1/2} u^{(\nu_1/2)-1}}{2^{(\nu_1+\nu_2)/2}\Gamma(\nu_1/2)\Gamma(\nu_2/2)} \int_0^\infty w^{[(\nu_1+\nu_2)/2]-1} e^{-[1+(\nu_1 u/\nu_2)](w/2)}\, dw$$

if $u > 0$ and 0 if $u \le 0$. But from 15, Appendix A,

$$\int_0^\infty w^{p-1} e^{-aw}\, dw = \frac{\Gamma(p)}{a^p}$$

Therefore, we have

$$h(u) = \frac{(\nu_1/\nu_2)^{\nu_1/2} u^{(\nu_1/2)-1} \Gamma\left(\frac{\nu_1+\nu_2}{2}\right)}{2^{(\nu_1+\nu_2)/2}\Gamma(\nu_1/2)\Gamma(\nu_2/2)\left[\frac{1}{2}\left(1 + \frac{\nu_1 u}{\nu_2}\right)\right]^{(\nu_1+\nu_2)/2}}$$

$$= \frac{\Gamma\left(\frac{\nu_1+\nu_2}{2}\right)}{\Gamma\left(\frac{\nu_1}{2}\right)\Gamma\left(\frac{\nu_2}{2}\right)} \nu_1^{\nu_1/2} \nu_2^{\nu_2/2} u^{(\nu_1/2)-1} (\nu_2 + \nu_1 u)^{-(\nu_1+\nu_2)/2}$$

if $u > 0$ and 0 if $u \le 0$, which is the required result.

4.46. Prove that the F distribution is unimodal at the value $\left(\frac{v_1 - 2}{v_1}\right)\left(\frac{v_2}{v_2 + 2}\right)$ if $v_1 > 2$.

The mode locates the maximum value of the density function. Apart from a constant, the density function of the F distribution is

$$u^{(v_1/2)-1}(v_2 + v_1 u)^{-(v_1+v_2)/2}$$

If this has a relative maximum, it will occur where the derivative is zero, i.e.,

$$\left(\frac{v_1}{2} - 1\right)u^{(v_1/2)-2}(v_2 + v_1 u)^{-(v_1+v_2)/2} - u^{(v_1/2)-1}v_1\left(\frac{v_1 + v_2}{2}\right)(v_2 + v_1 u)^{-[(v_1+v_2)/2]-1} = 0$$

Dividing by $u^{(v_1/2)-2}(v_2 + v_1 u)^{-[(v_1+v_2)/2]-1}$, $u \neq 0$, we find

$$\left(\frac{v_1}{2} - 1\right)(v_2 + v_1 u) - uv_1\left(\frac{v_1 + v_2}{2}\right) = 0 \qquad \text{or} \qquad u = \left(\frac{v_1 - 2}{v_1}\right)\left(\frac{v_2}{v_2 + 2}\right)$$

Using the second-derivative test, we can show that this actually gives the maximum.

4.47. Using the table for the F distribution in Appendix F, find (a) $F_{0.95,10,15}$, (b) $F_{0.99,15,9}$, (c) $F_{0.05,8,30}$, (d) $F_{0.01,15,9}$.

(a) From Appendix F, where $v_1 = 10$, $v_2 = 15$, we find $F_{0.95,10,15} = 2.54$.

(b) From Appendix F, where $v_1 = 15$, $v_2 = 9$, we find $F_{0.99,15,9} = 4.96$.

(c) By Theorem 4-8, page 117, $F_{0.05,8,30} = \dfrac{1}{F_{0.95,30,8}} = \dfrac{1}{3.08} = 0.325$.

(d) By Theorem 4-8, page 117, $F_{0.01,15,9} = \dfrac{1}{F_{0.99,9,15}} = \dfrac{1}{3.89} = 0.257$.

Relationships among F, χ^2, and t distributions

4.48. Verify that (a) $F_{0.95} = t^2_{0.975}$, (b) $F_{0.99} = t^2_{0.995}$.

(a) Compare the entries in the first column of the $F_{0.95}$ table in Appendix F with those in the t distribution under $t_{0.975}$. We see that

$$161 = (12.71)^2, \qquad 18.5 = (4.30)^2, \qquad 10.1 = (3.18)^2, \qquad 7.71 = (2.78)^2, \qquad \text{etc.}$$

(b) Compare the entries in the first column of the $F_{0.99}$ table in Appendix F with those in the t distribution under $t_{0.995}$. We see that

$$4050 = (63.66)^2, \qquad 98.5 = (9.92)^2, \qquad 34.1 = (5.84)^2, \qquad 21.2 = (4.60)^2, \qquad \text{etc.}$$

4.49. Prove Theorem 4-9, page 117, which can be briefly stated as

$$F_{1-p} = t^2_{1-(p/2)}$$

and therefore generalize the results of Problem 4.48.

Let $v_1 = 1$, $v_2 = v$ in the density function for the F distribution [(45), page 116]. Then

$$f(u) = \frac{\Gamma\left(\frac{v+1}{2}\right)}{\Gamma\left(\frac{1}{2}\right)\Gamma\left(\frac{v}{2}\right)} v^{v/2}u^{-1/2}(v + u)^{-(v+1)/2}$$

$$= \frac{\Gamma\left(\frac{v+1}{2}\right)}{\sqrt{\pi}\,\Gamma\left(\frac{v}{2}\right)} v^{v/2}u^{-1/2}v^{-(v+1)/2}\left(1 + \frac{u}{v}\right)^{-(v+1)/2}$$

$$= \frac{\Gamma\left(\frac{v+1}{2}\right)}{\sqrt{v\pi}\,\Gamma\left(\frac{v}{2}\right)} u^{-1/2}\left(1 + \frac{u}{v}\right)^{-(v+1)/2}$$

for $u > 0$, and $f(u) = 0$ for $u \le 0$. Now, by the definition of a percentile value, F_{1-p} is the number such that $P(U \le F_{1-p}) = 1 - p$. Therefore,

$$\frac{\Gamma\left(\frac{\nu+1}{2}\right)}{\sqrt{\nu\pi}\,\Gamma\left(\frac{\nu}{2}\right)}\int_0^{F_{1-p}} u^{-1/2}\left(1+\frac{u}{\nu}\right)^{-(\nu+1)/2} du = 1 - p$$

In the integral make the change of variable $t = +\sqrt{u}$:

$$2\frac{\Gamma\left(\frac{\nu+1}{2}\right)}{\sqrt{\nu\pi}\,\Gamma\left(\frac{\nu}{2}\right)}\int_0^{+\sqrt{F_{1-p}}} \left(1+\frac{t^2}{\nu}\right)^{-(\nu+1)/2} dt = 1 - p$$

Comparing with (42), page 115, we see that the left-hand side of the last equation equals

$$2 \cdot P(0 < T \le +\sqrt{F_{1-p}})$$

where T is a random variable having Student's t distribution with ν degrees of freedom. Therefore,

$$\begin{aligned}\frac{1-p}{2} &= P(0 < T \le +\sqrt{F_{1-p}})\\ &= P(T \le +\sqrt{F_{1-p}}) - P(T \le 0)\\ &= P(T \le +\sqrt{F_{1-p}}) - \tfrac{1}{2}\end{aligned}$$

where we have used the symmetry of the t distribution. Solving, we have

$$P(T \le +\sqrt{F_{1-p}}) = 1 - \frac{p}{2}$$

But, by definition, $t_{1-(p/2)}$ is the number such that

$$P(T \le t_{1-(p/2)}) = 1 - \frac{p}{2}$$

and this number is uniquely determined, since the density function of the t distribution is strictly positive. Therefore,

$$+\sqrt{F_{1-p}} = t_{1-(p/2)} \qquad \text{or} \qquad F_{1-p} = t^2_{1-(p/2)}$$

which was to be proved.

4.50. Verify Theorem 4-10, page 117, for (a) $p = 0.95$, (b) $p = 0.99$.

(a) Compare the entries in the last row of the $F_{0.95}$ table in Appendix F (corresponding to $\nu_2 = \infty$) with the entries under $\chi^2_{0.95}$ in Appendix E. Then we see that

$$3.84 = \frac{3.84}{1}, \quad 3.00 = \frac{5.99}{2}, \quad 2.60 = \frac{7.81}{3}, \quad 2.37 = \frac{9.49}{4}, \quad 2.21 = \frac{11.1}{5}, \quad \text{etc.}$$

which provides the required verification.

(b) Compare the entries in the last row of the $F_{0.99}$ table in Appendix F (corresponding to $\nu_2 = \infty$) with the entries under $\chi^2_{0.99}$ in Appendix E. Then we see that

$$6.63 = \frac{6.63}{1}, \quad 4.61 = \frac{9.21}{2}, \quad 3.78 = \frac{11.3}{3}, \quad 3.32 = \frac{13.3}{4}, \quad 3.02 = \frac{15.1}{5}, \quad \text{etc.}$$

which provides the required verification.

The general proof of Theorem 4-10 follows by letting $\nu_2 \to \infty$ in the F distribution on page 116.

The bivariate normal distribution

4.51. Suppose that X and Y are random variables whose joint density function is the bivariate normal distribution. Show that X and Y are independent if and only if their correlation coefficient is zero.

If the correlation coefficient $\rho = 0$, then the bivariate normal density function (49), page 117, becomes

$$f(x, y) = \left[\frac{1}{\sigma_1\sqrt{2\pi}} e^{-(x-\mu_1)^2/2\sigma_1^2}\right]\left[\frac{1}{\sigma_2\sqrt{2\pi}} e^{-(y-\mu_2)^2/2\sigma_2^2}\right]$$

and since this is a product of a function of x alone and a function of y alone for all values of x and y, it follows that X and Y are independent.

Conversely, if X and Y are independent, $f(x, y)$ given by (49) must for all values of x and y be the product of a function of x alone and a function of y alone. This is possible only if $\rho = 0$.

Miscellaneous distributions

4.52. Find the probability that in successive tosses of a fair die, a 3 will come up for the first time on the fifth toss.

Method 1

The probability of not getting a 3 on the first toss is 5/6. Similarly, the probability of not getting a 3 on the second toss is 5/6, etc. Then the probability of not getting a 3 on the first 4 tosses is $(5/6)(5/6)(5/6)(5/6) = (5/6)^4$. Therefore, since the probability of getting a 3 on the fifth toss is 1/6, the required probability is

$$\left(\frac{5}{6}\right)^4\left(\frac{1}{6}\right) = \frac{625}{7776}$$

Method 2 (using formula)

Using the geometric distribution, page 117, with $p = 1/6$, $q = 5/6$, $x = 5$, we see that the required probability is

$$\left(\frac{1}{6}\right)\left(\frac{5}{6}\right)^4 = \frac{625}{7776}$$

4.53. Verify the expressions given for (a) the mean, (b) the variance, of the Weibull distribution, page 118.

(a)

$$\begin{aligned}
\mu = E(X) &= \int_0^\infty abx^b e^{-ax^b}\,dx \\
&= \frac{ab}{a^{1/b}}\int_0^\infty \left(\frac{u}{a}\right) e^{-u} \frac{1}{b} u^{(1/b)-1}\,du \\
&= a^{-1/b}\int_0^\infty u^{1/b} e^{-u}\,du \\
&= a^{-1/b}\Gamma\left(1 + \frac{1}{b}\right)
\end{aligned}$$

where we have used the substitution $u = ax^b$ to evaluate the integral.

(b)

$$\begin{aligned}
E(X^2) &= \int_0^\infty abx^{b+1} e^{-ax^b}\,dx \\
&= \frac{ab}{a^{1/b}}\int_0^\infty \left(\frac{u}{a}\right)^{1+(1/b)} e^{-u} \frac{1}{b} u^{(1/b)-1}\,du \\
&= a^{-2/b}\int_0^\infty u^{2/b} e^{-u}\,du \\
&= a^{-2/b}\Gamma\left(1 + \frac{2}{b}\right)
\end{aligned}$$

Then

$$\begin{aligned}
\sigma^2 &= E[(X - \mu)^2] = E(X^2) - \mu^2 \\
&= a^{-2/b}\left[\Gamma\left(1 + \frac{2}{b}\right) - \Gamma^2\left(1 + \frac{1}{b}\right)\right]
\end{aligned}$$

Miscellaneous problems

4.54. The probability that an entering college student will graduate is 0.4. Determine the probability that out of 5 students (a) none, (b) 1, (c) at least 1, will graduate.

(a) $P(\text{none will graduate}) = {}_5C_0(0.4)^0(0.6)^5 = 0.07776$, or about 0.08

(b) $P(\text{1 will graduate}) = {}_5C_1(0.4)^1(0.6)^4 = 0.2592$, or about 0.26

(c) $P(\text{at least 1 will graduate}) = 1 - P(\text{none will graduate}) = 0.92224$, or about 0.92

4.55. What is the probability of getting a total of 9 (a) twice, (b) at least twice in 6 tosses of a pair of dice?

Each of the 6 ways in which the first die can fall can be associated with each of the 6 ways in which the second die can fall, so there are $6 \cdot 6 = 36$ ways in which both dice can fall. These are: 1 on the first die and 1 on the second die, 1 on the first die and 2 on the second die, etc., denoted by (1, 1), (1, 2), etc.

Of these 36 ways, all equally likely if the dice are fair, a total of 9 occurs in 4 cases: (3, 6), (4, 5), (5, 4), (6, 3). Then the probability of a total of 9 in a single toss of a pair of dice is $p = 4/36 = 1/9$, and the probability of not getting a total of 9 in a single toss of a pair of dice is $q = 1 - p = 8/9$.

(a) $$P(\text{two 9s in 6 tosses}) = {}_6C_2\left(\frac{1}{9}\right)^2\left(\frac{8}{9}\right)^{6-2} = \frac{61{,}440}{531{,}441}$$

(b) $$P(\text{at least two 9s}) = P(\text{two 9s}) + P(\text{three 9s}) + P(\text{four 9s}) + P(\text{five 9s}) + P(\text{six 9s})$$

$$= {}_6C_2\left(\frac{1}{9}\right)^2\left(\frac{8}{9}\right)^4 + {}_6C_3\left(\frac{1}{9}\right)^3\left(\frac{8}{9}\right)^3 + {}_6C_4\left(\frac{1}{9}\right)^4\left(\frac{8}{9}\right)^2 + {}_6C_5\left(\frac{1}{9}\right)^5\frac{8}{9} + {}_6C_6\left(\frac{1}{9}\right)^2$$

$$= \frac{61{,}440}{531{,}441} + \frac{10{,}240}{531{,}441} + \frac{960}{531{,}441} + \frac{48}{531{,}441} + \frac{1}{531{,}441} = \frac{72{,}689}{531{,}441}$$

Another method

$$P(\text{at least two 9s}) = 1 - P(\text{zero 9s}) - P(\text{one 9})$$

$$= 1 - {}_6C_0\left(\frac{1}{9}\right)^0\left(\frac{8}{9}\right)^6 - {}_6C_1\left(\frac{1}{9}\right)^1\left(\frac{8}{9}\right)^5 = \frac{72{,}689}{531{,}441}$$

4.56. If the probability of a defective bolt is 0.1, find (a) the mean, (b) the standard deviation for the distribution of defective bolts in a total of 400.

(a) Mean $= np = 400(0.1) = 40$, i.e., we can *expect* 40 bolts to be defective.

(b) Variance $= npq = 400(0.1)(0.9) = 36$. Hence the standard deviation $= \sqrt{36} = 6$.

4.57. Find the coefficients of (a) skewness, (b) kurtosis of the distribution in Problem 4.56.

(a) $$\text{Coefficient of skewness} = \frac{q - p}{\sqrt{npq}} = \frac{0.9 - 0.1}{6} = 0.133$$

Since this is positive, the distribution is skewed to the right.

(b) $$\text{Coefficient of kurtosis} = 3 + \frac{1 - 6pq}{npq} = 3 + \frac{1 - 6(0.1)(0.9)}{36} = 3.01$$

The distribution is slightly more peaked than the normal distribution.

4.58. The grades on a short quiz in biology were 0, 1, 2, . . . , 10 points, depending on the number answered correctly out of 10 questions. The mean grade was 6.7, and the standard deviation was 1.2. Assuming the grades to be normally distributed, determine (a) the percentage of students scoring 6 points, (b) the maximum grade of the lowest 10% of the class, (c) the minimum grade of the highest 10% of the class.

(a) To apply the normal distribution to discrete data, it is necessary to treat the data as if they were continuous. Thus a score of 6 points is considered as 5.5 to 6.5 points. See Fig. 4-20.

$$5.5 \text{ in standard units} = (5.5 - 6.7)/1.2 = -1.0$$

$$6.5 \text{ in standard units} = (6.5 - 6.7)/1.2 = -0.17$$

$$\text{Required proportion} = \text{area between } z = -1 \text{ and } z = -0.17$$
$$= (\text{area between } z = -1 \text{ and } z = 0)$$
$$- (\text{area between } z = -0.17 \text{ and } z = 0)$$
$$= 0.3413 - 0.0675 = 0.2738 = 27\%$$

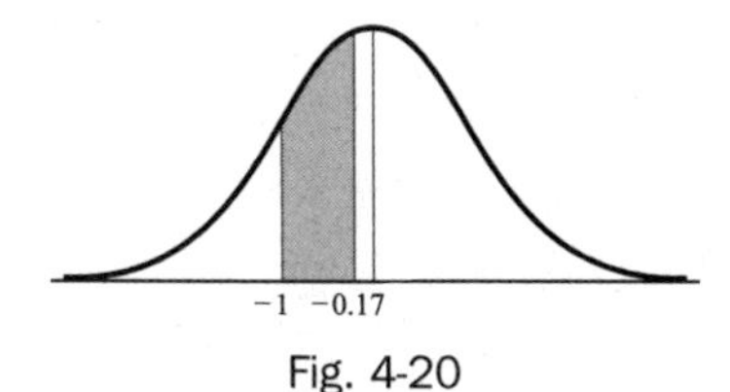

Fig. 4-20

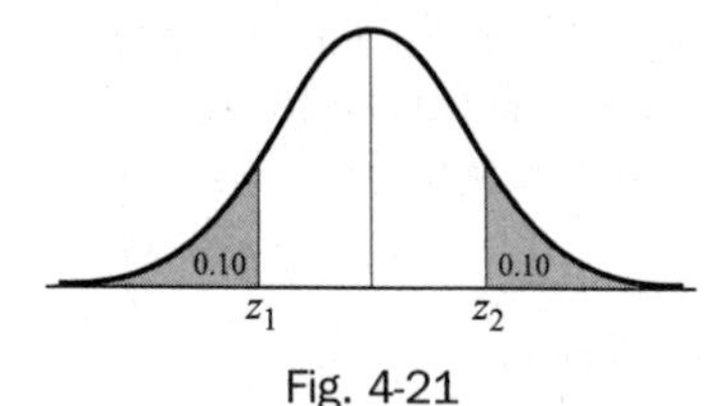

Fig. 4-21

(b) Let x_1 be the required maximum grade and z_1 its equivalent in standard units. From Fig. 4-21 the area to the left of z_1 is $10\% = 0.10$; hence,

$$\text{Area between } z_1 \text{ and } 0 = 0.40$$

and $z_1 = -1.28$ (very closely).
Then $z_1 = (x_1 - 6.7)/1.2 = -1.28$ and $x_1 = 5.2$ or 5 to the nearest integer.

(c) Let x_2 be the required minimum grade and z_2 the same grade in standard units. From (b), by symmetry, $z_2 = 1.28$. Then $(x_2 - 6.7)/1.2 = 1.28$, and $x_2 = 8.2$ or 8 to the nearest integer.

4.59. A Geiger counter is used to count the arrivals of radioactive particles. Find the probability that in time t no particles will be counted.

Let Fig. 4-22 represent the time axis with O as the origin. The probability that a particle is counted in a small time Δt is proportional to Δt and so can be written as $\lambda \Delta t$. Therefore, the probability of no count in time Δt is $1 - \lambda \Delta t$. More precisely, there will be additional terms involving $(\Delta t)^2$ and higher orders, but these are negligible if Δt is small.

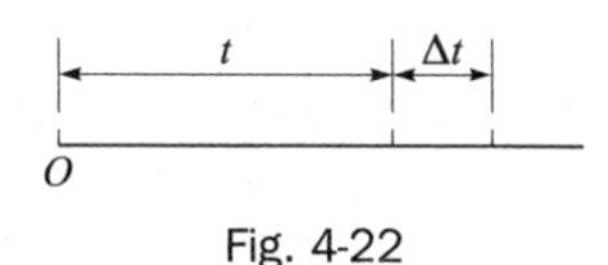

Fig. 4-22

Let $P_0(t)$ be the probability of no count in time t. Then $P_0(t + \Delta t)$ is the probability of no count in time $t + \Delta t$. If the arrivals of the particles are assumed to be independent events, the probability of no count in time $t + \Delta t$ is the product of the probability of no count in time t and the probability of no count in time Δt. Therefore, neglecting terms involving $(\Delta t)^2$ and higher, we have

(1) $$P_0(t + \Delta t) = P_0(t)[1 - \lambda \Delta t]$$

From (1) we obtain

(2) $$\lim_{\Delta t \to 0} \frac{P_0(t + \Delta t) - P_0(t)}{\Delta t} = -\lambda P_0(t)$$

i.e.,

(3) $$\frac{dP_0}{dt} = -\lambda P_0 \quad \text{or} \quad \frac{dP_0}{P_0} = -\lambda\, dt$$

Solving (3) by integration we obtain

$$\ln P_0 = -\lambda t + c_1 \quad \text{or} \quad P_0(t) = ce^{-\lambda t}$$

To determine c, note that if $t = 0$, $P_0(0) = c$ is the probability of no counts in time zero, which is of course 1. Thus $c = 1$ and the required probability is

(4) $$P_0(t) = e^{-\lambda t}$$

4.60. Referring to Problem 4.59, find the probability of exactly one count in time t.

Let $P_1(t)$ be the probability of 1 count in time t, so that $P_1(t + \Delta t)$ is the probability of 1 count in time $t + \Delta t$. Now we will have 1 count in time $t + \Delta t$ in the following two mutually exclusive cases:

(i) 1 count in time t and 0 counts in time Δt

(ii) 0 counts in time t and 1 count in time Δt

The probability of (i) is $P_1(t)(1 - \lambda\Delta t)$.
The probability of (ii) is $P_0(t)\,\lambda\Delta t$.
Thus, apart from terms involving $(\Delta t)^2$ and higher,

$$P_1(t + \Delta t) = P_1(t)(1 - \lambda\Delta t) + P_0(t)\lambda\Delta t \tag{1}$$

This can be written

$$\frac{P_1(t + \Delta t) - P_1(t)}{\Delta t} = \lambda P_0(t) - \lambda P_1(t) \tag{2}$$

Taking the limit as $\Delta t \to 0$ and using the expression for $P_0(t)$ obtained in Problem 4.59, this becomes

$$\frac{dP_1}{dt} = \lambda e^{-\lambda t} - \lambda P_1 \tag{3}$$

or

$$\frac{dP_1}{dt} + \lambda P_1 = \lambda e^{-\lambda t} \tag{4}$$

Multiplying by $e^{\lambda t}$, this can be written

$$\frac{d}{dt}(e^{\lambda t}P_1) = \lambda \tag{5}$$

which yields on integrating

$$P_1(t) = \lambda t e^{-\lambda t} + c_2 e^{-\lambda t} \tag{6}$$

If $t = 0$, $P_1(0)$ is the probability of 1 count in time 0, which is zero. Using this in (6), we find $c_2 = 0$. Therefore,

$$P_1(t) = \lambda t e^{-\lambda t} \tag{7}$$

By continuing in this manner, we can show that the probability of exactly n counts in time t is given by

$$P_n(t) = \frac{(\lambda t)^n e^{-\lambda t}}{n!} \tag{8}$$

which is the Poisson distribution.

SUPPLEMENTARY PROBLEMS

The binomial distribution

4.61. Find the probability that in tossing a fair coin 6 times, there will appear (a) 0, (b) 1, (c) 2, (d) 3, (e) 4, (f) 5, (g) 6 heads.

4.62. Find the probability of (a) 2 or more heads, (b) fewer than 4 heads, in a single toss of 6 fair coins.

4.63. If X denotes the number of heads in a single toss of 4 fair coins, find (a) $P(X = 3)$, (b) $P(X < 2)$, (c) $P(X \leq 2)$, (d) $P(1 < X \leq 3)$.

4.64. Out of 800 families with 5 children each, how many would you expect to have (a) 3 boys, (b) 5 girls, (c) either 2 or 3 boys? Assume equal probabilities for boys and girls.

4.65. Find the probability of getting a total of 11 (a) once, (b) twice, in two tosses of a pair of fair dice.

4.66. What is the probability of getting a 9 exactly once in 3 throws with a pair of dice?

4.67. Find the probability of guessing correctly at least 6 of the 10 answers on a true-false examination.

4.68. An insurance sales representative sells policies to 5 men, all of identical age and in good health. According to the actuarial tables, the probability that a man of this particular age will be alive 30 years hence is $\frac{2}{3}$. Find the probability that in 30 years (a) all 5 men, (b) at least 3 men, (c) only 2 men, (d) at least 1 man will be alive.

4.69. Compute the (a) mean, (b) standard deviation, (c) coefficient of skewness, (d) coefficient of kurtosis for a binomial distribution in which $p = 0.7$ and $n = 60$. Interpret the results.

4.70. Show that if a binomial distribution with $n = 100$ is symmetric; its coefficient of kurtosis is 2.9.

4.71. Evaluate (a) $\sum(x - \mu)^3 f(x)$, (b) $\sum(x - \mu)^4 f(x)$ the binomial distribution.

The normal distribution

4.72. On a statistics examination the mean was 78 and the standard deviation was 10. (a) Determine the standard scores of two students whose grades were 93 and 62, respectively, (b) Determine the grades of two students whose standard scores were -0.6 and 1.2, respectively.

4.73. Find (a) the mean, (b) the standard deviation on an examination in which grades of 70 and 88 correspond to standard scores of -0.6 and 1.4, respectively.

4.74. Find the area under the normal curve between (a) $z = -1.20$ and $z = 2.40$, (b) $z = 1.23$ and $z = 1.87$, (c) $z = -2.35$ and $z = -0.50$.

4.75. Find the area under the normal curve (a) to the left of $z = -1.78$, (b) to the left of $z = 0.56$, (c) to the right of $z = -1.45$, (d) corresponding to $z \geq 2.16$, (e) corresponding to $-0.80 \leq z \leq 1.53$, (f) to the left of $z = -2.52$ and to the right of $z = 1.83$.

4.76. If Z is normally distributed with mean 0 and variance 1, find: (a) $P(Z \geq -1.64)$, (b) $P(-1.96 \leq Z \leq 1.96)$, (c) $P(|Z| \geq 1)$.

4.77. Find the values of z such that (a) the area to the right of z is 0.2266, (b) the area to the left of z is 0.0314, (c) the area between -0.23 and z is 0.5722, (d) the area between 1.15 and z is 0.0730, (e) the area between $-z$ and z is 0.9000.

4.78. Find z_1 if $P(Z \geq z_1) = 0.84$, where z is normally distributed with mean 0 and variance 1.

4.79. If X is normally distributed with mean 5 and standard deviation 2, find $P(X > 8)$.

4.80. If the heights of 300 students are normally distributed with mean 68.0 inches and standard deviation 3.0 inches, how many students have heights (a) greater than 72 inches, (b) less than or equal to 64 inches, (c) between 65 and 71 inches inclusive, (d) equal to 68 inches? Assume the measurements to be recorded to the nearest inch.

4.81. If the diameters of ball bearings are normally distributed with mean 0.6140 inches and standard deviation 0.0025 inches, determine the percentage of ball bearings with diameters (a) between 0.610 and 0.618 inches inclusive, (b) greater than 0.617 inches, (c) less than 0.608 inches, (d) equal to 0.615 inches.

4.82. The mean grade on a final examination was 72, and the standard deviation was 9. The top 10% of the students are to receive *A*'s. What is the minimum grade a student must get in order to receive an *A*?

4.83. If a set of measurements are normally distributed, what percentage of these differ from the mean by (a) more than half the standard deviation, (b) less than three quarters of the standard deviation?

4.84. If μ is the mean and σ is the standard deviation of a set of measurements that are normally distributed, what percentage of the measurements are (a) within the range $\mu \pm 2\sigma$ (b) outside the range $\mu \pm 1.2\sigma$ (c) greater than $\mu - 1.5\sigma$?

4.85. In Problem 4.84 find the constant a such that the percentage of the cases (a) within the range $\mu \pm a\sigma$ is 75%, (b) less than $\mu - a\sigma$ is 22%.

Normal approximation to binomial distribution

4.86. Find the probability that 200 tosses of a coin will result in (a) between 80 and 120 heads inclusive, (b) less than 90 heads, (c) less than 85 or more than 115 heads, (d) exactly 100 heads.

4.87. Find the probability that a student can guess correctly the answers to (a) 12 or more out of 20, (b) 24 or more out of 40, questions on a true-false examination.

4.88. A machine produces bolts which are 10% defective. Find the probability that in a random sample of 400 bolts produced by this machine, (a) at most 30, (b) between 30 and 50, (c) between 35 and 45, (d) 65 or more, of the bolts will be defective.

4.89. Find the probability of getting more than 25 "sevens" in 100 tosses of a pair of fair dice.

The Poisson distribution

4.90. If 3% of the electric bulbs manufactured by a company are defective, find the probability that in a sample of 100 bulbs, (a) 0, (b) 1, (c) 2, (d) 3, (e) 4, (f) 5 bulbs will be defective.

4.91. In Problem 4.90, find the probability that (a) more than 5, (b) between 1 and 3, (c) less than or equal to 2, bulbs will be defective.

4.92. A bag contains 1 red and 7 white marbles. A marble is drawn from the bag, and its color is observed. Then the marble is put back into the bag and the contents are thoroughly mixed. Using (a) the binomial distribution, (b) the Poisson approximation to the binomial distribution, find the probability that in 8 such drawings, a red ball is selected exactly 3 times.

4.93. According to the National Office of Vital Statistics of the U.S. Department of Health and Human Services, the average number of accidental drownings per year in the United States is 3.0 per 100,000 population. Find the probability that in a city of population 200,000 there will be (a) 0, (b) 2, (c) 6, (d) 8, (e) between 4 and 8, (f) fewer than 3, accidental drownings per year.

4.94. Prove that if X_1 and X_2 are independent Poisson variables with respective parameters λ_1 and λ_2, then $X_1 + X_2$ has a Poisson distribution with parameter $\lambda_1 + \lambda_2$. (*Hint:* Use the moment generating function.) Generalize the result to n variables.

Multinomial distribution

4.95. A fair die is tossed 6 times. Find the probability that (a) 1 "one", 2 "twos" and 3 "threes" will turn up, (b) each side will turn up once.

4.96. A box contains a very large number of red, white, blue, and yellow marbles in the ratio 4:3:2:1. Find the probability that in 10 drawings (a) 4 red, 3 white, 2 blue, and 1 yellow marble will be drawn, (b) 8 red and 2 yellow marbles will be drawn.

4.97. Find the probability of not getting a 1, 2, or 3 in 4 tosses of a fair die.

The hypergeometric distribution

4.98. A box contains 5 red and 10 white marbles. If 8 marbles are to be chosen at random (without replacement), determine the probability that (a) 4 will be red, (b) all will be white, (c) at least one will be red.

4.99. If 13 cards are to be chosen at random (without replacement) from an ordinary deck of 52 cards, find the probability that (a) 6 will be picture cards, (b) none will be picture cards.

4.100. Out of 60 applicants to a university, 40 are from the East. If 20 applicants are to be selected at random, find the probability that (a) 10, (b) not more than 2, will be from the East.

The uniform distribution

4.101. Let X be uniformly distributed in $-2 \leq x \leq 2$ Find (a) $P(X < 1)$, (b) $P(|X - 1| \geq \frac{1}{2})$.

4.102. Find (a) the third, (b) the fourth moment about the mean of a uniform distribution.

4.103. Determine the coefficient of (a) skewness, (b) kurtosis of a uniform distribution.

4.104. If X and Y are independent and both uniformly distributed in the interval from 0 to 1, find $P(|X - Y| \geq \frac{1}{2})$.

The Cauchy distribution

4.105. Suppose that X is Cauchy distributed according to (29), page 114, with $a = 2$. Find (a) $P(X < 2)$, (b) $P(X^2 \geq 12)$.

4.106. Prove that if X_1 and X_2 are independent and have the same Cauchy distribution, then their arithmetic mean also has this distribution.

4.107. Let X_1 and X_2 be independent and normally distributed with mean 0 and variance 1. Prove that $Y = X_1/X_2$ is Cauchy distributed.

The gamma distribution

4.108. A random variable X is gamma distributed with $\alpha = 3$, $\beta = 2$. Find (a) $P(X \leq 1)$, (b) $P(1 \leq X \leq 2)$.

The chi-square distribution

4.109. For a chi-square distribution with 12 degrees of freedom, find the value of χ^2_c such that (a) the area to the right of χ^2_c is 0.05, (b) the area to the left of χ^2_c is 0.99, (c) the area to the right of χ^2_c is 0.025.

4.110. Find the values of χ^2 for which the area of the right-hand tail of the χ^2 distribution is 0.05, if the number of degrees of freedom v is equal to (a) 8, (b) 19, (c) 28, (d) 40.

4.111. Work Problem 4.110 if the area of the right-hand tail is 0.01.

4.112. (a) Find χ_1^2 and χ_2^2 such that the area under the χ^2 distribution corresponding to $v = 20$ between χ_1^2 and χ_2^2 is 0.95, assuming equal areas to the right of χ_2^2 and left of χ_1^2. (b) Show that if the assumption of equal areas in part (a) is not made, the values χ_1^2 and χ_2^2 are not unique.

4.113. If the variable U is chi-square distributed with $v = 7$, find χ_1^2 and χ_2^2 such that (a) $P(U > \chi_2^2) = 0.025$, (b) $P(U < \chi_1^2) = 0.50$, (c) $P(\chi_1^2 \le U \le \chi_2^2) = 0.90$.

4.114. Find (a) $\chi_{0.05}^2$ and (b) $\chi_{0.95}^2$ for $v = 150$.

4.115. Find (a) $\chi_{0.025}^2$ and (b) $\chi_{0.975}^2$ for $v = 250$.

Student's *t* distribution

4.116. For a Student's distribution with 15 degrees of freedom, find the value of t_1 such that (a) the area to the right of t_1 is 0.01, (b) the area to the left of t_1 is 0.95, (c) the area to the right of t_1 is 0.10, (d) the combined area to the right of t_1 and to the left of $-t_1$ is 0.01, (e) the area between $-t_1$ and t_1 is 0.95.

4.117. Find the values of t for which the area of the right-hand tail of the t distribution is 0.01, if the number of degrees of freedom v is equal to (a) 4, (b) 12, (c) 25, (d) 60, (e) 150.

4.118. Find the values of t_1 for Student's distribution that satisfy each of the following conditions: (a) the area between $-t_1$ and t_1 is 0.90 and $v = 25$, (b) the area to the left of $-t_1$ is 0.025 and $v = 20$, (c) the combined area to the right of t_1 and left of $-t_1$ is 0.01 and $v = 5$, (d) the area to the right of t_1 is 0.55 and $v = 16$.

4.119. If a variable U has a Student's distribution with $v = 10$, find the constant c such that (a) $P(U > c) = 0.05$, (b) $P(-c \le U \le c) = 0.98$, (c) $P(U \le c) = 0.20$, (d) $P(U \ge c) = 0.90$.

The *F* distribution

4.120. Evaluate each of the following:
(a) $F_{0.95,15,12}$; (b) $F_{0.99,120,60}$; (c) $F_{0.99,60,24}$; (d) $F_{0.01,30,12}$; (e) $F_{0.05,9,20}$; (f) $F_{0.01,8,8}$.

ANSWERS TO SUPPLEMENTARY PROBLEMS

4.61. (a) 1/64 (b) 3/32 (c) 15/64 (d) 5/16 (e) 15/64 (f) 3/32 (g) 1/64

4.62. (a) 57/64 (b) 21/32 **4.63.** (a) 1/4 (b) 5/16 (c) 11/16 (d) 5/8

4.64. (a) 250 (b) 25 (c) 500 **4.65.** (a) 17/162 (b) 1/324 **4.66.** 64/243

4.67. 193/512 **4.68.** (a) 32/243 (b) 192/243 (c) 40/243 (d) 242/243

4.69. (a) 42 (b) 3.550 (c) −0.1127 (d) 2.927

4.71. (a) $npq(q - p)$ (b) $npq(1 - 6pq) + 3n^2p^2q^2$ **4.72.** (a) 1.5, −1.6 (b) 72, 90

4.73. (a) 75.4 (b) 9 **4.74.** (a) 0.8767 (b) 0.0786 (c) 0.2991

4.75. (a) 0.0375 (b) 0.7123 (c) 0.9265 (d) 0.0154 (e) 0.7251 (f) 0.0395

4.76. (a) 0.9495 (b) 0.9500 (c) 0.6826

4.77. (a) 0.75 (b) -1.86 (c) 2.08 (d) 1.625 or 0.849 (e) ± 1.645

4.78. -0.995 **4.79.** 0.0668

4.80. (a) 20 (b) 36 (c) 227 (d) 40

4.81. (a) 93% (b) 8.1% (c) 0.47% (d) 15% **4.82.** 84

4.83. (a) 61.7% (b) 54.7% **4.84.** (a) 95.4% (b) 23.0% (c) 93.3%

4.85. (a) 1.15 (b) 0.77 **4.86.** (a) 0.9962 (b) 0.0687 (c) 0.0286 (d) 0.0558

4.87. (a) 0.2511 (b) 0.1342 **4.88.** (a) 0.0567 (b) 0.9198 (c) 0.6404 (d) 0.0079

4.89. 0.0089 **4.90.** (a) 0.04979 (b) 0.1494 (c) 0.2241 (d) 0.2241 (e) 0.1680 (f) 0.1008

4.91. (a) 0.0838 (b) 0.5976 (c) 0.4232 **4.92.** (a) 0.05610 (b) 0.06131

4.93. (a) 0.00248 (b) 0.04462 (c) 0.1607 (d) 0.1033 (e) 0.6964 (f) 0.0620

4.95. (a) 5/3888 (b) 5/324 **4.96.** (a) 0.0348 (b) 0.000295

4.97. 1/16 **4.98.** (a) 70/429 (b) 1/143 (c) 142/143

4.99. (a) $\binom{13}{6}\binom{39}{7}\Big/\binom{52}{13}$ (b) $\binom{13}{0}\binom{39}{13}\Big/\binom{52}{13}$

4.100. (a) $\binom{40}{10}\binom{20}{10}\Big/\binom{60}{20}$ (b) $[({}_{40}C_0)({}_{20}C_{20}) + ({}_{40}C_1)({}_{20}C_{19}) + ({}_{40}C_2)({}_{20}C_{18})]/{}_{60}C_{20}$

4.101. (a) 3/4 (b) 3/4 **4.102.** (a) 0 (b) $(b-a)^4/80$

4.103. (a) 0 (b) 9/5 **4.104.** 1/4

4.105. (a) 3/4 (b) 1/3 **4.108.** (a) $1 - \dfrac{13}{8\sqrt{e}}$ (b) $\dfrac{13}{8}e^{-1/2} - \dfrac{5}{2}e^{-1}$

4.109. (a) 21.0 (b) 26.2 (c) 23.3 **4.110.** (a) 15.5 (b) 30.1 (c) 41.3 (d) 55.8

4.111. (a) 20.1 (b) 36.2 (c) 48.3 (d) 63.7 **4.112.** (a) 9.59 and 34.2

4.113. (a) 16.0 (b) 6.35 (c) assuming equal areas in the two tails, $\chi_1^2 = 2.17$ and $\chi_2^2 = 14.1$

4.114. (a) 122.5 (b) 179.2 **4.115.** (a) 207.7 (b) 295.2

4.116. (a) 2.60 (b) 1.75 (c) 1.34 (d) 2.95 (e) 2.13

4.117. (a) 3.75 (b) 2.68 (c) 2.48 (d) 2.39 (e) 2.33

4.118. (a) 1.71 (b) 2.09 (c) 4.03 (d) −0.128

4.119. (a) 1.81 (b) 2.76 (c) −0.879 (d) −1.37

4.120. (a) 2.62 (b) 1.73 (c) 2.40 (d) 0.352 (e) 0.340 (f) 0.166

PART II

Statistics

Sampling Theory

Population and Sample. Statistical Inference

Often in practice we are interested in drawing valid conclusions about a large group of individuals or objects. Instead of examining the entire group, called the *population*, which may be difficult or impossible to do, we may examine only a small part of this population, which is called a *sample*. We do this with the aim of inferring certain facts about the population from results found in the sample, a process known as *statistical inference*. The process of obtaining samples is called *sampling*.

EXAMPLE 5.1 We may wish to draw conclusions about the heights (or weights) of 12,000 adult students (the population) by examining only 100 students (a sample) selected from this population.

EXAMPLE 5.2 We may wish to draw conclusions about the percentage of defective bolts produced in a factory during a given 6-day week by examining 20 bolts each day produced at various times during the day. In this case all bolts produced during the week comprise the population, while the 120 selected bolts constitute a sample.

EXAMPLE 5.3 We may wish to draw conclusions about the fairness of a particular coin by tossing it repeatedly. The population consists of all possible tosses of the coin. A sample could be obtained by examining, say, the first 60 tosses of the coin and noting the percentages of heads and tails.

EXAMPLE 5.4 We may wish to draw conclusions about the colors of 200 marbles (the population) in an urn by selecting a sample of 20 marbles from the urn, where each marble selected is returned after its color is observed.

Several things should be noted. First, the word *population* does not necessarily have the same meaning as in everyday language, such as "the population of Shreveport is 180,000." Second, the word *population* is often used to denote the observations or measurements rather than the individuals or objects. In Example 5.1 we can speak of the population of 12,000 heights (or weights) while in Example 5.4 we can speak of the population of all 200 colors in the urn (some of which may be the same). Third, the population can be finite or infinite, the number being called the *population size*, usually denoted by N. Similarly the number in the sample is called the *sample size*, denoted by n, and is generally finite. In Example 5.1, $N = 12{,}000$, $n = 100$, while in Example 5.3, N is infinite, $n = 60$.

Sampling With and Without Replacement

If we draw an object from an urn, we have the choice of replacing or not replacing the object into the urn before we draw again. In the first case a particular object can come up again and again, whereas in the second it can come up only once. Sampling where each member of a population may be chosen more than once is called *sampling with replacement*, while sampling where each member cannot be chosen more than once is called *sampling without replacement*.

A finite population that is sampled with replacement can theoretically be considered infinite since samples of any size can be drawn without exhausting the population. For most practical purposes, sampling from a finite population that is very large can be considered as sampling from an infinite population.

Random Samples. Random Numbers

Clearly, the reliability of conclusions drawn concerning a population depends on whether the sample is properly chosen so as to represent the population sufficiently well, and one of the important problems of statistical inference is just how to choose a sample.

One way to do this for finite populations is to make sure that each member of the population has the same chance of being in the sample, which is then often called a *random sample*. Random sampling can be accomplished for relatively small populations by drawing lots or, equivalently, by using a table of *random numbers* (Appendix H) specially constructed for such purposes. See Problem 5.43.

Because inference from sample to population cannot be certain, we must use the language of probability in any statement of conclusions.

Population Parameters

A population is considered to be known when we know the probability distribution $f(x)$ (probability function or density function) of the associated random variable X. For instance, in Example 5.1 if X is a random variable whose values are the heights (or weights) of the 12,000 students, then X has a probability distribution $f(x)$.

If, for example, X is normally distributed, we say that the population is *normally distributed* or that we have a *normal population*. Similarly, if X is binomially distributed, we say that the population is *binomially distributed* or that we have a *binomial population*.

There will be certain quantities that appear in $f(x)$, such as μ and σ in the case of the normal distribution or p in the case of the binomial distribution. Other quantities such as the median, moments, and skewness can then be determined in terms of these. All such quantities are often called *population parameters*. When we are given the population so that we know $f(x)$, then the population parameters are also known.

An important problem arises when the probability distribution $f(x)$ of the population is not known precisely, although we may have some idea of, or at least be able to make some hypothesis concerning, the general behavior of $f(x)$. For example, we may have some reason to suppose that a particular population is normally distributed. In that case we may not know one or both of the values μ and σ and so we might wish to draw statistical inferences about them.

Sample Statistics

We can take random samples from the population and then use these samples to obtain values that serve to estimate and test hypotheses about the population parameters.

By way of illustration, let us consider Example 5.1 where X is a random variable whose values are the various heights. To obtain a sample of size 100, we must first choose one individual at random from the population. This individual can have any one value, say, x_1, of the various possible heights, and we can call x_1 the value of a random variable X_1, where the subscript 1 is used since it corresponds to the first individual chosen. Similarly, we choose the second individual for the sample, who can have any one of the values x_2 of the possible heights, and x_2 can be taken as the value of a random variable X_2. We can continue this process up to X_{100} since the sample size is 100. For simplicity let us assume that the sampling is with replacement so that the same individual could conceivably be chosen more than once. In this case, since the sample size is much smaller than the population size, sampling without replacement would give practically the same results as sampling with replacement.

In the general case a sample of size n would be described by the values $x_1, x_2, \ldots, x_n$ of the random variables $X_1, X_2, \ldots, X_n$. In the case of sampling with replacement, $X_1, X_2, \ldots, X_n$ would be independent, identically distributed random variables having probability distribution $f(x)$. Their joint distribution would then be

$$P(X_1 = x_1, X_2 = x_2, \ldots, X_n = x_n) = f(x_1)f(x_2) \ldots f(x_n) \tag{1}$$

Any quantity obtained from a sample for the purpose of estimating a population parameter is called a *sample statistic*, or briefly *statistic*. Mathematically, a sample statistic for a sample of size n can be defined as a function of the random variables $X_1, \ldots, X_n$, i.e., $g(X_1, \ldots, X_n)$. The function $g(X_1, \ldots, X_n)$ is another random variable, whose values can be represented by $g(x_1, \ldots, x_n)$. The word *statistic* is often used for the random variable or for its values, the particular sense being clear from the context.

In general, corresponding to each population parameter there will be a statistic to be computed from the sample. Usually the method for obtaining this statistic from the sample is similar to that for obtaining the parameter from a finite population, since a sample consists of a finite set of values. As we shall see, however, this may not always produce the "best estimate," and one of the important problems of sampling theory is to decide how to form the proper sample statistic that will best estimate a given population parameter. Such problems are considered in later chapters.

Where possible we shall try to use Greek letters, such as μ and σ, for values of population parameters, and Roman letters, m, s, etc., for values of corresponding sample statistics.

Sampling Distributions

As we have seen, a sample statistic that is computed from $X_1, \ldots, X_n$ is a function of these random variables and is therefore itself a random variable. The probability distribution of a sample statistic is often called the *sampling distribution* of the statistic.

Alternatively we can consider all possible samples of size n that can be drawn from the population, and for each sample we compute the statistic. In this manner we obtain the distribution of the statistic, which is its sampling distribution.

For a sampling distribution, we can of course compute a mean, variance, standard deviation, moments, etc. The standard deviation is sometimes also called the *standard error.*

The Sample Mean

Let $X_1, X_2, \ldots, X_n$ denote the independent, identically distributed, random variables for a random sample of size n as described above. Then the *mean of the sample* or *sample mean* is a random variable defined by

$$\bar{X} = \frac{X_1 + X_2 + \cdots + X_n}{n} \tag{2}$$

in analogy with (3), page 75. If $x_1, x_2, \ldots, x_n$ denote values obtained in a particular sample of size n, then the mean for that sample is denoted by

$$\bar{x} = \frac{x_1 + x_2 + \cdots + x_n}{n} \tag{3}$$

EXAMPLE 5.5 If a sample of size 5 results in the sample values 7, 9, 1, 6, 2, then the sample mean is

$$\bar{x} = \frac{7 + 9 + 1 + 6 + 2}{5} = 5$$

Sampling Distribution of Means

Let $f(x)$ be the probability distribution of some given population from which we draw a sample of size n. Then it is natural to look for the probability distribution of the sample statistic $\bar{X}$, which is called the *sampling distribution for the sample mean*, or the *sampling distribution of means*. The following theorems are important in this connection.

Theorem 5-1 The mean of the sampling distribution of means, denoted by $\mu_{\bar{X}}$, is given by

$$E(\bar{X}) = \mu_{\bar{X}} = \mu \tag{4}$$

where μ is the mean of the population.

Theorem 5-1 states that the expected value of the sample mean is the population mean.

Theorem 5-2 If a population is infinite and the sampling is random or if the population is finite and sampling is with replacement, then the variance of the sampling distribution of means, denoted by $\sigma_{\bar{X}}^2$, is given by

$$E[(\bar{X} - \mu)^2] = \sigma_{\bar{X}}^2 = \frac{\sigma^2}{n} \tag{5}$$

where σ^2 is the variance of the population.

Theorem 5-3 If the population is of size N, if sampling is without replacement, and if the sample size is $n \le N$, then (5) is replaced by

$$\sigma_{\bar{X}}^2 = \frac{\sigma^2}{n}\left(\frac{N - n}{N - 1}\right) \tag{6}$$

while $\mu_{\bar{X}}$ is still given by (4).

Note that (6) reduces to (5) as $N \to \infty$.

Theorem 5-4 If the population from which samples are taken is normally distributed with mean μ and variance σ^2, then the sample mean is normally distributed with mean μ and variance σ^2/n.

Theorem 5-5 Suppose that the population from which samples are taken has a probability distribution with mean μ and variance σ^2 that is not necessarily a normal distribution. Then the standardized variable associated with $\bar{X}$, given by

$$Z = \frac{\bar{X} - \mu}{\sigma/\sqrt{n}} \tag{7}$$

is *asymptotically normal*, i.e.,

$$\lim_{n \to \infty} P(Z \le z) = \frac{1}{\sqrt{2\pi}} \int_{-\infty}^{z} e^{-u^2/2}\, du \tag{8}$$

Theorem 5-5 is a consequence of the central limit theorem, page 112. It is assumed here that the population is infinite or that sampling is with replacement. Otherwise, the above is correct if we replace $\sigma/\sqrt{n}$ in (7) by $\sigma_{\bar{X}}$ as given by (6).

Sampling Distribution of Proportions

Suppose that a population is infinite and binomially distributed, with p and $q = 1 - p$ being the respective probabilities that any given member exhibits or does not exhibit a certain property. For example, the population may be all possible tosses of a fair coin, in which the probability of the event heads is $p = \frac{1}{2}$.

Consider all possible samples of size n drawn from this population, and for each sample determine the statistic that is the proportion P of successes. In the case of the coin, P would be the proportion of heads turning up in n tosses. Then we obtain a *sampling distribution of proportions* whose mean μ_p and standard deviation σ_p are given by

$$\mu_p = p \qquad \sigma_p = \sqrt{\frac{pq}{n}} = \sqrt{\frac{p(1 - p)}{n}} \tag{9}$$

which can be obtained from (4) and (5) on placing $\mu = p, \sigma = \sqrt{pq}$.

For large values of n ($n \ge 30$), the sampling distribution is very nearly a normal distribution, as is seen from Theorem 5-5.

For finite populations in which sampling is without replacement, the second equation in (9) is replaced by $\sigma_{\bar{x}}$ as given by (6) with $\sigma = \sqrt{pq}$.

Note that equations (9) are obtained most easily on dividing by n the mean and standard deviation (np and $\sqrt{npq}$) of the binomial distribution.

Sampling Distribution of Differences and Sums

Suppose that we are given two populations. For each sample of size n_1 drawn from the first population, let us compute a statistic S_1. This yields a sampling distribution for S_1 whose mean and standard deviation we denote by μ_{S_1} and σ_{S_1}, respectively. Similarly for each sample of size n_2 drawn from the second population, let us compute a statistic S_2 whose mean and standard deviation are μ_{S_2} and σ_{S_2}, respectively.

Taking all possible combinations of these samples from the two populations, we can obtain a distribution of the differences, $S_1 - S_2$, which is called the *sampling distribution of differences* of the statistics. The mean and standard deviation of this sampling distribution, denoted respectively by $\mu_{S_1-S_2}$ and $\sigma_{S_1-S_2}$, are given by

$$\mu_{S_1-S_2} = \mu_{S_1} - \mu_{S_2} \qquad \sigma_{S_1-S_2} = \sqrt{\sigma_{S_1}^2 + \sigma_{S_2}^2}, \tag{10}$$

provided that the samples chosen do not in any way depend on each other, i.e., the samples are *independent* (in other words, the random variables S_1 and S_2 are independent).

If, for example, S_1 and S_2 are the sample means from two populations, denoted by $\bar{X}_1, \bar{X}_2$, respectively, then the sampling distribution of the differences of means is given for infinite populations with mean and standard deviation μ_1, σ_1 and μ_2, σ_2, respectively by

$$\mu_{\bar{X}_1-\bar{X}_2} = \mu_{\bar{X}_1} - \mu_{\bar{X}_2} = \mu_1 - \mu_2, \qquad \sigma_{\bar{X}_1-\bar{X}_2} = \sqrt{\sigma_{\bar{X}_1}^2 + \sigma_{\bar{X}_2}^2} = \sqrt{\frac{\sigma_1^2}{n_1} + \frac{\sigma_2^2}{n_2}} \tag{11}$$

using (4) and (5). This result also holds for finite populations if sampling is with replacement. The standardized variable

$$Z = \frac{(\bar{X}_1 - \bar{X}_2) - (\mu_1 - \mu_2)}{\sqrt{\dfrac{\sigma_1^2}{n_1} + \dfrac{\sigma_2^2}{n_2}}} \tag{12}$$

in that case is very nearly normally distributed if n_1 and n_2 are large ($n_1, n_2 \geq 30$). Similar results can be obtained for finite populations in which sampling is without replacement by using (4) and (6).

Corresponding results can be obtained for sampling distributions of differences of proportions from two binomially distributed populations with parameters p_1, q_1 and p_2, q_2, respectively. In this case S_1, and S_2 correspond to the proportions of successes P_1 and P_2, and equations (11) yield

$$\mu_{P_1-P_2} = \mu_{P_1} - \mu_{P_2} = p_1 - p_2, \qquad \sigma_{P_1-P_2} = \sqrt{\sigma_{P_1}^2 + \sigma_{P_2}^2} = \sqrt{\frac{p_1q_1}{n_1} + \frac{p_2q_2}{n_2}} \tag{13}$$

Instead of taking differences of statistics, we sometimes are interested in the sum of statistics. In that case the *sampling distribution of the sum of statistics* S_1 and S_2 has mean and standard deviation given by

$$\mu_{S_1+S_2} = \mu_{S_1} + \mu_{S_2} \qquad \sigma_{S_1+S_2} = \sqrt{\sigma_{S_1}^2 + \sigma_{S_2}^2} \tag{14}$$

assuming the samples are independent. Results similar to (11) can then be obtained.

The Sample Variance

If $X_1, X_2, \ldots, X_n$ denote the random variables for a random sample of size n, then the random variable giving the *variance of the sample* or the *sample variance* is defined in analogy with (14), page 77, by

$$S^2 = \frac{(X_1 - \bar{X})^2 + (X_2 - \bar{X})^2 + \cdots + (X_n - \bar{X})^2}{n} \tag{15}$$

Now in Theorem 5-1 we found that $E(\bar{X}) = \mu$, and it would be nice if we could also have $E(S^2) = \sigma^2$. Whenever the expected value of a statistic is equal to the corresponding population parameter, we call the statistic an *unbiased estimator*, and the value an *unbiased estimate*, of this parameter. It turns out, however (see Problem 5.20), that

$$E(S^2) = \mu_{S^2} = \frac{n-1}{n}\sigma^2 \tag{16}$$

which is very nearly σ^2 only for large values of n (say, $n \geq 30$). The desired unbiased estimator is defined by

$$\hat{S}^2 = \frac{n}{n-1}S^2 = \frac{(X_1 - \bar{X})^2 + (X_2 - \bar{X})^2 + \cdots + (X_n - \bar{X})^2}{n-1} \tag{17}$$

so that

$$E(\hat{S}^2) = \sigma^2 \tag{18}$$

Because of this, some statisticians choose to define the sample variance by $\hat{S}^2$ rather than S^2 and they simply replace n by $n - 1$ in the denominator of (15). We shall, however, continue to define the sample variance by (15) because by doing this, many later results are simplified.

EXAMPLE 5.6 Referring to Example 5.5, page 155, the sample variance has the value

$$s^2 = \frac{(4-6)^2 + (7-6)^2 + (5-6)^2 + (8-6)^2 + (6-6)^2}{5} = 2$$

while an unbiased estimate is given by

$$\hat{s}^2 = \frac{5}{4}s^2 = \frac{(4-6)^2 + (7-6)^2 + (5-6)^2 + (8-6)^2 + (6-6)^2}{4} = 2.5$$

The above results hold if sampling is from an infinite population or with replacement from a finite population. If sampling is without replacement from a finite population of size N, then the mean of the sampling distribution of variances is given by

$$E(S^2) = \mu_{S^2} = \left(\frac{N}{N-1}\right)\left(\frac{n-1}{n}\right)\sigma^2 \tag{19}$$

As $N \to \infty$, this reduces to (16).

Sampling Distribution of Variances

By taking all possible random samples of size n drawn from a population and then computing the variance for each sample, we can obtain the sampling distribution of variances. Instead of finding the sampling distribution of S^2 or $\hat{S}^2$ itself, it is convenient to find the sampling distribution of the related random variable

$$\frac{nS^2}{\sigma^2} = \frac{(n-1)\hat{S}^2}{\sigma^2} = \frac{(X_1 - \bar{X})^2 + (X_2 - \bar{X})^2 + \cdots + (X_n - \bar{X})^2}{\sigma^2} \tag{20}$$

The distribution of this random variable is described in the following theorem.

Theorem 5-6 If random samples of size n are taken from a population having a normal distribution, then the sampling variable (20) has a chi-square distribution with $n - 1$ degrees of freedom.

Because of Theorem 5-6, the variable in (20) is often denoted by χ^2. For a proof of this theorem see Problem 5.22.

Case Where Population Variance Is Unknown

In Theorems 5-4 and 5-5 we found that the standardized variable

$$Z = \frac{\bar{X} - \mu}{\sigma/\sqrt{n}} \tag{21}$$

is normally distributed if the population from which samples of size n are taken is normally distributed, while it is asymptotically normal if the population is not normal provided that $n \geq 30$. In (21) we have, of course, assumed that the population variance σ^2 is known.

It is natural to ask what would happen if we do not know the population variance. One possibility is to estimate the population variance by using the sample variance and then put the corresponding standard deviation in (21). A better idea is to replace the σ in (21) by the random variable $\hat{S}$ giving the sample standard deviation and to seek the distribution of the corresponding statistic, which we designate by

$$T = \frac{\bar{X} - \mu}{\hat{S}/\sqrt{n}} = \frac{\bar{X} - \mu}{S/\sqrt{n-1}} \tag{22}$$

We can then show by using Theorem 4-6, page 116, that T has Student's t distribution with $n - 1$ degrees of freedom whenever the population random variable is normally distributed. We state this in the following theorem, which is proved in Problem 5.24.

Theorem 5-7 If random samples of size n are taken from a normally distributed population, then the statistic (22) has Student's distribution with $n - 1$ degrees of freedom.

Sampling Distribution of Ratios of Variances

On page 157, we indicated how sampling distributions of differences, in particular differences of means, can be obtained. Using the same idea, we could arrive at the sampling distribution of differences of variances, say, $S_1^2 - S_2^2$. It turns out, however, that this sampling distribution is rather complicated. Instead, we may consider the statistic S_1^2/S_2^2, since a large or small ratio would indicate a large difference while a ratio nearly equal to 1 would indicate a small difference.

Theorem 5-8 Let two independent random samples of sizes m and n, respectively, be drawn from two normal populations with variances σ_1^2, σ_2^2, respectively. Then if the variances of the random samples are given by S_1^2, S_2^2, respectively, the statistic

$$F = \frac{mS_1^2/(m-1)\sigma_1^2}{nS_2^2/(n-1)\sigma_2^2} = \frac{\hat{S}_1^2/\sigma_1^2}{\hat{S}_2^2/\sigma_2^2} \tag{23}$$

has the F distribution with $m - 1, n - 1$ degrees of freedom.

Other Statistics

Many other statistics besides the mean and variance or standard deviation can be defined for samples. Examples are the median, mode, moments, skewness, and kurtosis. Their definitions are analogous to those given for populations in Chapter 3. Sampling distributions for these statistics, or at least their means and standard deviations (standard errors), can often be found. Some of these, together with ones already given, are shown in Table 5-1.

Table 5-1
Standard Errors for Some Sample Statistics

Sample Statistic	Standard Error	Remarks
Means	$\sigma_{\bar{X}} = \dfrac{\sigma}{\sqrt{n}}$	This is true for large or small samples where the population is infinite or sampling is with replacement. The sampling distribution of means is very nearly normal (asymptotically normal) for $n \geq 30$ even when the population is nonnormal. $\mu_{\bar{X}} = \mu$, the population mean in all cases.
Proportions	$\sigma_P = \sqrt{\dfrac{p(1-p)}{n}} = \sqrt{\dfrac{pq}{n}}$	The remarks made for means apply here as well. $\mu_P = p$ in all cases
Medians	$\sigma_{\text{med}} = \sigma\sqrt{\dfrac{\pi}{2n}} = \dfrac{1.2533\sigma}{\sqrt{n}}$	For $n \geq 30$, the sampling distribution of the medians is very nearly normal. The given result holds only if the population is normal or approximately normal. $\mu_{\text{med}} = \mu$
Standard deviations	(1) $\sigma_S = \dfrac{\sigma}{\sqrt{2n}}$ (2) $\sigma_S = \sqrt{\dfrac{\mu_4 - \sigma^4}{4n\sigma^2}}$	For $n \geq 100$, the sampling distribution of S is very nearly normal. σ_S is given by (1) only if the population is normal (or approximately normal). If the population is nonnormal, (2) can be used. Note that (2) reduces to (1) when $\mu_4 = 3\sigma^4$, which is true for normal populations. For $n \geq 100$, $\mu_S = \sigma$ very nearly.
Variances	(1) $\sigma_{S^2} = \sigma^2\sqrt{\dfrac{2}{n}}$ (2) $\sigma_{S^2} = \sqrt{\dfrac{\mu_4 - \sigma^2}{n}}$	The remarks made for standard deviations apply here as well. Note that (2) yields (1) in case the population is normal. $\mu_{S^2} = (n-1)\sigma^2/n$ which is very nearly σ^2 for large n ($n \geq 30$).

Frequency Distributions

If a sample (or even a population) is large, it is difficult to observe the various characteristics or to compute statistics such as mean or standard deviation. For this reason it is useful to organize or group the *raw data*. As an illustration, suppose that a sample consists of the heights of 100 male students at XYZ University. We arrange the data into *classes* or *categories* and determine the number of individuals belonging to each class, called the *class frequency*. The resulting arrangement, Table 5-2, is called a *frequency distribution* or *frequency table*.

The first class or category, for example, consists of heights from 60 to 62 inches, indicated by 60–62, which is called a *class interval*. Since 5 students have heights belonging to this class, the corresponding class frequency is 5. Since a height that is recorded as 60 inches is actually between 59.5 and 60.5 inches while one recorded as 62 inches is actually between 61.5 and 62.5 inches, we could just as well have recorded the class interval as 59.5–62.5. The next class interval would then be 62.5–65.5, etc. In the class interval 59.5–62.5 the numbers 59.5 and 62.5 are often called *class boundaries*. The width of the jth class interval, denoted by c_j, which is usually the same for all classes (in which case it is denoted by c), is the difference between the upper and lower class boundaries. In this case $c = 62.5 - 59.5 = 3$.

The midpoint of the class interval, which can be taken as representative of the class, is called the *class mark*. In the above table the class mark corresponding to the class interval 60–62 is 61.

A graph for the frequency distribution can be supplied by a histogram, as shown shaded in Fig. 5-1, or by a *polygon graph* (often called a *frequency polygon*) connecting the midpoints of the tops in the histogram. It is of

Table 5-2
Heights of 100 Male Students at XYZ University

Height (inches)	Number of Students
60–62	5
63–65	18
66–68	42
69–71	27
72–74	8
TOTAL	100

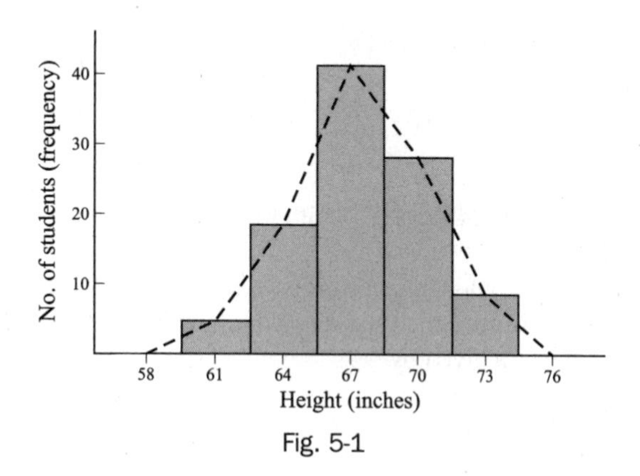

Fig. 5-1

interest that the shape of the graph seems to indicate that the sample is drawn from a population of heights that is normally distributed.

Relative Frequency Distributions

If in Table 5-2 we recorded the relative frequency or percentage rather than the number of students in each class, the result would be a *relative*, or *percentage, frequency distribution.* For example, the relative or percentage frequency corresponding to the class 63–65 is 18/100, or 18%. The corresponding histogram is then similar to that in Fig. 5-1 except that the vertical axis is relative frequency instead of frequency. The sum of the rectangular areas is then 1, or 100%.

We can consider a relative frequency distribution as a probability distribution in which probabilities are replaced by relative frequencies. Since relative frequencies can be thought of as *empirical probabilities* (see page 5), relative frequency distributions are known as *empirical probability distributions.*

Computation of Mean, Variance, and Moments for Grouped Data

We can represent a frequency distribution as in Table 5-3 by giving each class mark and the corresponding class frequency. The total frequency is n, i.e.,

$$n = f_1 + f_2 + \cdots + f_k = \sum f$$

Table 5-3

Class Mark	Class Frequency
x_1	f_1
x_2	f_2
$\vdots$	$\vdots$
x_k	f_k
TOTAL	n

Since there are f_1 numbers equal to x_1, f_2 numbers equal to $x_2, \ldots, f_k$ numbers equal to x_k, the mean is given by

$$\bar{x} = \frac{f_1 x_1 + f_2 x_2 + \cdots + f_k x_k}{n} = \frac{\sum fx}{n} \tag{24}$$

Similarly the variance is given by

$$s^2 = \frac{f_1(x_1 - \bar{x})^2 + f_2(x_2 - \bar{x})^2 + \cdots + f_k(x_k - \bar{x})^2}{n} = \frac{\sum f(x - \bar{x})^2}{n} \tag{25}$$

Note the analogy of (24) and (25) with the results (2), page 75, and (13), page 77, if f_j/n correspond to empirical probabilities.

In the case where class intervals all have equal size c, there are available short methods for computing the mean and variance. These are called *coding methods* and make use of the transformation from the class mark x to a corresponding integer u given by

$$x = a + cu \tag{26}$$

where a is an arbitrarily chosen class mark corresponding to $u = 0$. The coding formulas for the mean and variance are then given by

$$\bar{x} = a + \frac{c}{n}\sum fu = a + c\bar{u} \tag{27}$$

$$s^2 = c^2\left[\frac{\sum fu^2}{n} - \left(\frac{\sum fu}{n}\right)^2\right] = c^2(\overline{u^2} - \bar{u}^2) \tag{28}$$

Similar formulas are available for higher moments. The rth moments about the mean and the origin, respectively, are given by

$$m_r = \frac{f_1(x_1 - \bar{x})^r + \cdots + f_k(x_k - \bar{x})^r}{n} = \frac{\sum f(x - \bar{x})^r}{n} \tag{29}$$

$$m'_r = \frac{f_1x_1^r + \cdots + f_kx_k^r}{n} = \frac{\sum fx^r}{n} \tag{30}$$

The two kinds of moments are related by

$$\begin{aligned} m_1 &= 0 \\ m_2 &= m'_2 - m'^2_1 \\ m_3 &= m'_3 - 3m'_1m'_2 + 2m'^3_1 \\ m_4 &= m'_4 - 4m'_1m'_3 + 6m'^2_1m'_2 - 3m'^4_1 \end{aligned} \tag{31}$$

etc. If we write

$$M_r = \frac{\sum f(u - \bar{u})^r}{n} \qquad M'_r = \frac{\sum fu^r}{n}$$

where u is given by (26), then the relations (31) also hold between the M's. But

$$m_r = \frac{\sum f(x - \bar{x})^r}{n} = \frac{\sum f[(a + cu) - (a + c\bar{u})]^r}{n} = \frac{\sum fc^r(u - \bar{u})^r}{n} = c^rM_r$$

so that we obtain from (31) the coding formulas

$$\begin{aligned} m_1 &= 0 \\ m_2 &= c^2(M'_2 - M'^2_1) \\ m_3 &= c^3(M'_3 - 3M'_1M'_2 + 2M'^3_1) \\ m_4 &= c^4(M'_4 - 4M'_1M'_3 + 6M'^2_1M'_2 - 3M'^4_1) \end{aligned} \tag{32}$$

etc. The second equation of (32) is, of course, the same as (28).

In a similar manner other statistics, such as skewness and kurtosis, can be found for grouped samples.

SOLVED PROBLEMS

Sampling distribution of means

5.1. A population consists of the five numbers 2, 3, 6, 8, 11. Consider all possible samples of size two which can be drawn with replacement from this population. Find (a) the mean of the population, (b) the standard deviation of the population, (c) the mean of the sampling distribution of means, (d) the standard deviation of the sampling distribution of means, i.e., the standard error of means.

(a)
$$\mu = \frac{2+3+6+8+11}{5} = \frac{30}{5} = 6.0$$

(b)
$$\sigma^2 = \frac{(2-6)^2+(3-6)^2+(6-6)^2+(8-6)^2+(11-6)^2}{5} = \frac{16+9+0+4+25}{5} = 10.8$$

and $\sigma = 3.29$.

(c) There are 5(5) = 25 samples of size two which can be drawn with replacement (since any one of five numbers on the first draw can be associated with any one of the five numbers on the second draw). These are

(2, 2)	(2, 3)	(2, 6)	(2, 8)	(2, 11)
(3, 2)	(3, 3)	(3, 6)	(3, 8)	(3, 11)
(6, 2)	(6, 3)	(6, 6)	(6, 8)	(6, 11)
(8, 2)	(8, 3)	(8, 6)	(8, 8)	(8, 11)
(11, 2)	(11, 3)	(11, 6)	(11, 8)	(11, 11)

The corresponding sample means are

	2.0	2.5	4.0	5.0	6.5
	2.5	3.0	4.5	5.5	7.0
(1)	4.0	4.5	6.0	7.0	8.5
	5.0	5.5	7.0	8.0	9.5
	6.5	7.0	8.5	9.5	11.0

and the mean of the sampling distribution of means is

$$\mu_{\bar{X}} = \frac{\text{sum of all sample means in (1) above}}{25} = \frac{150}{25} = 6.0$$

illustrating the fact that $\mu_{\bar{X}} = \mu$. For a general proof of this, see Problem 5.6.

(d) The variance $\sigma_{\bar{X}}^2$ of the sampling distribution of means is obtained by subtracting the mean 6 from each number in (1), squaring the result, adding all 25 numbers obtained, and dividing by 25. The final result is

$$\sigma_{\bar{X}}^2 = \frac{135}{25} = 5.40 \quad \text{so that} \quad \sigma_{\bar{X}} = \sqrt{5.40} = 2.32$$

This illustrates the fact that for finite populations involving sampling with replacement (or infinite populations), $\sigma_{\bar{X}}^2 = \sigma^2/n$ since the right-hand side is $10.8/2 = 5.40$, agreeing with the above value. For a general proof of this, see Problem 5.7.

5.2. Solve Problem 5.1 in case sampling is without replacement.

As in (a) and (b) of Problem 5.1, $\mu = 6$ and $\sigma^2 = 10.8$, $\sigma = 3.29$.

(c) There are ${}_5C_2 = 10$ samples of size two which can be drawn without replacement (this means that we draw one number and then another number different from the first) from the population, namely,

(2, 3), (2, 6), (2, 8), (2, 11), (3, 6), (3, 8), (3, 11), (6, 8), (6, 11), (8, 11).

The selection (2, 3), for example, is considered the same as (3, 2).

The corresponding sample means are

2.5, 4.0, 5.0, 6.5, 4.5, 5.5, 7.0, 7.0, 8.5, 9.5

and the mean of sampling distribution of means is

$$\mu_{\bar{X}} = \frac{2.5 + 4.0 + 5.0 + 6.5 + 4.5 + 5.5 + 7.0 + 7.0 + 8.5 + 9.5}{10} = 6.0$$

illustrating the fact that $\mu_{\bar{X}} = \mu$.

(d) The variance of the sampling distribution of means is

$$\sigma_{\bar{X}}^2 = \frac{(2.5 - 6.0)^2 + (4.0 - 6.0)^2 + (5.0 - 6.0)^2 + \cdots + (9.5 - 6.0)^2}{10} = 4.05$$

and $\sigma_{\bar{X}} = 2.01$.

This illustrates $\sigma_{\bar{X}}^2 = \frac{\sigma^2}{n}\left(\frac{N-n}{N-1}\right)$, since the right side equals $\frac{10.8}{2}\left(\frac{5-2}{5-1}\right) = 4.05$, as obtained above. For a general proof of this result, see Problem 5.47.

5.3. Assume that the heights of 3000 male students at a university are normally distributed with mean 68.0 inches and standard deviation 3.0 inches. If 80 samples consisting of 25 students each are obtained, what would be the mean and standard deviation of the resulting sample of means if sampling were done (a) with replacement, (b) without replacement?

The numbers of samples of size 25 that could be obtained theoretically from a group of 3000 students with and without replacement are $(3000)^{25}$ and ${}_{3000}C_{25}$, which are much larger than 80. Hence, we do not get a true sampling distribution of means but only an *experimental* sampling distribution. Nevertheless, since the number of samples is large, there should be close agreement between the two sampling distributions. Hence, the mean and standard deviation of the 80 sample means would be close to those of the theoretical distribution. Therefore, we have

(a) $$\mu_{\bar{X}} = \mu = 68.0 \text{ inches} \quad \text{and} \quad \sigma_{\bar{X}} = \frac{\sigma}{\sqrt{n}} = \frac{3}{\sqrt{25}} = 0.6 \text{ inches}$$

(b) $$\mu_{\bar{X}} = \mu = 68.0 \text{ inches} \quad \text{and} \quad \sigma_{\bar{X}} = \frac{\sigma}{\sqrt{n}}\sqrt{\frac{N-n}{N-1}} = \frac{3}{\sqrt{25}}\sqrt{\frac{3000-25}{3000-1}}$$

which is only very slightly less than 0.6 inches and can for all practical purposes be considered the same as in sampling with replacement.

Thus we would expect the experimental sampling distribution of means to be approximately normally distributed with mean 68.0 inches and standard deviation 0.6 inches.

5.4. In how many samples of Problem 5.3 would you expect to find the mean (a) between 66.8 and 68.3 inches, (b) less than 66.4 inches?

The mean $\bar{X}$ of a sample in standard units is here given by $Z = \frac{\bar{X} - \mu_{\bar{X}}}{\sigma_{\bar{X}}} = \frac{\bar{X} - 68.0}{0.6}$.

(a) 66.8 in standard units $= (66.8 - 68.0)/0.6 = -2.0$

68.3 in standard units $= (68.3 - 68.0)/0.6 = 0.5$

Proportion of samples with means between 66.8 and 68.3 inches

$= $ (area under normal curve between $z = -2.0$ and $z = 0.5$)

$= $ (area between $z = -2$ and $z = 0$)

$+$ (area between $z = 0$ and $z = 0.5$)

$= 0.4772 + 0.1915 = 0.6687$

Then the expected number of samples $= (80)(0.6687)$ or 53 (Fig. 5-2).

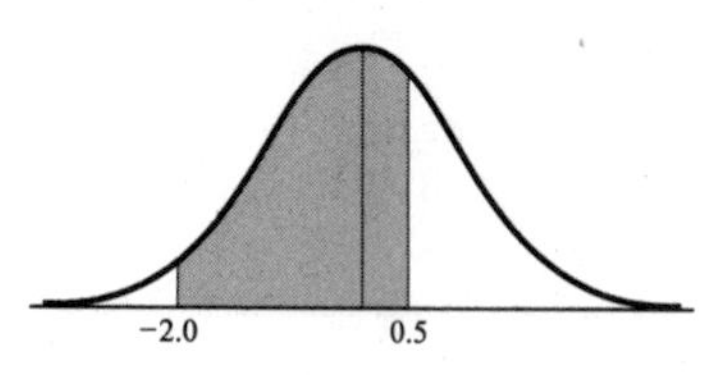

Fig. 5-2

(b) 66.4 in standard units $= (66.4 - 68.0)/0.6 = -2.67$

$$\begin{aligned}&\text{Proportion of samples with means less than 66.4 inches}\\&= (\text{area under normal curve to left of } z = -2.67)\\&= (\text{area to left of } z = 0)\\&\quad - (\text{area between } z = -2.67 \text{ and } z = 0)\\&= 0.5 - 0.4962 = 0.0038\end{aligned}$$

Then the expected number of samples $= (80)(0.0038) = 0.304$ or zero (Fig. 5-3).

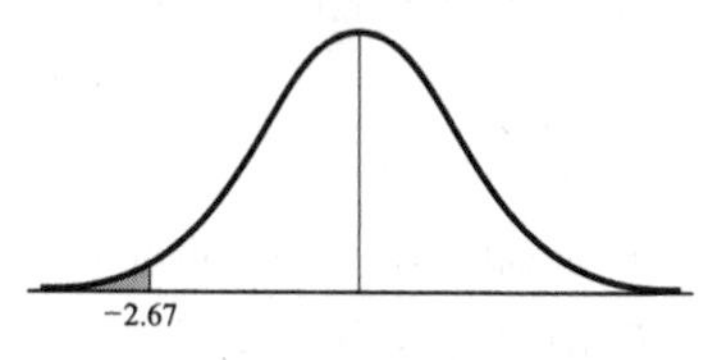

Fig. 5-3

5.5. Five hundred ball bearings have a mean weight of 5.02 oz and a standard deviation of 0.30 oz. Find the probability that a random sample of 100 ball bearings chosen from this group will have a combined weight, (a) between 496 and 500 oz, (b) more than 510 oz.

For the sampling distribution of means, $\mu_{\bar{X}} = \mu = 5.02$ oz.

$$\sigma_{\bar{X}} = \frac{\sigma}{\sqrt{n}}\sqrt{\frac{N-n}{N-1}} = \frac{0.30}{\sqrt{100}}\sqrt{\frac{500-100}{500-1}} = 0.027$$

(a) The combined weight will lie between 496 and 500 oz if the mean weight of the 100 ball bearings lies between 4.96 and 5.00 oz (Fig. 5-4).

$$4.96 \text{ in standard units} = \frac{4.96 - 5.02}{0.027} = -2.22$$

$$5.00 \text{ in standard units} = \frac{5.00 - 5.02}{0.027} = -0.74$$

$$\begin{aligned}&\text{Required probability}\\&= (\text{area between } z = -2.22 \text{ and } z = -0.74)\\&= (\text{area between } z = -2.22 \text{ and } z = 0)\\&\quad - (\text{area between } z = -0.74 \text{ and } z = 0)\\&= 0.4868 - 0.2704 = 0.2164\end{aligned}$$

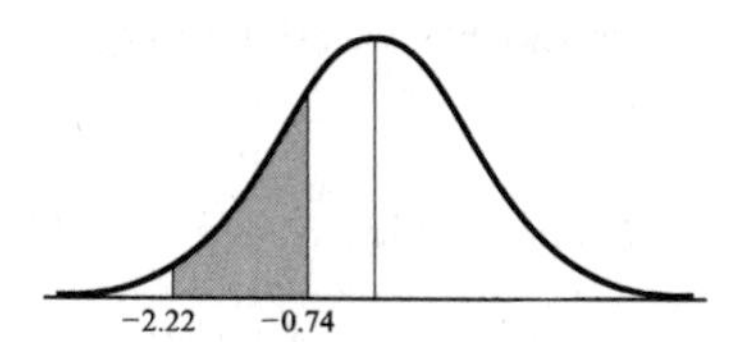

Fig. 5-4

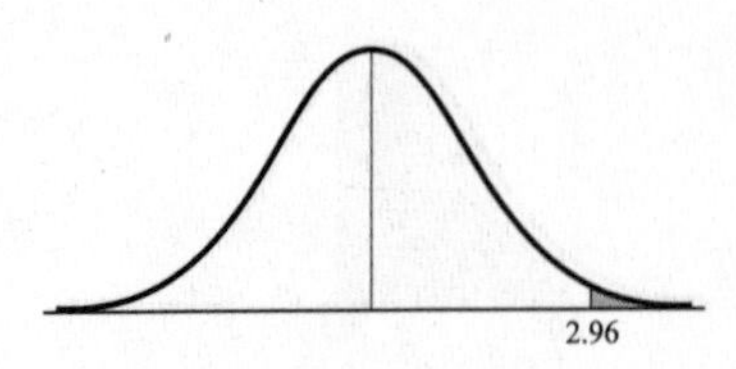

Fig. 5-5

(b) The combined weight will exceed 510 oz if the mean weight of the 100 bearings exceeds 5.10 oz (Fig. 5-5).

$$5.10 \text{ in standard units} = \frac{5.10 - 5.02}{0.027} = 2.96$$

Required probability

$$\begin{aligned} &= (\text{area to right of } z = 2.96) \\ &= (\text{area to right of } z = 0) \\ &\quad - (\text{area between } z = 0 \text{ and } z = 2.96) \\ &= 0.5 - 0.4985 = 0.0015 \end{aligned}$$

Therefore, there are only 3 chances in 2000 of picking a sample of 100 ball bearings with a combined weight exceeding 510 oz.

5.6. Theorem 5-1, page 155.

Since $X_1, X_2, \dots, X_n$ are random variables having the same distribution as the population, which has mean μ, we have

$$E(X_k) = \mu \qquad k = 1, 2, \dots, n$$

Then since the sample mean is defined as

$$\bar{X} = \frac{X_1 + \cdots + X_n}{n}$$

we have as required

$$E(\bar{X}) = \frac{1}{n}[E(X_1) + \cdots + E(X_n)] = \frac{1}{n}(n_\mu) = \mu$$

5.7. Prove Theorem 5-2, page 156.

We have

$$\bar{X} = \frac{X_1}{n} + \frac{X_2}{n} + \cdots + \frac{X_n}{n}$$

Then since $X_1, \dots, X_n$ are independent and have variance σ^2, we have by Theorems 3-5 and 3-7:

$$\text{Var}(\bar{X}) = \frac{1}{n^2}\text{Var}(X_1) + \cdots + \frac{1}{n^2}\text{Var}(X_n) = n\left(\frac{1}{n^2}\sigma^2\right) = \frac{\sigma^2}{n}$$

Sampling distribution of proportions

5.8. Find the probability that in 120 tosses of a fair coin (a) between 40% and 60% will be heads, (b) $\frac{5}{8}$ or more will be heads.

We consider the 120 tosses of the coin as a sample from the infinite population of all possible tosses of the coin. In this population the probability of heads is $p = \frac{1}{2}$, and the probability of tails is $q = 1 - p = \frac{1}{2}$.

(a) We require the probability that the number of heads in 120 tosses will be between 40% of 120, or 48, and 60% of 120, or 72. We proceed as in Chapter 4, using the normal approximation to the binomial distribution. Since the number of heads is a discrete variable, we ask for the probability that the number of heads lies between 47.5 and 72.5. (See Fig. 5-6.)

$$\mu = \text{expected number of heads} = np = 120\left(\frac{1}{2}\right) = 60$$

and

$$\sigma = \sqrt{npq} = \sqrt{(120)\left(\frac{1}{2}\right)\left(\frac{1}{2}\right)} = 5.48$$

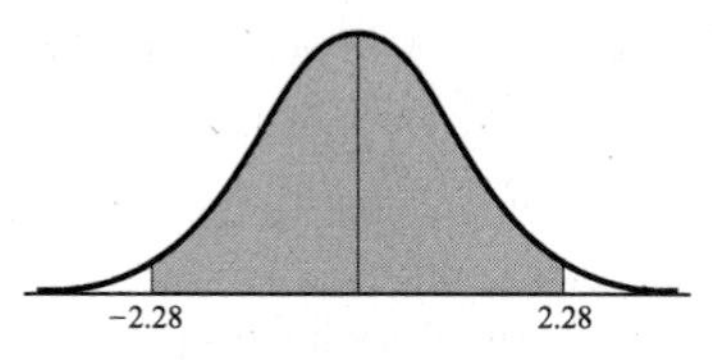

Fig. 5-6

$$47.5 \text{ in standard units} = \frac{47.5 - 60}{5.48} = -2.28$$

$$72.5 \text{ in standard units} = \frac{72.5 - 60}{5.48} = 2.28$$

Required probability

= (area under normal curve between $z = -2.28$ and $z = 2.28$)

= 2(area between $z = 0$ and $z = 2.28$)

= 2(0.4887) = 0.9774

Another method

$$\mu_P = p = \frac{1}{2} = 0.50 \qquad \sigma_P = \sqrt{\frac{pq}{n}} = \sqrt{\frac{\frac{1}{2}\left(\frac{1}{2}\right)}{120}} = 0.0456$$

$$40\% \text{ in standard units} = \frac{0.40 - 0.50}{0.0456} = -2.19$$

$$60\% \text{ in standard units} = \frac{0.60 - 0.50}{0.0456} = 2.19$$

Therefore, the required probability is the area under the normal curve between $z = -2.19$ and $z = 2.19$, i.e., $2(0.4857) = 0.9714$.

Although this result is accurate to two significant figures, it does not agree exactly since we have not used the fact that the proportion is actually a discrete variable. To account for this, we subtract $\frac{1}{2n} = \frac{1}{2(120)}$ from 0.40 and add $\frac{1}{2n} = \frac{1}{2(120)}$ to 0.60. Therefore, the required proportions in standard units are, since $1/240 = 0.00417$,

$$\frac{0.40 - 0.00417 - 0.50}{0.0456} = -2.28 \quad \text{and} \quad \frac{0.60 + 0.00417 - 0.50}{0.0456} = 2.28$$

so that agreement with the first method is obtained.

Note that $(0.40 - 0.00417)$ and $(0.60 + 0.00417)$ correspond to the proportions $47.5/120$ and $72.5/120$ in the first method above.

(b) Using the second method of (a), we find that since $\frac{5}{8} = 0.6250$,

$$(0.6250 - 0.00417) \text{ in standard units} = \frac{0.6250 - 0.00417 - 0.50}{0.0456} = 2.65$$

Required probability = (area under normal curve to right of $z = 2.65$)

= (area to right of $z = 0$)

− (area between $z = 0$ and $z = 2.65$)

= 0.5 − 0.4960 = 0.0040

5.9. Each person of a group of 500 people tosses a fair coin 120 times. How many people should be expected to report that (a) between 40% and 60% of their tosses resulted in heads, (b) $\frac{5}{8}$ or more of their tosses resulted in heads?

This problem is closely related to Problem 5.8. Here we consider 500 samples, of size 120 each, from the infinite population of all possible tosses of a coin.

(a) Part (a) of Problem 5.8 states that of all possible samples, each consisting of 120 tosses of a coin, we can expect to find 97.74% with a percentage of heads between 40% and 60%. In 500 samples we can expect to find about 97.74% of 500, or 489, samples with this property. It follows that about 489 people would be expected to report that their experiment resulted in between 40% and 60% heads.

It is interesting to note that $500 - 489 = 11$ people would be expected to report that the percentage of heads was not between 40% and 60%. These people might reasonably conclude that their coins were loaded, even though they were fair. This type of error is a *risk* that is always present whenever we deal with probability.

(b) By reasoning as in (a), we conclude that about $(500)(0.0040) = 2$ persons would report that $\frac{5}{8}$ or more of their tosses resulted in heads.

5.10. It has been found that 2% of the tools produced by a certain machine are defective. What is the probability that in a shipment of 400 such tools, (a) 3% or more, (b) 2% or less will prove defective?

$$\mu_P = p = 0.02 \qquad \text{and} \qquad \sigma_P = \sqrt{\frac{pq}{n}} = \sqrt{\frac{0.02(0.98)}{400}} = \frac{0.14}{20} = 0.007$$

(a) Using the correction for discrete variables, $1/2n = 1/800 = 0.00125$, we have

$$(0.03 - 0.00125) \text{ in standard units} = \frac{0.03 - 0.00125 - 0.02}{0.007} = 1.25$$

$$\text{Required probability} = (\text{area under normal curve to right } z = 1.25) = 0.1056$$

If we had not used the correction we would have obtained 0.0764.

Another method

(3% of 400) = 12 defective tools. On a continuous basis, 12 or more tools means 11.5 or more.

$$\mu = (2\% \text{ of } 400) = 8 \qquad \sigma = \sqrt{npq} = \sqrt{(400)(0.02)(0.98)} = 2.8$$

Then, 11.5 in standard units $= (11.5 - 8)/2.8 = 1.25$, and as before the required probability is 0.1056.

(b)
$$(0.02 + 0.00125) \text{ in standard units} = \frac{0.02 + 0.00125 - 0.02}{0.007} = 0.18$$

$$\text{Required probability} = (\text{area under normal curve to left of } z = 0.18)$$
$$= 0.5000 + 0.0714 = 0.5714$$

If we had not used the correction, we would have obtained 0.5000. The second method of part (a) can also be used.

5.11. The election returns showed that a certain candidate received 46% of the votes. Determine the probability that a poll of (a) 200, (b) 1000 people selected at random from the voting population would have shown a majority of votes in favor of the candidate.

(a)
$$\mu_P = p = 0.46 \qquad \text{and} \qquad \sigma_P = \sqrt{\frac{pq}{n}} = \sqrt{\frac{0.46(0.54)}{200}} = 0.0352$$

Since $1/2n = 1/400 = 0.0025$, a majority is indicated in the sample if the proportion in favor of the candidate is $0.50 + 0.0025 = 0.5025$ or more. (This proportion can also be obtained by realizing that 101 or more indicates a majority, but this as a continuous variable is 100.5; and so the proportion is $100.5/200 = 0.5025$.)

Then, 0.5025 in standard units $= (0.5025 - 0.46)/0.0352 = 1.21$ and

$$\text{Required probability} = (\text{area under normal curve to right of } z = 1.21)$$
$$= 0.5000 - 0.3869 = 0.1131$$

(b) $\mu_P = p = 0.46$, $\sigma_P = \sqrt{pq/n} = \sqrt{0.46(0.54)1000} = 0.0158$, and

$$0.5025 \text{ in standard units} = \frac{0.5025 - 0.46}{0.0158} = 2.69$$

$$\text{Required probability} = (\text{area under normal curve to right of } z = 2.69)$$
$$= 0.5000 - 0.4964 = 0.0036$$

Sampling distributions of differences and sums

5.12. Let U_1 be a variable that stands for any of the elements of the population 3, 7, 8 and U_2 a variable that stands for any of the elements of the population 2, 4. Compute (a) μ_{U_1}, (b) μ_{U_2}, (c) $\mu_{U_1-U_2}$, (d) σ_{U_1}, (e) σ_{U_2}, (f) $\sigma_{U_1-U_2}$.

(a)
$$\mu_{U_1} = \text{mean of population } U_1 = \frac{1}{3}(3 + 7 + 8) = 6$$

(b)
$$\mu_{U_2} = \text{mean of population } U_2 = \frac{1}{2}(2 + 4) = 3$$

(c) The population consisting of the differences of any member of U_1 and any member of U_2 is

$$\begin{matrix} 3-2 & 7-2 & 8-2 \\ 3-4 & 7-4 & 8-4 \end{matrix} \quad \text{or} \quad \begin{matrix} 1 & 5 & 6 \\ -1 & 3 & 4 \end{matrix}$$

Then
$$\mu_{U_1-U_2} = \text{mean of } (U_1 - U_2) = \frac{1 + 5 + 6 + (-1) + 3 + 4}{6} = 3$$

which illustrates the general result $\mu_{U_1-U_2} = \mu_{U_1-U_2}$, as is seen from (a) and (b).

(d)
$$\sigma^2_{U_1} = \text{variance of population } U_1 = \frac{(3-6)^2 + (7-6)^2 + (8-6)^2}{3} = \frac{14}{3}$$

or $\sigma_{U_1} = \sqrt{\frac{14}{3}}$.

(e)
$$\sigma^2_{U_2} = \text{variance of population } U_2 = \frac{(2-3)^2 + (4-3)^2}{2} = 1$$

or $\sigma_{U_2} = 1$.

(f)
$$\sigma^2_{U_1-U_2} = \text{variance of population } (U_1 - U_2)$$

$$= \frac{(1-3)^2 + (5-3)^2 + (6-3)^2 + (-1-3)^2 + (3-3)^2 + (4-3)^2}{6} = \frac{17}{3}$$

or $\sigma_{U_1-U_2} = \sqrt{\frac{17}{3}}$.

This illustrates the general result for independent samples, $\sigma_{U_1-U_2} = \sqrt{\sigma^2_{U_1} = \sigma^2_{U_2}}$, as is seen from (d) and (e). The proof of the general result follows from Theorem 3-7, page 78.

5.13. The electric light bulbs of manufacturer A have a mean lifetime of 1400 hours with a standard deviation of 200 hours, while those of manufacturer B have a mean lifetime of 1200 hours with a standard deviation of 100 hours. If random samples of 125 bulbs of each brand are tested, what is the probability that the brand A bulbs will have a mean lifetime that is at least (a) 160 hours, (b) 250 hours more than the brand B bulbs?

Let $\bar{X}_A$ and $\bar{X}_B$ denote the mean lifetimes of samples A and B, respectively. Then

$$\mu_{\bar{X}_A-\bar{X}_B} = \mu_{\bar{X}_A} - \mu_{\bar{X}_B} = 1400 - 1200 = 200 \text{ hours}$$

and
$$\sigma_{\bar{X}_A-\bar{X}_B} = \sqrt{\frac{\sigma_A^2}{n_A} + \frac{\sigma_B^2}{n_B}} = \sqrt{\frac{(100)^2}{125} + \frac{(200)^2}{125}} = 20 \text{ hours}$$

The standardized variable for the difference in means that

$$Z = \frac{(\bar{X}_A - \bar{X}_B) - (\mu_{\bar{X}_A-\bar{X}_B})}{\sigma_{\bar{X}_A-\bar{X}_B}} = \frac{(\bar{X}_A - \bar{X}_B) - 200}{20}$$

and is very nearly normally distributed.

(a) The difference 160 hours in standard units $= (160 - 200)/20 = -2$.

$$\text{Required probability} = (\text{area under normal curve to right of } z = -2)$$
$$= 0.5000 + 0.4772 = 0.9772$$

(b) The difference 250 hours in standard units $= (250 - 200)/20 = 2.50$.

$$\text{Required probability} = (\text{area under normal curve to right of } z = 2.50)$$
$$= 0.5000 - 0.4938 = 0.0062$$

5.14. Ball bearings of a given brand weigh 0.50 oz with a standard deviation of 0.02 oz. What is the probability that two lots, of 1000 ball bearings each, will differ in weight by more than 2 oz?

Let $\bar{X}_1$ and $\bar{X}_2$ denote the mean weights of ball bearings in the two lots. Then

$$\mu_{\bar{X}_1-\bar{X}_2} = \mu_{\bar{X}_1} - \mu_{\bar{X}_2} = 0.50 - 0.50 = 0$$

$$\sigma_{\bar{X}_1-\bar{X}_2} = \sqrt{\frac{\sigma_1^2}{n_1} + \frac{\sigma_2^2}{n_2}} = \sqrt{\frac{(0.02)^2}{1000} + \frac{(0.02)^2}{1000}} = 0.000895$$

The standardized variable for the difference in means is $Z = \dfrac{(\bar{X}_1 - \bar{X}_2) - 0}{0.000895}$ and is very nearly normally distributed.

A difference of 2 oz in the lots is equivalent to a difference of $2/1000 = 0.002$ oz in the means. This can occur either if $\bar{X}_1 - \bar{X}_2 \geq 0.002$ or $\bar{X}_1 - \bar{X}_2 \leq -0.002$, i.e.,

$$Z \geq \frac{0.002 - 0}{0.000895} = 2.23 \qquad \text{or} \qquad Z \leq \frac{-0.002 - 0}{0.000895} = -2.23$$

Then

$$P(Z \geq 2.23 \text{ or } Z \leq -2.23) = P(Z \geq 2.23) + P(Z \leq 2.23) = 2(0.5000 - 0.4871) = 0.0258$$

5.15. A and B play a game of heads and tails, each tossing 50 coins. A will win the game if he tosses 5 or more heads than B, otherwise B wins. Determine the odds against A winning any particular game.

Let P_A and P_B denote the proportion of heads obtained by A and B. If we assume the coins are all fair, the probability p of heads is $\frac{1}{2}$. Then

$$\mu_{P_A-P_B} = \mu_{P_A} - \mu_{P_B} = 0$$

and
$$\sigma_{P_A-P_B} = \sqrt{\sigma_{P_A}^2 + \sigma_{P_B}^2} = \sqrt{\frac{pq}{n_A} + \frac{pq}{n_B}} = \sqrt{\frac{2(\frac{1}{2})(\frac{1}{2})}{50}} = 0.10$$

The standardized variable for the difference in proportions is $Z = (P_A - P_B - 0)/0.10$.

On a continuous-variable basis, 5 or more heads means 4.5 or more heads, so that the difference in proportions should be $4.5/50 = 0.09$ or more, i.e., Z is greater than or equal to $(0.09 - 0)/0.10 = 0.9$

(or $Z \geq 0.9$). The probability of this is the area under the normal curve to the right of $Z = 0.9$, which is $0.5000 - 0.3159 = 0.1841$.

Therefore, the odds against A winning are $(1 - 0.1841) : 0.1841 = 0.8159 : 0.1841$, or 4.43 to 1.

5.16. Two distances are measured as 27.3 inches and 15.6 inches, with standard deviations (standard errors) of 0.16 inches and 0.08 inches, respectively. Determine the mean and standard deviation of (a) the sum, (b) the difference of the distances.

If the distances are denoted by D_1 and D_2, then

(a)
$$\mu_{D_1+D_2} = \mu_{D_1} + \mu_{D_2} = 27.3 + 15.6 = 42.9 \text{ inches}$$

$$\sigma_{D_1+D_2} = \sqrt{\sigma_{D_1}^2 + \sigma_{D_2}^2} = \sqrt{(0.16)^2 + (0.08)^2} = 0.18 \text{ inches}$$

(b)
$$\mu_{D_1-D_2} = \mu_{D_1} + \mu_{D_2} = 27.3 + 15.6 = 11.7 \text{ inches}$$

$$\sigma_{D_1-D_2} = \sqrt{\sigma_{D_1}^2 + \sigma_{D_2}^2} = \sqrt{(0.16)^2 + (0.08)^2} = 0.18 \text{ inches}$$

5.17. A certain type of electric light bulb has a mean lifetime of 1500 hours and a standard deviation of 150 hours. Three bulbs are connected so that when one burns out, another will go on. Assuming the lifetimes are normally distributed, what is the probability that lighting will take place for (a) at least 5000 hours, (b) at most 4200 hours?

Denote the lifetimes as L_1, L_2, and L_3. Then

$$\mu_{L_1+L_2+L_3} = \mu_{L_1} + \mu_{L_2} + \mu_{L_3} = 1500 + 1500 + 1500 = 4500 \text{ hours}$$

$$\sigma_{L_1+L_2+L_3} = \sqrt{\sigma_{L_1}^2 + \sigma_{L_2}^2 + \sigma_{L_3}^2} = \sqrt{3(150)^2} = 260 \text{ hours}$$

(a) 5000 hours in standard units $= (5000 - 4500)/260 = 1.92$.

$$\text{Required probability} = (\text{area under normal curve to right of } z = 1.92)$$
$$= 0.5000 - 0.4726 = 0.0274$$

(b) 4200 hours in standard units $= (4200 - 4500)/260 = -1.15$.

$$\text{Required probability} = (\text{area under normal curve to left of } z = -1.15)$$
$$= 0.5000 - 0.3749 = 0.1251$$

Sampling distribution of variances

5.18. With reference to Problem 5.1, find (a) the mean of the sampling distribution of variances, (b) the standard deviation of the sampling distribution of variances, i.e., the standard error of variances.

(a) The sample variances corresponding to each of the 25 samples in Problem 5.1 are

0	0.25	4.00	9.00	20.25
0.25	0	2.25	6.25	16.00
4.00	2.25	0	1.00	6.25
9.00	6.25	1.00	0	2.25
20.25	16.00	6.25	2.25	0

The mean of sampling distribution of variances is

$$\mu_{S^2} = \frac{\text{sum of all variances in the table above}}{25} = \frac{135}{25} = 5.40$$

This illustrates the fact that $\mu_{S^2} = (n - 1)(\sigma^2)/n$, since for $n = 2$ and $\sigma^2 = 10.8$ [see Problem 5.1(b)], the right-hand side is $\frac{1}{2}(10.8) = 5.4$.

The result indicates why a corrected variance for samples is often defined as $\hat{S}^2 = \frac{n}{n-1}S^2$, since it then follows that $\mu_{\hat{S}^2} = \sigma^2$ (see also remarks on page 158).

(b) The variance of the sampling distribution of variances $\sigma^2_{S^2}$ is obtained by subtracting the mean 5.40 from each of the 25 numbers in the above table, squaring these numbers, adding them, and then dividing the result by 25. Therefore, $\sigma^2_{S^2} = 575.75/25 = 23.03$ or $\sigma_{S^2} = 4.80$.

5.19. Work the previous problem if sampling is without replacement.

(a) There are 10 samples whose variances are given by the numbers above (or below) the diagonal of zeros in the table of Problem 5.18(a). Then

$$\mu_{S^2} = \frac{0.25 + 4.00 + 9.00 + 20.25 + 2.25 + 6.25 + 16.00 + 1.00 + 6.25 + 2.25}{10} = 6.75$$

This is a special case of the general result $\mu_{S^2} = \left(\frac{N}{N-1}\right)\left(\frac{n-1}{n}\right)\sigma^2$ [equation (*19*), page 158] as is verified by putting $N = 5$, $n = 2$, and $\sigma^2 = 10.8$ on the right-hand side to obtain $\mu_{S^2} = \left(\frac{5}{4}\right)\left(\frac{1}{2}\right)(10.8) = 6.75$.

(b) Subtracting 6.75 from each of the 10 numbers above the diagonal of zeros in the table of Problem 5.18(a), squaring these numbers, adding them, and dividing by 10, we find $\sigma^2_{S^2} = 39.675$, or $\sigma_{S^2} = 6.30$.

5.20. Prove that

$$E(S^2) = \frac{n-1}{n}\sigma^2$$

where S^2 is the sample variance for a random sample of size n, as defined on pages 157–158, and σ^2 is the variance of the population.

Method 1

We have

$$X_1 - \bar{X} = X_1 - \frac{1}{n}(X_1 + \cdots + X_n) = \frac{1}{n}[(n-1)X_1 - X_2 - \cdots - X_n]$$

$$= \frac{1}{n}[(n-1)(X_1 - \mu) - (X_2 - \mu) - \cdots - (X_n - \mu)]$$

Then

$$(X_1 - \bar{X})^2 = \frac{1}{n^2}[(n-1)^2(X_1 - \mu)^2 + (X_2 - \mu)^2 + \cdots + (X_n - \mu)^2 + \text{cross-product terms}]$$

Since the X's are independent, the expectation of each cross-product term is zero, and we have

$$E[(X_1 - \bar{X})^2] = \frac{1}{n^2}\{(n-1)^2E[(X_1 - \mu)^2] + E[(X_2 - \mu)^2] + \cdots + E[(X_n - \mu)^2]\}$$

$$= \frac{1}{n^2}\{(n-1)^2\sigma^2 + \sigma^2 + \cdots + \sigma^2\}$$

$$= \frac{1}{n^2}\{(n-1)^2\sigma^2 + (n-1)\sigma^2\} = \frac{n-1}{n}\sigma^2$$

Similarly, $E[(X_k - \bar{X})^2] = (n-1)\sigma^2/n$ for $k = 2, \ldots, n$. Therefore,

$$E(S^2) = \frac{1}{n}E[(X_1 - \bar{X})^2 + \cdots + (X_n - \bar{X})^2]$$

$$= \frac{1}{n}\left[\frac{n-1}{n}\sigma^2 + \cdots + \frac{n-1}{n}\sigma^2\right] = \frac{n-1}{n}\sigma^2$$

Method 2

We have $X_j - \bar{X} = (X_j - \mu) - (\bar{X} - \mu)$. Then

$$(X_j - \bar{X})^2 = (X_j - \mu)^2 - 2(X_j - \mu)(\bar{X} - \mu) + (\bar{X} - \mu)^2$$

and

$$\text{(1)} \qquad \sum(X_j - \bar{X})^2 = \sum(X_j - \mu)^2 - 2(\bar{X} - \mu)\sum(X_j - \mu) + \sum(\bar{X} - \mu)^2$$

where the sum is from $j = 1$ to n. This can be written

$$\text{(2)} \qquad \sum(X_j - \bar{X})^2 = \sum(X_j - \mu)^2 - 2n(\bar{X} - \mu)^2 + n(\bar{X} - \mu)^2$$
$$= \sum(X_j - \mu)^2 - n(\bar{X} - \mu)^2$$

since $\Sigma(X_j - \mu) = \Sigma X_j - n_\mu = n(\bar{X} - \mu)$. Taking the expectation of both sides of (2) and using Problem 5.7, we find

$$E\left[\sum(X_j - \bar{X})^2\right] = E\left[\sum(X_j - \mu)^2\right] - nE[(\bar{X} - \mu)^2]$$

$$= n\sigma^2 - n\left(\frac{\sigma^2}{n}\right) = (n - 1)\sigma^2$$

from which

5.21. Prove Theorem 5-4, page 156.

If $X_j, j = 1, 2, \ldots, n$, is normally distributed with mean μ and variance σ^2, then its characteristic function is (see Table 4-2, page 110)

$$\phi_j(\omega) = e^{i\mu\omega - (\sigma^2\omega^2)/2}$$

The characteristic function of $X_1 + X_2 + \cdots X_n$ is then, by Theorem 3-12,

$$\phi(\omega) = \phi_1(\omega)\phi_2(\omega) \cdots \phi_n(\omega) = e^{in\mu\omega - (n\sigma^2\omega^2)/2}$$

since the X_j are independent. Then, by Theorem 3-11, the characteristic function of

$$\bar{X} = \frac{X_1 + X_2 + \cdots + X_n}{n}$$

is

$$\phi_{\bar{X}}(\omega) = \phi\left(\frac{\omega}{n}\right) = e^{i\mu\omega - [(\sigma^2/n)\omega^2]/2}$$

But this is the characteristic function for a normal distribution with mean μ and variance σ^2/n, and the desired result follows from Theorem 3-13.

5.22. Prove Theorem 5-6, page 158.

By definition, $(n - 1)\hat{S}^2 = \Sigma_{j=1}^{n}(X_j - \bar{X})^2$. It then follows from (2) of Method 2 in Problem 5.20 that $V = V_1 + V_2$,where

$$V = \sum_{j=1}^{n}\frac{(X_j - \mu)^2}{\sigma^2}, \qquad V_1 = \frac{(n - 1)\hat{S}^2}{\sigma^2}, \qquad V_2 = \frac{(\bar{X} - \mu)^2}{\sigma^2/n}$$

Now by Theorem 4-3, page 115, V is chi-square distributed with n degrees of freedom [as is seen on replacing X_j by $(X_j - \mu)/\sigma$]. Also, by Problem 5.21, $\bar{X}$ is normally distributed with mean μ and variance σ^2/n. Therefore, from Theorem 4-3 with $\nu = 1$ and X_1 replaced by $(\bar{X} - \mu)/\sqrt{\sigma^2/n}$, we see that V_2 is chi-square distributed with 1 degree of freedom. It follows from Theorem 4-5, page 115, that if V_1 and V_2 are independent, then V_1, is chi-square distributed with $n - 1$ degrees of freedom. Since it can be shown that V_1 and V_2 are indeed independent, the required result follows.

5.23. (a) Use Theorem 5-6 to determine the expected number of samples in Problem 5.1 for which sample variances are greater than 7.2. (b) Check the result in (a) with the actual result.

(a) We have $n = 2$, $\sigma^2 = 10.8$ [from Problem 5.1(b)]. For $s_1^2 = 7.2$, we have

$$\frac{ns_1^2}{\sigma^2} = \frac{(2)(7.2)}{10.8} = 1.33$$

According to Theorem 5-6, $\chi^2 = nS^2/\sigma^2 = 2S^2/10.8$ has the chi-square distribution with 1 degree of freedom. From the table in Appendix E it follows that

$$P(S^2 \geq s_1^2) = P(\chi^2 \geq 1.33) = 0.25$$

Therefore, we would expect about 25% of the samples, or 6, to have variances greater than 7.2.

(b) From Problem 5.18 we find by counting that there are actually 6 variances greater than 7.2, so that there is agreement.

Case where population variance is unknown

5.24. Prove Theorem 5-7, page 159.

Let $Y = \dfrac{\bar{X} - \mu}{\sigma/\sqrt{n}}$, $Z = \dfrac{nS^2}{\sigma^2}$, $\nu = n - 1$. Then since the X_j are normally distributed with mean μ and variance σ^2, we know (Problem 5.21) that $\bar{X}$ is normally distributed with mean μ and variance σ^2/n, so that y is normally distributed with mean 0 and variance 1. Also, from Theorem 5-6, page 158, or Problem 5.22, Z is chi square distributed with $\nu = n - 1$ degrees of freedom. Furthermore, it can be shown that Y and Z are independent.

It follows from Theorem 4-6, page 116, that

$$T = \frac{Y}{\sqrt{Z/\nu}} = \frac{\bar{X} - \mu}{S/\sqrt{n-1}} = \frac{\bar{X} - \mu}{\hat{S}/\sqrt{n}}$$

has the t distribution with $n - 1$ degrees of freedom.

5.25. According to the table of Student's t distribution for 1 degree of freedom (Appendix D), we have $P(-1.376 \leq T \leq 1.376) = 0.60$. Check whether this is confirmed by the results obtained in Problem 5.1.

From the values of $\bar{X}$ in (1) on page 155, and the values of S^2 in Problem 5.18(a), we obtain the following values for $T = (\bar{X} - \mu)/(S/\sqrt{1})$:

$-\infty$	-7.0	-1.0	-0.33	0.11
-7.0	$-\infty$	-1.0	-0.20	0.25
-1.0	-1.0	$\cdots$	1.0	1.0
-0.33	-0.20	1.0	∞	2.33
0.11	0.25	1.0	2.33	∞

There are actually 16 values for which $-1.376 \leq T \leq 1.376$ whereas we would expect $(0.60)(25) = 15$. This is not too bad considering the small amount of data involved. This method of sampling was in fact the way "Student" originally obtained the t distribution.

Sampling distribution of ratios of variances

5.26. Prove Theorem 5-8, page 159.

Denote the samples of sizes m and n by $X_1, \ldots, X_m$ and $Y_1, \ldots, Y_n$, respectively. Then the sample variances are given by

$$S_1^2 = \frac{1}{m}\sum_{j=1}^{m}(X_j - \bar{X})^2, \qquad S_2^2 = \frac{1}{n}\sum_{j=1}^{n}(Y_j - \bar{Y})^2$$

where $\bar{X}$, $\bar{Y}$ are the sample means.

Now from Theorem 5-6, page 158, we know that mS_1^2/σ_1^2 and nS_2^2/σ_2^2 are chi-square distributed with $m - 1$ and $n - 1$ degrees of freedom, respectively. Therefore, from Theorem 4-7, page 117, it follows that

$$F = \frac{mS_1^2/(m-1)\sigma_1^2}{nS_2^2/(n-1)\sigma_2^2} = \frac{\hat{S}_1^2/\sigma_1^2}{\hat{S}_2^2/\sigma_2^2}$$

has the F distribution with $m - 1, n - 1$ degrees of freedom.

5.27. Two samples of sizes 8 and 10 are drawn from two normally distributed populations having variances 20 and 36, respectively. Find the probability that the variance of the first sample is more than twice the variance of the second.

We have $m = 8, n = 10, \sigma_1^2 = 20, \sigma_2^2 = 36$. Therefore,

$$F = \frac{8S_1^2/(7)(20)}{10S_2^2/(9)(36)} = 1.85\frac{S_1^2}{S_2^2}$$

The number of degrees of freedom for numerator and denominator are $\nu_1 = m - 1 = 7, \nu_2 = n - 1 = 9$. Now if S_1^2 is more than twice S_2^2, i.e., $S_1^2 > 2S_2^2$, then $F > 3.70$. Referring to the tables in Appendix F, we see that the probability is less than 0.05 but more than 0.01. For exact values we need a more extensive tabulation of the F distribution.

Frequency distributions

5.28. In Table 5-4 the weights of 40 male students at State University are recorded to the nearest pound. Construct a frequency distribution.

Table 5-4

138	164	150	132	144	125	149	157
146	158	140	147	136	148	152	144
168	126	138	176	163	119	154	165
146	173	142	147	135	153	140	135
161	145	135	142	150	156	145	128

The largest weight is 176 lb, and the smallest weight is 119 lb, so that the range is $176 - 119 = 57$ lb.

If 5 class intervals are used, the class interval size is $57/5 = 11$ approximately.

If 20 class intervals are used, the class interval size is $57/20 = 3$ approximately.

One convenient choice for the class interval size is 5 lb. Also, it is convenient to choose the class marks as 120, 125, 130, 135, . . . pounds. Therefore, the class intervals can be taken as 118–122, 123–127, 128–132, ... With this choice the class boundaries are 117.5, 122.5, 127.5, . . . , which do not coincide with observed data.

The required frequency distribution is shown in Table 5-5. The center column, called a *tally*, or *score, sheet*, is used to tabulate the class frequencies from the raw data and is usually omitted in the final presentation of the frequency distribution.

Another possibility

Of course, other possible frequency distributions exist. Table 5-6, for example, shows a frequency distribution with only 7 classes, in which the class interval is 9 lb.

Table 5-5

Weight (lb)	Tally	Frequency
118–122	/	1
123–127	//	2
128–132	//	2
133–137	////	4
138–142	~~////~~ /	6
143–147	~~////~~ ///	8
148–152	~~////~~	5
153–157	////	4
158–162	//	2
163–167	///	3
168–172	/	1
173–177	//	2
	TOTAL	40

Table 5-6

Weight (lb)	Tally	Frequency
118–126	///	3
127–135	~~////~~	5
136–144	~~////~~ ////	9
145–153	~~////~~ ~~////~~ //	12
154–162	~~////~~	5
163–171	////	4
172–180	//	2
	TOTAL	40

5.29. Construct a histogram and a frequency polygon for the weight distribution in Problem 5.28.

The histogram and frequency polygon for each of the two cases considered in Problem 5.28 are given in Figs. 5-7 and 5-8. Note that the centers of the bases of the rectangles are located at the class marks.

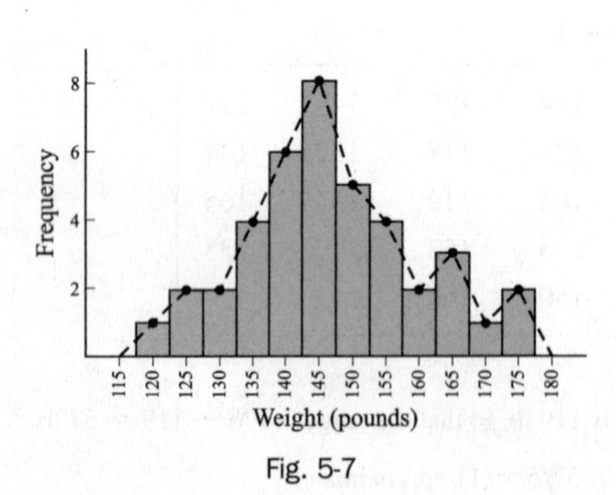

Fig. 5-7

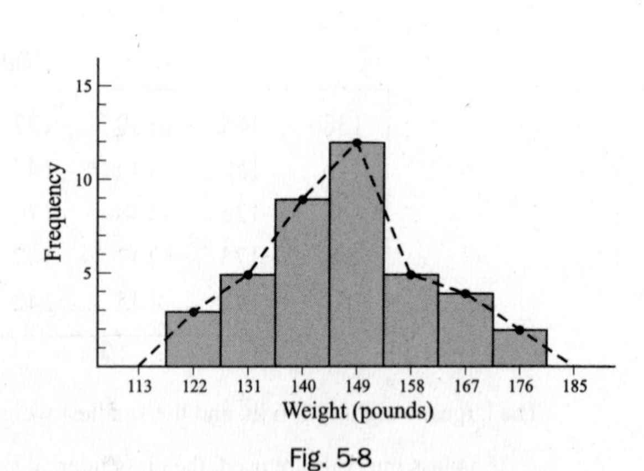

Fig. 5-8

5.30. Five pennies were simultaneously tossed 1000 times and at each toss the number of heads was observed. The numbers of tosses during which 0, 1, 2, 3, 4, and 5 heads were obtained are shown in Table 5-7. Graph the data.

The data can be shown graphically either as in Fig. 5-9 or Fig. 5-10.

Figure 5-9 seems to be a more natural graph to use. One reason is that the number of heads cannot be 1.5 or 3.2. This graph is a form of bar graph where the bars have zero width, and it is sometimes called a *rod graph.* It is especially useful when the data are discrete.

Figure 5-10 shows a histogram of the data. Note that the total area of the histogram is the total frequency 1000, as it should be.

Table 5-7

Number of Heads	Number of Tosses (frequency)
0	38
1	144
2	342
3	287
4	164
5	25
TOTAL	1000

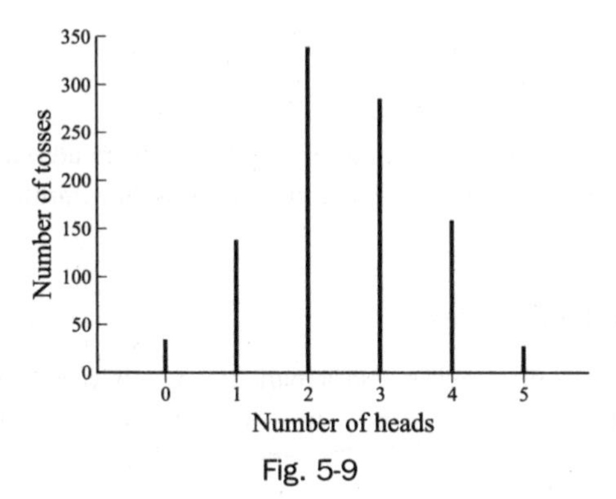

Fig. 5-9

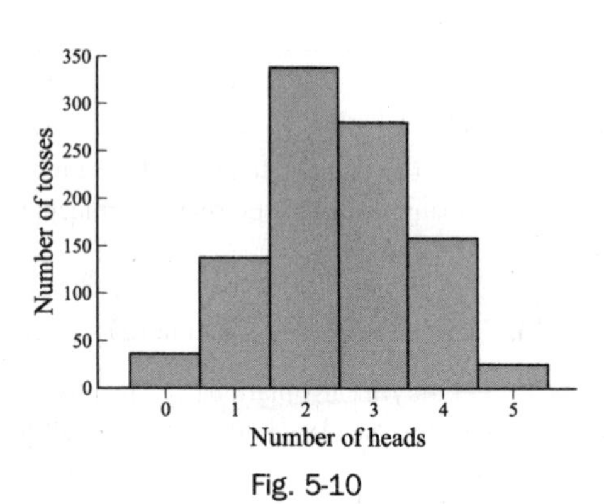

Fig. 5-10

Computation of mean, variance, and moments for samples

5.31. Find the arithmetic mean of the numbers 5, 3, 6, 5, 4, 5, 2, 8, 6, 5, 4, 8, 3, 4, 5, 4, 8, 2, 5, 4.

Method 1

$$\bar{x} = \frac{\sum x}{n} = \frac{5+3+6+5+4+5+2+8+6+5+4+8+3+4+5+4+8+2+5+4}{20}$$

$$= \frac{96}{20} = 4.8$$

Method 2

There are six 5s, two 3s, two 6s, five 4s, two 2s, and three 8s. Then

$$\bar{x} = \frac{\sum fx}{n} = \frac{(6)(5)+(2)(3)+(2)(6)+(5)(4)+(2)(2)+(3)(8)}{6+2+2+5+2+3} = \frac{96}{20} = 4.8$$

5.32. Four groups of students, consisting of 15, 20, 10, and 18 individuals, reported mean weights of 162, 148, 153, and 140 lb, respectively. Find the mean weight of all the students.

$$\bar{x} = \frac{\sum fx}{n} = \frac{(15)(162)+(20)(148)+(10)(153)+(18)(140)}{15+20+10+18} = 150 \text{ lb}$$

5.33. Use the frequency distribution of heights in Table 5-2, page 161, to find the mean height of the 100 male students at XYZ University.

The work is outlined in Table 5-8. Note that all students having heights 60–62 inches, 63–65 inches, etc., are considered as having heights 61, 64, etc., inches. The problem then reduces to finding the mean height of 100 students if 5 students have height 61 inches, 18 have height 64 inches, etc.

$$\bar{x} = \frac{\sum fx}{\sum f} = \frac{\sum fx}{n} = \frac{6745}{100} = 67.45 \text{ inches}$$

Table 5-8

Height (inches)	Class Mark (x)	Frequency (f)	fx
60–62	61	5	305
63–65	64	18	1152
66–68	67	42	2814
69–71	70	27	1890
72–74	73	8	584
		$n = \sum f = 100$	$\sum fx = 6745$

The computations involved can become tedious, especially for cases in which the numbers are large and many classes are present. Short techniques are available for lessening the labor in such cases. See Problem 5.35, for example.

5.34. Derive the coding formula (27), page 162, for the arithmetic mean.

Let the jth class mark be x_j. Then the deviation of x_j, from some specified class mark a, which is $x_j - a$, will be equal to the class interval size c multiplied by some integer u_j, i.e., $x_j - a = cu_j$ or $x_j = a + cu_j$ (also written briefly as $x = a + cu$).

The mean is then given by

$$\bar{x} = \frac{\sum f_j x_j}{n} = \frac{\sum f_j(a + cu_j)}{n} = \frac{a\sum f_j}{n} + c\frac{\sum f_j u_j}{n}$$
$$= a + c\frac{\sum f_j u_j}{n} = a + c\bar{u}$$

since $n = \sum f_j$.

5.35. Use the coding formula of Problem 5.34 to find the mean height of the 100 male students at XYZ University (see Problem 5.33).

The work may be arranged as in Table 5-9. The method is called the *coding method* and should be employed whenever possible.

$$\bar{x} = a + \left(\frac{\sum fu}{n}\right)c = 67 + \left(\frac{15}{100}\right)(3) = 67.45 \text{ inches}$$

Table 5-9

x	u	f	fu
61	−2	5	−10
64	−1	18	−18
$a \rightarrow$ 67	0	42	0
70	1	27	27
73	2	8	16
		$n = 100$	$\sum fu = 15$

5.36. Find (a) the variance, (b) the standard deviation for the numbers in Problem 5.31.

(a) **Method 1**

As in Problem 5.31, we have $\bar{x} = 4.8$. Then

$$s^2 = \frac{\sum(x - \bar{x})^2}{n} = \frac{(5 - 4.8)^2 + (3 - 4.8)^2 + (6 - 4.8)^2 + (5 - 4.8)^2 + \cdots + (4 - 4.8)^2}{20}$$

$$= \frac{59.20}{20} = 2.96$$

Method 2

$$s^2 = \frac{\sum f(x - \bar{x})^2}{n} = \frac{6(5 - 4.8)^2 + 2(3 - 4.8)^2 + 2(6 - 4.8)^2 + 5(4 - 4.8)^2 + 3(8 - 4.8)^2}{20}$$

$$= \frac{59.20}{20} = 2.96$$

(b) From (a), $s^2 = 2.96$ and $s = \sqrt{2.96} = 1.72$.

5.37. Find the standard deviation of the weights of students in Problem 5.32.

$$s^2 = \frac{\sum f(x - \bar{x})^2}{n} = \frac{15(162 - 150)^2 + 20(148 - 150)^2 + 10(153 - 150)^2 + 18(140 - 150)^2}{15 + 20 + 10 + 18}$$

$$= \frac{4130}{63} = 65.6 \text{ in units pounds square or (pounds)}^2$$

Then $s = \sqrt{65.6\ (\text{lb})^2} = \sqrt{65.6}$ lb $= 8.10$ lb, where we have used the fact that units follow the usual laws of algebra.

5.38. Find the standard deviation of the heights of the 100 male students at XYZ University. See Problem 5.33.

From Problem 5.33, $\bar{X} = 67.45$ inches. The work can be arranged as in Table 5-10.

$$s = \sqrt{\frac{\sum f(x - \bar{x})^2}{n}} = \sqrt{\frac{852.7500}{100}} = \sqrt{8.5275} = 2.92 \text{ inches}$$

Table 5-10

Height (inches)	Class Mark (x)	$x - \bar{x} = x - 67.45$	$(x - \bar{x})^2$	Frequency (f)	$f(x - \bar{x})^2$
60–62	61	−6.45	41.6025	5	208.0125
63–65	64	−3.45	11.9025	18	214.2450
66–68	67	−0.45	0.2025	42	8.5050
69–71	70	2.55	6.5025	27	175.5675
72–74	73	5.55	30.8025	8	246.4200
				$n = \sum f = 100$	$\sum f(x - \bar{x})^2 = 852.7500$

5.39. Derive the coding formula (28), page 162, for the variance.

As in Problem 5.34, we have $x_j = a + cu_j$ and

$$\bar{x} = a + c\frac{\sum f_j u_j}{n} = a + c\bar{u}$$

Then

$$s^2 = \frac{1}{n}\sum f_j(x_j - \bar{x})^2 = \frac{1}{n}\sum f_j(cu_j - c\bar{u})^2$$

$$= \frac{c^2}{n}\sum f_j(u_j - \bar{u})^2$$

$$= \frac{c^2}{n}\sum f_j(u_j^2 - 2u_j\bar{u} + \bar{u}^2)$$

$$= \frac{c^2}{n}\sum f_j u_j^2 - \frac{2\bar{u}c^2}{n}\sum f_j u_j + \frac{c^2}{n}\sum f_j \bar{u}^2$$

$$= c^2\frac{\sum f_j u_j^2}{n} - 2\bar{u}^2c^2 + c^2\bar{u}^2$$

$$= c^2\frac{\sum f_j u_j^2}{n} - c^2\left(\frac{\sum f_j u_j}{n}\right)^2$$

$$= c^2\left[\frac{\sum fu^2}{n} - \left(\frac{\sum fu}{n}\right)^2\right]$$

$$= c^2[\overline{u^2} - \bar{u}^2]$$

5.40. Use the coding formula of Problem 5.39 to find the standard deviation of heights in Problem 5.33.

The work may be arranged as in Table 5-11. This enables us to find $\bar{x}$ as in Problem 5.35. From the last column we then have

$$s^2 = c^2\left[\frac{\sum fu^2}{n} - \left(\frac{\sum fu}{n}\right)^2\right] = c^2(\overline{u^2} - \bar{u}^2)$$

$$= (3)^2\left[\frac{97}{100} - \left(\frac{15}{100}\right)^2\right] = 8.5275$$

and so $s = 2.92$ inches.

Table 5-11

x	u	f	fu	fu^2
61	−2	5	−10	20
64	−1	18	−18	18
$a \rightarrow$ 67	0	42	0	0
70	1	27	27	27
73	2	8	8	32
		$n = \sum f = 100$	$\sum fu = 15$	$\sum fu^2 = 97$

5.41. Find the first four moments about the mean for the height distribution of Problem 5.33.

Continuing the method of Problem 5.40, we obtain Table 5-12. Then, using the notation of page 162, we have

$$M_1' = \frac{\sum fu}{n} = 0.15 \qquad M_3' = \frac{\sum fu^3}{n} = 0.33$$

$$M_2' = \frac{\sum fu^2}{n} = 0.97 \qquad M_4' = \frac{\sum fu^4}{n} = 2.53$$

Table 5-12

x	u	f	fu	fu^2	fu^3	fu^4
61	−2	5	−10	20	−40	80
64	−1	18	−18	18	−18	18
67	0	42	0	0	0	0
70	1	27	27	27	27	27
73	2	8	16	32	64	128
		$n = \sum f = 100$	$\sum fu = 15$	$\sum fu^2 = 97$	$\sum fu^3 = 33$	$\sum fu^4 = 253$

and from (32),

$$
\begin{aligned}
m_1 &= 0 \\
m_2 &= c^2\left(M_2' - M_1'^2\right) = 9[0.97 - (0.15)^2] = 8.5275 \\
m_3 &= c^3\left(M_3' - 3M_1'M_2' + 2M_1'^3\right) = 27[0.33 - 3(0.15)(0.97) + 2(0.15)^3] = -2.6932 \\
m_4 &= c^4\left(M_4' - 4M_1'M_3' + 6M_1'^2M_2' - 3M_1'^4\right) \\
&= 81[2.53 - 4(0.15)(0.33) + 6(0.15)^2(0.97) - 3(0.15)^4] = 199.3759
\end{aligned}
$$

5.42. Find the coefficients of (a) skewness, (b) kurtosis for the height distribution of Problem 5.33.

(a) From Problem 5.41,

$$m_2 = s^2 = 8.5275 \qquad m_3 = -2.6932$$

Then

$$\text{Coefficient of skewness} = a_3 = \frac{m_3}{s^3} = \frac{-2.6932}{\sqrt{(8.5275)^3}} = -0.14$$

(b) From Problem 5.41,

$$m_4 = 199.3759 \qquad m_2 = s^2 = 8.5275$$

Then

$$\text{Coefficient of kurtosis} = a_4 = \frac{m_4}{s^4} = \frac{199.3759}{(8.5275)^2} = 2.74$$

From (a) we see that the distribution is moderately skewed to the left. From (b) we see that it is slightly less peaked than the normal distribution (which has coefficient of kurtosis = 3).

Miscellaneous problems

5.43. (a) Show how to select 30 random samples of 4 students each (with replacement) from the table of heights on page 161 by using random numbers, (b) Find the mean and standard deviation of the sampling distribution of means in (a). (c) Compare the results of (b) with theoretical values, explaining any discrepancies.

(a) Use two digits to number each of the 100 students: 00, 01, 02, . . . , 99 (see Table 5-13). Therefore, the 5 students with heights 60–62 inches are numbered 00–04, the 18 students with heights 63–65 inches are numbered 05–22, etc. Each student number is called a *sampling number.*

Table 5-13

Height (inches)	Frequency	Sampling Number
60–62	5	00–04
63–65	18	05–22
66–68	42	23–64
69–71	27	65–91
72–74	8	92–99

We now draw sampling numbers from the random number table (Appendix H). From the first line we find the sequence 51, 77, 27, 46, 40, etc., which we take as random sampling numbers, each of which yields the height of a particular student. Therefore, 51 corresponds to a student having height 66–68 inches, which we take as 67 inches (the class mark). Similarly 77, 27, 46 yield heights of 70, 67, 67 inches, respectively.

By this process we obtain Table 5-14, which shows the sampling numbers drawn, the corresponding heights, and the mean height for each of 30 samples. It should be mentioned that although we have entered the random number table on the first line, we could have started *anywhere* and chosen any specified pattern.

Table 5-14

	Sampling Numbers Drawn	Corresponding Heights	Mean Height
1.	51, 77, 27, 46	67, 70, 67, 67	67.75
2.	40, 42, 33, 12	67, 67, 67, 64	66.25
3.	90, 44, 46, 62	70, 67, 67, 67	67.75
4.	16, 28, 98, 93	64, 67, 73, 73	69.25
5.	58, 20, 41, 86	67, 64, 67, 70	67.00
6.	19, 64, 08, 70	64, 67, 64, 70	66.25
7.	56, 24, 03, 32	67, 67, 61, 67	65.50
8.	34, 91, 83, 58	67, 70, 70, 67	68.50
9.	70, 65, 68, 21	70, 70, 70, 64	68.50
10.	96, 02, 13, 87	73, 61, 64, 70	67.00
11.	76, 10, 51, 08	70, 64, 67, 64	66.25
12.	63, 97, 45, 39	67, 73, 67, 67	68.50
13.	05, 81, 45, 93	64, 70, 67, 73	68.50
14.	96, 01, 73, 52	73, 61, 70, 67	67.75
15.	07, 82, 54, 24	64, 70, 67, 67	67.00
16.	11, 64, 55, 58	64, 67, 67, 67	66.25
17.	70, 56, 97, 43	70, 67, 73, 67	69.25
18.	74, 28, 93, 50	70, 67, 73, 67	69.25
19.	79, 42, 71, 30	70, 67, 70, 67	68.50
20.	58, 60, 21, 33	67, 67, 64, 67	66.25
21.	75, 79, 74, 54	70, 70, 70, 67	69.25
22.	06, 31, 04, 18	64, 67, 61, 64	64.00
23.	67, 07, 12, 97	70, 64, 64, 73	67.75
24.	31, 71, 69, 88	67, 70, 70, 70	69.25
25.	11, 64, 21, 87	64, 67, 64, 70	66.25
26.	03, 58, 57, 93	61, 67, 67, 73	67.00
27.	53, 81, 93, 88	67, 70, 73, 70	70.00
28.	23, 22, 96, 79	67, 64, 73, 70	68.50
29.	98, 56, 59, 36	73, 67, 67, 67	68.50
30.	08, 15, 08, 84	64, 64, 64, 70	65.50

(b) Table 5-15 gives the frequency distribution of sample mean heights obtained in (a). This is a *sampling distribution of means.* The mean and the standard deviation are obtained as usual by the coding methods already described.

$$\text{Mean} = a + c\bar{u} = a + \frac{c\sum fu}{n} = 67.00 + \frac{(0.75)(23)}{30} = 67.58 \text{ inches}$$

$$\text{Standard deviation} = c\sqrt{\bar{u}^2 - \bar{u}^2} = c\sqrt{\frac{\sum fu^2}{n} - \left(\frac{\sum fu}{n}\right)^2}$$

$$= (0.75)\sqrt{\frac{123}{30} - \left(\frac{23}{30}\right)^2} = 1.41 \text{ inches}$$

Table 5-15

Sample Mean	Tally	f	u	fu	fu^2
64.00	/	1	−4	−4	16
64.75		0	−3	0	0
65.50	//	2	−2	−4	8
66.25	~~////~~ /	6	−1	−6	6
$a \rightarrow$ 67.00	////	4	0	0	0
67.75	////	4	1	4	4
68.50	~~////~~ //	7	2	14	28
69.25	~~////~~	5	3	15	45
70.00	/	1	4	4	16
		$\sum f = n = 30$		$\sum fu = 23$	$\sum fu^2 = 123$

(c) The theoretical mean of the sampling distribution of means, given by $\mu_{\bar{X}}$, equals the population mean μ, which is 67.45 inches (see Problem 5.33), in close agreement with the value 67.58 inches of part (b).

The theoretical standard deviation (standard error) of the sampling distribution of means, given by $\sigma_{\bar{X}}$, equals $\sigma/\sqrt{n}$, where the population standard deviation $\sigma = 2.92$ inches (see Problem 5.40) and the sample size $n = 4$. Since $\sigma/\sqrt{n} = 2.92/\sqrt{4} = 1.46$ inches, we have close agreement with the value 1.41 inches of part (b). Discrepancies are due to the fact that only 30 samples were selected and the sample size was small.

5.44. The standard deviation of the weights of a very large population of students is 10.0 lb. Samples of 200 students each are drawn from this population, and the standard deviations of the weights in each sample are computed. Find (a) the mean, (b) the standard deviation of the sampling distribution of standard deviations.

We can consider that sampling is either from an infinite population or with replacement from a finite population. From Table 5-1, page 160, we have:

(a)
$$\mu_S = \sigma = 10.0 \text{ lb}$$

(b)
$$\sigma_S = \frac{\sigma}{\sqrt{2n}} = \frac{10}{\sqrt{400}} = 0.50 \text{ lb}$$

5.45. What percentage of the samples in Problem 5.44 would have standard deviations (a) greater than 11.0 lb, (b) less than 8.8 lb?

The sampling distribution of standard deviations is approximately normal with mean 10.0 lb and standard deviation 0.50 lb.

(a) 11.0 lb in standard units $= (11.0 - 10.0)/0.50 = 2.0$. Area under normal curve to right of $z = 2.0$ is $(0.5 - 0.4772) = 0.0228$; hence, the required percentage is 2.3%.

(b) 8.8 lb in standard units $= (8.8 - 10.0)/0.50 = -2.4$. Area under normal curve to left of $z = -2.4$ is $(0.5 - 0.4918) = 0.0082$; hence, the required percentage is 0.8%.

5.46. A sample of 6 observations is drawn at random from a continuous population. What is the probability that the last 2 observations are less than the first 4?

Assume that the population has density function $f(x)$. The probability that 3 of the first 4 observations are greater than u while the 4th observation lies between u and $u + du$ is given by

(1)
$$_4C_3\left[\int_u^\infty f(x)\,dx\right]^3 f(u)\,du$$

The probability that the last 2 observations are less than u (and thus less than the first 4) is given by

(2) $$\left[\int_{-\infty}^{u} f(x)\,dx\right]^2$$

Then the probability that the first 4 are greater than u *and* the last 2 are less than u is the product of (1) and (2), i.e.,

(3) $${}_4C_3\left[\int_{u}^{\infty} f(x)\,dx\right]^3 f(u)\,du\left[\int_{-\infty}^{u} f(x)\,dx\right]^2$$

Since u can take on values between $-\infty$ and ∞, the total probability of the last 2 observations being less than the first 4 is the integral of (3) from $-\infty$ to ∞, i.e.,

(4) $${}_4C_3\int_{-\infty}^{\infty}\left[\int_{u}^{\infty} f(x)\,dx\right]^3\left[\int_{-\infty}^{u} f(x)\,dx\right]^2 f(u)\,du$$

To evaluate this, let

(5) $$v = \int_{-\infty}^{u} f(x)\,dx$$

Then

(6) $$dv = f(u)\,du \qquad 1 - v = \int_{u}^{\infty} f(x)\,dx$$

When $u = \infty$, $v = 1$, and when $u = -\infty$, $v = 0$. Therefore, (4) becomes

$${}_4C_3\int_{0}^{1} v^2(1 - v)^3\,dv = 4\frac{\Gamma(3)\Gamma(4)}{\Gamma(7)} = \frac{1}{15}$$

which is the required probability. It is of interest to note that the probability does not depend on the probability distribution $f(x)$. This is an example of *nonparametric statistics* since no population parameters have to be known.

Another method

Denote the observations by $x_1, x_2, \ldots, x_6$. Since the population is continuous, we may assume that the x_i's are distinct. There are 6! ways of arranging the subscripts 1, 2, . . . , 6, and any one of these is as likely as any other one to result in arranging the corresponding x_i's in increasing order. Out of the 6!, exactly $4! \times 2!$ arrangements would have x_1, x_2, x_3, x_4 as the smallest 4 observations and x_5, x_6 as the largest 2 observations. The required probability is, therefore,

$$\frac{4! \times 2!}{6!} = \frac{1}{15}$$

5.47. Let $\{X_1, X_2, \ldots, X_n\}$ be a random sample of size n drawn without replacement from a finite population of size N. Prove that if the population mean and variance are μ and σ^2, then (a) $E(X_j) = \mu$, (b) $\text{Cov}(X_j, X_k) = -\sigma^2/(N - 1)$.

Assume that the population consists of the set of numbers $(\alpha_1, \alpha_2, \ldots, \alpha_N)$, where the α's are not necessarily distinct. A random sampling procedure is one under which each selection of n out of N α's has the same probability (i.e., $1/{}_NC_n$). This means that the X_j are identically distributed:

$$X_j = \begin{cases} \alpha_1 & \text{prob. } 1/N \\ \alpha_2 & \text{prob. } 1/N \\ \vdots & \\ \alpha_N & \text{prob. } 1/N \end{cases} \qquad (j = 1, 2, \ldots, n)$$

They are not, however, mutually independent. Indeed, when $j \neq k$, the joint distribution of X_j and X_k is given by

$$P(X_j = \alpha_\lambda, X_k = \alpha_\nu) = P(X_j = \alpha_\lambda)P(X_k = \alpha_\nu | X_j = \alpha_\lambda)$$

$$= \frac{1}{N} P(X_k = \alpha_\nu | X_j = \alpha_\lambda)$$

$$= \begin{cases} \frac{1}{N}\left(\frac{1}{N-1}\right) & \lambda \neq \nu \\ 0 & \lambda = \nu \end{cases}$$

where λ and ν range from 1 to N.

(a)
$$E(X_j) = \sum_{\lambda=1}^{N} \alpha_\lambda P(X_j = \alpha_\lambda) = \frac{1}{N}\sum_{\lambda=1}^{N} \alpha_\lambda = \mu$$

(b)
$$\text{Cov}(X_j, X_k) = E[X_j - \mu)(X_k - \mu)]$$

$$= \sum_{\lambda=1}^{N}\sum_{\nu=1}^{N} (\alpha_\lambda - \mu)(\alpha_\nu - \mu)P(X_j = \alpha_\lambda, X_k = \alpha_\nu)$$

$$= \frac{1}{N}\left(\frac{1}{N-1}\right)\sum_{\lambda \neq \nu = 1}^{N} (\alpha_\lambda - \mu)(\alpha_\nu - \mu)$$

where the last sum contains a total of $N(N - 1)$ terms, corresponding to all possible pairs of unequal λ and ν.

Now, by elementary algebra,

$$[(\alpha_1 - \mu) + (\alpha_2 - \mu) + \cdots + (\alpha_N - \mu)]^2 = \sum_{\lambda=1}^{N} (\alpha_\lambda - \mu)^2 + \sum_{\lambda \neq \nu = 1}^{N} (\alpha_\lambda - \mu)(\alpha_\nu - \mu)$$

In this equation, the left-hand side is zero, since by definition

$$\alpha_1 + \alpha_2 + \cdots + \alpha_N = N\mu$$

and the first sum on the right-hand side equals, by definition, $N\sigma^2$. Hence,

$$\sum_{\lambda \neq \nu = 1}^{N} (\alpha_\lambda - \mu)(\alpha_\nu - \mu) = -N\sigma^2$$

and

$$\text{Cov}(X_j, X_k) = \frac{1}{N}\left(\frac{1}{N-1}\right)(-N\sigma^2) = -\frac{\sigma^2}{N-1}$$

5.48. Prove that (a) the mean, (b) the variance of the sample mean in Problem 5.47 are given, respectively, by

$$\mu_{\bar{X}} = \mu \qquad \sigma^2_{\bar{X}} = \frac{\sigma^2}{n}\left(\frac{N-n}{N-1}\right)$$

(a)
$$E(\bar{X}) = E\left(\frac{X_1 + \cdots + X_n}{n}\right) = \frac{1}{n}[E(X_1) + \cdots + E(X_n)]$$

$$= \frac{1}{n}(\mu + \cdots + \mu) = \mu$$

where we have used Problem 5.47(a).

(b) Using Theorems 3-5 and 3-16 (generalized), and Problem 5.47, we obtain

$$\mathrm{Var}(\bar{X}) = \frac{1}{n^2}\mathrm{Var}\left(\sum_{j=1}^{n} X_j\right) = \frac{1}{n^2}\left[\sum_{j=1}^{n}\mathrm{Var}(X_j) + \sum_{j\neq k=1}^{n}\mathrm{Cov}(X_j, X_k)\right]$$

$$= \frac{1}{n^2}\left[n\sigma^2 + n(n-1)\left(-\frac{\sigma^2}{N-1}\right)\right]$$

$$= \frac{\sigma^2}{n}\left[1 - \frac{n-1}{N-1}\right] = \frac{\sigma^2}{n}\left(\frac{N-n}{N-1}\right)$$

SUPPLEMENTARY PROBLEMS

Sampling distribution of means

5.49. A population consists of the four numbers 3, 7, 11, 15. Consider all possible samples of size two that can be drawn with replacement from this population. Find (a) the population mean, (b) the population standard deviation, (c) the mean of the sampling distribution of means, (d) the standard deviation of the sampling distribution of means. Verify (c) and (d) directly from (a) and (b) by use of suitable formulas.

5.50. Solve Problem 5.49 if sampling is without replacement.

5.51. The weights of 1500 ball bearings are normally distributed with a mean of 22.40 oz and a standard deviation of 0.048 oz. If 300 random samples of size 36 are drawn from this population, determine the expected mean and standard deviation of the sampling distribution of means if sampling is done (a) with replacement, (b) without replacement.

5.52. Solve Problem 5.51 if the population consists of 72 ball bearings.

5.53. In Problem 5.51, how many of the random samples would have their means (a) between 22.39 and 22.41 oz, (b) greater than 22.42 oz, (c) less than 22.37 oz, (d) less than 22.38 or more than 22.41 oz?

5.54. Certain tubes manufactured by a company have a mean lifetime of 800 hours and a standard deviation of 60 hours. Find the probability that a random sample of 16 tubes taken from the group will have a mean lifetime (a) between 790 and 810 hours, (b) less than 785 hours, (c) more than 820 hours, (d) between 770 and 830 hours.

5.55. Work Problem 5.54 if a random sample of 64 tubes is taken. Explain the difference.

5.56. The weights of packages received by a department store have a mean of 300 lb and a standard deviation of 50 lb. What is the probability that 25 packages received at random and loaded on an elevator will exceed the safety limit of the elevator, listed as 8200 lb?

Sampling distribution of proportions

5.57. Find the probability that of the next 200 children born, (a) less than 40% will be boys, (b) between 43% and 57% will be girls, (c) more than 54% will be boys. Assume equal probabilities for births of boys and girls.

5.58. Out of 1000 samples of 200 children each, in how many would you expect to find that (a) less than 40% are boys, (b) between 40% and 60% are girls, (c) 53% or more are girls?

5.59. Work Problem 5.57 if 100 instead of 200 children are considered, and explain the differences in results.

5.60. An urn contains 80 marbles of which 60% are red and 40% are white. Out of 50 samples of 20 marbles, each selected with replacement from the urn, how many samples can be expected to consist of (a) equal numbers of red and white marbles, (b) 12 red and 8 white marbles, (c) 8 red and 12 white marbles, (d) 10 or more white marbles?

5.61. Design an experiment intended to illustrate the results of Problem 5.60. Instead of red and white marbles, you may use slips of paper on which *R* or *W* are written in the correct proportions. What errors might you introduce by using two different sets of marbles?

5.62. A manufacturer sends out 1000 lots, each consisting of 100 electric bulbs. If 5% of the bulbs are normally defective, in how many of the lots should we expect (a) fewer than 90 good bulbs, (b) 98 or more good bulbs?

Sampling distributions of differences and sums

5.63. *A* and *B* manufacture two types of cables, having mean breaking strengths of 4000 and 4500 lb and standard deviations of 300 and 200 lb, respectively. If 100 cables of brand *A* and 50 cables of brand *B* are tested, what is the probability that the mean breaking strength of *B* will be (a) at least 600 lb more than *A*, (b) at least 450 lb more than *A*?

5.64. What are the probabilities in Problem 5.63 if 100 cables of both brands are tested? Account for the differences.

5.65. The mean score of students on an aptitude test is 72 points with a standard deviation of 8 points. What is the probability that two groups of students, consisting of 28 and 36 students, respectively, will differ in their mean scores by (a) 3 or more points, (b) 6 or more points, (c) between 2 and 5 points?

5.66. An urn holds 60 red marbles and 40 white marbles. Two sets of 30 marbles each are drawn with replacement from the urn, and their colors are noted. What is the probability that the two sets differ by 8 or more red marbles?

5.67. Solve Problem 5.66 if sampling is without replacement in obtaining each set.

5.68. The election returns showed that a certain candidate received 65% of the votes. Find the probability that two random samples, each consisting of 200 voters, indicated a greater than 10% difference in the proportions that voted for the candidate.

5.69. If U_1 and U_2 are the sets of numbers in Problem 5.12, verify that (a) $\mu_{U_1+U_2} = \mu_{U_1} + \mu_{U_2}$, (b) $\sigma_{U_1+U_2} = \sqrt{\sigma^2_{U_1} + \sigma^2_{U_2}}$.

5.70. Three weights are measured as 20.48, 35.97, and 62.34 lb with standard deviations of 0.21, 0.46, and 0.54 lb, respectively. Find the (a) mean, (b) standard deviation of the sum of the weights.

5.71. The voltage of a battery is very nearly normal with mean 15.0 volts and standard deviation 0.2 volts. What is the probability that four such batteries connected in series will have a combined voltage of 60.8 or more volts?

Sampling distribution of variances

5.72. With reference to Problem 5.49, find (a) the mean of the sampling distribution of variances, (b) the standard error of variances.

5.73. Work Problem 5.72 if sampling is without replacement.

5.74. A normal population has a variance of 15. If samples of size 5 are drawn from this population, what percentage can be expected to have variances (a) less than 10, (b) more than 20, (c) between 5 and 10?

5.75. It is found that the lifetimes of television tubes manufactured by a company have a normal distribution with a mean of 2000 hours and a standard deviation of 60 hours. If 10 tubes are selected at random, find the probability that the sample standard deviation will (a) not exceed 50 hours, (b) lie between 50 and 70 hours.

Case where population variance is unknown

5.76. According to the table of Student's t distribution for 1 degree of freedom (Appendix D), we have $P(-1 \leq T \leq 1) = 0.50$. Check whether the results of Problem 5.1 are confirmed by this value, and explain any difference.

5.77. Check whether the results of Problem 5.49 are confirmed by using (a) $P(-1 \leq T \leq 1) = 0.50$, (b) $P(-1.376 \leq T \leq 1.376) = 0.60$, where T has Student's t distribution with $\nu = 1$.

5.78. Explain how you could use Theorem 5-7, page 159, to design a table of Student's t distribution such as that in Appendix D.

Sampling distribution of ratios of variances

5.79. Two samples of sizes 4 and 8 are drawn from a normally distributed population. Is the probability that one variance is greater than 1.5 times the other greater than 0.05, between 0.05 and 0.01, or less than 0.01?

5.80. Two companies, A and B, manufacture light bulbs. The lifetimes of both are normally distributed. Those for A have a standard deviation of 40 hours while the lifetimes for B have a standard deviation of 50 hours. A sample of 8 bulbs is taken from A and 16 bulbs from B. Determine the probability that the variance of the first sample is more than (a) twice, (b) 1.2 times, that of the second.

5.81. Work Problem 5.80 if the standard deviations of lifetimes are (a) both 40 hours, (b) both 50 hours.

Frequency distribution

5.82. Table 5-16 shows a frequency distribution of the lifetimes of 400 radio tubes tested at the L & M Tube Company. With reference to this table, determine the

(a) upper limit of the fifth class

(b) lower limit of the eighth class

(c) class mark of the seventh class

Table 5-16

Lifetime (hours)	Number of Tubes
300–399	14
400–499	46
500–599	58
600–699	76
700–799	68
800–899	62
900–999	48
1000–1099	22
1100–1199	6
TOTAL	400

(d) class boundaries of the last class

(e) class interval size

(f) frequency of the fourth class

(g) relative frequency of the sixth class

(h) percentage of tubes whose lifetimes do not exceed 600 hours

(i) percentage of tubes with lifetimes greater than or equal to 900 hours

(j) percentage of tubes whose lifetimes are at least 500 but less than 1000 hours

5.83. Construct (a) a histogram, (b) a frequency polygon corresponding to the frequency distribution of Problem 5.82.

5.84. For the data of Problem 5.82, construct (a) a relative, or percentage, frequency distribution, (b) a relative frequency histogram, (c) a relative frequency polygon.

5.85. Estimate the percentage of tubes of Problem 5.82 with lifetimes of (a) less than 560 hours, (b) 970 or more hours, (c) between 620 and 890 hours.

5.86. The inner diameters of washers produced by a company can be measured to the nearest thousandth of an inch. If the class marks of a frequency distribution of these diameters are given in inches by 0.321, 0.324, 0.327, 0.330, 0.333, and 0.336, find (a) the class interval size, (b) the class boundaries, (c) the class limits.

5.87. Table 5-17 shows the diameters in inches of a sample of 60 ball bearings manufactured by a company. Construct a frequency distribution of the diameters using appropriate class intervals.

Table 5-17

0.738	0.729	0.743	0.740	0.736	0.741	0.735	0.731	0.726	0.737
0.728	0.737	0.736	0.735	0.724	0.733	0.742	0.736	0.739	0.735
0.745	0.736	0.742	0.740	0728	0.738	0.725	0.733	0.734	0.732
0.733	0.730	0.732	0.730	0.739	0.734	0.738	0.739	0.727	0.735
0.735	0.732	0.735	0.727	0.734	0.732	0.736	0.741	0.736	0.744
0.732	0.737	0.731	0.746	0.735	0.735	0.729	0.734	0.730	0.740

5.88. For the data of Problem 5.87, construct (a) a histogram, (b) a frequency polygon, (c) a relative frequency distribution, (d) a relative frequency histogram, (e) a relative frequency polygon.

5.89. From the results in Problem 5-88, determine the percentage of ball bearings having diameters (a) exceeding 0.732 inch, (b) no more than 0.736 inch, (c) between 0.730 and 0.738 inch. Compare your results with those obtained directly from the raw data of Table 5-17.

5.90. Work Problem 5.88 for the data of Problem 5.82.

Computation of mean, standard deviation, and moments for samples

5.91. A student received grades of 85, 76, 93, 82, and 96 in five subjects. Determine the arithmetic mean of the grades.

5.92. The reaction times of an individual to certain stimuli were measured by a psychologist to be 0.53, 0.46, 0.50, 0.49, 0.52, 0.53, 0.44, and 0.55 seconds. Determine the mean reaction time of the individual to the stimuli.

5.93. A set of numbers consists of six 6s, seven 7s, eight 8s, nine 9s, and ten 10s. What is the arithmetic mean of the numbers?

5.94. A student's grades in the laboratory, lecture, and recitation parts of a physics course were 71, 78, and 89, respectively, (a) If the weights accorded these grades are 2, 4, and 5, respectively, what is an appropriate average grade? (b) What is the average grade if equal weights are used?

5.95. Three teachers of economics reported mean examination grades of 79, 74, and 82 in their classes, which consisted of 32, 25, and 17 students, respectively. Determine the mean grade for all the classes.

5.96. The mean annual salary paid to all employees in a company was \$5000. The mean annual salaries paid to male and female employees of the company were \$5200 and \$4200, respectively. Determine the percentages of males and females employed by the company.

5.97. Table 5-18 shows the distribution of the maximum loads in short tons (1 short ton $=$ 2000 lb) supported by certain cables produced by a company. Determine the mean maximum loading using (a) the "long method," (b) the coding method.

Table 5-18

Maximum Load (short tons)	Number of Cables
9–9.7	2
9.8–10.2	5
10.3–10.7	12
10.8–11.2	17
11.3–11.7	14
11.8–12.2	6
12.3–12.7	3
12.8–13.2	1
TOTAL	60

Table 5-19

x	462	480	498	516	534	552	570	588	606	624
f	98	75	56	42	30	21	15	11	6	2

5.98. Find $\bar{x}$ for the data in Table 5-19 using (a) the long method (b) the coding method.

5.99. Table 5-20 shows the distribution of the diameters of the heads of rivets manufactured by a company. Compute the mean diameter.

Table 5-20

Diameter (inches)	Frequency
0.7247–0.7249	2
0.7250–0.7252	6
0.7253–0.7255	8
0.7256–0.7258	15
0.7259–0.7261	42
0.7262–0.7264	68
0.7265–0.7267	49
0.7268–0.7270	25
0.7271–0.7273	18
0.7274–0.7276	12
0.7277–0.7279	4
0.7280–0.7282	1
TOTAL	250

5.100. Compute the mean for the data in Table 5-21.

Table 5-21

Class	Frequency
10–under 15	3
15–under 20	7
20–under 25	16
25–under 30	12
30–under 35	9
35–under 40	5
40–under 45	2
TOTAL	54

5.101. Find the standard deviation of the numbers:
(a) 3, 6, 2, 1, 7, 5; (b) 3.2, 4.6, 2.8, 5.2, 4.4; (c) 0, 0, 0, 0, 0, 1, 1, 1.

5.102. (a) By adding 5 to each of the numbers in the set 3, 6, 2, 1, 7, 5, we obtain the set 8, 11, 7, 6, 12, 10. Show that the two sets have the same standard deviation but different means. How are the means related?

(b) By multiplying each of the numbers 3, 6, 2, 1, 7, 5 by 2 and then adding 5, we obtain the set 11, 17, 9, 7, 19, 15. What is the relationship between the standard deviations and between the means for the two sets?

(c) What properties of the mean and standard deviation are illustrated by the particular sets of numbers in (a) and (b)?

5.103. Find the standard deviation of the set of numbers in the arithmetic progression 4, 10, 16, 22, . . . , 154.

5.104. Find the standard deviations for the distributions of: (a) Problem 5-97, (b) Problem 5.98.

5.105. Find (a) the mean, (b) the standard deviation for the distribution of Problem 5.30, explaining the significance of the results obtained.

5.106. (a) Find the standard deviation s of the rivet diameters in Problem 5.99 (b) What percentage of rivet diameters lie in $(\bar{x} \pm s)$, $(\bar{x} \pm 2s)$, $(\bar{x} \pm 3s)$? (c) Compare the percentages in (b) with those that would theoretically be expected if the distribution were normal, and account for any observed differences.

5.107. (a) Find the mean and standard deviation for the data of Problem 5.28.

(b) Construct a frequency distribution for the data, and find the standard deviation.

(c) Compare the result of (b) with that of (a).

5.108. Work Problem 5.107 for the data of Problem 5.87.

5.109. (a) Of a total of n numbers, the fraction p are ones while the fraction $q = 1 - p$ are zeros. Prove that the standard deviation of the set of numbers is $\sqrt{pq}$. (b) Apply the result of (a) to Problem 5.101(c).

5.110. Find the (a) first, (b) second, (c) third, (d) fourth moment about the origin for the set of numbers 4, 7, 5, 9, 8, 3, 6.

5.111. Find the (a) first, (b) second, (c) third, (d) fourth moment about the mean for the set of numbers in Problem 5.110.

5.112. Find the (a) first, (b) second, (c) third, (d) fourth moment about the number 7 for the set of numbers in Problem 5.110.

5.113. Using the results of Problems 5.110 and 5.111, verify the following relations between the moments: (a) $m_2 = m'_2 - m'^2_1$, (b) $m_3 = m'_3 - 3m'_1m'_2 + 2m'^3_1$, (c) $m_4 = m'_4 - 4m'_1m'_3 + 6m'^2_1m'_2 - 3m'^4_1$.

5.114. Find the first four moments about the mean of the set of numbers in the arithmetic progression 2, 5, 8, 11, 14, 17.

5.115. If the first moment about the number 2 is equal to 5, what is the mean?

5.116. If the first four moments of a set of numbers about the number 3 are equal to -2, 10, -25, and 50, determine the corresponding moments (a) about the mean, (b) about the number 5, (c) about zero.

5.117. Find the first four moments about the mean of the numbers 0, 0, 0, 1, 1, 1, 1, 1.

5.118. (a) Prove that $m_5 = m'_5 - 5m'_1m'_4 + 10m'^2_1m'_3 - 10m'^3_1m'_2 + 4m'^5_1$. (b) Derive a similar formula for m_6.

5.119. Of a total of n numbers, the fraction p are ones while the fraction $q = 1 - p$ are zeros. Find (a) m_1, (b) m_2, (c) m_3, (d) m_4 for the set of numbers. Compare with Problem 5.117.

5.120. Calculate the first four moments about the mean for the distribution of Table 5-22.

Table 5-22

x	f
12	1
14	4
16	6
18	10
20	7
22	2
TOTAL	30

5.121. Calculate the first four moments about the mean for the distribution of Problem 5.97.

5.122. Find (a) m_1, (b) m_2, (c) m_3, (d) m_4, (e) $\bar{x}$, (f) s, (g) $\overline{x^2}$ (h) $\overline{x^3}$, (i) $\overline{x^4}$, (j) $\overline{(x+1)^3}$ for the distribution of Problem 5.100.

5.123. Find the coefficient of (a) skewness, (b) kurtosis for the distribution Problem 5.120.

5.124. Find the coefficient of (a) skewness, (b) kurtosis for the distribution of Problem 5.97. See Problem 5.121.

5.125. The second moments about the mean of two distributions are 9 and 16, while the third moments about the mean are -8.1 and -12.8, respectively. Which distribution is more skewed to the left?

5.126. The fourth moments about the mean of the two distributions of Problem 5.125 are 230 and 780, respectively. Which distribution more nearly approximates the normal distribution from the viewpoint of (a) peakedness, (b) skewness?

Miscellaneous problems

5.127. A population of 7 numbers has a mean of 40 and a standard deviation of 3. If samples of size 5 are drawn from this population and the variance of each sample is computed, find the mean of the sampling distribution of variances if sampling is (a) with replacement, (b) without replacement.

5.128. Certain tubes produced by a company have a mean lifetime of 900 hours and a standard deviation of 80 hours. The company sends out 1000 lots of 100 tubes each. In how many lots can we expect (a) the mean lifetimes to exceed 910 hours, (b) the standard deviations of the lifetimes to exceed 95 hours? What assumptions must be made?

5.129. In Problem 5.128 if the median lifetime is 900 hours, in how many lots can we expect the median lifetimes to exceed 910 hours? Compare your answer with Problem 5.128(a) and explain the results.

5.130. On a citywide examination the grades were normally distributed with mean 72 and standard deviation 8. (a) Find the minimum grade of the top 20% of the students. (b) Find the probability that in a random sample of 100 students, the minimum grade of the top 20% will be less than 76.

5.131. (a) Prove that the variance of the set of n numbers $a, a + d, a + 2d, \ldots, a + (n - 1)d$ (i.e., an arithmetic progression with first term a and common difference d) is given by $\frac{1}{12}(n^2 - 1)d^2$. [*Hint:* Use $1 + 2 + 3 + \cdots + (n - 1) = \frac{1}{2}n(n - 1)$, $1^2 + 2^2 + 3^2 + \cdots + (n - 1)^2 = \frac{1}{6}n(n - 1)(2n - 1)$.]
(b) Use (a) in Problem 5.103.

5.132. Prove that the first four moments about the mean of the arithmetic progression $a, a + d, a + 2d, \ldots, a + (n - 1)d$ are

$$m_1 = 0, \qquad m_2 = \frac{1}{12}(n^2 - 1)d^2, \qquad m_3 = 0, \qquad m_4 = \frac{1}{240}(n^2 - 1)(3n^2 - 7)d^4$$

Compare with Problem 5.114. [*Hint:* $1^4 + 2^4 + 3^4 + \cdots + (n - 1)^4 = \frac{1}{30}n(n - 1)(2n - 1)(3n^2 - 3n - 1)$.]

ANSWERS TO SUPPLEMENTARY PROBLEMS

5.49. (a) 9.0 (b) 4.47 (c) 9.0 (d) 3.16 **5.50.** (a) 9.0 (b) 4.47 (c) 9.0 (d) 2.58

5.51. (a) $\mu_{\bar{X}} = 22.40$ oz, $\sigma_{\bar{X}} = 0.008$ oz (b) $\mu_{\bar{X}} = 22.40$ oz, $\sigma_{\bar{X}}$ is slightly less than 0.008 oz

5.52. (a) $\mu_{\bar{X}} = 22.40$ oz, $\sigma_{\bar{X}} = 0.008$ oz (b) $\mu_{\bar{X}} = 22.40$ oz, $\sigma_{\bar{X}} = 0.0057$ oz

5.53. (a) 237 (b) 2 (c) none (d) 24 **5.54.** (a) 0.4972 (b) 0.1587 (c) 0.0918 (d) 0.9544

5.55. (a) 0.8164 (b) 0.0228 (c) 0.0038 (d) 1.0000 **5.56.** 0.0026

5.57. (a) 0.0029 (b) 0.9596 (c) 0.1446 **5.58.** (a) 2 (b) 996 (c) 218

5.59. (a) 0.0179 (b) 0.8664 (c) 0.1841 **5.60.** (a) 6 (b) 9 (c) 2 (d) 12

5.62. (a) 19 (b) 125 **5.63.** (a) 0.0077 (b) 0.8869 **5.64.** (a) 0.0028 (b) 0.9172

5.65. (a) 0.2150 (b) 0.0064 (c) 0.4504 **5.66.** 0.0482 **5.67.** 0.0188 **5.68.** 0.0410

5.70. (a) 118.79 lb (b) 0.74 lb **5.71.** 0.0228 **5.72.** (a) 10.00 (b) 11.49

5.73. (a) 40/3 (b) 28.10 **5.74.** (a) 0.50 (b) 0.17 (c) 0.28 **5.75.** (a) 0.36 (b) 0.49

5.80. (a) between 0.01 and 0.05 (b) greater than 0.05 **5.81.** (a) greater than 0.05 (b) greater than 0.05

5.82. (a) 799 (b) 1000 (c) 949.5 (d) 1099.5, 1199.5 (e) 100 (hours) (f) 76 (g) $62/400 = 0.155$ or 15.5% (h) 29.5% (i) 19.0% (j) 78.0%

5.85. (a) 24% (b) 11% (c) 46%

5.86. (a) 0.003 inch (b) 0.3195, 0.3225, 0.3255, . . . ,0.3375 inch
(c) 0.320–0.322, 0.323–0.325, 0.326–0.328, . . . ,0.335–0.337

5.91. 86 **5.92.** 0.50 s **5.93.** 8.25 **5.94.** (a) 82 (b) 79 **5.95.** 78 **5.96.** 80%, 20%

5.97. 11.09 tons **5.98.** 501.0 **5.99.** 0.72642 inch **5.100.** 26.2

5.101. (a) 2.16 (b) 0.90 (c) 0.484 **5.103.** 45 **5.104.** (a) 0.733 ton (b) 38.60

5.105. (a) $\bar{x} = 2.47$ (b) $s = 1.11$ **5.106.** (a) 0.000576 inch (b) 72.1%, 93.3%, 99.76%

5.107. (a) 146.8 lb, 12.9 lb **5.108.** (a) 0.7349 inch, 0.00495 inch

5.110. (a) 6 (b) 40 (c) 288 (d) 2188 **5.111.** (a) 0 (b) 4 (c) 0 (d) 25.86

5.112. (a) -1 (b) 5 (c) -91 (d) 53 **5.114.** 0, 26.25, 0, 1193.1 **5.115.** 7

5.116. (a) 0, 6, 19, 42 (b) -4, 22, -117, 560 (c) 1, 7, 38, 155

5.117. 0, 0.2344, -0.0586, 0.0696 **5.120.** $m_1 = 0$, $m_2 = 5.97$, $m_3 = -3.97$, $m_4 = 89.22$

5.121. $m_1 = 0$, $m_2 = 0.53743$, $m_3 = 0.36206$, $m_4 = 0.84914$

5.122. (a) 0 (b) 52.95 (c) 92.35 (d) 7158.20 (e) 26.2 (f) 7.28 (g) 739.38 (h) 22,247 (i) 706,428 (j) 24,545

5.123. (a) -0.2464 (b) 2.62 **5.124.** (a) 0.9190 (b) 2.94

5.125. first distribution **5.126.** (a) second (b) first **5.127.** (a) 7.2 (b) 8.4

5.128. (a) 106 (b) 4 **5.129.** 159 **5.130.** (a) 78.7 (b) 0.0090

Estimation Theory

Unbiased Estimates and Efficient Estimates

As we remarked in Chapter 5 (see page 158), a statistic is called an *unbiased estimator* of a population parameter if the mean or expectation of the statistic is equal to the parameter. The corresponding value of the statistic is then called an *unbiased estimate* of the parameter.

EXAMPLE 6.1 The mean $\bar{X}$ and variance $\hat{S}^2$ as defined on pages 155 and 158 are unbiased estimators of the population mean μ and variance σ^2, since $E(\bar{X}) = \mu$, $E(\hat{S}^2) = \sigma^2$. The values $\bar{x}$ and $\hat{s}^2$ are then called *unbiased estimates.* However, $\hat{S}$ is actually a biased estimator of σ, since in general $E(\hat{S}) \neq \sigma$.

If the sampling distributions of two statistics have the same mean, the statistic with the smaller variance is called a *more efficient estimator* of the mean. The corresponding value of the efficient statistic is then called an *efficient estimate.* Clearly one would in practice prefer to have estimates that are both efficient and unbiased, but this is not always possible.

EXAMPLE 6.2 For a normal population, the sampling distribution of the mean and median both have the same mean, namely, the population mean. However, the variance of the sampling distribution of means is smaller than that of the sampling distribution of medians. Therefore, the mean provides a more efficient estimate than the median. See Table 5-1, page 160.

Point Estimates and Interval Estimates: Reliability

An estimate of a population parameter given by a single number is called a *point estimate* of the parameter. An estimate of a population parameter given by two numbers between which the parameter may be considered to lie is called an *interval estimate* of the parameter.

EXAMPLE 6.3 If we say that a distance is 5.28 feet, we are giving a point estimate. If, on the other hand, we say that the distance is 5.28 ± 0.03 feet, i.e., the distance lies between 5.25 and 5.31 feet, we are giving an interval estimate.

A statement of the error or precision of an estimate is often called its *reliability.*

Confidence Interval Estimates of Population Parameters

Let μ_S and σ_S be the mean and standard deviation (standard error) of the sampling distribution of a statistic S. Then, if the sampling distribution of S is approximately normal (which as we have seen is true for many statistics if the sample size $n \geq 30$), we can expect to find S lying in the intervals $\mu_S - \sigma_S$ to $\mu_S + \sigma_S$, $\mu_S - 2\sigma_S$ to $\mu_S + 2\sigma_S$ or $\mu_S - 3\sigma_S$ to $\mu_S + 3\sigma_S$ about 68.27%, 95.45%, and 99.73% of the time, respectively.

Equivalently we can expect to find, or we can be *confident* of finding, μ_S in the intervals $S - \sigma_S$ to $S + \sigma_S$, $S - 2\sigma_S$ to $S + 2\sigma_S$ or $S - 3\sigma_S$ to $S + 3\sigma_S$ about 68.27%, 95.45%, and 99.73% of the time, respectively. Because of this, we call these respective intervals the 68.27%, 95.45%, and 99.73% *confidence intervals* for estimating μ_S (i.e., for estimating the population parameter, in the case of an unbiased S). The end numbers of these intervals ($S \pm \sigma_S$, $S \pm 2\sigma_S$, $S \pm 3\sigma_S$) are then called the 68.27%, 95.45%, and 99.73% *confidence limits.*

Similarly, $S \pm 1.96\sigma_S$ and $S \pm 2.58\sigma_S$ are 95% and 99% (or 0.95 and 0.99) confidence limits for μ_S. The percentage confidence is often called the *confidence level.* The numbers 1.96, 2.58, etc., in the confidence limits are called *critical values*, and are denoted by z_c. From confidence levels we can find critical values, and conversely.

In Table 6-1 we give values of z_c corresponding to various confidence levels used in practice. For confidence levels not presented in the table, the values of z_c can be found from the normal curve area table in Appendix C.

Table 6-1

Confidence Level	99.73%	99%	98%	96%	95.45%	95%	90%	80%	68.27%	50%
z_c	3.00	2.58	2.33	2.05	2.00	1.96	1.645	1.28	1.00	0.6745

In cases where a statistic has a sampling distribution that is different from the normal distribution (such as chi square, t, or F), appropriate modifications to obtain confidence intervals have to be made.

Confidence Intervals for Means

1. **LARGE SAMPLES ($n \geq 30$).** If the statistic S is the sample mean $\bar{X}$, then 95% and 99% confidence limits for estimation of the population mean μ are given by $\bar{X} \pm 1.96\sigma_{\bar{X}}$ and $\bar{X} \pm 2.58\sigma_{\bar{X}}$, respectively. More generally, the confidence limits are given by $\bar{X} \pm z_c\sigma_{\bar{X}}$ where z_c, which depends on the particular level of confidence desired, can be read from the above table. Using the values of $\sigma_{\bar{X}}$ obtained in Chapter 5, we see that the confidence limits for the population mean are given by

$$\bar{X} \pm z_c \frac{\sigma}{\sqrt{n}} \tag{1}$$

in case sampling is from an infinite population or if sampling is with replacement from a finite population, and by

$$\bar{X} \pm z_c \frac{\sigma}{\sqrt{n}} \sqrt{\frac{N-n}{N-1}} \tag{2}$$

if sampling is without replacement from a population of finite size N.

In general, the population standard deviation σ is unknown, so that to obtain the above confidence limits, we use the estimator $\hat{S}$ or S.

2. **SMALL SAMPLES ($n < 30$) AND POPULATION NORMAL.** In this case we use the t distribution to obtain confidence levels. For example, if $-t_{0.975}$ and $t_{0.975}$ are the values of T for which 2.5% of the area lies in each tail of the t distribution, then a 95% confidence interval for T is given by (see page 159)

$$-t_{0.975} < \frac{(\bar{X} - \mu)\sqrt{n}}{\hat{S}} < t_{0.975} \tag{3}$$

from which we see that μ can be estimated to lie in the interval

$$\bar{X} - t_{0.975}\frac{\hat{S}}{\sqrt{n}} < \mu < \bar{X} + t_{0.975}\frac{\hat{S}}{\sqrt{n}} \tag{4}$$

with 95% confidence. In general the confidence limits for population means are given by

$$\bar{X} \pm t_c \frac{\hat{S}}{\sqrt{n}} \tag{5}$$

where the value t_c can be read from Appendix D.

A comparison of (5) with (1) shows that for small samples we replace z_c by t_c. For $n \geq 30$, z_c and t_c are practically equal. It should be noted that an advantage of the small sampling theory (which can of course be used for large samples as well, i.e., it is *exact*) is that $\hat{S}$ appears in (5) so that the sample standard deviation can be used instead of the population standard deviation (which is usually unknown) as in (1).

Confidence Intervals for Proportions

Suppose that the statistic S is the proportion of "successes" in a sample of size $n \geq 30$ drawn from a binomial population in which p is the proportion of successes (i.e., the probability of success). Then the confidence limits for p are given by $P \pm z_c\sigma_P$, where P denotes the proportion of successes in the sample of size n. Using the values of σ_P obtained in Chapter 5, we see that the confidence limits for the population proportion are given by

$$P \pm z_c\sqrt{\frac{pq}{n}} = P \pm z_c\sqrt{\frac{p(1-p)}{n}} \tag{6}$$

in case sampling is from an infinite population or if sampling is with replacement from a finite population. Similarly, the confidence limits are

$$P \pm z_c\sqrt{\frac{pq}{n}}\sqrt{\frac{N-n}{N-1}} \tag{7}$$

if sampling is without replacement from a population of finite size N. Note that these results are obtained from (1) and (2) on replacing $\bar{X}$ by P and σ by $\sqrt{pq}$.

To compute the above confidence limits, we use the sample estimate P for p. A more exact method is given in Problem 6.27.

Confidence Intervals for Differences and Sums

If S_1 and S_2 are two sample statistics with approximately normal sampling distributions, confidence limits for the differences of the population parameters corresponding to S_1 and S_2 are given by

$$S_1 - S_2 \pm z_c\sigma_{S_1-S_2} = S_1 - S_2 \pm z_c\sqrt{\sigma_{S_1}^2 + \sigma_{S_2}^2} \tag{8}$$

while confidence limits for the sum of the population parameters are given by

$$S_1 + S_2 \pm z_c\sigma_{S_1+S_2} = S_1 + S_2 \pm z_c\sqrt{\sigma_{S_1}^2 + \sigma_{S_2}^2} \tag{9}$$

provided that the samples are independent.

For example, confidence limits for the difference of two population means, in the case where the populations are infinite and have known standard deviations σ_1, σ_2, are given by

$$\bar{X}_1 - \bar{X}_2 \pm z_c\sigma_{\bar{X}_1-\bar{X}_2} = \bar{X}_1 - \bar{X}_2 \pm z_c\sqrt{\frac{\sigma_1^2}{n_1} + \frac{\sigma_2^2}{n_2}} \tag{10}$$

where $\bar{X}_1$, n_1 and $\bar{X}_2$, n_2 are the respective means and sizes of the two samples drawn from the populations.

Similarly, confidence limits for the difference of two population proportions, where the populations are infinite, are given by

$$P_1 - P_2 \pm z_c\sqrt{\frac{P_1(1-P_1)}{n_1} + \frac{P_2(1-P_2)}{n_2}} \tag{11}$$

where P_1 and P_2 are the two sample proportions and n_1 and n_2 are the sizes of the two samples drawn from the populations.

Confidence Intervals for the Variance of a Normal Distribution

The fact that $nS^2/\sigma^2 = (n-1)\hat{S}^2/\sigma^2$ has a chi-square distribution with $n - 1$ degrees of freedom enables us to obtain confidence limits for σ^2 or σ. For example, if $\chi^2_{0.025}$ and $\chi^2_{0.975}$ are the values of χ^2 for which 2.5% of the area lies in each tail of the distribution, then a 95% confidence interval is

$$\chi^2_{0.025} \leq \frac{nS^2}{\sigma^2} \leq \chi^2_{0.975} \tag{12}$$

or equivalently

$$\chi^2_{0.025} \leq \frac{(n-1)\hat{S}^2}{\sigma^2} \leq \chi^2_{0.975} \tag{13}$$

From these we see that σ can be estimated to lie in the interval

$$\frac{S\sqrt{n}}{\chi_{0.975}} \leq \sigma \leq \frac{S\sqrt{n}}{\chi_{0.025}} \tag{14}$$

or equivalently

$$\frac{\hat{S}\sqrt{n-1}}{\chi_{0.975}} \leq \sigma \leq \frac{\hat{S}\sqrt{n-1}}{\chi_{0.025}} \tag{15}$$

with 95% confidence. Similarly, other confidence intervals can be found.

It is in general desirable that the expected width of a confidence interval be as small as possible. For statistics with symmetric sampling distributions, such as the normal and t distributions, this is achieved by using tails of equal areas. However, for nonsymmetric distributions, such as the chi-square distribution, it may be desirable to adjust the areas in the tails so as to obtain the smallest interval. The process is illustrated in Problem 6.28.

Confidence Intervals for Variance Ratios

In Chapter 5, page 159, we saw that if two independent random samples of sizes m and n having variances S_1^2, S_2^2 are drawn from two normally distributed populations of variances σ_1^2, σ_2^2, respectively, then the random variable $\dfrac{\hat{S}_1^2/\sigma_1^2}{\hat{S}_2^2/\sigma_2^2}$ has an F distribution with $m-1$, $n-1$, degrees of freedom. For example, if we denote by $F_{0.01}$ and $F_{0.99}$ the values of F for which 1% of the area lies in each tail of the F distribution, then with 98% confidence we have

$$F_{0.01} \leq \frac{\hat{S}_1^2/\sigma_1^2}{\hat{S}_2^2/\sigma_2^2} \leq F_{0.99} \tag{16}$$

From this we can see that a 98% confidence interval for the variance ratio σ_1^2/σ_2^2 of the two populations is given by

$$\frac{1}{F_{0.99}}\frac{\hat{S}_1^2}{\hat{S}_2^2} \leq \frac{\sigma_1^2}{\sigma_2^2} \leq \frac{1}{F_{0.01}}\frac{\hat{S}_1^2}{\hat{S}_2^2} \tag{17}$$

Note that $F_{0.99}$ is read from one of the tables in Appendix F. The value $F_{0.01}$ is the reciprocal of $F_{0.99}$ with the degrees of freedom for numerator and denominator reversed, in accordance with Theorem 4-8, page 117.

In a similar manner we could find a 90% confidence interval by use of the appropriate table in Appendix F. This would be given by

$$\frac{1}{F_{0.95}}\frac{\hat{S}_1^2}{\hat{S}_2^2} \leq \frac{\sigma_1^2}{\sigma_2^2} \leq \frac{1}{F_{0.05}}\frac{\hat{S}_1^2}{\hat{S}_2^2} \tag{18}$$

Maximum Likelihood Estimates

Although confidence limits are valuable for estimating a population parameter, it is still often convenient to have a single or point estimate. To obtain a "best" such estimate, we employ a technique known as the *maximum likelihood method* due to Fisher.

To illustrate the method, we assume that the population has a density function that contains a population parameter, say, θ, which is to be estimated by a certain statistic. Then the density function can be denoted by $f(x, \theta)$. Assuming that there are n independent observations, $X_1, \ldots, X_n$, the joint density function for these observations is

$$L = f(x_1, \theta)f(x_2, \theta) \cdots f(x_n, \theta) \tag{19}$$

which is called the *likelihood*. The *maximum likelihood* can then be obtained by taking the derivative of L with respect to θ and setting it equal to zero. For this purpose it is convenient to first take logarithms and then take the derivative. In this way we find

$$\frac{1}{f(x_1, \theta)} \frac{\partial f(x_1, \theta)}{\partial \theta} + \cdots + \frac{1}{f(x_n, \theta)} \frac{\partial f(x_n, \theta)}{\partial \theta} = 0 \tag{20}$$

The solution of this equation, for θ in terms of the x_k, is known as the *maximum likelihood estimator of* θ.

The method is capable of generalization. In case there are several parameters, we take the partial derivatives with respect to each parameter, set them equal to zero, and solve the resulting equations simultaneously.

SOLVED PROBLEMS

Unbiased and efficient estimates

6.1. Give examples of estimators (or estimates) which are (a) unbiased and efficient, (b) unbiased and inefficient, (c) biased and inefficient.

Assume that the population is normal. Then

(a) The sample mean $\bar{X}$ and the modified sample variance $\hat{S}^2 = \frac{n}{n-1}S^2$ are two such examples.

(b) The sample median and the sample statistic $\frac{1}{2}(Q_1 + Q_3)$, where Q_1 and Q_3 are the lower and upper sample quartiles, are two such examples. Both statistics are unbiased estimates of the population mean, since the mean of their sampling distributions can be shown to be the population mean. However, they are both inefficient compared with $\bar{X}$.

(c) The sample standard deviation S, the modified standard deviation $\hat{S}$, the mean deviation, and the semi-interquartile range are four such examples for evaluating the population standard deviation, σ.

6.2. A sample of five measurements of the diameter of a sphere were recorded by a scientist as 6.33, 6.37, 6.36, 6.32, and 6.37 cm. Determine unbiased and efficient estimates of (a) the true mean, (b) the true variance. Assume that the measured diameter is normally distributed.

(a) An unbiased and efficient estimate of the true mean (i.e., the population mean) is

$$\bar{x} = \frac{\sum x}{n} = \frac{6.33 + 6.37 + 6.36 + 6.32 + 6.37}{5} = 6.35 \text{ cm}$$

(b) An unbiased and efficient estimate of the true variance (i.e., the population variance) is

$$\hat{s}^2 = \frac{n}{n-1}s^2 = \frac{\sum(x - \bar{x})^2}{n-1}$$

$$= \frac{(6.33 - 6.35)^2 + (6.37 - 6.35)^2 + (6.36 - 6.35)^2 + (6.32 - 6.35)^2 + (6.37 - 6.35)^2}{5 - 1}$$

$$= 0.00055 \text{ cm}^2$$

Note that $\hat{s} = \sqrt{0.00055} = 0.023$ is an estimate of the true standard deviation, but this estimate is neither unbiased nor efficient.

6.3. Suppose that the heights of 100 male students at XYZ University represent a random sample of the heights of all 1546 male students at the university. Determine unbiased and efficient estimates of (a) the true mean, (b) the true variance.

(a) From Problem 5.33:

$$\text{Unbiased and efficient estimate of true mean height} = \bar{x} = 67.45 \text{ inch}$$

(b) From Problem 5.38:

$$\text{Unbiased and efficient estimate of true variance} = \hat{s}^2 = \frac{n}{n-1}s^2 = \frac{100}{99}(8.5275) = 8.6136$$

Therefore, $\hat{s} = \sqrt{8.6136} = 2.93$. Note that since n is large there is essentially no difference between s^2 and $\hat{s}^2$ or between s and $\hat{s}$.

6.4. Give an unbiased and inefficient estimate of the true (mean) diameter of the sphere of Problem 6.2.

The median is one example of an unbiased and inefficient estimate of the population mean. For the five measurements arranged in order of magnitude, the median is 6.36 cm.

Confidence interval estimates for means (large samples)

6.5. Find (a) 95%, (b) 99% confidence intervals for estimating the mean height of the XYZ University students in Problem 6.3.

(a) The 95% confidence limits are $\bar{X} \pm 1.96\sigma/\sqrt{n}$.

Using $\bar{x} = 67.45$ inches and $\hat{s} = 2.93$ inches as an estimate of σ (see Problem 6.3), the confidence limits are $67.45 \pm 1.96(2.93/\sqrt{100})$, or 67.45 ± 0.57, inches. Then the 95% confidence interval for the population mean μ is 66.88 to 68.02 inches, which can be denoted by $66.88 < \mu < 68.02$.

We can therefore say that the probability that the population mean height lies between 66.88 and 68.02 inches is about 95%, or 0.95. In symbols we write $P(66.88 < \mu < 68.02) = 0.95$. This is equivalent to saying that we are 95% *confident* that the population mean (or true mean) lies between 66.88 and 68.02 inches.

(b) The 99% confidence limits are $\bar{X} \pm 2.58\sigma/\sqrt{n}$. For the given sample,

$$\bar{x} \pm 2.58\frac{\hat{s}}{\sqrt{n}} = 67.45 \pm 2.58\frac{2.93}{\sqrt{100}} = 67.45 \pm 0.76 \text{ inches}$$

Therefore, the 99% confidence interval for the population mean μ is 66.69 to 68.21 inches, which can be denoted by $66.69 < \mu < 68.21$.

In obtaining the above confidence intervals, we assumed that the population was infinite or so large that we could consider conditions to be the same as sampling with replacement. For finite populations where sampling is without replacement, we should use $\frac{\sigma}{\sqrt{n}}\sqrt{\frac{N-n}{N-1}}$ in place of $\frac{\sigma}{\sqrt{n}}$. However, we can consider the factor $\sqrt{\frac{N-n}{N-1}} = \sqrt{\frac{1546-100}{1546-1}} = 0.967$ as essentially 1.0, so that it need not be used. If it is used, the above confidence limits become 67.45 ± 0.56 and 67.45 ± 0.73 inches, respectively.

6.6. Measurements of the diameters of a random sample of 200 ball bearings made by a certain machine during one week showed a mean of 0.824 inch and a standard deviation of 0.042 inch. Find (a) 95%, (b) 99% confidence limits for the mean diameter of all the ball bearings.

Since $n = 200$ is large, we can assume that $\bar{X}$ is very nearly normal.

(a) The 95% confidence limits are

$$\bar{X} \pm 1.96\frac{\sigma}{\sqrt{n}} = \bar{x} \pm 1.96\frac{\hat{s}}{\sqrt{n}} = 0.824 \pm 1.96\frac{0.042}{\sqrt{200}} = 0.824 \pm 0.0058 \text{ inch}$$

or 0.824 ± 0.006 inch.

(b) The 99% confidence limits are

$$\bar{X} \pm 2.58\frac{\sigma}{\sqrt{n}} = \bar{x} \pm 2.58\frac{\hat{s}}{\sqrt{n}} = 0.824 \pm 2.58\frac{0.042}{\sqrt{200}} = 0.824 \pm 0.0077 \text{ inch}$$

or 0.824 ± 0.008 inches.

Note that we have assumed the reported standard deviation to be the *modified* standard deviation $\hat{s}$. If the standard deviation had been s, we would have used $\hat{s} = \sqrt{n/(n-1)}s = \sqrt{200/199}\,s$ which can be taken as s for all practical purposes. In general, for $n \geq 30$, we may take s and $\hat{s}$ as practically equal.

6.7. Find (a) 98%, (b) 90%, (c) 99.73% confidence limits for the mean diameter of the ball bearings in Problem 6.6.

(a) Let z_c be such that the area under the normal curve to the right of $z = z_c$ is 1%. Then by symmetry the area to the left of $z = -z_c$ is also 1%, so that the shaded area is 98% of the total area (Fig. 6-1).

Since the total area under the curve is one, the area from $z = 0$ is $z = z_c$ is 0.49; hence, $z_c = 2.33$. Therefore, 98% confidence limits are

$$\bar{x} \pm 2.33\frac{\sigma}{\sqrt{n}} = 0.824 \pm 2.33\frac{0.042}{\sqrt{200}} = 0.824 \pm 0.0069 \text{ inch}$$

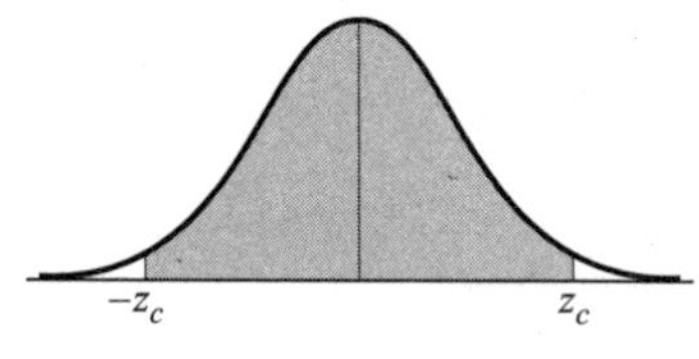

Fig. 6-1

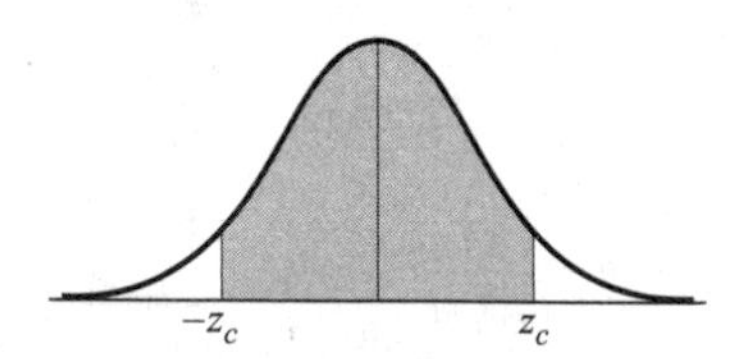

Fig. 6-2

(b) We require z_c such that the area from $z = 0$ to $z = z_c$ is 0.45; then $z_c = 1.645$ (Fig. 6-2).

Therefore, 90% confidence limits are

$$\bar{x} \pm 1.645\frac{\sigma}{\sqrt{n}} = 0.824 \pm 1.645\frac{0.042}{\sqrt{200}} = 0.824 \pm 0.0049 \text{ inch}$$

(c) The 99.73% confidence limits are

$$\bar{x} \pm 3\frac{\sigma}{\sqrt{n}} = 0.824 \pm 3\frac{0.042}{\sqrt{200}} = 0.824 \pm 0.0089 \text{ inch}$$

6.8. In measuring reaction time, a psychologist estimates that the standard deviation is 0.05 second. How large a sample of measurements must he take in order to be (a) 95%, (b) 99% confident that the error in his estimate of mean reaction time will not exceed 0.01 second?

(a) The 95% confidence limits are $\bar{X} \pm 1.96\sigma/\sqrt{n}$, the error of the estimate being $1.96\sigma/\sqrt{n}$. Taking $\sigma = s = 0.05$ second, we see that this error will be equal to 0.01 second if $(1.96)(0.05)/\sqrt{n} = 0.01$, i.e., $\sqrt{n} = (1.96)(0.05)/0.01 = 9.8$, or $n = 96.04$. Therefore, we can be 95% confident that the error in the estimate will be less than 0.01 if n is 97 or larger.

(b) The 99% confidence limits are $\bar{X} \pm 2.58\sigma/\sqrt{n}$. Then $(2.58)(0.05)/\sqrt{n} = 0.01$, or $n = 166.4$. Therefore, we can be 99% confident that the error in the estimate will be less than 0.01 only if n if 167 or larger.

Note that the above solution assumes a nearly normal distribution for $\bar{X}$, which is justified since the n obtained is large.

6.9. A random sample of 50 mathematics grades out of a total of 200 showed a mean of 75 and a standard deviation of 10. (a) What are the 95% confidence limits for the mean of the 200 grades? (b) With what degree of confidence could we say that the mean of all 200 grades is 75 ± 1?

(a) Since the population size is not very large compared with the sample size, we must adjust for *sampling without replacement.* Then the 95% confidence limits are

$$\bar{X} \pm 1.96\sigma_{\bar{X}} = \bar{X} \pm 1.96\frac{\sigma}{\sqrt{n}}\sqrt{\frac{N-n}{N-1}} = 75 \pm 1.96\frac{10}{\sqrt{50}}\sqrt{\frac{200-50}{200-1}} = 75 \pm 2.4$$

(b) The confidence limits can be represented by

$$\bar{X} \pm z_c\sigma_{\bar{X}} = \bar{X} \pm z_c\frac{\sigma}{\sqrt{n}}\sqrt{\frac{N-n}{N-1}} = 75 \pm z_c\frac{(10)}{\sqrt{50}}\sqrt{\frac{200-50}{200-1}} = 75 \pm 1.23z_c$$

Since this must equal 75 ± 1, we have $1.23z_c = 1$ or $z_c = 0.81$. The area under the normal curve from $z = 0$ to $z = 0.81$ is 0.2910; hence, the required degree of confidence is $2(0.2919) = 0.582$ or 58.2%.

Confidence interval estimates for means (small samples)

6.10. The 95% critical values (two-tailed) for the normal distribution are given by ± 1.96. What are the corresponding values for the t distribution if the number of degrees of freedom is (a) $\nu = 9$, (b) $\nu = 20$, (c) $\nu = 30$, (d) $\nu = 60$?

For 95% critical values (two-tailed) the total shaded area in Fig. 6-3 must be 0.05. Therefore, the shaded area in the right tail is 0.025, and the corresponding critical value is $t_{0.975}$. Then the required critical values are $\pm t_{0.975}$. For the given values of ν these are (a) ± 2.26, (b) ± 2.09, (c) ± 2.04, (d) ± 2.00.

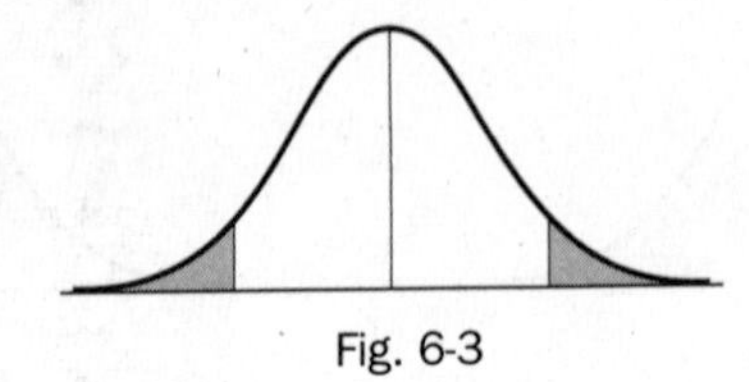

Fig. 6-3

6.11. A sample of 10 measurements of the diameter of a sphere gave a mean $\bar{x} = 4.38$ inches and a standard deviation $s = 0.06$ inch. Find (a) 95%, (b) 99% confidence limits for the actual diameter.

(a) The 95% confidence limits are given by $\bar{X} \pm t_{0.975}(S/\sqrt{n-1})$.

Since $\nu = n - 1 = 10 - 1 = 9$, we find $t_{0.975} = 2.26$ [see also Problem 6.10(a)]. Then using $\bar{x} = 4.38$ and $s = 0.06$, the required 95% confidence limits are

$$4.38 \pm 2.26 \frac{0.06}{\sqrt{10-1}} = 4.38 \pm 0.0452 \text{ inch}$$

Therefore, we can be 95% confident that the true mean lies between $4.38 - 0.045 = 4.335$ inches and $4.38 + 0.045 = 4.425$ inches.

(b) For $\nu = 9$, $t_{0.995} = 3.25$. Then the 99% confidence limits are

$$\bar{X} \pm t_{0.995}(S/\sqrt{n-1}) = 4.38 \pm 3.25(0.06/\sqrt{10-1}) = 4.38 \pm 0.0650 \text{ inch}$$

and the 99% confidence interval is 4.315 to 4.445 inches.

6.12. (a) Work Problem 6.11 assuming that the methods of large sampling theory are valid.
(b) Compare the results of the two methods.

(a) Using large sampling theory, the 95% confidence limits are

$$\bar{X} \pm 1.96 \frac{\sigma}{\sqrt{n}} = 4.38 \pm 1.96 \frac{0.06}{\sqrt{10}} = 4.38 \pm 0.037 \text{ inch}$$

where we have used the sample standard deviation 0.06 as estimate of σ. Similarly, the 99% confidence limits are $4.38 \pm (2.58)(0.06)/\sqrt{10} = 4.38 \pm 0.049$ inch.

(b) In each case the confidence intervals using the small or exact sampling methods are wider than those obtained by using large sampling methods. This is to be expected since less precision is available with small samples than with large samples.

Confidence interval estimates for proportions

6.13. A sample poll of 100 voters chosen at random from all voters in a given district indicated that 55% of them were in favor of a particular candidate. Find (a) 95%, (b) 99%, (c) 99.73% confidence limits for the proportion of all the voters in favor of this candidate.

(a) The 95% confidence limits for the population p are

$$P \pm 1.96\sigma_P = P \pm 1.96\sqrt{\frac{p(1-p)}{n}} = 0.55 \pm 1.96\sqrt{\frac{(0.55)(0.45)}{100}} = 0.55 \pm 0.10$$

where we have used the sample proportion 0.55 to estimate p.

(b) The 99% confidence limits for p are $0.55 \pm 2.58\sqrt{(0.55)(0.45)/100} = 0.55 \pm 0.13$.

(c) The 99.73% confidence limits for p are $0.55 \pm 3\sqrt{(0.55)(0.45)/100} = 0.55 \pm 0.15$.

For a more exact method of working this problem, see Problem 6.27.

6.14. How large a sample of voters should we take in Problem 6.13 in order to be 95% confident that the candidate will be elected?

The candidate is elected if $p > 0.50$, and to be 95% confident of his being elected, we require that Prob. $(p > 0.50) = 0.95$. Since $(P - p)/\sqrt{p(1-p)/n}$ is asymptotically normal,

$$\text{Prob.}\left(\frac{P-p}{\sqrt{p(1-p)/n}} < \beta\right) = \frac{1}{\sqrt{2\pi}}\int_{-\infty}^{\beta} e^{-u^2/2}du$$

or

$$\text{Prob.}\,(p > P - \beta\sqrt{p(1-p)/n}) = \frac{1}{\sqrt{2\pi}}\int_{-\infty}^{\beta} e^{-u^2/2}du$$

Comparison with Prob.$(p > 0.50) = 0.95$, using Appendix C, shows that

$$P - \beta\sqrt{p(1-p)/n} = 0.50 \qquad \text{where} \qquad \beta = 1.645$$

Then, using $P = 0.55$ and the estimate $p = 0.55$ from Problem 6.13, we have

$$0.55 - 1.645\sqrt{(0.55)(0.45)/n} = 0.50 \qquad \text{or} \qquad n = 271$$

6.15. In 40 tosses of a coin, 24 heads were obtained. Find (a) 95%, (b) 99.73% confidence limits for the proportion of heads that would be obtained in an unlimited number of tosses of the coin.

(a) At the 95% level, $z_c = 1.96$. Substituting the values $P = 24/40 = 0.6$ and $n = 40$ in the formula $p = P \pm z_c\sqrt{P(1-P)/n}$, we find $p = 0.60 \pm 0.15$, yielding the interval 0.45 to 0.75.

(b) At the 99.73% level, $z_c = 3$. Using the formula $p = P \pm z_c\sqrt{P(1-P)/n}$, we find $p = 0.60 \pm 0.23$, yielding the interval 0.37 to 0.83.

The more exact formula of Problem 6.27 gives the 95% confidence interval as 0.45 to 0.74 and the 99.73% confidence interval as 0.37 to 0.79.

Confidence intervals for differences and sums

6.16. A sample of 150 brand A light bulbs showed a mean lifetime of 1400 hours and a standard deviation of 120 hours. A sample of 200 brand B light bulbs showed a mean lifetime of 1200 hours and a standard deviation of 80 hours. Find (a) 95%, (b) 99% confidence limits for the difference of the mean lifetimes of the populations of brands A and B.

Confidence limits for the difference in means of brands A and B are given by

$$\bar{X}_A - \bar{X}_B \pm z_c\sqrt{\frac{\sigma_A^2}{n_A} + \frac{\sigma_B^2}{n_B}}$$

(a) The 95% confidence limits are $1400 - 1200 \pm 1.96\sqrt{(120)^2/150 + (80)^2/100} = 200 \pm 24.8$.

Therefore, we can be 95% confident that the difference of population means lies between 175 and 225 hours.

(b) The 99% confidence limits are $1400 - 1200 \pm 2.58\sqrt{(120)^2/150 + (80)^2/100} = 200 \pm 32.6$.

Therefore, we can be 99% confident that the difference of population means lies between 167 and 233 hours.

6.17. In a random sample of 400 adults and 600 teenagers who watched a certain television program, 100 adults and 300 teenagers indicated that they liked it. Construct (a) 95%, (b) 99% confidence limits for the difference in proportions of all adults and all teenagers who watched the program and liked it.

Confidence limits for the difference in proportions of the two groups are given by

$$P_1 - P_2 \pm z_c\sqrt{\frac{P_1Q_1}{n_1} + \frac{P_2Q_2}{n_2}}$$

where subscripts 1 and 2 refer to teenagers and adults, respectively, and $Q_1 = 1 - P_1$, $Q_2 = 1 - P_2$. Here $P_1 = 300/600 = 0.50$ and $P_2 = 100/400 = 0.25$ are, respectively, the proportion of teenagers and adults who liked the program.

(a) 95% confidence limits: $0.50 - 0.25 \pm 1.96\sqrt{(0.50)(0.50)/600 + (0.25)(0.75)/400} = 0.25 \pm 0.06$.

Therefore we can be 95% confident that the true difference in proportions lies between 0.19 and 0.31.

(b) 99% confidence limits: $0.50 - 0.25 \pm 2.58\sqrt{(0.50)(0.50)/600 + (0.25)(0.75)/400} = 0.25 \pm 0.08$.

Therefore, we can be 99% confident that the true difference in proportions lies between 0.17 and 0.33.

6.18. The electromotive force (emf) of batteries produced by a company is normally distributed with mean 45.1 volts and standard deviation 0.04 volt. If four such batteries are connected in series, find (a) 95%, (b) 99%, (c) 99.73%, (d) 50% confidence limits for the total electromotive force.

If E_1, E_2, E_3, and E_4 represent the emfs of the four batteries, we have

$$\mu_{E_1+E_2+E_3+E_4} = \mu_{E_1} + \mu_{E_2} + \mu_{E_3} + \mu_{E_4} \quad \text{and} \quad \sigma_{E_1+E_2+E_3+E_4} = \sqrt{\sigma_{E_1}^2 + \sigma_{E_2}^2 + \sigma_{E_3}^2 + \sigma_{E_4}^2}$$

Then, since $\mu_{E_1} = \mu_{E_2} = \mu_{E_3} = \mu_{E_4} = 45.1$ volts and $\sigma_{E_1} = \sigma_{E_2} = \sigma_{E_3} = \sigma_{E_4} = 0.04$ volt,

$$\mu_{E_1+E_2+E_3+E_4} = 4(45.1) = 180.4 \quad \text{and} \quad \sigma_{E_1+E_2+E_3+E_4} = \sqrt{4(0.04)^2} = 0.08$$

(a) 95% confidence limits are $180.4 \pm 1.96(0.08) = 180.4 \pm 0.16$ volts.

(b) 99% confidence limits are $180.4 \pm 2.58(0.08) = 180.4 \pm 0.21$ volts.

(c) 99.73% confidence limits are $180.4 \pm 3(0.08) = 180.4 \pm 0.24$ volts.

(d) 50% confidence limits are $180.4 \pm 0.6745(0.08) = 180.4 \pm 0.054$ volts.

The value 0.054 volts is called the *probable error.*

Confidence intervals for variances

6.19. The standard deviation of the lifetimes of a sample of 200 electric light bulbs was computed to be 100 hours. Find (a) 95%, (b) 99% confidence limits for the standard deviation of all such electric light bulbs.

In this case large sampling theory applies. Therefore (see Table 5-1, page 160) confidence limits for the population standard deviation σ are given by $S \pm z_c\sigma/\sqrt{2n}$, where z_c indicates the level of confidence. We use the sample standard deviation to estimate σ.

(a) The 95% confidence limits are $100 \pm 1.96(100)/\sqrt{400} = 100 \pm 9.8$.

Therefore, we can be 95% confident that the population standard deviation will lie between 90.2 and 109.8 hours.

(b) The 99% confidence limits are $100 \pm 2.58(100)/\sqrt{400} = 100 \pm 12.9$.

Therefore, we can be 99% confident that the population standard deviation will lie between 87.1 and 112.9 hours.

6.20. How large a sample of the light bulbs in Problem 6.19 must we take in order to be 99.73% confident that the true population standard deviation will not differ from the sample standard deviation by more than (a) 5%, (b) 10%?

As in Problem 6.19, 99.73% confidence limits for σ are $S \pm 3\sigma/\sqrt{2n} = s \pm 3s/\sqrt{2n}$, using s as an estimate of σ. Then the percentage error in the standard deviation is

$$\frac{3s/\sqrt{2n}}{s} = \frac{300}{\sqrt{2n}}\%$$

(a) If $300/\sqrt{2n} = 5$, then $n = 1800$. Therefore, the sample size should be 1800 or more.

(b) If $300/\sqrt{2n} = 10$, then $n = 450$. Therefore, the sample size should be 450 or more.

6.21. The standard deviation of the heights of 16 male students chosen at random in a school of 1000 male students is 2.40 inches. Find (a) 95%, (b) 99% confidence limits of the standard deviation for all male students at the school. Assume that height is normally distributed.

(a) 95% confidence limits are given by $S\sqrt{n}/\chi_{0.975}$ and $S\sqrt{n}/\chi_{0.025}$.

For $\nu = 16 - 1 = 15$ degrees of freedom, $\chi^2_{0.975} = 27.5$ or $\chi_{0.975} = 5.24$ and $\chi^2_{0.025} = 6.26$ or $\chi_{0.025} = 2.50$.

Then the 95% confidence limits are $2.40\sqrt{16}/5.24$ and $2.40\sqrt{16}/2.50$, i.e., 1.83 and 3.84 inches. Therefore, we can be 95% confident that the population standard deviation lies between 1.83 and 3.84 inches.

(b) 99% confidence limits are given by $S\sqrt{n}/\chi_{0.995}$ and $S\sqrt{n}/\chi_{0.005}$.

For $\nu = 16 - 1 = 15$ degrees of freedom, $\chi^2_{0.995} = 32.8$ or $\chi_{0.995} = 5.73$ and $\chi^2_{0.005} = 4.60$ or $\chi_{0.005} = 21.4$.

Then the 99% confidence limits are $2.40\sqrt{16}/5.73$ and $2.40\sqrt{16}/2.14$, i.e., 1.68 and 4.49 inches. Therefore, we can be 99% confident that the population standard deviation lies between 1.68 and 4.49 inches.

6.22. Work Problem 6.19 using small or exact sampling theory.

(a) 95% confidence limits are given by $S\sqrt{n}/\chi_{0.975}$ and $S\sqrt{n}/\chi_{0.025}$.

For $\nu = 200 - 1 = 199$ degrees of freedom, we find as in Problem 4.41, page 136,

$$\chi^2_{0.975} = \frac{1}{2}(z_{0.975} + \sqrt{2(199) - 1})^2 = \frac{1}{2}(1.96 + 19.92)^2 = 239$$

$$\chi^2_{0.025} = \frac{1}{2}(z_{0.025} + \sqrt{2(199) - 1})^2 = \frac{1}{2}(-1.96 + 19.92)^2 = 161$$

from which $\chi_{0.975} = 15.5$ and $\chi_{0.025} = 12.7$.

Then the 95% confidence limits are $100\sqrt{200}/15.5 = 91.2$ and $100\sqrt{200}/12.7 = 111.3$ hours respectively. Therefore, we can be 95% confident that the population standard deviation will lie between 91.2 and 111.3 hours.

This should be compared with the result of Problem 6.19(a).

(b) 99% confidence limits are given by $S\sqrt{n}/\chi_{0.995}$ and $S\sqrt{n}/\chi_{0.005}$.

For $\nu = 200 - 1 = 199$ degrees of freedom,

$$\chi^2_{0.995} = \frac{1}{2}(z_{0.995} + \sqrt{2(199) - 1})^2 = \frac{1}{2}(2.58 + 19.92)^2 = 253$$

$$\chi^2_{0.005} = \frac{1}{2}(z_{0.005} + \sqrt{2(199) - 1})^2 = \frac{1}{2}(-2.58 + 19.92)^2 = 150$$

from which $\chi_{0.995} = 15.9$ and $\chi_{0.005} = 12.2$.

Then the 99% confidence limits are $100\sqrt{200}/15.9 = 88.9$ and $100\sqrt{200}/12.2 = 115.9$ hours respectively. Therefore, we can be 99% confident that the population standard deviation will lie between 88.9 and 115.9 hours.

This should be compared with the result of Problem 6.19(b).

Confidence intervals for variance ratios

6.23. Two samples of sizes 16 and 10, respectively, are drawn at random from two normal populations. If their variances are found to be 24 and 18, respectively, find (a) 98%, (b) 90% confidence limits for the ratio of the variances.

(a) We have $m = 16, n = 10, s_1^2 = 20, s_2^2 = 18$ so that

$$\hat{s}_1^2 = \frac{m}{m-1}s_1^2 = \left(\frac{16}{15}\right)(24) = 25.2$$

$$\hat{s}_2^2 = \frac{n}{n-1}s_2^2 = \left(\frac{10}{9}\right)(18) = 20.0$$

From Problem 4.47(b), page 139, we have $F_{0.99} = 4.96$ for $\nu_1 = 16 - 1 = 15$ and $\nu_2 = 10 - 1 = 9$ degrees of freedom. Also, from Problem 4.47(d), we have for $\nu_1 = 15$ and $\nu_2 = 9$ degrees of freedom $F_{0.01} = 1/3.89$ so that $1/F_{0.01} = 3.89$. Then using (17), page 198, we find for the 98% confidence interval

$$\left(\frac{1}{4.96}\right)\left(\frac{25.2}{20.0}\right) \le \frac{\sigma_1^2}{\sigma_2^2} \le (3.89)\left(\frac{25.2}{20.0}\right)$$

or

$$0.283 \le \frac{\sigma_1^2}{\sigma_2^2} \le 4.90$$

(b) As in (a) we find from Appendix F that $F_{0.95} = 3.01$ and $F_{0.05} = 1/2.59$. Therefore, the 90% confidence interval is

$$\frac{1}{3.01}\left(\frac{25.2}{20.0}\right) \le \frac{\sigma_1^2}{\sigma_2^2} \le (2.59)\left(\frac{25.2}{20.0}\right)$$

or

$$0.4186 \le \frac{\sigma_1^2}{\sigma_2^2} \le 3.263$$

Note that the 90% confidence interval is much smaller than the 98% confidence interval, as we would of course expect.

6.24. Find the (a) 98%, (b) 90% confidence limits for the ratio of the standard deviations in Problem 6.23.

By taking square roots of the inequalities in Problem 6.23, we find for the 98% and 90% confidence limits

(a)
$$0.53 \le \frac{\sigma_1}{\sigma_2} \le 2.21$$

(b)
$$0.65 \le \frac{\sigma_1}{\sigma_2} \le 1.81$$

Maximum likelihood estimates

6.25. Suppose that n observations, $X_1, \ldots, X_n$, are made from a normally distributed population of which the mean is unknown and the variance is known. Find the maximum likelihood estimate of the mean.

Since

$$f(x_k, \mu) = \frac{1}{\sqrt{2\pi\sigma^2}}e^{-(x_k-\mu)^2/2\sigma^2}$$

we have

(1)
$$L = f(x_1, \mu) \cdots f(x_n, \mu) = (2\pi\sigma^2)^{-n/2}e^{-\sum(x_k-\mu)^2/2\sigma^2}$$

Therefore,

(2)
$$\ln L = -\frac{n}{2}\ln(2\pi\sigma^2) - \frac{1}{2\sigma^2}\sum(x_k - \mu)^2$$

Taking the partial derivative with respect to μ yields

(3)
$$\frac{1}{L}\frac{\partial L}{\partial \mu} = \frac{1}{\sigma^2}\sum(x_k - \mu)$$

Setting $\partial L/\partial\mu = 0$ gives

(4) $$\sum(x_k - \mu) = 0 \qquad \text{i.e.} \qquad \sum x_k - n\mu = 0$$

or

(5) $$\mu = \frac{\sum x_k}{n}$$

Therefore, the maximum likelihood estimate is the sample mean.

6.26. If in Problem 6.25 the mean is known but the variance is unknown, find the maximum likelihood estimate of the variance.

If we write $f(x_k, \sigma^2)$ instead of $f(x_k, \mu)$, everything done in Problem 6.25 through equation (2) still applies. Then, taking the partial derivative with respect to σ^2, we have

$$\frac{1}{L}\frac{\partial L}{\partial \sigma^2} = -\frac{n}{2\sigma^2} + \frac{1}{2(\sigma^2)^2}\sum(x_k - \mu)^2$$

Setting $\partial L/\partial \sigma^2 = 0$, we find

$$\sigma^2 = \frac{\sum(x_k - \mu)^2}{n}$$

Miscellaneous problems

6.27. (a) If P is the observed proportion of successes in a sample of size n, show that the confidence limits for estimating the population proportion of successes p at the level of confidence determined by z_c are given by

$$\frac{P + \dfrac{z_c^2}{2n} \pm z_c\sqrt{\dfrac{P(1-P)}{n} + \dfrac{z_c^2}{4n^2}}}{1 + \dfrac{z_c^2}{n}}$$

(b) Use the formula derived in (a) to obtain the 99.73% confidence limits of Problem 6.13. (c) Show that for large n the formula in (a) reduces to $P \pm z_c\sqrt{P(1-P)/n}$, as used in Problem 6.13.

(a) The sample proportion P in standard units is $\dfrac{P - p}{\sigma_P} = \dfrac{P - p}{\sqrt{p(1-p)/n}}$

The largest and smallest values of this standardized variable are $\pm z_c$, where z_c determines the level of confidence. At these extreme values we must therefore have

$$P - p = \pm z_c\sqrt{\frac{p(1-p)}{n}}$$

Squaring both sides,

$$P^2 - 2pP + p^2 = z_c^2\frac{p(1-p)}{n}$$

Multiplying both sides by n and simplifying, we find

$$\left(n + z_c^2\right)p^2 - \left(2nP + z_c^2\right)p + nP^2 = 0$$

If $a = n + z_c^2$, $b = -\left(2nP + z_c^2\right)$ and $c = nP^2$, this equation becomes $ap^2 + bp + c = 0$, whose solution for p is given by the quadratic formula as

$$p = \frac{-b \pm \sqrt{b^2 - 4ac}}{2a} = \frac{2nP + z_c^2 \pm \sqrt{\left(2nP + z_c^2\right)^2 - 4\left(n + z_c^2\right)(nP^2)}}{2\left(n + z_c^2\right)}$$

$$= \frac{2nP + z_c^2 \pm z_c\sqrt{4nP(1-P) + z_c^2}}{2\left(n + z_c^2\right)}$$

Dividing the numerator and denominator by $2n$, this becomes

$$p = \frac{P + \dfrac{z_c^2}{2n} \pm z_c\sqrt{\dfrac{P(1-P)}{n} + \dfrac{z_c^2}{4n^2}}}{1 + \dfrac{z_c^2}{n}}$$

(b) For 99.73% confidence limits, $z_c = 3$. Then using $P = 0.55$ and $n = 100$ in the formula derived in (a), we find $p = 0.40$ and 0.69, agreeing with Problem 6.13(c).

(c) If n is large, then $z_c^2/2n$, $z_c^2/4n^2$, and z_c^2/n are all negligibly small and can essentially be replaced by zero, so that the required result is obtained.

6.28. Is it possible to obtain a 95% confidence interval for the population standard deviation whose expected width is smaller than that found in Problem 6.22(a)?

The 95% confidence limits for the population standard deviation as found in Problem 6.22(a) were obtained by choosing critical values of χ^2 such that the area in each tail was 2.5%. It is possible to find other 95% confidence limits by choosing critical values of χ^2 for which the sum of the areas in the tails is 5%, or 0.05, but such that the areas in each tail are not equal.

In Table 6-2 several such critical values have been obtained and the corresponding 95% confidence intervals shown.

Table 6-2

Critical Values	95% Confidence Interval	Width
$\chi_{0.01} = 12.44$, $\chi_{0.96} = 15.32$	92.3 to 113.7	21.4
$\chi_{0.02} = 12.64$, $\chi_{0.97} = 15.42$	91.7 to 111.9	20.2
$\chi_{0.03} = 12.76$, $\chi_{0.98} = 15.54$	91.0 to 110.8	19.8
$\chi_{0.04} = 12.85$, $\chi_{0.99} = 15.73$	89.9 to 110.0	20.1

From this table it is seen that a 95% interval of width only 19.8 is 91.0 to 110.8.

An interval with even smaller width can be found by continuing the same method of approach, using critical values such as $\chi_{0.031}$ and $\chi_{0.981}$, $\chi_{0.032}$ and $\chi_{0.982}$, etc.

In general, however, the decrease in the interval that is thereby obtained is usually negligible and is not worth the labor involved.

SUPPLEMENTARY PROBLEMS

Unbiased and efficient estimates

6.29. Measurements of a sample of weights were determined as 8.3, 10.6, 9.7, 8.8, 10.2, and 9.4 lb, respectively. Determine unbiased and efficient estimates of (a) the population mean, and (b) the population variance. (c) Compare the sample standard deviation with the estimated population standard deviation.

6.30. A sample of 10 television tubes produced by a company showed a mean lifetime of 1200 hours and a standard deviation of 100 hours. Estimate (a) the mean, (b) the standard deviation of the population of all television tubes produced by this company.

6.31. (a) Work Problem 6.30 if the same results are obtained for 30, 50, and 100 television tubes, (b) What can you conclude about the relation between sample standard deviations and estimates of population standard deviations for different sample sizes?

Confidence interval estimates for means (large samples)

6.32. The mean and standard deviation of the maximum loads supported by 60 cables (see Problem 5.98) are 11.09 tons and 0.73 tons, respectively. Find (a) 95%, (b) 99% confidence limits for the mean of the maximum loads of all cables produced by the company.

6.33. The mean and standard deviation of the diameters of a sample of 250 rivet heads manufactured by a company are 0.72642 inch and 0.00058 inch, respectively (see Problem 5.99). Find (a) 99%, (b) 98%, (c) 95%, (d) 90% confidence limits for the mean diameter of all the rivet heads manufactured by the company.

6.34. Find (a) the 50% confidence limits, (b) the probable error for the mean diameter in Problem 6.33.

6.35. If the standard deviation of the lifetimes of television tubes is estimated as 100 hours, how large a sample must we take in order to be (a) 95%, (b) 90%, (c) 99%, (d) 99.73% confident that the error in the estimated mean lifetime will not exceed 20 hours.

6.36. What are the sample sizes in Problem 6.35 if the error in the estimated mean lifetime must not exceed 10 hours?

Confidence interval estimates for means (small samples)

6.37. A sample of 12 measurements of the breaking strengths of cotton threads gave a mean of 7.38 oz and a standard deviation of 1.24 oz. Find (a) 95%, (b) 99% confidence limits for the actual mean breaking strength.

6.38. Work Problem 6.37 assuming that the methods of large sampling theory are applicable, and compare the results obtained.

6.39. Five measurements of the reaction time of an individual to certain stimuli were recorded as 0.28, 0.30, 0.27, 0.33, 0.31 second. Find (a) 95%, (b) 99% confidence limits for the actual mean reaction time.

Confidence interval estimates for proportions

6.40. An urn contains red and white marbles in an unknown proportion. A random sample of 60 marbles selected with replacement from the urn showed that 70% were red. Find (a) 95%, (b) 99%, (c) 99.73% confidence limits for the actual proportion of red marbles in the urn. Present the results using both the approximate formula and the more exact formula of Problem 6.27.

6.41. How large a sample of marbles should one take in Problem 6.40 in order to be (a) 95%, (b) 99%, (c) 99.73% confident that the true and sample proportions do not differ more than 5%?

6.42. It is believed that an election will result in a very close vote between two candidates. Explain by means of an example, stating all assumptions, how you would determine the least number of voters to poll in order to be (a) 80%, (b) 95%, (c) 99% confident of a decision in favor of either one of the candidates.

Confidence intervals for differences and sums

6.43. Of two similar groups of patients, A and B, consisting of 50 and 100 individuals, respectively, the first was given a new type of sleeping pill and the second was given a conventional type. For patients in group A the mean number of hours of sleep was 7.82 with a standard deviation of 0.24 hour. For patients in group B the mean number of hours of sleep was 6.75 with a standard deviation of 0.30 hour. Find (a) 95% and (b) 99% confidence limits for the difference in the mean number of hours of sleep induced by the two types of sleeping pills.

6.44. A sample of 200 bolts from one machine showed that 15 were defective, while a sample of 100 bolts from another machine showed that 12 were defective. Find (a) 95%, (b) 99%, (c) 99.73% confidence limits for the difference in proportions of defective bolts from the two machines. Discuss the results obtained.

6.45. A company manufactures ball bearings having a mean weight of 0.638 oz and a standard deviation of 0.012 oz. Find (a) 95%, (b) 99% confidence limits for the weights of lots consisting of 100 ball bearings each.

Confidence intervals for variances or standard deviations

6.46. The standard deviation of the breaking strengths of 100 cables tested by a company was 1800 lb. Find (a) 95%, (b) 99%, (c) 99.73% confidence limits for the standard deviation of all cables produced by the company.

6.47. How large a sample should one take in order to be (a) 95%, (b) 99%, (c) 99.73% confident that a population standard deviation will not differ from a sample standard deviation by more than 2%?

6.48. The standard deviation of the lifetimes of 10 electric light bulbs manufactured by a company is 120 hours. Find (a) 95%, (b) 99% confidence limits for the standard deviation of all bulbs manufactured by the company.

6.49. Work Problem 6.48 if 25 electric light bulbs show the same standard deviation of 120 hours.

6.50. Work Problem 6.48 by using the χ^2 distribution if a sample of 100 electric bulbs shows the same standard deviation of 120 hours.

Confidence intervals for variance ratios

6.51. The standard deviations of the diameters of ball bearings produced by two machines were found to be 0.042 cm and 0.035 cm, respectively, based on samples of sizes 10 each. Find (a) 98%, (b) 90% confidence intervals for the ratio of the variances.

6.52. Determine the (a) 98%, (b) 90% confidence intervals for the ratio of the standard deviations in Problem 6.51.

6.53. Two samples of sizes 6 and 8, respectively, turn out to have the same variance. Find (a) 98%, (b) 90% confidence intervals for the ratio of the variances of the populations from which they were drawn.

6.54. Work (a) Problem 6.51, (b) Problem 6.53 if the samples have sizes 120 each.

Maximum likelihood estimates

6.55. Suppose that n observations, $X_1, \ldots, X_n$, are made from a Poisson distribution with unknown parameter λ. Find the maximum likelihood estimate of λ.

6.56. A population has a density function given by $f(x) = 2\nu\sqrt{\nu/\pi}x^2e^{-\nu x^2}, -\infty < x < \infty$. For n observations, $X_1, \ldots, X_n$, made from this population, find the maximum likelihood estimate of ν.

6.57. A population has a density function given by

$$f(x) = \begin{cases} (k+1)x^k & 0 \le x \le 1 \\ 0 & \text{otherwise} \end{cases}$$

For n observations $X_1, \ldots, X_n$ made from this population, find the maximum likelihood estimate of k.

Miscellaneous problems

6.58. The 99% confidence coefficients (two-tailed) for the normal distribution are given by ±2.58. What are the corresponding coefficients for the t distribution if (a) $\nu = 4$, (b) $\nu = 12$, (c) $\nu = 25$, (d) $\nu = 30$, (e) $\nu = 40$?

6.59. A company has 500 cables. A test of 40 cables selected at random showed a mean breaking strength of 2400 lb and a standard deviation of 150 lb. (a) What are the 95% and 99% confidence limits for estimating the mean breaking strength of the remaining 460 cables? (b) With what degree of confidence could we say that the mean breaking strength of the remaining 460 cables is 2400 ± 35 lb?

ANSWERS TO SUPPLEMENTARY PROBLEMS

6.29. (a) 9.5 lb (b) 0.74 lb^2 (c) 0.78 and 0.86 lb, respectively.

6.30. (a) 1200 hours (b) 105.4 hours

6.31. (a) Estimates of population standard deviations for sample sizes 30, 50, and 100 tubes are, respectively, 101.7, 101.0, and 100.5 hours. Estimates of population means are 1200 hours in all cases.

6.32. (a) 11.09 ± 0.18 tons (b) 11.09 ± 0.24 tons

6.33. (a) 0.72642 ± 0.000095 inch (c) 0.72642 ± 0.000072 inch
(b) 0.72642 ± 0.000085 inch (d) 0.72642 ± 0.000060 inch

6.34. (a) 0.72642 ± 0.000025 inch (b) 0.000025 inch

6.35. (a) at least 97 (b) at least 68 (c) at least 167 (d) at least 225

6.36. (a) at least 385 (b) at least 271 (c) at least 666 (d) at least 900

6.37. (a) 7.38 ± 0.82 oz (b) 7.38 ± 1.16 oz **6.38.** (a) 7.38 ± 0.70 oz (*b)* 7.38 ± 0.96 oz

6.39. (a) 0.298 ± 0.030 second (b) 0.298 ± 0.049 second

6.40. (a) 0.70 ± 0.12, 0.69 ± 0.11 (b) 0.70 ± 0.15, 0.68 ± 0.15 (c) 0.70 ± 0.18, 0.67 ± 0.17

6.41. (a) at least 323 (b) at least 560 (c) at least 756

6.43. (a) 1.07 ± 0.09 hours (b) 1.07 ± 0.12 hours

6.44. (a) 0.045 ± 0.073 (b) 0.045 ± 0.097 (c) 0.045 ± 0.112

6.45. (a) 63.8 ± 0.24 oz (b) 63.8 ± 0.31 oz

6.46. (a) 1800 ± 249 lb (b) 1800 ± 328 lb (c) 1800 ± 382 lb

6.47. (a) at least 4802 (b) at least 8321 (c) at least 11,250

6.48. (a) 87.0 to 230.9 hours (b) 78.1 to 288.5 hours

6.49. (a) 95.6 to 170.4 hours (b) 88.9 to 190.8 hours

6.50. (a) 106.1 to 140.5 hours (b) 102.1 to 148.1 hours

6.51. (a) 0.269 to 7.70 (b) 0.453 to 4.58 **6.52.** (a) 0.519 to 2.78 (b) 0.673 to 2.14

6.53. (a) 0.140 to 11.025 (b) 0.264 to 5.124

6.54. (a) 0.941 to 2.20, 1.067 to 1.944 (b) 0.654 to 1.53, 0.741 to 1.35

6.55. $\lambda = \left(\sum x_k\right)/n$ **6.56.** $\nu = \dfrac{3n}{2(x_1^2 + \cdots + x_n^2)}$

6.57. $k = -1 - \dfrac{n}{\ln(x_1 \cdots x_n)}$

6.58. (a) ± 4.60 (b) ± 3.06 (c) ± 2.79 (d) ± 2.75 (e) ± 2.70

6.59. (a) 2400 $\pm$ 45 lb, 2400 $\pm$ 59 lb (b) 87.6%

CHAPTER 7

Tests of Hypotheses and Significance

Statistical Decisions

Very often in practice we are called upon to make decisions about populations on the basis of sample information. Such decisions are called *statistical decisions.* For example, we may wish to decide on the basis of sample data whether a new serum is really effective in curing a disease, whether one educational procedure is better than another, or whether a given coin is loaded.

Statistical Hypotheses. Null Hypotheses

In attempting to reach decisions, it is useful to make assumptions or guesses about the populations involved. Such assumptions, which may or may not be true, are called *statistical hypotheses* and in general are statements about the probability distributions of the populations.

For example, if we want to decide whether a given coin is loaded, we formulate the hypothesis that the coin is fair, i.e., $p = 0.5$, where p is the probability of heads. Similarly, if we want to decide whether one procedure is better than another, we formulate the hypothesis that there is *no difference* between the procedures (i.e., any observed differences are merely due to fluctuations in sampling from the *same* population). Such hypotheses are often called *null hypotheses* or simply *hypotheses*, are denoted by H_0.

Any hypothesis that differs from a given null hypothesis is called an *alternative hypothesis*. For example, if the null hypothesis is $p = 0.5$, possible alternative hypotheses are $p = 0.7$, $p \neq 0.5$, or $p > 0.5$. A hypothesis alternative to the null hypothesis is denoted by H_1.

Tests of Hypotheses and Significance

If on the supposition that a particular hypothesis is true we find that results observed in a random sample differ markedly from those expected under the hypothesis on the basis of pure chance using sampling theory, we would say that the observed differences are *significant* and we would be inclined to reject the hypothesis (or at least not accept it on the basis of the evidence obtained). For example, if 20 tosses of a coin yield 16 heads, we would be inclined to reject the hypothesis that the coin is fair, although it is conceivable that we might be wrong.

Procedures that enable us to decide whether to accept or reject hypotheses or to determine whether observed samples differ significantly from expected results are called *tests of hypotheses, tests of significance*, or *decision rules.*

Type I and Type II Errors

If we reject a hypothesis when it happens to be true, we say that a *Type I error* has been made. If, on the other hand, we accept a hypothesis when it should be rejected, we say that a *Type II error* has been made. In either case a wrong decision or error in judgment has occurred.

In order for any tests of hypotheses or decision rules to be good, they must be designed so as to minimize errors of decision. This is not a simple matter since, for a given sample size, an attempt to decrease one type of error is accompanied in general by an increase in the other type of error. In practice one type of error may be more serious than the other, and so a compromise should be reached in favor of a limitation of the more serious error. The only way to reduce both types of error is to increase the sample size, which may or may not be possible.

Level of Significance

In testing a given hypothesis, the maximum probability with which we would be willing to risk a Type I error is called the *level of significance* of the test. This probability is often specified before any samples are drawn, so that results obtained will not influence our decision.

In practice a level of significance of 0.05 or 0.01 is customary, although other values are used. If for example a 0.05 or 5% level of significance is chosen in designing a test of a hypothesis, then there are about 5 chances in 100 that we would reject the hypothesis when it should be accepted, i.e., whenever the null hypotheses is true, we are about 95% *confident* that we would make the right decision. In such cases we say that the hypothesis has been *rejected at a* 0.05 *level of significance*, which means that we could be wrong with probability 0.05.

Tests Involving the Normal Distribution

To illustrate the ideas presented above, suppose that under a given hypothesis the sampling distribution of a statistic S is a normal distribution with mean μ_S and standard deviation σ_S. Also, suppose we decide to reject the hypothesis if S is either too small or too large. The distribution of the standardized variable $Z = (S - \mu_S)/\sigma_S$ is the standard normal distribution (mean 0, variance 1) shown in Fig. 7-1, and extreme values of Z would lead to the rejection of the hypothesis.

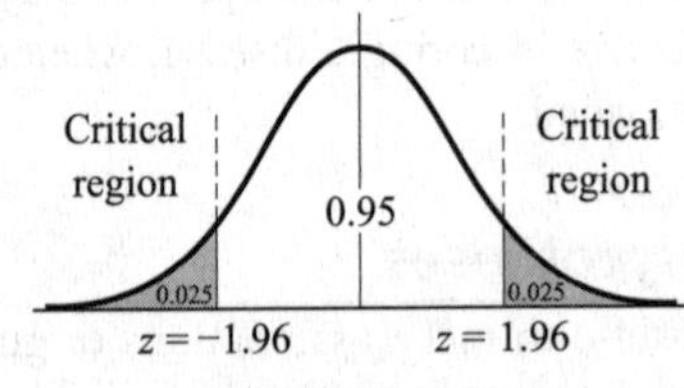

Fig. 7-1

As indicated in the figure, we can be 95% confident that, if the hypothesis is true, the z score of an actual sample statistic S will lie between -1.96 and 1.96 (since the area under the normal curve between these values is 0.95).

However, if on choosing a single sample at random we find that the z score of its statistic lies *outside* the range -1.96 to 1.96, we would conclude that such an event could happen with the probability of only 0.05 (total shaded area in the figure) if the given hypothesis were true. We would then say that this z score differed *significantly* from what would be expected under the hypothesis, and we would be inclined to reject the hypothesis.

The total shaded area 0.05 is the level of significance of the test. It represents the probability of our being wrong in rejecting the hypothesis, i.e., the probability of making a Type I error. Therefore, we say that the hypothesis is *rejected at a* 0.05 *level of significance* or that the z score of the given sample statistic is *significant* at a 0.05 *level of significance*.

The set of z scores outside the range -1.96 to 1.96 constitutes what is called the *critical region* or *region of rejection of the hypothesis* or *the region of significance*. The set of z scores inside the range -1.96 to 1.96 could then be called the *region of acceptance of the hypothesis* or the *region of nonsignificance*.

On the basis of the above remarks, we can formulate the following decision rule:

(a) Reject the hypothesis at a 0.05 level of significance if the z score of the statistic S lies outside the range -1.96 to 1.96 (i.e., either $z > 1.96$ or $z < -1.96$). This is equivalent to saying that the observed sample statistic is significant at the 0.05 level.

(b) Accept the hypothesis (or, if desired, make no decision at all) otherwise.

It should be noted that other levels of significance could have been used. For example, if a 0.01 level were used we would replace 1.96 everywhere above by 2.58 (see Table 7-1). Table 6-1, page 196, can also be used since the sum of the level of significance and level of confidence is 100%.

One-Tailed and Two-Tailed Tests

In the above test we displayed interest in extreme values of the statistic S or its corresponding z score on both sides of the mean, i.e., in both tails of the distribution. For this reason such tests are called *two-tailed tests* or *two-sided tests*.

Often, however, we may be interested only in extreme values to one side of the mean, i.e., in one tail of the distribution, as, for example, when we are testing the hypothesis that one process is better than another (which is different from testing whether one process is better or worse than the other). Such tests are called *one-tailed tests* or *one-sided tests*. In such cases the critical region is a region to one side of the distribution, with area equal to the level of significance.

Table 7-1, which gives critical values of z for both one-tailed and two-tailed tests at various levels of significance, will be useful for reference purposes. Critical values of z for other levels of significance are found by use of the table of normal curve areas.

Table 7-1

Level of Significance α	0.10	0.05	0.01	0.005	0.002
Critical Values of z for One-Tailed Tests	-1.28 *or* 1.28	-1.645 *or* 1.645	-2.33 *or* 2.33	-2.58 *or* 2.58	-2.88 *or* 2.88
Critical Values of z for Two-Tailed Tests	-1.645 *and* 1.645	-1.96 *and* 1.96	-2.58 *and* 2.58	-2.81 *and* 2.81	-3.08 *and* 3.08

P Value

In most of the tests we will consider, the null hypothesis H_0 will be an assertion that a population parameter has a specific value, and the alternative hypothesis H_1 will be one of the following assertions:

(i) The parameter is greater than the stated value (right-tailed test).

(ii) The parameter is less than the stated value (left-tailed test).

(iii) The parameter is either greater than or less than the stated value (two-tailed test).

In cases (i) and (ii), H_1 has a single direction with respect to the parameter, and in case (iii), H_1 is bidirectional. After the test has been performed and the test statistic S computed, the P value of the test is the probability that a value of S in the direction(s) of H_1 and as extreme as the one that actually did occur would occur if H_0 were true.

For example, suppose the standard deviation σ of a normal population is known to be 3, and H_0 asserts that the mean μ is equal to 12. A random sample of size 36 drawn from the population yields a sample mean $\bar{x} = 12.95$. The test statistic is chosen to be

$$Z = \frac{\bar{X} - 12}{\sigma/\sqrt{n}} = \frac{\bar{X} - 12}{0.5},$$

which, if H_0 is true, is the standard normal random variable. The test value of Z is $(12.95 - 12)/0.5 = 1.9$. The P value for the test then depends on the alternative hypothesis H_1 as follows:

(i) For H_1: $\mu > 12$ [case (i) above], the P value is the probability that a random sample of size 36 would yield a sample mean of 12.95 or more if the true mean were 12, i.e., $P(Z \geq 1.9) = 0.029$. In other words, the chances are about 3 in 100 that $\bar{x} > 12.95$ if $\mu = 12$.

(ii) For H_1: $\mu < 12$ [case (ii) above], the P value of the test is the probability that a random sample of size 36 would yield a sample mean of 12.95 or less if the true mean were 12, i.e., $P(Z \leq 1.9) = 0.97$, or the chances are about 97 in 100 that $\bar{x} \leq 12.95$ if $\mu = 12$.

(iii) For H_1: $\mu \neq 12$ [case (iii) above], the P value is the probability that a random sample of size 36 would yield a sample mean 0.95 or more units away from 12, i.e., $\bar{x} \geq 12.95$ or $\bar{x} \leq 11.05$, if the true mean were 12. Here the P value is $P(Z \geq 1.9) + P(Z \leq -1.9) = 0.057$, which says the chances are about 6 in 100 that $|\bar{x} - 12| \geq 0.095$ if $\mu = 12$.

Small P values provide evidence for rejecting the null hypothesis in favor of the alternative hypothesis, and large P values provide evidence for not rejecting the null hypothesis in favor of the alternative hypothesis. In case (i) of the above example, the small P value 0.029 is a fairly strong indicator that the population mean is greater than 12, whereas in case (ii), the large P value 0.97 strongly suggests that H_0: $\mu = 12$ should not be rejected in favor of H_1: $\mu < 12$. In case (iii), the P value 0.057 provides evidence for rejecting H_0 in favor of H_1: $\mu \neq 12$ but not as much evidence as is provided for rejecting H_0 in favor of H_1: $\mu > 12$.

It should be kept in mind that the P value and the level of significance do not provide criteria for rejecting or not rejecting the null hypothesis by itself, but for rejecting or not rejecting the null hypothesis in favor of the alternative hypothesis. As the previous example illustrates, identical test results and significance levels can lead to different conclusions regarding the same null hypothesis in relation to different alternative hypotheses.

When the test statistic S is the standard normal random variable, the table in Appendix C is sufficient to compute the P value, but when S is one of the t, F, or chi-square random variables, all of which have different distributions depending on their degrees of freedom, either computer software or more extensive tables than those in Appendices D, E, and F will be needed to compute the P value.

Special Tests of Significance for Large Samples

For large samples, many statistics S have nearly normal distributions with mean μ_S and standard deviation σ_S. In such cases we can use the above results to formulate decision rules or tests of hypotheses and significance. The following special cases are just a few of the statistics of practical interest. In each case the results hold for infinite populations or for sampling with replacement. For sampling without replacement from finite populations, the results must be modified. See pages 156 and 158.

1. MEANS. Here $S = \bar{X}$, the sample mean; $\mu_S = \mu_{\bar{X}} = \mu$, the population mean; $\sigma_S = \sigma_{\bar{X}} = \sigma/\sqrt{n}$, where σ is the population standard deviation and n is the sample size. The standardized variable is given by

$$Z = \frac{\bar{X} - \mu}{\sigma/\sqrt{n}} \tag{1}$$

When necessary the observed sample standard deviation, s (or $\hat{s}$), is used to estimate σ.

To test the null hypothesis H_0 that the population mean is $\mu = a$, we would use the statistic (1). Then, if the alternative hypothesis is $\mu \neq a$, using a two-tailed test, we would accept H_0 (or at least not reject it) at the 0.05 level if for a particular sample of size n having mean $\bar{x}$

$$-1.96 \leq \frac{\bar{x} - a}{\sigma/\sqrt{n}} \leq 1.96 \tag{2}$$

and would reject it otherwise. For other significance levels we would change (2) appropriately. To test H_0 against the alternative hypothesis that the population mean is greater than a, we would use a one-tailed test and accept H_0 (or at least not reject it) at the 0.05 level if

$$\frac{\bar{x} - a}{\sigma/\sqrt{n}} < 1.645 \tag{3}$$

(see Table 7-1) and reject it otherwise. To test H_0 against the alternative hypothesis that the population mean is less than a, we would accept H_0 at the 0.05 level if

$$\frac{\bar{x} - a}{\sigma/\sqrt{n}} > -1.645 \tag{4}$$

2. PROPORTIONS. Here $S = P$, the proportion of "successes" in a sample; $\mu_S = \mu_P = p$, where p is the population proportion of successes and n is the sample size; $\sigma_S = \sigma_P = \sqrt{pq/n}$, where $q = 1 - p$. The standardized variable is given by

$$Z = \frac{P - p}{\sqrt{pq/n}} \tag{5}$$

In case $P = X/n$, where X is the actual number of successes in a sample, (5) becomes

$$Z = \frac{X - np}{\sqrt{npq}} \tag{6}$$

Remarks similar to those made above about one- and two-tailed tests for means can be made.

3. **DIFFERENCES OF MEANS.** Let $\bar{X}_1$ and $\bar{X}_2$ be the sample means obtained in large samples of sizes n_1 and n_2 drawn from respective populations having means μ_1 and μ_2 and standard deviations σ_1 and σ_2. Consider the null hypothesis that there is *no difference* between the population means, i.e., $\mu_1 = \mu_2$. From (11), page 157, on placing $\mu_1 = \mu_2$ we see that the sampling distribution of differences in means is approximately normal with mean and standard deviation given by

$$\mu_{\bar{X}_1-\bar{X}_2} = 0 \qquad \sigma_{\bar{X}_1-\bar{X}_2} = \sqrt{\frac{\sigma_1^2}{n_1} + \frac{\sigma_2^2}{n_2}} \tag{7}$$

where we can, if necessary, use the observed sample standard deviations s_1 and s_2 (or $\hat{s}_1$ and $\hat{s}_2$) as estimates of σ_1 and σ_2.

By using the standardized variable given by

$$Z = \frac{\bar{X}_1 - \bar{X}_2 - 0}{\sigma_{\bar{X}_1-\bar{X}_2}} = \frac{\bar{X}_1 - \bar{X}_2}{\sigma_{\bar{X}_1-\bar{X}_2}} \tag{8}$$

in a manner similar to that described in Part 1 above, we can test the null hypothesis against alternative hypotheses (or the significance of an observed difference) at an appropriate level of significance.

4. **DIFFERENCES OF PROPORTIONS.** Let P_1 and P_2 be the sample proportions obtained in large samples of sizes n_1 and n_2 drawn from respective populations having proportions p_1 and p_2. Consider the null hypothesis that there is *no difference* between the population proportions, i.e., $p_1 = p_2$, and thus that the samples are really drawn from the same population.

From (13), page 157, on placing $p_1 = p_2 = p$, we see that the sampling distribution of differences in proportions is approximately normal with mean and standard deviation given by

$$\mu_{P_1-P_2} = 0 \qquad \sigma_{P_1-P_2} = \sqrt{p(1-p)\left(\frac{1}{n_1} + \frac{1}{n_2}\right)} \tag{9}$$

where $\bar{P} = \dfrac{n_1P_1 + n_2P_2}{n_1 + n_2}$ is used as an estimate of the population proportion p.

By using the standardized variable

$$Z = \frac{P_1 - P_2 - 0}{\sigma_{P_1-P_2}} = \frac{P_1 - P_2}{\sigma_{P_1-P_2}} \tag{10}$$

we can test observed differences at an appropriate level of significance and thereby test the null hypothesis.

Tests involving other statistics (see Table 5-1, page 160) can similarly be designed.

Special Tests of Significance for Small Samples

In case samples are small ($n < 30$), we can formulate tests of hypotheses and significance using other distributions besides the normal, such as Student's t, chi-square, and F. These involve exact sampling theory and so, of course, hold even when samples are large, in which case they reduce to those given above. The following are some examples.

1. **MEANS.** To test the hypothesis H_0 that a normal population has mean, μ, we use

$$T = \frac{\bar{X} - \mu}{S}\sqrt{n-1} = \frac{\bar{X} - \mu}{\hat{S}}\sqrt{n} \tag{11}$$

where $\bar{X}$ is the mean of a sample of size n. This is analogous to using the standardized variable $Z = \dfrac{\bar{X} - \mu}{\sigma/\sqrt{n}}$ for large n except that $\hat{S} = \sqrt{n/(n-1)}\, S$ is used in place of σ. The difference is that while Z is normally distributed, T has Student's t distribution. As n increases, these tend toward agreement. Tests of hypotheses similar to those for means on page 216, can be made using critical t values in place of critical z values.

2. DIFFERENCES OF MEANS. Suppose that two random samples of sizes n_1 and n_2 are drawn from normal (or approximately normal) populations whose standard deviations are equal, i.e., $\sigma_1 = \sigma_2$. Suppose further that these two samples have means and standard deviations given by $\bar{X}_1, \bar{X}_2$ and S_1, S_2, respectively. To test the hypothesis H_0 that the samples come from the same population (i.e., $\mu_1 = \mu_2$ as well as $\sigma_1 = \sigma_2$), we use the variable given by

$$T = \frac{\bar{X}_1 - \bar{X}_2}{\sigma\sqrt{\dfrac{1}{n_1} + \dfrac{1}{n_2}}} \qquad \text{where} \qquad \sigma = \sqrt{\frac{n_1 S_1^2 + n_2 S_2^2}{n_1 + n_2 - 2}} \tag{12}$$

The distribution of T is Student's t distribution with $\nu = n_1 + n_2 - 2$ degrees of freedom. Use of (12) is made plausible on placing $\sigma_1 = \sigma_2 = \sigma$ in (12), page 157, and then using as an estimator of σ_2 the weighted average

$$\frac{(n_1 - 1)\hat{S}_1^2 + (n_2 - 1)\hat{S}_2^2}{(n_1 - 1) + (n_2 - 1)} = \frac{n_1 S_1^2 + n_2 S_2^2}{n_1 + n_2 - 2}$$

where $\hat{S}_1^2$ and $\hat{S}_2^2$ are the unbiased estimators of σ_1^2 and σ_2^2. This is the *pooled variance* obtained by combining the data.

3. VARIANCES. To test the hypothesis H_0 that a normal population has variance σ^2, we consider the random variables

$$\chi^2 = \frac{nS^2}{\sigma^2} = \frac{(n - 1)\hat{S}^2}{\sigma^2} \tag{13}$$

which (see pages 158–159) has the chi-square distribution with $n - 1$ degrees of freedom. Then if a random sample of size n turns out to have variance s^2, we would, on the basis of a two-tailed test, accept H_0 (or at least not reject it) at the 0.05 level if

$$\chi^2_{0.025} \le \frac{ns^2}{\sigma^2} \le \chi^2_{0.975} \tag{14}$$

and reject it otherwise. A similar result can be obtained for the 0.01 or other level.

To test the hypothesis H_1 that the population variance is greater than σ^2, we would still use the null hypothesis H_0 but would now employ a one-tailed test. Thus we would reject H_0 at the 0.05 level (and thereby conclude that H_1 is correct) if the particular sample variance s^2 were such that

$$\frac{ns^2}{\sigma^2} > \chi^2_{0.95} \tag{15}$$

and would accept H_0 (or at least not reject it) otherwise.

4. RATIOS OF VARIANCES. In some problems we wish to decide whether two samples of sizes m and n, respectively, whose measured variances are s_1^2 and s_2^2, do or do not come from normal populations with the same variance. In such cases, we use the statistic (see page 159).

$$F = \frac{\hat{S}_1^2/\sigma_1^2}{\hat{S}_2^2/\sigma_2^2} \tag{16}$$

where σ_1^2, σ_2^2 are the variances of the two normal populations from which the samples are drawn. Suppose that H_0 denotes the null hypothesis that there is no difference between population variances, i.e., $\sigma_1^2 = \sigma_2^2$. Then under this hypothesis (16) becomes

$$F = \frac{\hat{S}_1^2}{\hat{S}_2^2} \tag{17}$$

To test this hypothesis at the 0.10 level, for example, we first note that F in (16) has the F distribution with $m - 1, n - 1$ degrees of freedom. Then, using a two-tailed test, we would accept H_0 (or not reject it) at the 0.10 level if

$$F_{0.05} \leq \frac{\hat{s}_1^2}{\hat{s}_2^2} \leq F_{0.95} \tag{18}$$

and reject it otherwise.

Similar approaches using one-tailed tests can be formulated in case we wish to test the hypothesis that one particular population variance is in fact greater than the other.

Relationship Between Estimation Theory and Hypothesis Testing

From the above remarks one cannot help but notice that there is a relationship between estimation theory involving confidence intervals and the theory of hypothesis testing. For example, we note that the result (2) for accepting H_0 at the 0.05 level is equivalent to the result (1) on page 196, leading to the 95% confidence interval

$$\bar{x} - \frac{1.96\sigma}{\sqrt{n}} \leq \mu \leq \bar{x} + \frac{1.96\sigma}{\sqrt{n}} \tag{19}$$

Thus, at least in the case of two-tailed tests, we could actually employ the confidence intervals of Chapter 6 to test hypotheses. A similar result for one-tailed tests would require one-sided confidence intervals (see Problem 6.14).

Operating Characteristic Curves. Power of a Test

We have seen how the Type I error can be limited by properly choosing a level of significance. It is possible to avoid risking Type II errors altogether by simply not making them, which amounts to never accepting hypotheses. In many practical cases, however, this cannot be done. In such cases use is often made of *operating characteristic curves*, or *OC curves*, which are graphs showing the probabilities of Type II errors under various hypotheses. These provide indications of how well given tests will enable us to minimize Type II errors, i.e., they indicate the *power of a test* to avoid making wrong decisions. They are useful in designing experiments by showing, for instance, what sample sizes to use.

Quality Control Charts

It is often important in practice to know if a process has changed sufficiently to make remedial steps necessary. Such problems arise, for example, in *quality control* where one must, often quickly, decide whether observed changes are due simply to chance fluctuations or to actual changes in a manufacturing process because of deterioration of machine parts, or mistakes of employees. *Control charts* provide a useful and simple method for dealing with such problems (see Problem 7.29).

Fitting Theoretical Distributions to Sample Frequency Distributions

When one has some indication of the distribution of a population by probabilistic reasoning or otherwise, it is often possible to fit such theoretical distributions (also called "model" or "expected" distributions) to frequency distributions obtained from a sample of the population. The method used in general consists of employing the mean and standard deviation of the sample to estimate the mean and standard deviation of the population. See Problems 7.30, 7.32, and 7.33.

The problem of testing the *goodness of fit* of theoretical distributions to sample distributions is essentially the same as that of deciding whether there are significant differences between population and sample values. An important significance test for the goodness of fit of theoretical distributions, the *chi-square test*, is described below.

In attempting to determine whether a normal distribution represents a good fit for given data, it is convenient to use *normal curve graph paper*, or *probability graph paper* as it is sometimes called (see Problem 7.31).

The Chi-Square Test for Goodness of Fit

To determine whether the proportion P of "successes" in a sample of size n drawn from a binomial population differs significantly from the population proportion p of successes, we have used the statistic given by (5) or (6) on page 216. In this simple case only two possible events A_1, A_2 can occur, which we have called "success" and "failure" and which have probabilities p and $q = 1 - p$, respectively. A particular sample value of the random variable $X = nP$ is often called the *observed frequency* for the event A_1, while np is called the *expected frequency*.

EXAMPLE 7.1 If we obtain a sample of 100 tosses of a fair coin, so that $n = 100, p = \frac{1}{2}$, then the expected frequency of heads (successes) is $np = (100)(\frac{1}{2}) = 50$. The observed frequency in the sample could of course be different.

A natural generalization is to the case where k possible events $A_1, A_2, \ldots, A_k$ can occur, the respective probabilities being $p_1, p_2, \ldots, p_k$. In such cases, we have a *multinomial population* (see page 112). If we draw a sample of size n from this population, the observed frequencies for the events $A_1, \ldots, A_k$ can be described by random variables $X_1, \ldots, X_k$ (whose specific values $x_1, x_2, \ldots, x_k$ would be the observed frequencies for the sample), while the expected frequencies would be given by $np_1, \ldots, np_k$, respectively. The results can be indicated as in Table 7-2.

Table 7-2

Event	A_1	A_2	$\cdots$	A_k
Observed Frequency	x_1	x_2	$\cdots$	x_k
Expected Frequency	np_1	np_2	$\cdots$	np_k

EXAMPLE 7.2 If we obtain a sample of 120 tosses of a fair die, so that $n = 120$, then the probabilities of the faces 1, 2, . . . , 6 are denoted by $p_1, p_2, \ldots, p_6$, respectively, and are all equal to $\frac{1}{6}$. The corresponding expected frequencies are $np_1, np_2, \ldots, np_6$ and are all equal to $(120)(\frac{1}{6}) = 20$. The observed frequencies of the various faces that come up in the sample can of course be different.

A clue as to the possible generalization of the statistic (6) which could measure the discrepancies existing between observed and expected frequencies in Table 7-2 is obtained by squaring the statistic (6) and writing it as

$$Z^2 = \frac{(X - np)^2}{npq} = \frac{(X_1 - np)^2}{np} + \frac{(X_2 - nq)^2}{nq} \tag{20}$$

where $X_1 = X$ is the random variable associated with "success" and $X_2 = n - X_1$ is the random variable associated with "failure." Note that nq in (20) is the expected frequency of failures.

The form of the result (20) suggests that a measure of the discrepancy between observed and expected frequencies for the general case is supplied by the statistic

$$\chi^2 = \frac{(X_1 - np_1)^2}{np_1} + \frac{(X_2 - np_2)^2}{np_2} + \cdots + \frac{(X_k - np_k)^2}{np_k} = \sum_{j=1}^{k} \frac{(X_j - np_j)^2}{np_j} \tag{21}$$

where the total frequency (i.e., the sample size) is n, so that

$$X_1 + X_2 + \cdots + X_k = n \tag{22}$$

An expression equivalent to (21) is

$$\chi^2 = \sum_{j=1}^{k} \frac{X_j^2}{np_j} - n \tag{23}$$

If $\chi^2 = 0$, the observed and expected frequencies agree exactly while if $\chi^2 > 0$, they do not agree exactly. The larger the value of χ^2, the greater is the discrepancy between observed and expected frequencies.

As is shown in Problem 7.62, the sampling distribution of χ^2 as defined by (21) is approximated very closely by the chi-square distribution [hence the choice of symbol in (21)] if the expected frequencies np_j are at least equal to 5, the approximation improving for larger values. The number of degrees of freedom for this chi-square distribution is given by:

(a) $\nu = k - 1$ if expected frequencies can be computed without having to estimate population parameters from sample statistics. Note that we subtract 1 from k because of the constraint condition (22), which states that if we know $k - 1$ of the expected frequencies, the remaining frequency can be determined.

(b) $\nu = k - 1 - m$ if the expected frequencies can be computed only by estimating m population parameters from sample statistics.

In practice, expected frequencies are computed on the basis of a hypothesis H_0. If under this hypothesis the computed value of χ^2 given by (21) or (23) is greater than some critical value (such as $\chi^2_{0.95}$ or $\chi^2_{0.99}$, which are the critical values at the 0.05 and 0.01 significance levels, respectively), we would conclude that observed frequencies differ *significantly* from expected frequencies and would reject H_0 at the corresponding level of significance. Otherwise, we would accept it or at least not reject it. This procedure is called the *chi-square test* of hypotheses or significance.

Besides applying to the multinomial distribution, the chi-square test can be used to determine how well other theoretical distributions, such as the normal or Poisson, fit empirical distributions, i.e., those obtained from sample data. See Problem 7.44.

Contingency Tables

Table 7-2 above, in which observed frequencies occupy a single row, is called a *one-way classification table.* Since the number of columns is k, this is also called a $1 \times k$ (read "1 by k") *table*. By extending these ideas, we can arrive at *two-way classification tables* or $h \times k$ *tables* in which the observed frequencies occupy h rows and k columns. Such tables are often called *contingency tables*.

Corresponding to each observed frequency in an $h \times k$ contingency table, there is an *expected* or *theoretical frequency*, which is computed subject to some hypothesis according to rules of probability. These frequencies that occupy the *cells* of a contingency table are called *cell frequencies*. The total frequency in each row or each column is called the *marginal frequency*.

To investigate agreement between observed and expected frequencies, we compute the statistic

$$\chi^2 = \sum_j \frac{(X_j - np_j)^2}{np_j} \tag{24}$$

where the sum is taken over all cells in the contingency table, the symbols X_j and np_j representing, respectively, the observed and expected frequencies in the jth cell. This sum, which is analogous to (21), contains hk terms. The sum of all observed frequencies is denoted n and is equal to the sum of all expected frequencies [compare with equation (22)].

As before, the statistic (24) has a sampling distribution given very closely by the chi-square distribution provided expected frequencies are not too small. The number of degrees of freedom ν of this chi-square distribution is given for $h > 1, k < 1$ by

(a) $\nu = (h - 1)(k - 1)$ if the expected frequencies can be computed without having to estimate population parameters from sample statistics. For a proof of this see Problem 7.48.

(b) $\nu = (h - 1)(k - 1) - m$ if the expected frequencies can be computed only by estimating m population parameters from sample statistics.

Significance tests for $h \times k$ tables are similar to those for $1 \times k$ tables. Expected frequencies are found subject to a particular hypothesis H_0. A hypothesis commonly tested is that the two classifications are independent of each other.

Contingency tables can be extended to higher dimensions. For example, we can have $h \times k \times l$ tables where 3 classifications are present.

Yates' Correction for Continuity

When results for continuous distributions are applied to discrete data, certain corrections for continuity can be made as we have seen in previous chapters. A similar correction is available when the chi-square distribution is used. The correction consists in rewriting (21) as

$$\chi^2\,(\text{corrected}) = \frac{(|X_1 - np_1| - 0.5)^2}{np_1} + \frac{(|X_2 - np_2| - 0.5)^2}{np_2} + \cdots + \frac{(|X_k - np_k| - 0.5)^2}{np_k} \tag{25}$$

and is often referred to as *Yates' correction*. An analogous modification of (24) also exists.

In general, the correction is made only when the number of degrees of freedom is $\nu = 1$. For large samples this yields practically the same results as the uncorrected χ^2, but difficulties can arise near critical values (see Problem 7.41). For small samples where each expected frequency is between 5 and 10, it is perhaps best to compare both the corrected and uncorrected values of χ^2. If both values lead to the same conclusion regarding a hypothesis, such as rejection at the 0.05 level, difficulties are rarely encountered. If they lead to different conclusions, one can either resort to increasing sample sizes or, if this proves impractical, one can employ exact methods of probability involving the multinomial distribution.

Coefficient of Contingency

A measure of the degree of relationship, association, or dependence of the classifications in a contingency table is given by

$$C = \sqrt{\frac{\chi^2}{\chi^2 + n}} \tag{26}$$

which is called the *coefficient of contingency*. The larger the value of C, the greater is the degree of association. The number of rows and columns in the contingency table determines the maximum value of C, which is never greater than one. For a $k \times k$ table the maximum value of C is given by $\sqrt{(k-1)/k}$. See Problems 7.52 and 7.53.

SOLVED PROBLEMS

Tests of means and proportions using normal distributions

7.1. Find the probability of getting between 40 and 60 heads inclusive in 100 tosses of a fair coin.

According to the binomial distribution the required probability is

$${}_{100}C_{40}\left(\frac{1}{2}\right)^{40}\left(\frac{1}{2}\right)^{60} + {}_{100}C_{41}\left(\frac{1}{2}\right)^{41}\left(\frac{1}{2}\right)^{59} + \cdots + {}_{100}C_{60}\left(\frac{1}{2}\right)^{60}\left(\frac{1}{2}\right)^{40}$$

The mean and standard deviation of the number of heads in 100 tosses are given by

$$\mu = np = 100\left(\frac{1}{2}\right) = 50 \qquad \sigma = \sqrt{npq} = \sqrt{(100)\left(\frac{1}{2}\right)\left(\frac{1}{2}\right)} = 5$$

Since np and nq are both greater than 5, the normal approximation to the binomial distribution can be used in evaluating the above sum.

On a continuous scale, between 40 and 60 heads inclusive is the same as between 39.5 and 60.5 heads.

$$39.5 \text{ in standard units} = \frac{39.5 - 50}{5} = -2.10 \qquad 60.5 \text{ in standard units} = \frac{60.5 - 50}{5} = 2.10$$

$$\begin{aligned}\text{Required probability} &= \text{area under normal curve between } z = -2.10 \text{ and } z = 2.10\\ &= 2(\text{area between } z = 0 \text{ and } z = 2.10) = 2(0.4821) = 0.9642\end{aligned}$$

7.2. To test the hypothesis that a coin is fair, the following decision rules are adopted: (1) Accept the hypothesis if the number of heads in a single sample of 100 tosses is between 40 and 60 inclusive, (2) reject the hypothesis otherwise.

(a) Find the probability of rejecting the hypothesis when it is actually correct.

(b) Interpret graphically the decision rule and the result of part (a).

(c) What conclusions would you draw if the sample of 100 tosses yielded 53 heads? 60 heads?

(d) Could you be wrong in your conclusions to (c)? Explain.

(a) By Problem 7.1, the probability of not getting between 40 and 60 heads inclusive if the coin is fair equals $1 - 0.9642 = 0.0358$. Then the probability of rejecting the hypothesis when it is correct equals 0.0358.

(b) The decision rule is illustrated by Fig. 7-2, which shows the probability distribution of heads in 100 tosses of a fair coin.

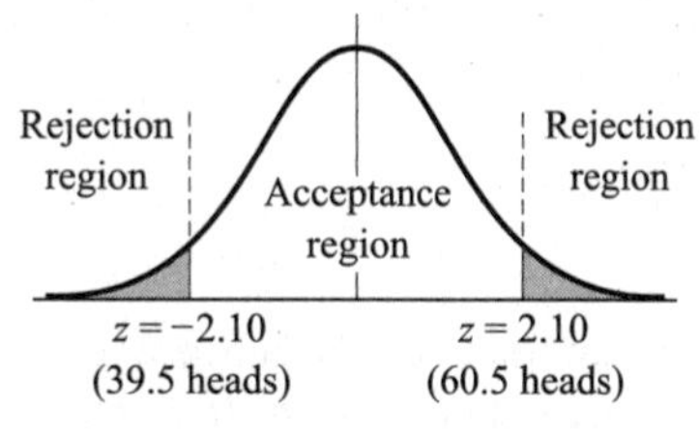

Fig. 7-2

If a single sample of 100 tosses yields a z score between -2.10 and 2.10, we accept the hypothesis; otherwise we reject the hypothesis and decide that the coin is not fair.

The error made in rejecting the hypothesis when it should be accepted is the *Type I error* of the decision rule; and the probability of making this error, equal to 0.0358 from part (a), is represented by the total shaded area of the figure.

If a single sample of 100 tosses yields a number of heads whose z score lies in the shaded regions, we could say that this z score differed *significantly* from what would be expected if the hypothesis were true. For this reason the total shaded area (i.e., probability of a Type I error) is called the *level of significance* of the decision rule; it equals 0.0358 in this case. We therefore speak of rejecting the hypothesis at a 0.0358, or 3.58%, level of significance.

(c) According to the decision rule, we would have to accept the hypothesis that the coin is fair in both cases. One might argue that if only one more head had been obtained, we would have rejected the hypothesis. This is what one must face when any sharp line of division is used in making decisions.

(d) Yes. We could accept the hypothesis when it actually should be rejected, as would be the case, for example, when the probability of heads is really 0.7 instead of 0.5.

The error made in accepting the hypothesis when it should be rejected is the *Type II error* of the decision. For further discussion see Problems 7.23–7.25.

7.3. Design a decision rule to test the hypothesis that a coin is fair if a sample of 64 tosses of the coin is taken and if a level of significance of (a) 0.05, (b) 0.01 is used.

(a) **First method**

If the level of significance is 0.05, each shaded area in Fig. 7-3 is 0.025 by symmetry. Then the area between 0 and z_1 is $0.5000 - 0.0250 = 0.4750$, and $z_1 = 1.96$.

Thus a possible decision rule is:

(1) Accept the hypothesis that the coin is fair if Z is between -1.96 and 1.96.

(2) Reject the hypothesis otherwise.

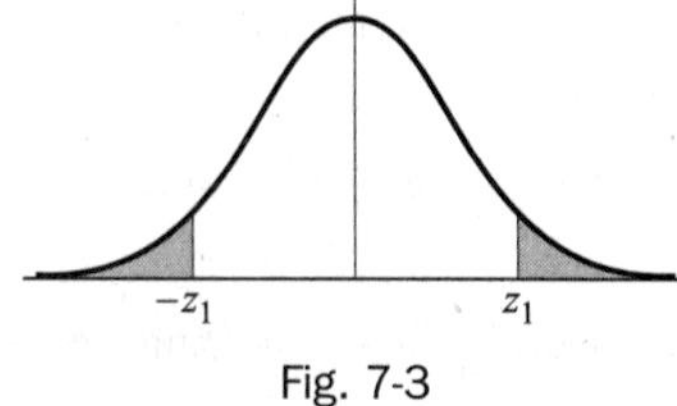

Fig. 7-3

The critical values -1.96 and 1.96 can also be read from Table 7-1.

To express this decision rule in terms of the number of heads to be obtained in 64 tosses of the coin, note that the mean and standard deviation of the exact binomial distribution of heads are given by

$$\mu = np = 64(0.5) = 32 \quad \text{and} \quad \sigma = \sqrt{npq} = \sqrt{64(0.5)(0.5)} = 4$$

under the hypothesis that the coin is fair. Then $Z = (X - \mu)/\sigma = (X - 32)/4$.

If $Z = 1.96$, $(X - 32)/4 = 1.96$ or $X = 39.84$. If $Z = -1.96$, $(X - 32)/4 = -1.96$ or $X = 24.16$. Therefore, the decision rule becomes:

(1) Accept the hypothesis that the coin is fair if the number of heads is between 24.16 and 39.84, i.e., between 25 and 39 inclusive.

(2) Reject the hypothesis otherwise.

Second method

With probability 0.95, the number of heads will lie between $\mu - 1.96\sigma$ and $\mu + 1.96\sigma$, i.e., $np - 1.96\sqrt{npq}$ and $np + 1.96\sqrt{npq}$ or between $32 - 1.96(4) = 24.16$ and $32 + 1.96(4) = 39.84$, which leads to the above decision rule.

Third method

$-1.96 < Z < 1.96$ is equivalent to $-1.96 < (X - 32)/4 < 1.96$. Consequently $-1.96(4) < (X - 32) < 1.96(4)$ or $32 - 1.96(4) < X < 32 + 1.96(4)$, i.e., $24.16 < X < 39.84$, which also leads to the above decision rule.

(b) If the level of significance is 0.01, each shaded area in Fig. 7-3 is 0.005. Then the area between 0 and z_1 is $0.5000 - 0.0050 = 0.4950$, and $z_1 = 2.58$ (more exactly, 2.575). This can also be read from Table 7-1.

Following the procedure in the second method of part (a), we see that with probability 0.99 the number of heads will lie between $\mu - 2.58\sigma$ and $\mu + 2.58\sigma$, i.e., $32 - 2.58(4) = 21.68$ and $32 + 2.58(4) = 42.32$.

Therefore, the decision rule becomes:

(1) Accept the hypothesis if the number of heads is between 22 and 42 inclusive.

(2) Reject the hypothesis otherwise.

7.4. How could you design a decision rule in Problem 7.3 so as to avoid a Type II error?

A Type II error is made by accepting a hypothesis when it should be rejected. To avoid this error, instead of accepting the hypothesis we simply do not reject it, which could mean that we are withholding any decision in this case. For example, we could word the decision rule of Problem 7.3(b) as:

(1) Do not reject the hypothesis if the number of heads is between 22 and 42 inclusive.

(2) Reject the hypothesis otherwise.

In many practical instances, however, it is important to decide whether a hypothesis should be accepted or rejected. A complete discussion of such cases requires consideration of Type II errors (see Problems 7.23 to 7.25).

7.5. In an experiment on extrasensory perception (ESP) a subject in one room is asked to state the color (red or blue) of a card chosen from a deck of 50 well-shuffled cards by an individual in another room. It is unknown to the subject how many red or blue cards are in the deck. If the subject identifies 32 cards correctly, determine whether the results are significant at the (a) 0.05, (b) 0.01 level of significance. (c) Find and interpret the P value of the test.

If p is the probability of the subject stating the color of a card correctly, then we have to decide between the following two hypotheses:

H_0: $p = 0.5$, and the subject is simply guessing, i.e., results are due to chance

H_1: $p > 0.5$, and the subject has powers of ESP.

We choose a one-tailed test, since we are not interested in ability to obtain extremely low scores but rather in ability to obtain high scores.

If the hypothesis H_0 is true, the mean and standard deviation of the number of cards identified correctly is given by

$$\mu = np = 50(0.5) = 25 \quad \text{and} \quad \sigma = \sqrt{npq} = \sqrt{50(0.5)(0.5)} = \sqrt{12.5} = 3.54$$

(a) For a one-tailed test at a level of significance of 0.05, we must choose z_1 in Fig. 7-4 so that the shaded area in the critical region of high scores is 0.05. Then the area between 0 and z_1 is 0.4500, and $z_1 = 1.645$. This can also be read from Table 7-1.

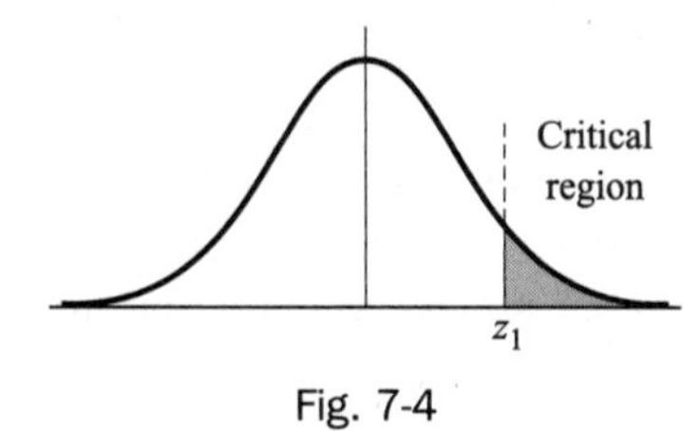

Fig. 7-4

Therefore, our decision rule or test of significance is:

(1) If the z score observed is greater than 1.645, the results are significant at the 0.05 level and the individual has powers of ESP.

(2) If the z score is less than 1.645, the results are due to chance, i.e., not significant at the 0.05 level.

Since 32 in standard units is $(32 - 25)/3.54 = 1.98$, which is greater than 1.645, decision (1) holds, i.e., we conclude at the 0.05 level that the individual has powers of ESP.

Note that we should really apply a continuity correction, since 32 on a continuous scale is between 31.5 and 32.5. However, 31.5 has a standard score of $(31.5 - 25)/3.54 = 1.84$, and so the same conclusion is reached.

(b) If the level of significance is 0.01, then the area between 0 and z_1 is 0.4900, and $z_1 = 2.33$. Since 32 (or 31.5) in standard units is 1.98 (or 1.84), which is less than 2.33, we conclude that the results are *not significant* at the 0.01 level.

Some statisticians adopt the terminology that results significant at the 0.01 level are *highly significant*, results significant at the 0.05 level but not at the 0.01 level are *probably significant*, while results significant at levels larger than 0.05 are *not significant*.

According to this terminology, we would conclude that the above experimental results are *probably significant*, so that further investigations of the phenomena are probably warranted.

(c) The P value of the test is the probability that the colors of 32 or more cards would, in a random selection, be identified correctly. The standard score of 32, taking into account the continuity correction is $z = 1.84$. Therefore the P value is $P(Z \geq 1.84) = 0.032$. The statistician could say that on the basis of the experiment, the chances of being wrong in concluding that the individual has powers of ESP are about 3 in 100.

7.6. The manufacturer of a patent medicine claimed that it was 90% effective in relieving an allergy for a period of 8 hours. In a sample of 200 people who had the allergy, the medicine provided relief for 160 people. (a) Determine whether the manufacturer's claim is legitimate by using 0.01 as the level of significance. (b) Find the P value of the test.

(a) Let p denote the probability of obtaining relief from the allergy by using the medicine. Then we must decide between the two hypotheses:

$$H_0: p = 0.9, \text{ and the claim is correct}$$

$$H_1: p < 0.9, \text{ and the claim is false}$$

We choose a one-tailed test, since we are interested in determining whether the proportion of people relieved by the medicine is too low.

If the level of significance is taken as 0.01, i.e., if the shaded area in Fig. 7-5 is 0.01, then $z_1 = -2.33$ as can be seen from Problem 7.5(b) using the symmetry of the curve, or from Table 7-1.

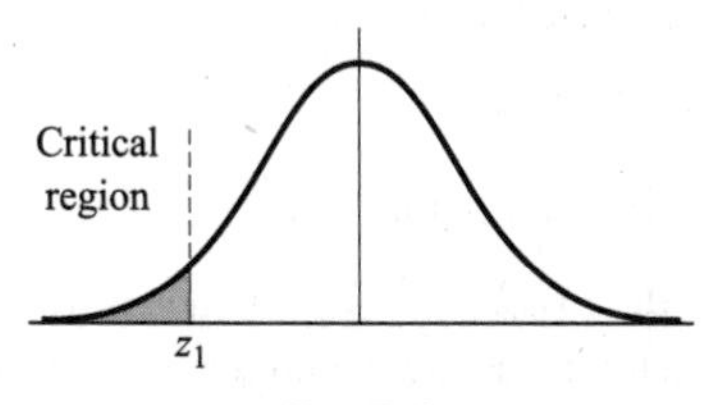

Fig. 7-5

We take as our decision rule:

(1) The claim is not legitimate if Z is less than -2.33 (in which case we reject H_0).

(2) Otherwise, the claim is legitimate, and the observed results are due to chance (in which case we accept H_0).

If H_0 is true, $\mu = np = 200(0.9) = 180$ and $\sigma = \sqrt{npq} = \sqrt{(200)(0.9)(0.1)} = 4.23$.

Now 160 in standard units is $(160 - 180)/4.23 = -4.73$, which is much less than -2.33. Thus by our decision rule we conclude that the claim is not legitimate and that the sample results are *highly significant* (see end of Problem 7.5).

(b) The P value of the test is $P(Z \leq -4.73) \approx 0$, which shows that the claim is almost certainly false. That is, if H_0 were true, it is almost certain that a random sample of 200 allergy sufferers who used the medicine would include more than 160 people who found relief.

7.7. The mean lifetime of a sample of 100 fluorescent light bulbs produced by a company is computed to be 1570 hours with a standard deviation of 120 hours. If μ is the mean lifetime of all the bulbs produced by the company, test the hypothesis $\mu = 1600$ hours against the alternative hypothesis $\mu \neq 1600$ hours, using a level of significance of (a) 0.05 and (b) 0.01. (c) Find the P value of the test.

We must decide between the two hypotheses

$$H_0: \mu = 1600 \text{ hours} \qquad H_1: \mu \neq 1600 \text{ hours}$$

A two-tailed test should be used here since $\mu \neq 1600$ includes values both larger and smaller than 1600.

(a) For a two-tailed test at a level of significance of 0.05, we have the following decision rule:

(1) Reject H_0 if the z score of the sample mean is outside the range -1.96 to 1.96.

(2) Accept H_0 (or withhold any decision) otherwise.

The statistic under consideration is the sample mean $\bar{X}$. The sampling distribution of X has a mean $\mu_{\bar{X}} = \mu$ and standard deviation $\sigma_{\bar{X}} = \sigma/\sqrt{n}$, where μ and σ are the mean and standard deviation of the population of all bulbs produced by the company.

Under the hypothesis H_0, we have $\mu = 1600$ and $\sigma_{\bar{X}} = \sigma/\sqrt{n} = 120/\sqrt{100} = 12$, using the sample standard deviation as an estimate of σ. Since $Z = (\bar{X} - 1600)/12 = (1570 - 1600)/12 = -2.50$ lies outside the range -1.96 to 1.96, we reject H_0 at a 0.05 level of significance.

(b) If the level of significance is 0.01, the range -1.96 to 1.96 in the decision rule of part (a) is replaced by -2.58 to 2.58. Then since the z score of -2.50 lies inside this range, we accept H_0 (or withhold any decision) at a 0.01 level of significance.

(c) The P value of the two-tailed test is $P(Z \leq -2.50) + P(Z \geq 2.50) = 0.0124$, which is the probability that a mean lifetime of less than 1570 hours or more than 1630 hours would occur by chance if H_0 were true.

7.8. In Problem 7.7 test the hypothesis $\mu = 1600$ hours against the alternative hypothesis $\mu < 1600$ hours, using a level of significance of (a) 0.05, (b) 0.01. (c) Find the P value of the test.

We must decide between the two hypotheses

$$H_0: \mu = 1600 \text{ hours} \qquad H_1: \mu < 1600 \text{ hours}$$

A one-tailed test should be used here (see Fig. 7-5).

(a) If the level of significance is 0.05, the shaded region of Fig. 7-5 has an area of 0.05, and we find that $z_1 = -1.645$. We therefore adopt the decision rule:

(1) Reject H_0 if Z is less than -1.645.

(2) Accept H_0 (or withhold any decision) otherwise.

Since, as in Problem 7.7(a), the z score is -2.50, which is less than -1.645, we reject H_0 at a 0.05 level of significance. Note that this decision is identical with that reached in Problem 7.7(a) using a two-tailed test.

(b) If the level of significance is 0.01, the z_1 value in Fig. 7-5 is -2.33. Hence we adopt the decision rule:

(1) Reject H_0 if Z is less than -2.33.

(2) Accept H_0 (or withhold any decision) otherwise.

Since, as in Problem 7.7(a), the z score is -2.50, which is less than -2.33, we reject H_0 at a 0.01 level of significance. Note that this decision is not the same as that reached in Problem 7.7(b) using a two-tailed test.

It follows that decisions concerning a given hypothesis H_0 based on one-tailed and two-tailed tests are not always in agreement. This is, of course, to be expected since we are testing H_0 against a different alternative in each case.

(c) The P value of the test is $P(Z < 1570) = 0.0062$, which is the probability that a mean lifetime of less than 1570 hours would occur by chance if H_0 were true.

7.9. The breaking strengths of cables produced by a manufacturer have mean 1800 lb and standard deviation 100 lb. By a new technique in the manufacturing process it is claimed that the breaking strength can be increased. To test this claim, a sample of 50 cables is tested, and it is found that the mean breaking strength is 1850 lb. (a) Can we support the claim at a 0.01 level of significance? (b) What is the P value of the test?

(a) We have to decide between the two hypotheses

H_0: $\mu = 1800$ lb, and there is really no change in breaking strength

H_1: $u > 1800$ lb, and there is a change in breaking strength

A one-tailed test should be used here (see Fig. 7-4). At a 0.01 level of significance the decision rule is:

(1) If the z score observed is greater than 2.33, the results are significant at the 0.01 level and H_0 is rejected.

(2) Otherwise H_0 is accepted (or the decision is withheld).

Under the hypothesis that H_0 is true, we find

$$Z = \frac{\bar{X} - \mu}{\sigma/\sqrt{n}} = \frac{1850 - 1800}{100/\sqrt{50}} = 3.55$$

which is greater than 2.33. Hence we conclude that the results are *highly significant* and the claim should be supported.

(b) The P value of the test is $P(Z \geq 3.55) = 0.0002$, which is the probability that a mean breaking strength of 1850 lb or more would occur by chance if H_0 were true.

Tests involving differences of means and proportions

7.10. An examination was given to two classes consisting of 40 and 50 students, respectively. In the first class the mean grade was 74 with a standard deviation of 8, while in the second class the mean grade was 78 with a standard deviation of 7. Is there a significant difference between the performance of the two classes at a level of significance of (a) 0.05, (b) 0.01? (c) What is the P value of the test?

Suppose the two classes come from two populations having the respective means μ_1 and μ_2. Then we have to decide between the hypotheses

H_0: $\mu_1 = \mu_2$, and the difference is merely due to chance

H_1: $\mu_1 \neq \mu_2$, and there is a significant difference between classes

Under the hypothesis H_0, both classes come from the same population. The mean and standard deviation of the difference in means are given by

$$\mu_{\bar{X}_1 - \bar{X}_2} = 0 \qquad \sigma_{\bar{X}_1 - \bar{X}_2} = \sqrt{\frac{\sigma_1^2}{n_1} + \frac{\sigma_2^2}{n_2}} = \sqrt{\frac{8^2}{40} + \frac{7^2}{50}} = 1.606$$

where we have used the sample standard deviations as estimates of σ_1 and σ_2.

Then
$$Z = \frac{\bar{X}_1 - \bar{X}_2}{\sigma_{\bar{X}_1-\bar{X}_2}} = \frac{74 - 78}{1.606} = -2.49$$

(a) For a two-tailed test, the results are significant at a 0.05 level if Z lies outside the range -1.96 to 1.96. Hence we conclude that at a 0.05 level there is a significant difference in performance of the two classes and that the second class is probably better.

(b) For a two-tailed test the results are significant at a 0.01 level if Z lies outside the range -2.58 and 2.58. Hence we conclude that at a 0.01 level there is no significant difference between the classes.

Since the results are significant at the 0.05 level but not at the 0.01 level, we conclude that the results are *probably significant*, according to the terminology used at the end of Problem 7.5.

(c) The P value of the two-tailed test is $P(Z \leq -2.49) + P(Z \geq 2.49) = 0.0128$, which is the probability that the observed statistics would occur in the same population.

7.11. The mean height of 50 male students who showed above-average participation in college athletics was 68.2 inches with a standard deviation of 2.5 inches, while 50 male students who showed no interest in such participation had a mean height of 67.5 inches with a standard deviation of 2.8 inches. (a) Test the hypothesis that male students who participate in college athletics are taller than other male students. (b) What is the P value of the test?

(a) We must decide between the hypotheses

H_0: $\mu_1 = \mu_2$, and there is no difference between the mean heights

H_1: $\mu_1 > \mu_2$, and mean height of first group is greater than that of second group

Under the hypothesis H_0,

$$\mu_{\bar{X}_1-\bar{X}_2} = 0 \qquad \sigma_{\bar{X}_1-\bar{X}_2} = \sqrt{\frac{\sigma_1^2}{n_1} + \frac{\sigma_2^2}{n_2}} = \sqrt{\frac{(2.5)^2}{50} + \frac{(2.8)^2}{50}} = 0.53$$

where we have used the sample standard deviations as estimates of σ_1 and σ_2.

Then
$$Z = \frac{\bar{X}_1 - \bar{X}_2}{\sigma_{\bar{X}_1-\bar{X}_2}} = \frac{68.2 - 67.5}{0.53} = 1.32$$

On the basis of a one-tailed test at a level of significance of 0.05, we would reject the hypothesis H_0 if the z score were greater than 1.645. We therefore cannot reject the hypothesis at this level of significance.

It should be noted, however, that the hypothesis can be rejected at a level of 0.10 if we are willing to take the risk of being wrong with a probability of 0.10, i.e., 1 chance in 10.

(b) The P value of the test is $P(Z \geq 1.32) = 0.0934$, which is the probability that the observed positive difference between mean heights of male athletes and other male students would occur by chance if H_0 were true.

7.12. By how much should the sample size of each of the two groups in Problem 7.11 be increased in order that the observed difference of 0.7 inch in the mean heights be significant at the level of significance (a) 0.05, (b) 0.01?

Suppose the sample size of each group is n and that the standard deviations for the two groups remain the same. Then under the hypothesis H_0 we have $\mu_{\bar{X}_1-\bar{X}_2} = 0$ and

$$\sigma_{\bar{X}_1-\bar{X}_2} = \sqrt{\frac{\sigma_1^2}{n} + \frac{\sigma_2^2}{n}} = \sqrt{\frac{(2.5)^2 + (2.8)^2}{n}} = \sqrt{\frac{14.09}{n}} = \frac{3.75}{\sqrt{n}}$$

For an observed difference in mean heights of 0.7 inch,

$$Z = \frac{\bar{X}_1 - \bar{X}_2}{\sigma_{\bar{X}_1-\bar{X}_2}} = \frac{0.7}{3.75/\sqrt{n}} = \frac{0.7\sqrt{n}}{3.75}$$

(a) The observed difference will be significant at a 0.05 level if

$$\frac{0.7\sqrt{n}}{3.75} \geq 1.645 \quad \text{or} \quad \sqrt{n} \geq 8.8 \quad \text{or} \quad n \geq 78$$

Therefore, we must increase the sample size in each group by at least $78 - 50 = 28$.

(b) The observed difference will be significant at a 0.01 level if

$$\frac{0.7\sqrt{n}}{3.75} \geq 2.33 \quad \text{or} \quad \sqrt{n} \geq 12.5 \quad \text{or} \quad n \geq 157$$

Hence, we must increase the sample size in each group by at least $157 - 50 = 107$.

7.13. Two groups, A and B, each consist of 100 people who have a disease. A serum is given to Group A but not to Group B (which is called the *control group*); otherwise, the two groups are treated identically. It is found that in Groups A and B, 75 and 65 people, respectively, recover from the disease. Test the hypothesis that the serum helps to cure the disease using a level of significance of (a) 0.01, (b) 0.05, (c) 0.10. (d) Find the P value of the test.

Let p_1 and p_2 denote, respectively, the population proportions cured by (1) using the serum, (2) not using the serum. We must decide between the two hypotheses

H_0: $p_1 = p_2$, and observed differences are due to chance, i.e., the serum is ineffective

H_1: $p_1 > p_2$, and the serum is effective

Under the hypothesis H_0,

$$\mu_{P_1-P_2} = 0 \qquad \sigma_{P_1-P_2} = \sqrt{pq\left(\frac{1}{n_1} + \frac{1}{n_2}\right)} = \sqrt{(0.70)(0.30)\left(\frac{1}{100} + \frac{1}{100}\right)} = 0.0648$$

where we have used as an estimate of p the average proportion of cures in the two sample groups, given by $(75 + 65)/200 = 0.70$, and where $q = 1 - p = 0.30$. Then

$$Z = \frac{P_1 - P_2}{\sigma_{P_1-P_2}} = \frac{0.750 - 0.650}{0.0648} = 1.54$$

(a) On the basis of a one-tailed test at a 0.01 level of significance, we would reject the hypothesis H_0 only if the z score were greater than 2.33. Since the z score is only 1.54, we must conclude that the results are due to chance at this level of significance.

(b) On the basis of a one-tailed test at a 0.05 level of significance, we would reject H_0 only if the z score were greater than 1.645. Hence, we must conclude that the results are due to chance at this level also.

(c) If a one-tailed test at a 0.10 level of significance were used, we would reject H_0 only if the z score were greater than 1.28. Since this condition is satisfied, we would conclude that the serum is effective at a 0.10 level of significance.

(d) The P value of the test is $P(Z \geq 1.54) = 0.0618$, which is the probability that a z score of 1.54 or higher in favor of the user group would occur by chance if H_0 were true.

Note that our conclusions above depended on how much we were willing to risk being wrong. If results are actually due to chance and we conclude that they are due to the serum (Type I error), we might proceed to give the serum to large groups of people, only to find then that it is actually ineffective. This is a risk that we are not always willing to assume.

On the other hand, we could conclude that the serum does not help when it actually does help (Type II error). Such a conclusion is very dangerous, especially if human lives are at stake.

7.14. Work Problem 7.13 if each group consists of 300 people and if 225 people in Group A and 195 people in Group B are cured.

In this case the proportions of people cured in the two groups are, respectively, $225/300 = 0.750$ and $195/300 = 0.650$, which are the same as in Problem 7.13. Under the hypothesis H_0,

$$\mu_{P_1-P_2} = 0 \qquad \sigma_{P_1-P_2} = \sqrt{pq\left(\frac{1}{n_1} + \frac{1}{n_2}\right)} = \sqrt{(0.70)(0.30)\left(\frac{1}{300} + \frac{1}{300}\right)} = 0.0374$$

where $(225 + 195)/600 = 0.70$ is used as an estimate of p. Then

$$Z = \frac{P_1 - P_2}{\sigma_{P_1-P_2}} = \frac{0.750 - 0.650}{0.0374} = 2.67$$

Since this value of z is greater than 2.33, we can reject the hypothesis at a 0.01 level of significance, i.e., we can conclude that the serum is effective with only a 0.01 probability of being wrong. Here the P value of the test is $P(Z \geq 2.67) = 0.0038$.

This shows how increasing the sample size can increase the reliability of decisions. In many cases, however, it may be impractical to increase sample sizes. In such cases we are forced to make decisions on the basis of available information and so must contend with greater risks of incorrect decisions.

7.15. A sample poll of 300 voters from district A and 200 voters from district B showed that 56% and 48%, respectively, were in favor of a given candidate. At a level of significance of 0.05, test the hypothesis that (a) there is a difference between the districts, (b) the candidate is preferred in district A. (c) Find the respective P values of the test.

Let p_1 and p_2 denote the proportions of all voters of districts A and B, respectively, who are in favor of the candidate.

Under the hypothesis H_0: $p_1 = p_2$, we have

$$\mu_{P_1-P_2} = 0 \qquad \sigma_{P_1-P_2} = \sqrt{pq\left(\frac{1}{n_1} + \frac{1}{n_2}\right)} = \sqrt{(0.528)(0.472)\left(\frac{1}{300} + \frac{1}{200}\right)} = 0.0456$$

where we have used as estimates of p and q the values $[(0.56)(300) + (0.48)(200)]/500 = 0.528$ and $1 - 0.528 = 0.472$. Then

$$Z = \frac{P_1 - P_2}{\sigma_{P_1-P_2}} = \frac{0.560 - 0.480}{0.0456} = 1.75$$

(a) If we wish to determine only whether there is a difference between the districts, we must decide between the hypotheses H_0: $p_1 = p_2$ and H_1: $p_1 \neq p_2$, which involves a two-tailed test.

On the basis of a two-tailed test at a 0.05 level of significance, we would reject H_0 if Z were outside the interval -1.96 to 1.96. Since $Z = 1.75$ lies inside this interval, we cannot reject H_0 at this level, i.e., there is no significant difference between the districts.

(b) If we wish to determine whether the candidate is preferred in district A, we must decide between the hypotheses H_0: $p_1 = p_2$ and H_0: $p_1 > p_2$, which involves a one-tailed test.

On the basis of a one-tailed test at a 0.05 level of significance, we would reject H_0 if Z were greater than 1.645. Since this is the case, we can reject H_0 at this level and conclude that the candidate is preferred in district A.

(c) In part (a), the P value is $P(Z \leq -1.75) + P(Z \geq 1.75) = 0.0802$, and the P value in part (b) is $P(Z \geq 1.75) = 0.0401$.

Tests involving student's *t* distribution

7.16. In the past a machine has produced washers having a mean thickness of 0.050 inch. To determine whether the machine is in proper working order a sample of 10 washers is chosen for which the mean thickness is 0.053 inch and the standard deviation is 0.003 inch. Test the hypothesis that the machine is in proper working order using a level of significance of (a) 0.05, (b) 0.01. (c) Find the P value of the test.

We wish to decide between the hypotheses

H_0: $\mu = 0.050$, and the machine is in proper working order

H_1: $\mu \neq 0.050$, and the machine is not in proper working order

so that a two-tailed test is required.

Under the hypothesis H_0, we have $T = \dfrac{\bar{X} - \mu}{S}\sqrt{n-1} = \dfrac{0.053 - 0.050}{0.003}\sqrt{10-1} = 3.00$.

(a) For a two-tailed test at a 0.05 level of significance, we adopt the decision rule:

(1) Accept H_0 if T lies inside the interval $-t_{0.975}$ to $t_{0.975}$, which for $10 - 1 = 9$ degrees of freedom is the interval -2.26 to 2.26.

(2) Reject H_0 otherwise.

Since $T = 3.00$, we reject H_0 at the 0.05 level.

(b) For a two-tailed test at a 0.01 level of significance, we adopt the decision rule:

(1) Accept H_0 if T lies inside the interval $-t_{0.995}$ to $t_{0.995}$, which for $10 - 1 = 9$ degrees of freedom is the interval -3.25 to 3.25.

(2) Reject H_0 otherwise.

Since $T = 3.00$, we accept H_0 at the 0.01 level.

Because we can reject H_0 at the 0.05 level but not at the 0.01 level, we can say that the sample result is *probably significant* (see terminology at the end of Problem 7.5). It would therefore be advisable to check the machine or at least to take another sample.

(c) The P value is $P(T \geq 3) + P(T \leq -3)$. The table in Appendix D shows that $0.01 < P < 0.02$. Using computer software, we find $P = 0.015$.

7.17. A test of the breaking strengths of 6 ropes manufactured by a company showed a mean breaking strength of 7750 lb and a standard deviation of 145 lb, whereas the manufacturer claimed a mean breaking strength of 8000 lb. Can we support the manufacturér's claim at a level of significance of (a) 0.05, (b) 0.01? (c) What is the P value of the test?

We must decide between the hypotheses

H_0: $\mu = 8000$ lb, and the manufacturer's claim is justified

H_1: $\mu < 8000$ lb, and the manufacturer's claim is not justified

so that a one-tailed test is required.

Under the hypothesis H_0, we have

$$T = \frac{\bar{X} - \mu}{S}\sqrt{n-1} = \frac{7750 - 8000}{145}\sqrt{6-1} = -3.86.$$

(a) For a one-tailed test at a 0.05 level of significance, we adopt the decision rule:

(1) Accept H_0 if T is greater than $-t_{0.95}$, which for $6 - 1 = 5$ degrees of freedom means $T > -2.01$.

(2) Reject H_0 otherwise.

Since $T = -3.86$, we reject H_0.

(b) For a one-tailed test at a 0.01 level of significance, we adopt the decision rule:

(1) Accept H_0 if T is greater than $-t_{0.99}$, which for 5 degrees of freedom means $T > -3.36$.

(2) Reject H_0 otherwise.

Since $T = -3.86$, we reject H_0.

We conclude that it is extremely unlikely that the manufacturer's claim is justified.

(c) The P value is $P(T \leq -3.86)$. The table in Appendix D shows $0.005 < P < 0.01$. By computer software, $P = 0.006$.

7.18. The IQs (intelligence quotients) of 16 students from one area of a city showed a mean of 107 with a standard deviation of 10, while the IQs of 14 students from another area of the city showed a mean of 112 with a standard deviation of 8. Is there a significant difference between the IQs of the two groups at a (a) 0.01, (b) 0.05 level of significance? (c) What is the P value of the test?

If μ_1 and μ_2 denote population mean IQs of students from the two areas, we have to decide between the hypotheses

$$H_0: \mu_1 = \mu_2, \text{ and there is essentially no difference between the groups}$$
$$H_1: \mu_1 \neq \mu_2, \text{ and there is a significant difference between the groups}$$

Under the hypothesis H_0,

$$T = \frac{\bar{X}_1 - \bar{X}_2}{\sigma\sqrt{1/n_1 + 1/n_2}} \quad \text{where} \quad \sigma = \sqrt{\frac{n_1 S_1^2 + n_2 S_2^2}{n_1 + n_2 - 2}}$$

Then

$$\sigma = \sqrt{\frac{16(10)^2 + 14(8)^2}{16 + 14 - 2}} = 9.44 \quad \text{and} \quad T = \frac{112 - 107}{9.44\sqrt{1/16 + 1/14}} = 1.45$$

(a) On the basis of a two-tailed test at a 0.01 level of significance, we would reject H_0 if T were outside the range $-t_{0.995}$ to $t_{0.995}$, which, for $n_1 + n_2 - 2 = 16 + 14 - 2 = 28$ degrees of freedom, is the range -2.76 to 2.76.

Therefore, we cannot reject H_0 at a 0.01 level of significance.

(b) On the basis of a two-tailed rest at a 0.05 level of significance, we would reject H_0 if T were outside the range $-t_{0.975}$ to $t_{0.975}$, which for 28 degrees of freedom is the range -2.05 to 2.05.

Therefore, we cannot reject H_0 at a 0.05 level of significance. We conclude that there is no significant difference between the IQs of the two groups.

(c) The P value is $P(T \geq 1.45) + P(T \leq -1.45)$. The table in Appendix D shows $0.1 < P < 0.2$. By computer software, $P = 0.158$.

7.19. At an agricultural station it was desired to test the effect of a given fertilizer on wheat production. To accomplish this, 24 plots of land having equal areas were chosen; half of these were treated with the fertilizer and the other half were untreated (control group). Otherwise the conditions were the same. The mean yield of wheat on the untreated plots was 4.8 bushels with a standard deviation of 0.40 bushels, while the mean yield on the treated plots was 5.1 bushels with a standard deviation of 0.36 bushels. Can we conclude that there is a significant improvement in wheat production because of the fertilizer if a significance level of (a) 1%, (b) 5% is used? (c) What is the P value of the test?

If μ_1 and μ_2 denote population mean yields of wheat on treated and untreated land, respectively, we have to decide between the hypotheses

$$H_0: \mu_1 = \mu_2, \text{ and the difference is due to chance}$$
$$H_1: \mu_1 > \mu_2, \text{ and the fertilizer improves the yield}$$

Under the hypothesis H_0,

$$T = \frac{\bar{X}_1 - \bar{X}_2}{\sigma\sqrt{1/n_1 + 1/n_2}} \quad \text{where} \quad \sigma = \sqrt{\frac{n_1 S_1^2 + n_2 S_2^2}{n_1 + n_2 - 2}}$$

Then

$$\sigma = \sqrt{\frac{12(0.40)^2 + 12(0.36)^2}{12 + 12 - 2}} = 0.397 \quad \text{and} \quad T = \frac{5.1 - 4.8}{0.397\sqrt{1/12 + 1/12}} = 1.85$$

(a) On the basis of a one-tailed test at a 0.01 level of significance, we would reject H_0 if T were greater than $t_{0.99}$, which, for $n_1 + n_2 - 2 = 12 + 12 - 2 = 22$ degrees of freedom, is 2.51.

Therefore, we cannot reject H_0 at a 0.01 level of significance.

(b) On the basis of one-tailed test at a 0.05 level of significance, we would reject H_0 if T were greater than $t_{0.95}$, which for 22 degrees of freedom is 1.72.

Therefore, we can reject H_0 at a 0.05 level of significance.

We conclude that the improvement in yield of wheat by use of the fertilizer is *probably significant.* However before definite conclusions are drawn concerning the usefulness of the fertilizer, it may be desirable to have some further evidence.

(c) The P value is $P(T \geq 1.85)$. The table in Appendix D shows $0.025 < P < 0.05$. By computer software, $P = 0.039$.

Tests involving the chi-square distribution

7.20. In the past the standard deviation of weights of certain 40.0 oz packages filled by a machine was 0.25 oz. A random sample of 20 packages showed a standard deviation of 0.32 oz. Is the apparent increase in variability significant at the (a) 0.05, (b) 0.01 level of significance? (c) What is the P value of the test?

We have to decide between the hypotheses

$$H_0: \sigma = 0.25 \text{ oz and the observed result is due to chance}$$

$$H_1: \sigma > 0.25 \text{ oz and the variability has increased}$$

The value of χ^2 for the sample is $\chi^2 = ns^2/\sigma^2 = 20(0.32)^2/(0.25)^2 = 32.8$.

(a) Using a one-tailed test, we would reject H_0 at a 0.05 level of significance if the sample value of χ^2 were greater than $\chi^2_{0.95}$, which equals 30.1 for $\nu = 20 - 1 = 19$ degrees of freedom. Therefore, we would reject H_0 at a 0.05 level of significance.

(b) Using a one-tailed test, we would reject H_0 at a 0.01 level of significance if the sample value of χ^2 were greater than $\chi^2_{0.99}$, which equals 36.2 for 19 degrees of freedom. Therefore, we would not reject H_0 at a 0.01 level of significance.

We conclude that the variability has probably increased. An examination of the machine should be made.

(c) The P value is $P(\chi^2 \geq 32.8)$. The table in Appendix E shows $0.025 < P < 0.05$. By computer software, $P = 0.0253$.

Tests involving the *F* distribution

7.21. An instructor has two classes, A and B, in a particular subject. Class A has 16 students while class B has 25 students. On the same examination, although there was no significant difference in mean grades, class A had a standard deviation of 9 while class B had a standard deviation of 12. Can we conclude at the (a) 0.01, (b) 0.05 level of significance that the variability of class B is greater than that of A? (c) What is the P value of the test?

(a) We have, on using subscripts 1 and 2 for classes A and B, respectively, $s_1 = 9$, $s_2 = 12$ so that

$$\hat{s}_1^2 = \frac{n_1}{n_1 - 1} s_1^2 = \frac{16}{15}(9)^2 = 86.4, \qquad \hat{s}_2^2 = \frac{n_2}{n_2 - 1} s_2^2 = \frac{25}{24}(12)^2 = 150$$

We have to decide between the hypotheses

$$H_0: \sigma_1 = \sigma_2, \text{ and any observed variability is due to chance}$$

$$H_1: \sigma_2 > \sigma_1, \text{ and the variability of class } B \text{ is greater than that of } A$$

The decision must therefore be based on a one-tailed test of the F distribution. For the samples in question,

$$F = \frac{\hat{s}_2^2}{\hat{s}_1^2} = \frac{150}{86.4} = 1.74$$

The number of degrees of freedom associated with the numerator is $r_2 = 25 - 1 = 24$; for the denominator, $r_1 = 16 - 1 = 15$. At the 0.01 level for 24, 15 degrees of freedom we have from Appendix F, $F_{0.99} = 3.29$. Then, since $F < F_{0.99}$, we cannot reject H_0 at the 0.01 level.

(b) Since $F_{0.95} = 2.29$ for 24, 15 degrees of freedom (see Appendix F), we see that $F < F_{0.95}$. Thus we cannot reject H_0 at the 0.05 level either.

(c) The P value of the test is $P(F \geq 1.74)$. The tables in Appendix F show that $P > 0.05$. By computer software, $P = 0.134$.

7.22. In Problem 7.21 would your conclusions be changed if it turned out that there was a significant difference in the mean grades of the classes? Explain your answer.

Since the actual mean grades were not used at all in Problem 7.21, it makes no difference what they are. This is to be expected in view of the fact that we are not attempting to decide whether there is a difference in mean grades, but only whether there is a difference in variability of the grades.

Operating characteristic curves

7.23. Referring to Problem 7.2, what is the probability of accepting the hypothesis that the coin is fair when the actual probability of heads is $p = 0.7$?

The hypothesis H_0 that the coin is fair, i.e., $p = 0.5$, is accepted when the number of heads in 100 tosses lies between 39.5 and 60.5. The probability of rejecting H_0 when it should be accepted (i.e., the probability of a Type I error) is represented by the total area α of the shaded region under the normal curve to the left in Fig. 7-6. As computed in Problem 7.2(a), this area α, which represents the level of significance of the test of H_0, is equal to 0.0358.

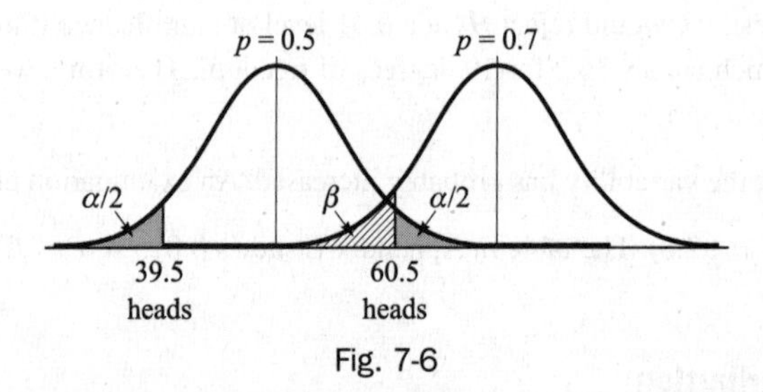

Fig. 7-6

If the probability of heads is $p = 0.7$, then the distribution of heads in 100 tosses is represented by the normal curve to the right in Fig. 7-6. The probability of accepting H_0 when actually $p = 0.7$ (i.e., the probability of a Type II error) is given by the cross-hatched area β. To compute this area, we observe that the distribution under the hypothesis $p = 0.7$ has mean and standard deviation given by

$$\mu = np = (100)(0.7) = 70 \qquad \sigma = \sqrt{npq} = \sqrt{(100)(0.7)(0.3)} = 4.58$$

$$60.5 \text{ in standard units} = \frac{60.5 - 70}{4.58} = -2.07$$

$$39.5 \text{ in standard units} = \frac{39.5 - 70}{4.58} = -6.66$$

Then β = area under the standard normal curve between $z = -6.66$ and $z = -2.07 = 0.0192$.

Therefore, with the given decision rule there is very little chance of accepting the hypothesis that the coin is fair when actually $p = 0.7$.

Note that in this problem we were given the decision rule from which we computed α and β. In practice two other possibilities may arise:

(1) We decide on α (such as 0.05 or 0.01), arrive at a decision rule, and then compute β.

(2) We decide on α and β and then arrive at a decision rule.

7.24. Work Problem 7.23 if (a) $p = 0.6$, (b) $p = 0.8$, (c) $p = 0.9$, (d) $p = 0.4$.

(a) If $p = 0.6$, the distribution of heads has mean and standard deviation given by

$$\mu = np = (100)(0.6) = 60 \qquad \sigma = \sqrt{npq} = \sqrt{(100)(0.6)(0.4)} = 4.90$$

$$60.5 \text{ in standard units} = \frac{60.5 - 60}{4.90} = 0.102$$

$$39.5 \text{ in standard units} = \frac{39.5 - 60}{4.90} = -4.18$$

Then β = area under the standard normal curve between $z = -4.18$ and $z = 0.102 = 0.5405$

Therefore, with the given decision rule there is a large chance of accepting the hypothesis that the coin is fair when actually $p = 0.6$.

(b) If $p = 0.8$, then $\mu = np = (100)(0.8) = 80$ and $\sigma = \sqrt{npq} = \sqrt{(100)(0.08)(0.2)} = 4$.

$$60.5 \text{ in standard units} = \frac{60.5 - 80}{4} = -4.88$$

$$39.5 \text{ in standard units} = \frac{39.5 - 80}{4} = -10.12$$

Then β = area under the standard curve between $z = -10.12$ and $z = -4.88 = 0.0000$, very closely.

(c) From comparison with (b) or by calculation, we see that if $p = 0.9$, $\beta = 0$ for all practical purposes.

(d) By symmetry, $p = 0.4$ yields the same value of β as $p = 0.6$, i.e., $\beta = 0.5405$.

7.25. Represent the results of Problems 7.23 and 7.24 by constructing a graph of (a) β vs. p, (b) $(1 - \beta)$ vs. p. Interpret the graphs obtained.

Table 7-3 shows the values of β corresponding to given values of p as obtained in Problems 7.23 and 7.24.

Note that β represents the probability of accepting the hypothesis $p = 0.5$ when actually p is a value other than 0.5. However, if it is actually true that $p = 0.5$, we can interpret β as the probability of accepting $p = 0.5$ when it should be accepted. This probability equals $1 - 0.0358 = 0.9642$ and has been entered into Table 7-3.

Table 7-3

p	0.1	0.2	0.3	0.4	0.5	0.6	0.7	0.8	0.9
β	0.0000	0.0000	0.0192	0.5405	0.9642	0.5405	0.0192	0.0000	0.0000

(a) The graph of β vs. p, shown in Fig. 7-7(a), is called the *operating characteristic curve*, or *OC curve*, of the decision rule or test of hypotheses.

The distance from the maximum point of the OC curve to the line $\beta = 1$ is equal to $\alpha = 0.0358$, the level of significance of the test.

In general, the sharper the peak of the OC curve the better is the decision rule for rejecting hypotheses that are not valid.

(b) The graph of $(1 - \beta)$ vs. p, shown in Fig. 7-7(b), is called the *power curve* of the decision rule or test of hypotheses. This curve is obtained simply by inverting the OC curve, so that actually both graphs are equivalent.

The quantity $1 - \beta$ is often called a *power function* since it indicates the ability or *power of a test* to reject hypotheses which are false, i.e., should be rejected. The quantity β is also called the *operating characteristic function* of a test.

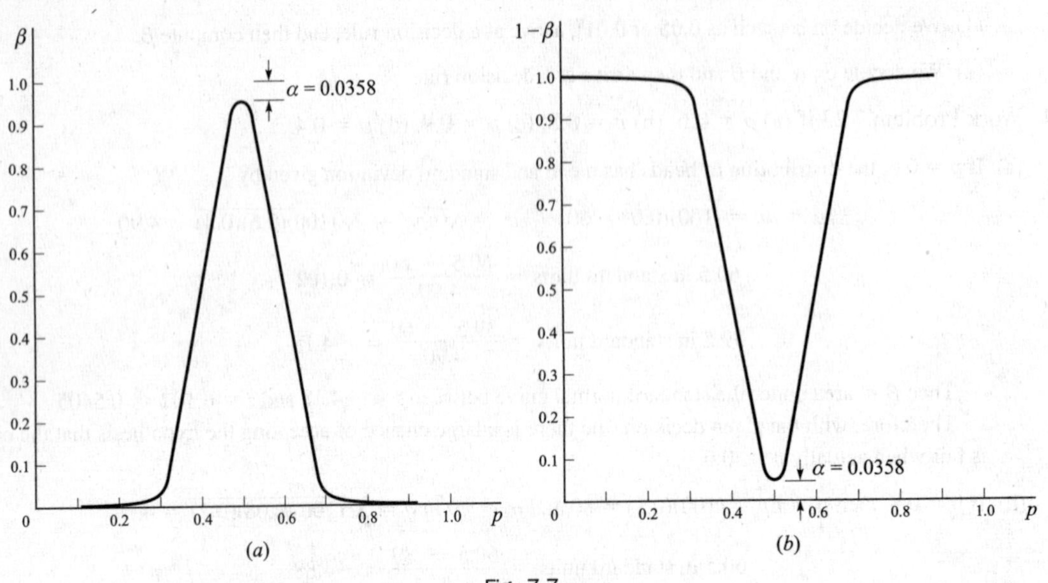

Fig. 7-7

7.26. A company manufactures rope whose breaking strengths have a mean of 300 lb and standard deviation 24 lb. It is believed that by a newly developed process the mean breaking strength can be increased, (a) Design a decision rule for rejecting the old process at a 0.01 level of significance if it is agreed to test 64 ropes, (b) Under the decision rule adopted in (a), what is the probability of accepting the old process when in fact the new process has increased the mean breaking strength to 310 lb? Assume that the standard deviation is still 24 lb.

(a) If μ is the mean breaking strength, we wish to decide between the hypotheses

H_0: $\mu = 300$ lb, and the new process is equivalent to the old one

H_1: $\mu > 300$ lb, and the new process is better than the old one

For a one-tailed test at a 0.01 level of significance, we have the following decision rule (refer to Fig. 7-8):

(1) Reject H_0 if the z score of the sample mean breaking strength is greater than 2.33.

(2) Accept H_0 otherwise.

Since $Z = \dfrac{\bar{X} - \mu}{\sigma/\sqrt{n}} = \dfrac{\bar{X} - 300}{24/\sqrt{64}}$, $\bar{X} = 300 + 3z$. Then if $Z > 2.33$, $\bar{X} > 300 + 3(2.33) = 307.0$ lb.

Therefore, the above decision rule becomes:

(1) Reject H_0 if the mean breaking strength of 64 ropes exceeds 307.0 lb.

(2) Accept H_0 otherwise.

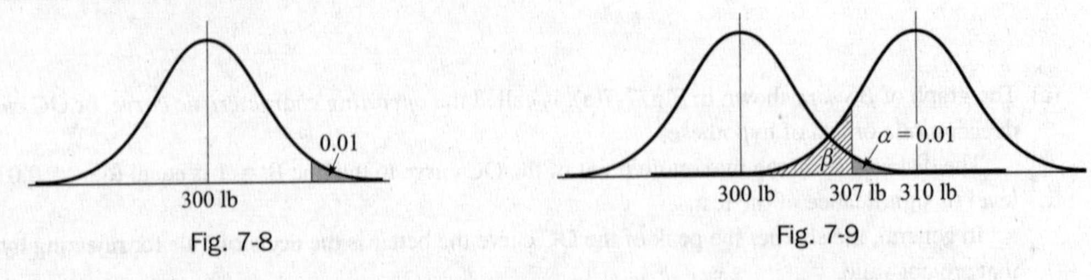

Fig. 7-8

Fig. 7-9

(b) Consider the two hypotheses (H_0: $\mu = 300$ lb) and (H_1: $\mu = 310$ lb). The distributions of breaking strengths corresponding to these two hypotheses are represented respectively by the left and right normal distributions of Fig. 7-9.

The probability of accepting the old process when the new mean breaking strength is actually 310 lb is represented by the region of area β in Fig. 7-9. To find this, note that 307.0 lb in standard units is $(307.0 - 310)/3 = -1.00$; hence

$$\beta = \text{area under right-hand normal curve to left of } z = -1.00 = 0.1587$$

This is the probability of accepting (H_0: $\mu = 300$ lb) when actually (H_1: $\mu = 310$ 1b) is true, i.e., it is the probability of making a Type II error.

7.27. Construct (a) an OC curve, (b) a power curve for Problem 7.26, assuming that the standard deviation of breaking strengths remains at 24 lb.

By reasoning similar to that used in Problem 7.26(b), we can find β for the cases where the new process yields mean breaking strengths μ equal to 305 lb, 315 lb, etc. For example, if $\mu = 305$ lb, then 307.0 lb in standard units is $(307.0 - 305)/3 = 0.67$, and hence

$$\beta = \text{area under right hand normal curve to left of } z = 0.67 = 0.7486$$

In this manner Table 7-4 is obtained.

Table 7-4

μ	290	295	300	305	310	315	320
β	1.0000	1.0000	0.9900	0.7486	0.1587	0.0038	0.0000

(a) The OC curve is shown in Fig. 7-10(a). From this curve we see that the probability of keeping the old process if the new breaking strength is less than 300 lb is practically 1 (except for the level of significance of 0.01 when the new process gives a mean of 300 lb). It then drops rather sharply to zero so that there is practically no chance of keeping the old process when the mean breaking strength is greater than 315 lb.

(b) The power curve shown in Fig. 7-10(b) is capable of exactly the same interpretation as that for the OC curve. In fact the two curves are essentially equivalent.

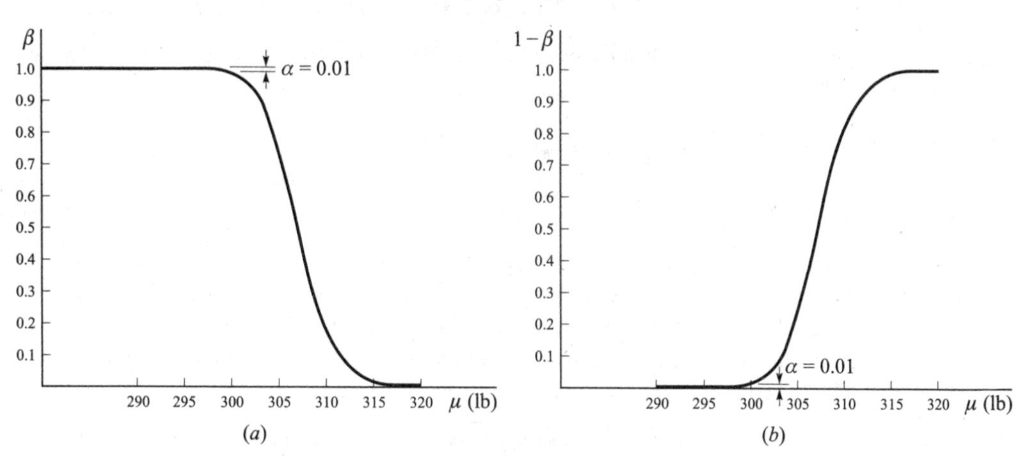

Fig. 7-10

7.28. To test the hypothesis that a coin is fair (i.e., $p = 0.5$) by a number of tosses of the coin, we wish to impose the following restrictions: (A) the probability of rejecting the hypothesis when it is actually correct must be 0.05 at most; (B) the probability of accepting the hypothesis when actually p differs from 0.5 by 0.1 or more (i.e., $p \geq 0.6$ or $p \leq 0.4$) must be 0.05 at most. Determine the minimum sample size that is necessary and state the resulting decision rule.

Here we have placed limits on the risks of Type I and Type II errors. For example, the imposed restriction (A) requires that the probability of a Type I error is $\alpha = 0.05$ at most, while restriction (B) requires that the probability of a Type II error is $\beta = 0.05$ at most. The situation is illustrated graphically in Fig. 7-11.

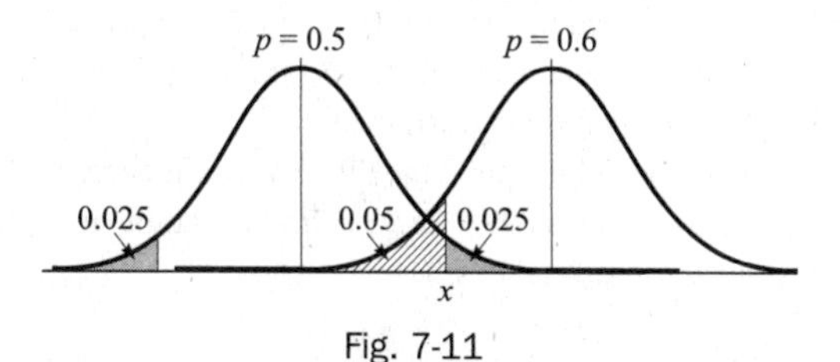

Fig. 7-11

Let n denote the required sample size and x the number of heads in n tosses above which we reject the hypothesis $p = 0.5$. From Fig. 7-11,

(1) Area under normal curve for $p = 0.5$ to right of $\dfrac{x - np}{\sqrt{npq}} = \dfrac{x - 0.5n}{0.5\sqrt{n}}$ is 0.025.

(2) Area under normal curve for $p = 0.6$ to left of $\dfrac{x - np}{\sqrt{npq}} = \dfrac{x - 0.6n}{0.49\sqrt{n}}$ is 0.05.

Actually, we should have equated the area between

$$\frac{(n - x) - 0.6n}{0.49\sqrt{n}} \quad \text{and} \quad \frac{x - 0.6n}{0.49\sqrt{n}}$$

to 0.05; however (2) is a close approximation. Notice that by making the acceptance probability 0.05 in the "worst case," $p = 0.6$, we automatically make it 0.05 or less when p has any other value outside the range 0.4 to 0.6. Hence, a weighted average of all these probabilities, which represents the probability of a Type II error, will also be 0.05 or less.

From (1), $\dfrac{x - 0.5n}{0.5\sqrt{n}} = 1.96$ or (3) $x = 0.5n + 0.980\sqrt{n}$.

From (2), $\dfrac{x - 0.6n}{0.49\sqrt{n}} = -1.645$ or (4) $x = 0.6n - 0.806\sqrt{n}$.

Then from (3) and (4), $n = 318.98$. It follows that the sample size must be at least 319, i.e., we must toss the coin at least 319 times. Putting $n = 319$ in (3) or (4), $x = 177$.

For $p = 0.5$, $x - np = 177 - 159.5 = 17.5$. Therefore, we adopt the following decision rule:

(a) Accept the hypothesis $p = 0.5$ if the number of heads in 319 tosses is in the range 159.5 ± 17.5, i.e., between 142 and 177.

(b) Reject the hypothesis otherwise.

Quality control charts

7.29. A machine is constructed to produce ball bearings having a mean diameter of 0.574 inch and a standard deviation of 0.008 inch. To determine whether the machine is in proper working order, a sample of 6 ball bearings is taken every 2 hours and the mean diameter is computed from this sample, (a) Design a decision rule whereby one can be fairly certain that the quality of the products is conforming to required standards, (b) Show how to represent graphically the decision rule in (a).

(a) With 99.73% confidence we can say that the sample mean $\bar{X}$ must lie in the range $(\mu_{\bar{X}} - 3\sigma_{\bar{X}})$ to $(\mu_{\bar{X}} + 3\sigma_{\bar{X}})$ or $(\mu - 3\sigma/\sqrt{n})$ to $(\mu + 3\sigma/\sqrt{n})$. Since $\mu = 0.574$, $\sigma = 0.008$ and $n = 6$, it follows that with 99.73% confidence the sample mean should lie between $(0.574 - 0.024/\sqrt{6})$ and $(0.574 + 0.024/\sqrt{6})$ or between 0.564 and 0.584 inches.

Hence, our decision rule is as follows:

(1) If a sample mean falls inside the range 0.564 to 0.584 inches, assume the machine is in proper working order.

(2) Otherwise conclude that the machine is not in proper working order and seek to determine the reason.

(b) A record of the sample means can be kept by means of a chart such as shown in Fig. 7-12, called a *quality control chart.* Each time a sample mean is computed, it is represented by a point. As long as the points lie between the lower limit 0.564 inch and upper limit 0.584 inch, the process is under control. When a point goes outside of these control limits (such as in the third sample taken on Thursday), there is a possibility that something is wrong and investigation is warranted.

The control limits specified above are called the 99.73% confidence limits, or briefly the 3σ limits. However, other confidence limits, such as 99% or 95% limits, can be determined as well. The choice in each case depends on particular circumstances.

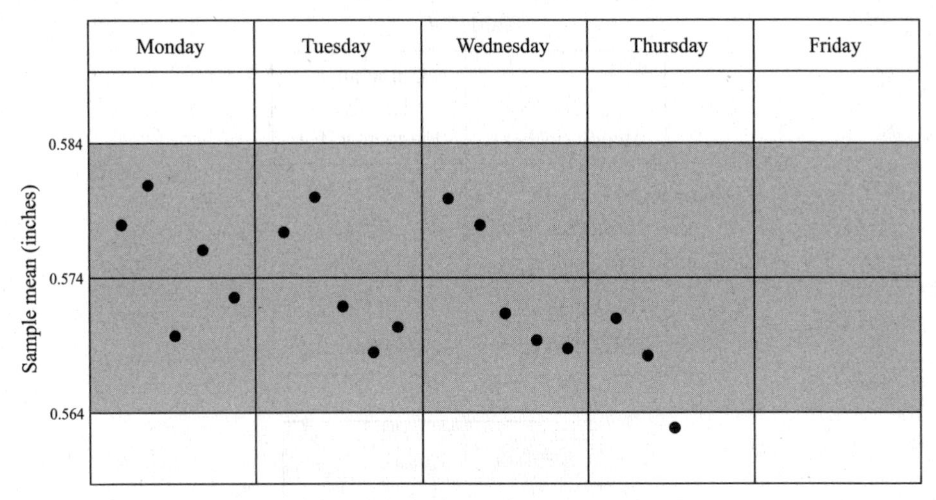

Fig. 7-12

Fitting of data by theoretical distributions

7.30. Fit a binomial distribution to the data of Problem 5.30, page 176.

We have $P(x \text{ heads in a toss of 5 pennies}) = f(x) = {}_5C_x p^x q^{5-x}$, where p and q are the respective probabilities of a head and tail on a single toss of a penny. The mean or expected number of heads is $\mu = np = 5p$.

For the actual or observed frequency distribution, the mean number of heads is

$$\frac{\sum fx}{\sum f} = \frac{(38)(0) + (144)(1) + (342)(2) + (287)(3) + (164)(4) + (25)(5)}{1000} = \frac{2470}{1000} = 2.47$$

Equating the theoretical and actual means, $5p = 2.47$ or $p = 0.494$. Therefore, the fitted binomial distribution is given by $f(x) = {}_5C_x = (0.494)^x(0.506)^{5-x}$.

In Table 7-5 these probabilities have been listed as well as the expected (theoretical) and actual frequencies. The fit is seen to be fair. The goodness of fit is investigated in Problem 7.43.

Table 7-5

Number of Heads (x)	$P(x \text{ heads})$	Expected Frequency	Observed Frequency
0	0.0332	33.2 or 33	38
1	0.1619	161.9 or 162	144
2	0.3162	316.2 or 316	342
3	0.3087	308.7 or 309	287
4	0.1507	150.7 or 151	164
5	0.0294	29.4 or 29	25

7.31. Use probability graph paper to determine whether the frequency distribution of Table 5-2, page 161, can be closely approximated by a normal distribution.

First the given frequency distribution is converted into a cumulative relative frequency distribution, as shown in Table 7-6. Then the cumulative relative frequencies expressed as percentages are plotted against upper class boundaries on special probability graph paper as shown in Fig. 7-13. The degree to which all plotted points lie on a straight line determines the closeness of fit of the given distribution to a normal distribution. It is seen that there is a normal distribution which fits the data closely. See Problem 7.32.

Table 7-6

Height (inches)	Cumulative Relative Frequency (%)
Less than 61.5	5.0
Less than 64.5	23.0
Less than 67.5	65.0
Less than 70.5	92.0
Less than 73.5	100.0

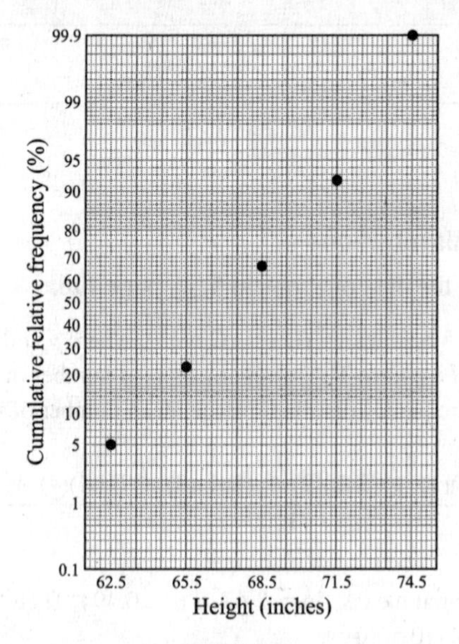

Fig. 7-13

7.32. Fit a normal curve to the data of Table 5-2, page 161.

$$\bar{x} = 67.45 \text{ inches}, \qquad s = 2.92 \text{ inches}$$

The work may be organized as in Table 7-7. In calculating z for the class boundaries, we use $z = (x - \bar{x})/s$ where the mean $\bar{x}$ and standard deviations s have been obtained respectively in Problems 5.35 and 5.40.

Table 7-7

Heights (inches)	Class Boundaries (x)	z for Class Boundaries	Area under Normal Curve from 0 to z	Area for Each Class	Expected Frequency	Observed Frequency
	59.5	−2.72	0.4967			
60–62				0.0413	4.13 or 4	5
	62.5	−1.70	0.4554			
63–65				0.2086	20.68 or 21	18
	65.5	−0.67	0.2486			
66–68			Add →	0.3892	38.92 or 39	42
	68.5	0.36	0.1406			
69–71				0.2771	27.71 or 28	27
	71.5	1.39	0.4177			
72–74				0.0743	7.43 or 7	8
	74.5	2.41	0.4920			

In the fourth column, the areas under the normal curve from 0 to z have been obtained by using the table in Appendix C. From this we find the areas under the normal curve between successive values of z as in the fifth column. These are obtained by subtracting the successive areas in the fourth column when the corresponding

z s have the same sign, and adding them when the z s have opposite signs (which occurs only once in the table). The reason for this is at once clear from a diagram.

Multiplying the entries in the fifth column (which represent relative frequencies) by the total frequency n (in this case $n = 100$) yields the theoretical or expected frequencies as in the sixth column. It is seen that they agree well with the actual or observed frequencies of the last column.

The goodness of fit of the distribution is considered in Problem 7.44.

7.33. Table 7-8 shows the number of days f in a 50-day period during which x automobile accidents occurred in a city. Fit a Poisson distribution to the data.

Table 7-8

Number of Accidents (x)	Number of Days (f)
0	21
1	18
2	7
3	3
4	1
TOTAL	50

The mean number of accidents is

$$\lambda = \frac{\sum fx}{\sum f} = \frac{(21)(0) + (18)(1) + (7)(2) + (3)(3) + (1)(4)}{50}$$

$$= \frac{45}{50} = 0.90$$

Then, according to the Poisson distribution,

$$P(x \text{ accidents}) = \frac{(0.90)^x e^{-0.90}}{x!}$$

In Table 7-9 are listed the probabilities for 0, 1, 2, 3, and 4 accidents as obtained from this Poisson distribution, as well as the theoretical number of days during which z accidents take place (obtained by multiplying the respective probabilities by 50). For convenience of comparison, the fourth column giving the actual number of days has been repeated.

Table 7-9

Number of Accidents (x)	P (x accidents)	Expected Number of Days	Actual Number of Days
0	0.4066	20.33 or 20	21
1	0.3659	18.30 or 18	18
2	0.1647	8.24 or 8	7
3	0.0494	2.47 or 2	3
4	0.0111	0.56 or 1	1

Note that the fit of the Poisson distribution to the data is good.

For a true Poisson distribution, $\sigma^2 = \lambda$. Computation of the variance of the given distribution gives 0.97. This compares favorably with the value 0.90 for λ, which can be taken as further evidence for the suitability of the Poisson distribution in approximating the sample data.

The chi-square test

7.34. In 200 tosses of a coin, 115 heads and 85 tails were observed. Test the hypothesis that the coin is fair using a level of significance of (a) 0.05, (b) 0.01. (c) Find the P value of the test.

Observed frequencies of heads and tails are, respectively, $x_1 = 115$, $x_2 = 85$.

Expected frequencies of heads and tails if the coin is fair are $np_1 = 100$, $np_2 = 100$, respectively. Then

$$x^2 = \frac{(x_1 - np_1)^2}{np_1} + \frac{(x_2 - np_2)^2}{np_2} = \frac{(115 - 100)^2}{100} + \frac{(85 - 100)^2}{100} = 4.50$$

Since the number of categories or classes (heads, tails) is $k = 2$, $\nu = k - 1 = 2 - 1 = 1$.

(a) The critical value $\chi^2_{0.95}$ for 1 degree of freedom is 3.84. Then since $4.50 > 3.84$, we reject the hypothesis that the coin is fair at a 0.05 level of significance.

(b) The critical value $\chi^2_{0.99}$ for 1 degree of freedom is 6.63. Then since $4.50 < 6.63$, we cannot reject the hypothesis that the coin is fair at a 0.01 level of significance.

We conclude that the observed results are *probably significant* and the coin is *probably not fair*. For a comparison of this method with previous methods used, see Method 1 of Problem 7.36.

(c) The P value is $P(\chi^2 \geq 4.50)$. The table in Appendix E shows $0.025 < P < 0.05$. By computer software, $P = 0.039$.

7.35. Work Problem 7.34 using Yates' correction.

$$\chi^2\text{(corrected)} = \frac{(|x - 1 - np_1| - 0.5)^2}{np_1} + \frac{(|x_2 - np_2| - 0.5)^2}{np_2}$$

$$= \frac{(|115 - 100| - 0.5)^2}{100} + \frac{(|85 - 100| - 0.5)^2}{100} = \frac{(14.5)^2}{100} + \frac{(14.5)^2}{100} = 4.205$$

The corrected P value is 0.04

Since $4.205 > 3.84$ and $4.205 < 6.63$, the conclusions arrived at in Problem 7.34 are valid. For a comparison with previous methods, see Method 2 of Problem 7.36.

7.36. Work Problem 7.34 by using the normal approximation to the binomial distribution.

Under the hypothesis that the coin is fair, the mean and standard deviation of the number of heads in 200 tosses of a coin are $\mu = np = (200)(0.5) = 100$ and $\sigma = \sqrt{npq} = \sqrt{(200)(0.5)(0.5)} = 7.07$.

Method 1

$$115 \text{ heads in standard units} = (115 - 100)/7.07 = 2.12.$$

Using a 0.05 significance level and a two-tailed test, we would reject the hypothesis that the coin is fair if the z score were outside the interval -1.96 to 1.96. With a 0.01 level the corresponding interval would be -2.58 to 2.58. It follows as in Problem 7.34 that we can reject the hypothesis at a 0.05 level but cannot reject it at a 0.01 level. The P value of the test is 0.034.

Note that the square of the above standard score, $(2.12)^2 = 4.50$, is the same as the value of χ^2 obtained in Problem 7.34. This is always the case for a chi-square test involving two categories. See Problem 7.60.

Method 2

Using the correction for continuity, 115 or more heads is equivalent to 114.5 or more heads. Then 114.5 in standard units $= (114.5 - 100)/7.07 = 2.05$. This leads to the same conclusions as in the first method. The corrected P value is 0.04.

Note that the square of this standard score is $(2.05)^2 = 4.20$, agreeing with the value of χ^2 corrected for continuity using Yates' correction of Problem 7.35. This is always the case for a chi-square test involving two categories in which Yates' correction is applied, again in consequence of Problem 7.60.

7.37. Table 7-10 shows the observed and expected frequencies in tossing a die 120 times. (a) Test the hypothesis that the die is fair, using a significance level of 0.05. (b) Find the P value of the test.

(a)

Table 7-10

Face	1	2	3	4	5	6
Observed Frequency	25	17	15	23	24	16
Expected Frequency	20	20	20	20	20	20

$$\chi^2 = \frac{(x_1 - np_1)^2}{np_1} + \frac{(x_2 - np_2)^2}{np_2} + \frac{(x_3 - np_3)^2}{np_3} + \frac{(x_4 - np_4)^2}{np_4} + \frac{(x_5 - np_5)^2}{np_5} + \frac{(x_6 - np_6)^2}{np_6}$$

$$= \frac{(25 - 20)^2}{20} + \frac{(17 - 20)^2}{20} + \frac{(15 - 20)^2}{20} + \frac{(23 - 20)^2}{20} + \frac{(24 - 20)^2}{20} + \frac{(16 - 20)^2}{20} = 5.00$$

Since the number of categories or classes (faces 1, 2, 3, 4, 5, 6) is $k = 6$, $\nu = k - 1 = 6 - 1 = 5$.

The critical value $\chi^2_{0.95}$ for 5 degrees of freedom is 11.1. Then since $5.00 < 11.1$, we cannot reject the hypothesis that the die is fair.

For 5 degrees of freedom $\chi^2_{0.05} = 1.15$, so that $\chi^2 = 5.00 > 1.15$. It follows that the agreement is not so exceptionally good that we would look upon it with suspicion.

(b) The P value of the test is $P(\chi^2 \geq 5.00)$. The table in Appendix E shows $0.25 < P < 0.5$. By computer software, $P = 0.42$.

7.38. A random number table of 250 digits had the distribution of the digits 0, 1, 2, . . . , 9 shown in Table 7-11. (a) Does the observed distribution differ significantly from the expected distribution? (b) What is the P value of the observation?

(a)

Table 7-11

Digit	0	1	2	3	4	5	6	7	8	9
Observed Frequency	17	31	29	18	14	20	35	30	20	36
Expected Frequency	25	25	25	25	25	25	25	25	25	25

$$\chi^2 = \frac{(17 - 25)^2}{25} + \frac{(31 - 25)^2}{25} + \frac{(29 - 25)^2}{25} + \frac{(18 - 25)^2}{25} + \cdots + \frac{(36 - 25)^2}{25} = 23.3$$

The critical value $\chi^2_{0.99}$ for $\nu = k - 1 = 9$ degrees of freedom is 21.7, and $23.3 > 21.7$. Hence we conclude that the observed distribution differs significantly from the expected distribution at the 0.01 level of significance. Some suspicion is therefore upon the table.

(b) The P value is $P(\chi^2 \geq 23.3)$. The table in Appendix E shows that $0.005 < P < 0.01$. By computer software, $P = 0.0056$.

7.39. In Mendel's experiments with peas he observed 315 round and yellow, 108 round and green, 101 wrinkled and yellow, and 32 wrinkled and green. According to his theory of heredity the numbers should be in the proportion 9:3:3:1. Is there any evidence to doubt his theory at the (a) 0.01, (b) 0.05 level of significance? (c) What is the P value of the observation?

The total number of peas is $315 + 108 + 101 + 32 = 556$. Since the expected numbers are in the proportion 9:3:3:1 (and $9 + 3 + 3 + 1 = 16$), we would expect

$\frac{9}{16}(556) = 312.75$ round and yellow $\qquad$ $\frac{3}{16}(556) = 104.25$ wrinkled and yellow

$\frac{3}{16}(556) = 104.25$ round and green $\qquad$ $\frac{1}{16}(556) = 34.75$ wrinkled and green

Then

$$\chi^2 = \frac{(315 - 312.75)^2}{312.75} + \frac{(108 - 104.25)^2}{104.25} + \frac{(101 - 104.25)^2}{104.25} + \frac{(32 - 34.75)^2}{37.75} = 0.470$$

Since there are four categories, $k = 4$ and the number of degrees of freedom is $\nu = 4 - 1 = 3$.

(a) For $\nu = 3$, $\chi^2_{0.99} = 11.3$ so that we cannot reject the theory at the 0.01 level.

(b) For $\nu = 3$, $\chi^2_{0.95} = 7.81$ so that we cannot reject the theory at the 0.05 level.

We conclude that the theory and experiment are in agreement.

Note that for 3 degrees of freedom, $\chi^2_{0.05} = 0.352$ and $\chi^2 = 0.470 > 0.352$. Therefore, although the agreement is good, the results obtained are subject to a reasonable amount of sampling error.

(c) The P value is $P(\chi^2 \geq 0.470)$. The table in Appendix E shows that $0.9 < P < 0.95$. By computer software, $P = 0.93$.

7.40. An urn consists of a very large number of marbles of four different colors: red, orange, yellow, and green. A sample of 12 marbles drawn at random from the urn revealed 2 red, 5 orange, 4 yellow, and 1 green marble. Test the hypothesis that the urn contains equal proportions of the differently colored marbles, and find the P value of the sample results.

Under the hypothesis that the urn contains equal proportions of the differently colored marbles, we would expect 3 of each kind in a sample of 12 marbles.

Since these expected numbers are less than 5, the chi-square approximation will be in error. To avoid this, we combine categories so that the expected number in each category is at least 5.

If we wish to reject the hypothesis, we should combine categories in such a way that the evidence against the hypothesis shows up best. This is achieved in our case by considering the categories "red or green" and "orange or yellow," for which the sample revealed 3 and 9 marbles, respectively. Since the expected number in each category under the hypothesis of equal proportions is 6, we have

$$\chi^2 = \frac{(3 - 6)^2}{6} + \frac{(9 - 6)^2}{6} = 3$$

For $\nu = 2 - 1 = 1$, $\chi^2_{0.95} = 3.84$. Therefore, we cannot reject the hypothesis at the 0.05 level of significance (although we can at the 0.10 level). Conceivably the observed results could arise on the basis of chance even when equal proportions of the colors are present. The P value is $P(\chi^2 \geq 3) = 0.083$.

Another method

Using Yates' correction, we find

$$\chi^2 = \frac{(|3 - 6| - 0.5)^2}{6} + \frac{(|9 - 6| - 0.5)^2}{6} = \frac{(2.5)^2}{6} + \frac{(2.5)^2}{6} = 2.1$$

which leads to the same conclusion given above. This is of course to be expected, since Yates' correction always *reduces* the value of χ^2. Here the P value is $P(\chi^2 \geq 2.1) = 0.15$.

It should be noted that if the χ^2 approximation is used despite the fact that the frequencies are too small, we would obtain

$$\chi^2 = \frac{(2 - 3)^2}{3} + \frac{(5 - 3)^2}{3} + \frac{(4 - 3)^2}{3} + \frac{(1 - 3)^2}{3} = 3.33$$

with a P value of 0.34.

Since for $\nu = 4 - 1 = 3$, $\chi^2_{0.95} = 7.81$, we would arrive at the same conclusions as above. Unfortunately the χ^2 approximation for small frequencies is poor; hence when it is not advisable to combine frequencies we must resort to exact probability methods involving the multinomial distribution.

7.41. In 360 tosses of a pair of dice, 74 "sevens" and 24 "elevens" are observed. Using a 0.05 level of significance, test the hypothesis that the dice are fair, and find the P value of the observed results.

A pair of dice can fall in 36 ways. A seven can occur in 6 ways, an eleven in 2 ways.

Then $P(\text{seven}) = \frac{6}{36} = \frac{1}{6}$ and $P(\text{eleven}) = \frac{2}{36} = \frac{1}{18}$. Therefore, in 360 tosses we would expect $\frac{1}{6}(360) = 60$ sevens and $\frac{1}{18}(360) = 20$ elevens, so that

$$\chi^2 = \frac{(74 - 60)^2}{60} + \frac{(24 - 20)^2}{20} = 4.07$$

with a P value of 0.044.

For $\nu = 2 - 1 = 1$, $\chi^2_{0.95} = 3.84$. Then since $4.07 > 3.84$, we would be inclined to reject the hypothesis that the dice are fair. Using Yates' correction, however, we find

$$\chi^2\,(\text{corrected}) = \frac{(|74 - 60| - 0.5)^2}{60} + \frac{(|24 - 20| - 0.5)^2}{20} = \frac{(13.5)^2}{60} + \frac{(3.5)^2}{20} = 3.65$$

with a P value of 0.056.

Therefore, on the basis of the corrected χ^2, we could not reject the hypothesis at the 0.05 level.

In general, for large samples such as we have here, results using Yates' correction prove to be more reliable than uncorrected results. However, since even the corrected value of χ^2 lies so close to the critical value, we are hesitant about making decisions one way or the other. In such cases it is perhaps best to increase the sample size by taking more observations if we are interested especially in the 0.05 level. Otherwise, we could reject the hypothesis at some other level (such as 0.10).

7.42. A survey of 320 families with 5 children each revealed the distribution of boys and girls shown in Table 7-12. (a) Is the result consistent with the hypothesis that male and female births are equally probable? (b) What is the P value of the sample results?

(a)

Table 7-12

Number of Boys and Girls	5 boys 0 girls	4 boys 1 girl	3 boys 2 girls	2 boys 3 girls	1 boy 4 girls	0 boys 5 girls	TOTAL
Number of Families	18	56	110	88	40	8	320

Let p = probability of a male birth, and $q = 1 - p$ = probability of a female birth. Then the probabilities of (5 boys), (4 boys and 1 girl), . . . , (5 girls) are given by the terms in the binomial expansion

$$(p + q)^5 = p^5 + 5p^4q + 10p^3q^2 + 10p^2q^3 + 5pq^4 + q^5$$

If $p = q = \frac{1}{2}$, we have

$$P(5 \text{ boys and } 0 \text{ girls}) = \left(\frac{1}{2}\right)^5 = \frac{1}{32} \qquad P(2 \text{ boys and } 3 \text{ girls}) = 10\left(\frac{1}{2}\right)^2\left(\frac{1}{2}\right)^3 = \frac{10}{32}$$

$$P(4 \text{ boys and } 1 \text{ girl}) = 5\left(\frac{1}{2}\right)^4\left(\frac{1}{2}\right) = \frac{5}{32} \qquad P(1 \text{ boy and } 4 \text{ girls}) = 5\left(\frac{1}{2}\right)\left(\frac{1}{2}\right)^4 = \frac{5}{32}$$

$$P(3 \text{ boys and } 2 \text{ girls}) = 10\left(\frac{1}{2}\right)^3\left(\frac{1}{2}\right)^2 = \frac{10}{32} \qquad P(0 \text{ boys and } 5 \text{ girls}) = \left(\frac{1}{2}\right)^5 = \frac{1}{32}$$

Then the expected number of families with 5, 4, 3, 2, 1, and 0 boys are obtained, respectively, by multiplying the above probabilities by 320, and the results are 10, 50, 100, 100, 50, 10. Hence,

$$\chi^2 = \frac{(18 - 10)^2}{10} + \frac{(56 - 50)^2}{50} + \frac{(100 - 100)^2}{100} + \frac{(88 - 100)^2}{100} + \frac{(40 - 50)^2}{50} + \frac{(8 - 10)^2}{10} = 12.0$$

Since $\chi^2_{0.95} = 11.1$ and $\chi^2_{0.99} = 15.1$ for $\nu = 6 - 1 = 5$ degrees of freedom, we can reject the hypothesis at the 0.05 but not at the 0.01 significance level. Therefore, we conclude that the results are probably significant, and male and female births are not equally probable.

(b) The P value is $P(\chi^2 \geq 12.0) = 0.035$.

Goodness of fit

7.43. Use the chi-square test to determine the goodness of fit of the data in Problem 7.30.

$$\chi^2 = \frac{(38-33.2)^2}{33.2} + \frac{(144-161.9)^2}{161.9} + \frac{(342-316.2)^2}{316.2} + \frac{(287-308.7)^2}{308.7} + \frac{(164-150.7)^2}{150.7} + \frac{(25-29.4)^2}{29.4}$$

$$= 7.45$$

Since the number of parameters used in estimating the expected frequencies is $m = 1$ (namely, the parameter p of the binomial distribution), $\nu = k - 1 - m = 6 - 1 - 1 = 4$.

For $\nu = 4$, $\chi^2_{0.95} = 9.49$. Hence the fit of the data is good.

For $\nu = 4$, $\chi^2_{0.05} = 0.711$. Therefore, since $\chi^2 = 7.54 > 0.711$, the fit is not so good as to be incredible.

The P value is $P(\chi^2 \geq 7.45) = 0.11$.

7.44. Determine the goodness of fit of the data in Problem 7.32.

$$\chi^2 = \frac{(5-4.13)^2}{4.13} + \frac{(18-20.68)^2}{20.68} + \frac{(42-38.92)^2}{38.92} + \frac{(27-27.71)^2}{27.71} + \frac{(8-7.43)^2}{7.43} = 0.959$$

Since the number of parameters used in estimating the expected frequencies is $m = 2$ (namely, the mean μ and the standard deviation σ of the normal distribution), $\nu = k - 1 - m = 5 - 1 - 2 = 2$.

For $\nu = 2$, $\chi^2_{0.95} = 5.99$. Therefore, we conclude that the fit of the data is very good.

For $\nu = 2$, $\chi^2_{0.05} = 0.103$. Then, since $\chi^2 = 0.959 > 0.103$, the fit is not "too good."

The P value is $P(\chi^2 \geq 0.959) = 0.62$.

Contingency tables

7.45. Work Problem 7.13 by using the chi-square test.

The conditions of the problem are presented in Table 7-13. Under the null hypothesis H_0 that the serum has no effect, we would expect 70 people in each of the groups to recover and 30 in each group not to recover, as indicated in Table 7-14. Note that H_0 is equivalent to the statement that recovery is independent of the use of the serum, i.e., the classifications are independent.

Table 7-13
Frequencies Observed

	Recover	Do Not Recover	TOTAL
Group *A* (using serum)	75	25	100
Group *B* (not using serum)	65	35	100
TOTAL	140	60	200

Table 7-14
Frequencies Expected under H_0

	Recover	Do Not Recover	TOTAL
Group *A* (using serum)	70	30	100
Group *B* (not using serum)	70	30	100
TOTAL	140	60	200

$$\chi^2 = \frac{(75-70)^2}{70} + \frac{(65-70)^2}{70} + \frac{(25-30)^2}{30} + \frac{(35-30)^2}{30} = 2.38$$

To determine the number of degrees of freedom, consider Table 7-15, which is the same as Tables 7-13 and 7-14 except that only totals are shown. It is clear that we have the freedom of placing only one number in any of the four empty cells, since once this is done the numbers in the remaining cells are uniquely determined from the indicated totals. Therefore, there is 1 degree of freedom.

Table 7-15

	Recover	Do Not Recover	TOTAL
Group A			100
Group B			100
TOTAL	140	60	200

Since $\chi^2_{0.95} = 3.84$ for 1 degree of freedom, and since $\chi^2 = 2.38 < 3.84$, we conclude that the results are *not significant* at a 0.05 level. We are therefore unable to reject H_0 at this level, and we conclude either that the serum is not effective or else withhold decision pending further tests. The P value of the observed frequencies is $P(\chi^2 \geq 2.38) = 0.12$.

Note that $\chi^2 = 2.38$ is the square of the z score, $z = 1.54$, obtained in Problem 7.13. In general, the chi-square test involving sample proportions in a 2×2 contingency table is equivalent to a test of significance of differences in proportions using the normal approximation as on page 217.

Note also that the P value 0.12 here is twice the P value 0.0618 in Problem 7.13. In general a one-tailed test using χ^2 is equivalent to a two-tailed test using χ since, for example, $\chi^2 > \chi^2_{0.95}$ corresponds to $\chi > \chi_{0.95}$ or $\chi < -\chi_{0.95}$. Since for 2×2 tables χ^2 is the square of the z score, χ is the same as z for this case. Therefore, a rejection of a hypothesis at the 0.05 level using χ^2 is equivalent to a rejection in a two-tailed test at the 0.10 level using z.

7.46. Work Problem 7.45 by using Yates' correction.

$$\chi^2(\text{corrected}) = \frac{(|75-70|-0.5)^2}{70} + \frac{(|65-70|-0.5)^2}{70} + \frac{(|25-30|-0.5)^2}{30} + \frac{(|35-30|-0.5)^2}{30} = 1.93$$

with a P value of 0.16.

Therefore, the conclusions given in Problem 7.45 are valid. This could have been realized at once by noting Yates' correction always decreases the value of χ^2 and increases the P value.

7.47. In Table 7-16 are indicated the numbers of students passed and failed by three instructors: Mr. X, Mr. Y, and Mr. Z. Test the hypothesis that the proportions of students failed by the three instructors are equal.

Table 7-16
Frequencies Observed

	Mr. X	Mr. Y	Mr. Z	TOTAL
Passed	50	47	56	153
Failed	5	14	8	27
TOTAL	55	61	64	180

Under the hypothesis H_0 that the proportions of students failed by the three instructors are the same, they would have failed $27/180 = 15\%$ of the students and passed 85% of the students. The frequencies expected under H_0 are shown in Table 7-17.

Then

$$\chi^2 = \frac{(50 - 46.75)^2}{46.75} + \frac{(47 - 51.85)^2}{51.85} + \frac{(56 - 54.40)^2}{54.40} + \frac{(5 - 8.25)^2}{8.25} + \frac{(14 - 9.15)^2}{9.15} + \frac{(8 - 9.60)^2}{9.60} = 4.84$$

Table 7-17
Frequencies Expected under H_0

	Mr. X	Mr. Y	Mr. Z	TOTAL
Passed	85% of 55 = 46.75	85% of 61 = 51.85	85% of 64 = 54.40	153
Failed	15% of 55 = 8.25	15% of 61 = 9.15	15% of 64 = 9.60	27
TOTAL	55	61	64	180

To determine the number of degrees of freedom, consider Table 7-18, which is the same as Tables 7-16 and 7-17 except that only totals are shown. It is clear that we have the freedom of placing only one number into an empty cell of the first column and one number into an empty cell of the second or third column, after which all numbers in the remaining cells will be uniquely determined from the indicated totals. Therefore, there are 2 degrees of freedom in this case.

Table 7-18

	Mr. X	Mr. Y	Mr. Z	TOTAL
Passed				153
Failed				27
TOTAL	55	61	64	180

Since $\chi^2_{0.95} = 5.99$, we cannot reject H_0 at the 0.05 level. Note, however, that since $\chi^2_{0.90} = 4.61$, we can reject H_0 at the 0.10 level if we are willing to take the risk of 1 chance in 10 of being wrong. The P value of the observed frequencies is $P(\chi^2 \geq 4.84) = 0.089$.

7.48. Show that for an $h \times k$ contingency table $(h > 1, k > 1)$, the number of degrees of freedom is given by $(h - 1)(k - 1)$.

There are $h + k - 1$ independent totals of the hk entries. It follows that the number of degrees of freedom is

$$hk - (h + k - 1) = (h - 1)(k - 1)$$

as required. The result holds if the population parameters needed in obtaining theoretical frequencies are known; otherwise adjustment is needed as described in (b), page 220.

7.49. Table 7-19 represents a general 2×2 contingency table. Show that

$$\chi^2 = \frac{n(a_1 b_2 - a_2 b_1)^2}{n_1 n_2 n_A n_B}$$

Table 7-19
Results Observed

	I	II	TOTAL
A	a_1	a_2	n_A
B	b_1	b_2	n_B
TOTAL	n_1	n_2	n

Table 7-20
Results Expected

	I	II	TOTAL
A	n_1n_A/n	n_2n_A/n	n_A
B	n_1n_B/n	n_2n_B/n	n_B
TOTAL	n_1	n_2	n

As in Problem 7.45, the results expected under a null hypothesis are shown in Table 7-20. Then

$$\chi^2 = \frac{(a_1 - n_1n_A/n)^2}{n_1n_A/n} + \frac{(a_2 - n_2n_A/n)^2}{n_2n_A/n} + \frac{(b_1 - n_1n_B/n)^2}{n_1n_B/n} + \frac{(b_2 - n_2n_B/n)^2}{n_2n_B/n}$$

But
$$a_1 - \frac{n_1n_A}{n} = a_1 - \frac{(a_1 + b_1)(a_1 + a_2)}{a_1 + b_1 + a_2 + b_2} = \frac{a_1b_2 - a_2b_1}{n}$$

Similarly
$$a_2 - \frac{n_2n_A}{n} = b_1 - \frac{n_1n_B}{n} = b_2 - \frac{n_2n_B}{n} = \frac{a_1b_2 - a_2b_1}{n}$$

We can therefore write

$$\chi^2 = \frac{n}{n_1n_A}\left(\frac{a_1b_2 - a_2b_1}{n}\right)^2 + \frac{n}{n_2n_A}\left(\frac{a_1b_2 - a_2b_1}{n}\right)^2 + \frac{n}{n_1n_B}\left(\frac{a_1b_2 - a_2b_1}{n}\right)^2 + \frac{n}{n_2n_B}\left(\frac{a_1b_2 - a_2b_1}{n}\right)^2$$

which simplifies to

(1)
$$\chi^2 = \frac{n(a_1b_2 - a_2b_1)^2}{n_1n_2n_An_B} = \frac{n\Delta^2}{n_1n_2n_An_B}$$

where $\Delta = a_1b_2 - a_2b_1$, $n = a_1 + a_2 + b_1 + b_2$, $n_1 = a_1 + b_1$, $n_2 = a_2 + b_2$, $n_A = a_1 + a_2$, $n_B = b_1 + b_2$. If Yates' correction is applied, (1) is replaced by

(2)
$$\chi^2(\text{corrected}) = \frac{n\left(|\Delta| - \frac{1}{2}n\right)^2}{n_1n_2n_An_B}$$

7.50. Illustrate the result of Problem 7.49 for the data of Problem 7.45.

In Problem 7.45, $a_1 = 75$, $a_2 = 25$, $b_1 = 65$, $b_2 = 35$, $n_1 = 140$, $n_2 = 60$, $n_A = 100$, $n_B = 100$, and $n = 200$; then (1) of Problem 7.49 gives

$$\chi^2 = \frac{200[(75)(35) - (25)(65)]^2}{(140)(60)(100)(100)} = 2.38$$

Using Yates' correction, the result is the same as in Problem 7.46:

$$\chi^2(\text{corrected}) = \frac{n\left(|a_1b_2 - a_2b_1| - \frac{1}{2}n\right)^2}{n_1n_2n_An_B} = \frac{200[|(75)(35) - (25)(65)| - 100]^2}{(140)(60)(100)(100)} = 1.93$$

7.51. Show that a chi-square test involving two sample proportions is equivalent to a significance test of differences in proportions using the normal approximation (see page 217).

Let P_1 and P_2 denote the two sample proportions and p the population proportion. With reference to Problem 7.49, we have

(1)
$$P_1 = \frac{a_1}{n_1}, \quad P_2 = \frac{a_2}{n_2}, \quad 1 - P_1 = \frac{b_1}{n_1}, \quad 1 - P_2 = \frac{b_2}{n_2}$$

(2)
$$p = \frac{n_A}{n}, \quad 1 - p = q = \frac{n_B}{n}$$

so that

$$a_1 = n_1P_1, \quad a_2 = n_2P_2, \quad b_1 = n_1(1 - P_1), \quad b_2 = n_2(1 - P_2) \tag{3}$$

$$n_A = np, \quad n_B = nq \tag{4}$$

Using (3) and (4), we have from Problem 7.49

$$\chi^2 = \frac{n(a_1b_2 - a_2b_1)^2}{n_1n_2n_An_B} = \frac{n[n_1P_1n_2(1 - P_2) - n_2P_2n_1(1 - P_1)]^2}{n_1n_2npnq}$$

$$= \frac{n_1n_2(P_1 - P_2)^2}{npq} = \frac{(P_1 - P_2)^2}{pq(1/n_1 + 1/n_2)} \qquad (\text{since } n = n_1 + n_2)$$

which is the square of the Z statistic given in (10) on page 217.

Coefficient of contingency

7.52. Find the coefficient of contingency for the data in the contingency table of Problem 7.45.

$$C = \sqrt{\frac{\chi^2}{\chi^2 + n}} = \sqrt{\frac{2.38}{2.38 + 200}} = \sqrt{0.01176} = 0.1084$$

7.53. Find the maximum value of C for all 2×2 tables that could arise in Problem 7.13.

The maximum value of C occurs when the two classifications are perfectly dependent or associated. In such cases, all those who take the serum will recover and all those who do not take the serum will not recover. The contingency table then appears as in Table 7-21.

Table 7-21

	Recover	Do Not Recover	TOTAL
Group A (using serum)	100	0	100
Group B (not using serum)	0	100	100
TOTAL	100	100	200

Since the expected cell frequencies, assuming complete independence, are all equal to 50,

$$\chi^2 = \frac{(100 - 50)^2}{50} + \frac{(0 - 50)^2}{50} + \frac{(0 - 50)^2}{50} + \frac{(100 - 50)^2}{50} = 200$$

Then the maximum value of C is $\sqrt{\chi^2/(\chi^2 + n)} = \sqrt{200/(200 + 200)} = 0.7071$.

In general, for perfect dependence in a contingency table where the numbers of rows and columns are both equal to k, the only nonzero cell frequencies occur in the diagonal from upper left to lower right. For such cases, $C_{\max} = \sqrt{(k - 1)/k}$.

Miscellaneous problems

7.54. An instructor gives a short quiz involving 10 true-false questions. To test the hypothesis that the student is guessing, the following decision rule is adopted: (*i*) If 7 or more are correct, the student is not guessing; (*ii*) if fewer than 7 are correct, the student is guessing. Find the probability of rejecting the hypothesis when it is correct.

Let p = probability that a question is answered correctly.

The probability of getting x questions out of 10 correct is ${}_{10}C_x p^x q^{10-x}$, where $q = 1 - p$.

Then under the hypothesis $p = 0.5$ (i.e., the student is guessing),

$P(7 \text{ or more correct}) = P(7 \text{ correct}) + P(8 \text{ correct}) + P(9 \text{ correct}) + P(10 \text{ correct})$

$$= {}_{10}C_7\left(\frac{1}{2}\right)^7\left(\frac{1}{2}\right)^3 + {}_{10}C_8\left(\frac{1}{2}\right)^8\left(\frac{1}{2}\right)^2 + {}_{10}C_9\left(\frac{1}{2}\right)^9\left(\frac{1}{2}\right) + {}_{10}C_{10}\left(\frac{1}{2}\right)^{10} = 0.1719$$

Therefore, the probability of concluding that the student is not guessing when in fact he is guessing is 0.1719. Note that this is the probability of a Type I error.

7.55. In Problem 7.54, find the probability of accepting the hypothesis $p = 0.5$ when actually $p = 0.7$.

Under the hypothesis $p = 0.7$,

$P(\text{less than 7 correct}) = 1 - P(7 \text{ or more correct})$

$$= 1 - [{}_{10}C_7(0.7)^7(0.3)^3 + {}_{10}C_8(0.7)^8(0.3)^2 + {}_{10}C_9(0.7)^9(0.3) + {}_{10}C_{10}(0.7)^{10}] = 0.3504$$

7.56. In Problem 7.54, find the probability of accepting the hypothesis $p = 0.5$ when actually (a) $p = 0.6$, (b) $p = 0.8$, (c) $p = 0.9$, (d) $p = 0.4$, (e) $p = 0.3$, (f) $p = 0.2$, (g) $p = 0.1$.

(a) If $p = 0.6$, the required probability is given by

$$1 - [P(7 \text{ correct}) + P(8 \text{ correct}) + P(9 \text{ correct}) + P(10 \text{ correct})]$$

$$= 1 - [{}_{10}C_7(0.6)^7(0.4)^3 + {}_{10}C_8(0.6)^8(0.4)^2 + {}_{10}C_9(0.6)^9(0.4) + {}_{10}C_{10}(0.6)^{10}] = 0.618$$

The results for (b), (c), . . . , (g) can be similarly found and are indicated in Table 7-22 together with the value corresponding to $p = 0.7$ found in Problem 7.55. Note that the probability is denoted by β (probability of a Type II error). We have also included the entry for $p = 0.5$, given by $\beta = 1 - 0.1719 = 0.828$ from Problem 7.54.

Table 7-22

p	0.1	0.2	0.3	0.4	0.5	0.6	0.7	0.8	0.9
β	1.000	0.999	0.989	0.945	0.828	0.618	0.350	0.121	0.013

7.57. Use Problem 7.56 to construct the graph of β vs. p, the operating characteristic curve of the decision rule in Problem 7.54.

The required graph is shown in Fig. 7-14. Note the similarity with the OC curve of Problem 7.27.

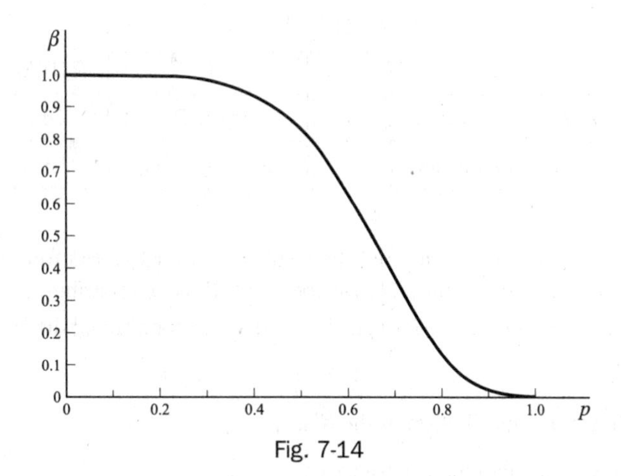

Fig. 7-14

If we had plotted $(1 - \beta)$ vs. p, the *power curve* of the decision rule would have been obtained.
The graph indicates that the given decision rule is *powerful* for rejecting $p = 0.5$ when actually $p \geq 0.8$.

7.58. A coin that is tossed 6 times comes up heads 6 times. Can we conclude at (a) 0.05, (b) 0.01 significance level that the coin is not fair? Consider both a one-tailed and a two-tailed test.

Let p = probability of heads in a single toss of the coin.
Under the hypothesis (H_0: $p = 0.5$) (i.e., the coin is fair),

$$f(x) = P(x \text{ heads in 6 tosses}) = {}_6C_x\left(\frac{1}{2}\right)^x\left(\frac{1}{2}\right)^{6-x} = \frac{{}_6C_x}{64}$$

Then the probabilities of 0, 1, 2, 3, 4, 5, and 6 heads are given, respectively, by $\frac{1}{64}$, $\frac{6}{64}$, $\frac{15}{64}$, $\frac{20}{64}$, $\frac{15}{64}$, $\frac{6}{64}$, and $\frac{1}{64}$.

One-tailed test

Here we wish to decide between the hypotheses (H_0: $p = 0.5$) and (H_1: $p > 0.5$). Since $P(6 \text{ heads}) = \frac{1}{64} = 0.01562$ and $P(5 \text{ or } 6 \text{ heads}) = \frac{6}{64} + \frac{1}{64} = 0.1094$, we can reject H_0 at a 0.05 but not a 0.01 level (i.e., the result observed is significant at a 0.05 but not a 0.01 level).

Two-tailed test

Here we wish to decide between the hypotheses (H_0: $p = 0.5$) and (H_1: $p \neq 0.5$). Since $P(0 \text{ or } 6 \text{ heads}) = \frac{1}{64} + \frac{1}{64} = 0.03125$, we can reject H_0 at a 0.05 but not a 0.01 level.

7.59. Work Problem 7.58 if the coin comes up heads 5 times.

One-tailed test

Since $P(5 \text{ or } 6 \text{ heads}) = \frac{6}{64} + \frac{1}{64} = \frac{7}{64} = 0.1094$, we cannot reject H_0 at a level of 0.05 or 0.01.

Two-tailed test

Since $P(0 \text{ or } 1 \text{ or } 5 \text{ or } 6 \text{ heads}) = 2\left(\frac{7}{64}\right) = 0.2188$, we cannot reject H_0 at a level of 0.05 or 0.01.

7.60. Show that a chi-square test involving only two categories is equivalent to the significance test for proportions (page 216).

If P is the sample proportion for category I, p is the population proportion, and n is the total frequency, we can describe the situation by means of Table 7-23. Then by definition,

$$\chi^2 = \frac{(nP - np)^2}{np} + \frac{[n(1-P) - n(1-p)]^2}{nq}$$

$$= \frac{n^2(P-p)^2}{np} + \frac{n^2(P-p)^2}{nq} = n(P-p)^2\left(\frac{1}{p} + \frac{1}{q}\right) = \frac{n(P-p)^2}{pq} = \frac{(P-p)^2}{pq/n}$$

which is the square of the Z statistic (5) on page 216.

Table 7-23

	I	II	TOTAL
Observed Frequency	nP	$n(1-P)$	n
Expected Frequency	np	$n(1-p) = nq$	n

7.61. Suppose $X_1, X_2, \ldots, X_k$ have a multinomial distribution, with expected frequencies $np_1, np_2, \ldots, np_k$, respectively. Let $Y_1, Y_2, \ldots, Y_k$ be mutually independent, Poisson-distributed variables, with parameters $\lambda_1 = np_1, \lambda_2 = np_2, \ldots, \lambda_k = np_k$, respectively. Prove that the conditional distribution of the Y's given that

$$Y_1 + Y_2 + \cdots + Y_k = n$$

is precisely the multinomial distribution of the X's.

For the joint probability function of the Y's, we have

$$(1) \qquad P(Y_1 = y_1, Y_2 = y_2, \ldots, Y_k = y_k) = \left[\frac{(np_1)^{y_1}e^{-np_1}}{y_1!}\right]\left[\frac{(np_2)^{y_2}e^{-np_2}}{y_2!}\right]\cdots\left[\frac{(np_k)^{y_k}e^{-np_k}}{y_k!}\right]$$

$$= \frac{n^{y_1+y_2+\cdots+y_k}p_1^{y_1}p_2^{y_2}\cdots p_k^{y_k}}{y_1!y_2!\cdots y_k!}e^{-n}$$

where we have used the fact that $p_1 + p_2 + \cdots + p_k = 1$. The conditional distribution we are looking for is given by

(2) $$P(Y_1 = y_1, Y_2 = y_2, \ldots, Y_k = y_k | Y_1 + Y_2 + \cdots + Y_k = n)$$
$$= \frac{P(Y_1 = y_1, Y_2 = y_2, \ldots, Y_k = y_k \text{ and } Y_1 + Y_2 + \cdots + Y_k = n)}{P(Y_1 + Y_2 + \cdots + Y_k = n)}$$

Now, the numerator in (2) has, from (1), the value

$$\frac{n^n p_1^{y_1} p_2^{y_2} \cdots p_k^{y_k}}{y_1! y_2! \cdots y_k!} e^{-n}$$

As for the denominator, we know from Problem 4.94, page 146, that $Y_1 + Y_2 + \cdots + Y_k$ is itself a Poisson variable with parameter $np_1 + np_2 + \cdots + np_k = n$. Hence, the denominator has the value

$$\frac{n^n e^{-n}}{n!}$$

Therefore, (2) becomes

$$P(Y_1 = y_1, Y_2 = y_2, \ldots, Y_k = y_k | Y_1 + Y_2 + \cdots + Y_k = n) = \frac{n!}{y_1! y_2! \cdots y_k!} p_1^{y_1} p_2^{y_2} \cdots p_k^{y_k}$$

which is just the multinomial distribution of the X's [compare (16), page 112].

7.62. Use the result of Problem 7.61 to show that χ^2, as defined by (21), page 220, is approximately chi-square distributed.

As it stands, (21) is difficult to deal with because the multinomially distributed X's are dependent, in view of the restriction (22). However, Problem 7.61 shows that we can replace the X's by the *independent*, Poisson-distributed Y's if it is given that $Y_1 + Y_2 + \cdots + Y_k = n$..Therefore, we rewrite (21) as

(1) $$\chi^2 = \left(\frac{Y_1 - \lambda_1}{\sqrt{\lambda_1}}\right)^2 + \left(\frac{Y_2 - \lambda_2}{\sqrt{\lambda_2}}\right)^2 + \cdots + \left(\frac{Y_k - \lambda_k}{\sqrt{\lambda_k}}\right)^2$$

As $n \to \infty$, all the λ's approach ∞, and the central limit theorem for the Poisson distribution [(14), page 112] gives

(2) $$\chi^2 \approx Z_1^2 + Z_2^2 + \cdots + Z_k^2$$

where the Z's are *independent* normal variables having mean 0 and variance 1 whose distribution is conditional upon the event

(3) $$\sqrt{\lambda_1} Z_1 + \sqrt{\lambda_2} Z_2 + \cdots + \sqrt{\lambda_k} Z_k = 0 \quad \text{or} \quad \sqrt{p_1} Z_1 + \sqrt{p_2} Z_2 + \cdots + \sqrt{p_k} Z_k = 0$$

or, since the random variables are continuous,

(4) $$|\sqrt{p_1} Z_1 + \sqrt{p_2} Z_2 + \cdots + \sqrt{p_k} Z_k| < \epsilon$$

Let us denote by $F_\nu(x)$ the cumulative distribution function for a chi-square variable with ν degrees of freedom. Then what we want to prove is

(5) $$P\left(Z_1^2 + Z_2^2 + \cdots + Z_k^2 \le x \,\middle|\, |\sqrt{p_1} Z_1 + \sqrt{p_2} Z_2 + \cdots + \sqrt{p_k} Z_k| < \epsilon\right)$$
$$= \frac{P\left(Z_1^2 + Z_2^2 + \cdots + Z_k^2 \le x \text{ and } |\sqrt{p_1} Z_1 + \sqrt{p_2} Z_2 + \cdots + \sqrt{p_k} Z_k| < \epsilon\right)}{P(|\sqrt{p_1} Z_1 + \sqrt{p_2} Z_2 + \cdots + \sqrt{p_k} Z_k| < \epsilon)}$$
$$= F_\nu(x)$$

for a suitable value of ν.

It is easy to establish (5) if we use our geometrical intuition. First of all, Theorem 4-3 shows that the unconditional distribution of $Z_1^2 + Z_2^2 + \cdots + Z_k^2$ is chi-square with k degrees of freedom. Hence, since the density function for each Z_j is $(2\pi)^{-1/2} e^{-z^2/2}$,

(6) $$F_k(x) = (2\pi)^{-k/2} \int \cdots \int_{z_1^2+z_2^2+\cdots+z_k^2 \le x} e^{-(z_1^2+z_2^2+\cdots+z_k^2)/2} dz_1 dz_2 \cdots dz_k$$

Furthermore, we have for the numerator in (5):

(7) $$\text{Numerator} = (2\pi)^{-k/2} \int \cdots \int_{\substack{z_1^2+z_2^2+\cdots+z_k^2 \le x, \\ |\sqrt{p_1}z_1+\sqrt{p_2}z_2+\cdots+\sqrt{p_k}z_k| < \epsilon}} e^{-(z_1^2+z_2^2+\cdots+z_k^2)/2} dz_1 dz_2 \cdots dz_k$$

We recall from analytic geometry that in three-dimensional space, $x_1^2 + x_2^2 + x_3^2 \le a^2$ represents a spherical solid of radius a centered at the origin, while $\alpha_1 x_1 + \alpha_2 x_2 + \alpha_3 x_3 = 0$ is a plane through the origin whose normal is the unit vector $(\alpha_1, \alpha_2, \alpha_3)$. Figure 7-15 shows the intersection of the two bodies. It is obvious that when a function which depends only on distance from the origin, i.e.,

$$f(r) \qquad \text{where} \qquad r = \sqrt{x_1^2 + x_2^2 + x_3^2}$$

is integrated over the circular area—or throughout a thin slab lying on that area—the value of the integral is completely independent of the direction-cosines $\alpha_1, \alpha_2, \alpha_3$. In other words, all cutting planes through the origin give the same integral.

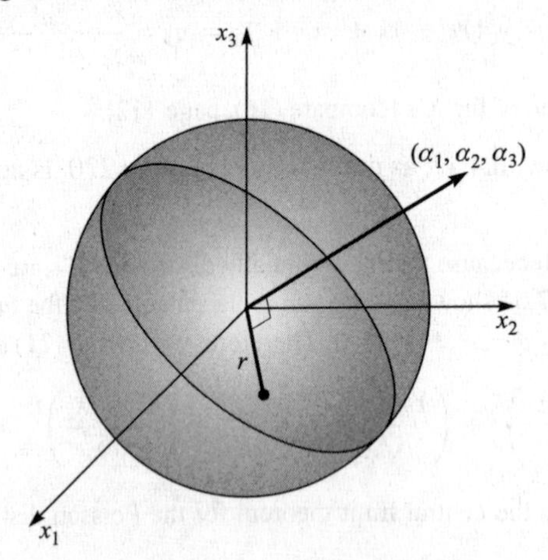

Fig. 7-15

By analogy we conclude that in (7), where $e^{-r^2/2}$ is integrated over the intersection of a *hypersphere* about the origin and a *hyperplane* through the origin, the p's may be given any convenient values. We choose

$$p_1 = p_2 = \cdots = p_{k-1} = 0, \qquad p_k = 1$$

and obtain

(8) $$\text{Numerator} = (2\pi)^{-k/2} \int \cdots \int_{z_1^2+z_2^2+\cdots+z_{k-1}^2 \le x} e^{-(z_1^2+z_2^2+\cdots+z_{k-1}^2)/2} dz_1 dz_2 \cdots dz_{k-1}(2\epsilon)$$
$$= (2\pi)^{-1/2} F_{k-1}(x)(2\epsilon)$$

using (6). The factor 2ϵ is the thickness of the slab.

To evaluate the denominator in (5), we note that the random variable

$$W = \sqrt{p_1}Z_1 + \sqrt{p_2}Z_2 + \cdots + \sqrt{p_k}Z_k$$

is normal (because it is a linear combination of the independent, normal Z's), and that

$$E(W) = \sqrt{p_1}(0) + \sqrt{p_2}(0) + \cdots + \sqrt{p_k}(0) = 0$$
$$\text{Var}(W) = p_1(1) + p_2(1) + \cdots + p_k(1) = 1$$

Therefore, the density function for W is $\phi(w) = (2\pi)^{-1/2}e^{-w^2/2}$, and

(9) $$\text{Denominator} = P(|W| < \epsilon) = \phi(0)(2\epsilon) = (2\pi)^{-1/2}(2\epsilon)$$

Dividing (8) by (9), we obtain the desired result, where $\nu = k - 1$.

The above "proof" (which can be made rigorous) shows incidentally that every linear constraint placed on the Z's, and hence on the Y's or X's, reduces the number of degrees of freedom in χ^2 by 1. This provides the basis for the rules given on page 221.

SUPPLEMENTARY PROBLEMS

Tests of means and proportions using normal distributions

7.63. An urn contains marbles that are either red or blue. To test the hypothesis of equal proportions of these colors, we agree to sample 64 marbles with replacement, noting the colors drawn and adopt the following decision rule: (1) accept the hypothesis if between 28 and 36 red marbles are drawn; (2) reject the hypothesis otherwise. (a) Find the probability of rejecting the hypothesis when it is actually correct. (b) Interpret graphically the decision rule and the result obtained in (a).

7.64. (a) What decision rule would you adopt in Problem 7.63 if you require the probability of rejecting the hypothesis when it is actually correct to be at most 0.01, i.e., you want a 0.01 level of significance? (b) At what level of confidence would you accept the hypothesis? (c) What would be the decision rule if a 0.05 level of significance were adopted?

7.65. Suppose that in Problem 7.63 you wish to test the hypothesis that there is a greater proportion of red than blue marbles. (a) What would you take as the null hypothesis and what would be the alternative? (b) Should you use a one- or two-tailed test? Why? (c) What decision rule should you adopt if the level of significance is 0.05? (d) What is the decision rule if the level of significance is 0.01?

7.66. A pair of dice is tossed 100 times, and it is observed that sevens appear 23 times. Test the hypothesis that the dice are fair, i.e., not loaded, using (a) a two-tailed test and (b) a one-tailed test, both with a significance level of 0.05. Discuss your reasons, if any, for preferring one of these tests over the other.

7.67. Work Problem 7.66 if the level of significance is 0.01.

7.68. A manufacturer claimed that at least 95% of the equipment which he supplied to a factory conformed to specifications. An examination of a sample of 200 pieces of equipment revealed that 18 were faulty. Test his claim at a significance level of (a) 0.01, (b) 0.05.

7.69. It has been found from experience that the mean breaking strength of a particular brand of thread is 9.72 oz with a standard deviation of 1.4 oz. Recently a sample of 36 pieces of thread showed a mean breaking strength of 8.93 oz. Can one conclude at a significance level of (a) 0.05, (b) 0.01 that the thread has become inferior?

7.70. On an examination given to students at a large number of different schools, the mean grade was 74.5 and the standard deviation was 8.0. At one particular school where 200 students took the examination, the mean grade was 75.9. Discuss the significance of this result at a 0.05 level from the viewpoint of (a) a one-tailed test, (b) a two-tailed test, explaining carefully your conclusions on the basis of these tests.

7.71. Answer Problem 7.70 if the significance level is 0.01.

Tests involving differences of means and proportions

7.72. A sample of 100 electric light bulbs produced by manufacturer A showed a mean lifetime of 1190 hours and a standard deviation of 90 hours. A sample of 75 bulbs produced by manufacturer B showed a mean lifetime of 1230 hours with a standard deviation of 120 hours. Is there a difference between the mean lifetimes of the two brands of bulbs at a significance level of (a) 0.05, (b) 0.01?

7.73. In Problem 7.72 test the hypothesis that the bulbs of manufacturer B are superior to those of manufacturer A using a significance level of (a) 0.05, (b) 0.01. Explain the differences between this and what was asked in Problem 7.72. Do the results contradict those of Problem 7.72?

7.74. On an elementary school examination in spelling, the mean grade of 32 boys was 72 with a standard deviation of 8, while the mean grade of 36 girls was 75 with a standard deviation of 6. Test the hypothesis at a (a) 0.05, (b) 0.01 level of significance that the girls are better in spelling than the boys.

7.75. To test the effects of a new fertilizer on wheat production, a tract of land was divided into 60 squares of equal areas, all portions having identical qualities as to soil, exposure to sunlight, etc. The new fertilizer was applied to 30 squares, and the old fertilizer was applied to the remaining squares. The mean number of bushels of wheat harvested per square of land using the new fertilizer was 18.2, with a standard deviation of 0.63 bushels. The corresponding mean and standard deviation for the squares using the old fertilizer were 17.8 and 0.54 bushels, respectively. Using a significance level of (a) 0.05, (b) 0.01, test the hypothesis that the new fertilizer is better than the old one.

7.76. Random samples of 200 bolts manufactured by machine A and 100 bolts manufactured by machine B showed 19 and 5 defective bolts, respectively. Test the hypothesis that (a) the two machines are showing different qualities of performance, (b) machine B is performing better than A. Use a 0.05 level of significance.

Tests involving student's *t* distribution

7.77. The mean lifetime of electric light bulbs produced by a company has in the past been 1120 hours with a standard deviation of 125 hours. A sample of 8 electric light bulbs recently chosen from a supply of newly produced bulbs showed a mean lifetime of 1070 hours. Test the hypothesis that the mean lifetime of the bulbs has not changed, using a level of significance of (a) 0.05, (b) 0.01.

7.78. In Problem 7.77 test the hypothesis $\mu = 1120$ hours against the alternative hypothesis $\mu < 1120$ hours, using a significance level of (a) 0.05, (b) 0.01.

7.79. The specifications for the production of a certain alloy call for 23.2% copper. A sample of 10 analyses of the product showed a mean copper content of 23.5% and a standard deviation of 0.24%. Can we conclude at a (a) 0.01, (b) 0.05 significance level that the product meets the required specifications?

7.80. In Problem 7.79 test the hypothesis that the mean copper content is higher than in required specifications, using a significance level of (a) 0.01, (b) 0.05.

7.81. An efficiency expert claims that by introducing a new type of machinery into a production process, he can decrease substantially the time required for production. Because of the expense involved in maintenance of the machines, management feels that unless the production time can be decreased by at least 8.0%, they cannot afford to introduce the process. Six resulting experiments show that the time for production is decreased by 8.4% with standard deviation of 0.32%. Using a level of significance of (a) 0.01, (b) 0.05, test the hypothesis that the process should be introduced.

7.82. Two types of chemical solutions, A and B, were tested for their pH (degree of acidity of the solution). Analysis of 6 samples of A showed a mean pH of 7.52 with a standard deviation of 0.024. Analysis of 5 samples of B showed a mean pH of 7.49 with a standard deviation of 0.032. Using a 0.05 significance level, determine whether the two types of solutions have different pH values.

7.83. On an examination in psychology 12 students in one class had a mean grade of 78 with a standard deviation of 6, while 15 students in another class had a mean grade of 74 with a standard deviation of 8. Using a significance level of 0.05, determine whether the first group is superior to the second group.

Tests involving the chi-square distribution

7.84. The standard deviation of the breaking strengths of certain cables produced by a company is given as 240 lb. After a change was introduced in the process of manufacture of these cables, the breaking strengths of a sample of 8 cables showed a standard deviation of 300 lb. Investigate the significance of the apparent increase in variability, using a significance level of (a) 0.05, (b) 0.01.

7.85. The annual temperature of a city is obtained by finding the mean of the mean temperatures on the 15th day of each month. The standard deviation of the annual temperatures of the city over a period of 100 years was 16° Fahrenheit. During the last 15 years a standard deviation of annual temperatures was computed as 10° Fahrenheit. Test the hypothesis that the temperatures in the city have become less variable than in the past, using a significance level of (a) 0.05, (b) 0.01.

7.86. In Problem 7.77 a sample of 20 electric light bulbs revealed a standard deviation in the lifetimes of 150 hours. Would you conclude that this is unusual? Explain.

Tests involving the *F* distribution

7.87. Two samples consisting of 21 and 9 observations have variances given by $s_1^2 = 16$ and $s_2^2 = 8$, respectively. Test the hypothesis that the first population variance is greater than the second at a (a) 0.05, (b) 0.01 level of significance.

7.88. Work Problem 7.87 if the two samples consist of 60 and 120 observations, respectively.

7.89. In Problem 7.82 can we conclude that there is a significant difference in the variability of the pH values for the two solutions at a 0.10 level of significance?

Operating characteristic curves

7.90. Referring to Problem 7.63, determine the probability of accepting the hypothesis that there are equal proportions of red and blue marbles when the actual proportion p of red marbles is (a) 0.6, (b) 0.7, (c) 0.8, (d) 0.9, (e) 0.3.

7.91. Represent the results of Problem 7.90 by constructing a graph of (a) β vs. p, (b) $(1 - \beta)$ vs. p. Compare these graphs with those of Problem 7.25 by considering the analogy of red and blue marbles to heads and tails, respectively.

Quality control charts

7.92. In the past a certain type of thread produced by a manufacturer has had a mean breaking strength of 8.64 oz and a standard deviation of 1.28 oz. To determine whether the product is conforming to standards, a sample of 16 pieces of thread is taken every 3 hours and the mean breaking strength is determined. Find the (a) 99.73% or 3σ (b) 99% and (c) 95% control limits on a quality control chart and explain their applications.

7.93. On the average about 3% of the bolts produced by a company are defective. To maintain this quality of performance, a sample of 200 bolts produced is examined every 4 hours. Determine (a) 99%, (b) 95% control limits for the number of defective bolts in each sample. Note that only *upper control limits* are needed in this case.

Fitting of data by theoretical distributions

7.94. Fit a binomial distribution to the data of Table 7-24.

Table 7-24

x	0	1	2	3	4
f	30	62	46	10	2

7.95. Fit a normal distribution to the data of Problem 5.98.

7.96. Fit a normal distribution to the data of Problem 5.100.

7.97. Fit a Poisson distribution to the data of Problem 7.44, and compare with the fit obtained by using the binomial distribution.

7.98. In 10 Prussian army corps over a period of 20 years from 1875 throughout 1894, the number of deaths per army corps per year resulting from the kick of a horse are given in Table 7-25. Fit a Poisson distribution to the data.

Table 7-25

x	0	1	2	3	4
f	109	65	22	3	1

The chi-square test

7.99. In 60 tosses of a coin, 37 heads and 23 tails were observed. Test the hypothesis that the coin is fair using a significance level of (a) 0.05, (b) 0.01.

7.100. Work Problem 7.99 using Yates' correction.

7.101. Over a long period of time the grades given by a group of instructors in a particular course have averaged 12% A's, 18% B's, 40% C's, 18% D's, and 12% F's. A new instructor gives 22 A's, 34 B's, 66 C's, 16 D's, and 12 F's during two semesters. Determine at a 0.05 significance level whether the new instructor is following the grade pattern set by the others.

7.102. Three coins were tossed together a total of 240 times, and each time the number of heads turning up was observed. The results are shown in Table 7-26 together with results expected under the hypothesis that the coins are fair. Test this hypothesis at a significance level of 0.05.

Table 7-26

	0 heads	1 head	2 heads	3 heads
Observed Frequency	24	108	95	23
Expected Frequency	30	90	90	30

7.103. The number of books borrowed from a public library during a particular week is given in Table 7-27. Test the hypothesis that the number of books borrowed does not depend on the day of the week, using a significance level of (a) 0.05, (b) 0.01.

Table 7-27

	Mon.	Tues.	Wed.	Thurs.	Fri.
Number of Books Borrowed	135	108	120	114	146

7.104. An urn consists of 6 red marbles and 3 white ones. Two marbles are selected at random from the urn, their colors are noted, and then the marbles are replaced in the urn. This process is performed a total of 120 times, and the results obtained are shown in Table 7-28. (a) Determine the expected frequencies. (b) Determine at a level of significance of 0.05 whether the results obtained are consistent with those expected.

Table 7-28

	0 red, 2 white	1 red, 1 white	2 red, 0 white
Number of Drawings	6	53	61

7.105. Two hundred bolts were selected at random from the production of each of 4 machines. The numbers of defective bolts found were 2, 9, 10, 3. Determine whether there is a significant difference between the machines using a significance level of 0.05.

Goodness of fit

7.106. (a) Use the chi-square test to determine the goodness of fit of the data of Problem 7.94. (b) Is the fit "too good"? Use a 0.05 level of significance.

7.107. Use the chi-square test to determine the goodness of fit of the data referred to in (a) Problem 7.95, (b) Problem 7.96. Use a level of significance of 0.05 and in each case determine whether the fit is "too good."

7.108. Use the chi-square test to determine the goodness of fit of the data referred to in (a) Problem 7.97, (b) Problem 7.98. Is your result in (a) consistent with that of Problem 7.106?

Contingency tables

7.109. Table 7-29 shows the result of an experiment to investigate the effect of vaccination of laboratory animals against a particular disease. Using an (a) 0.01, (b) 0.05 significance level, test the hypothesis that there is no difference between the vaccinated and unvaccinated groups, i.e., vaccination and this disease are independent.

Table 7-29

	Got Disease	Did Not Get Disease
Vaccinated	9	42
Not Vaccinated	17	28

7.110. Work Problem 7.109 using Yates' correction.

7.111. Table 7-30 shows the numbers of students in each of two classes, *A* and *B*, who passed and failed an examination given to both groups. Using an (a) 0.05, (b) 0.01 significance level, test the hypothesis that there is no difference between the two classes. Work the problem with and without Yates' correction.

Table 7-30

	Passed	Failed
Class A	72	17
Class B	64	23

7.112. Of a group of patients who complained that they did not sleep well, some were given sleeping pills while others were given sugar pills (although they all *thought* they were getting sleeping pills). They were later asked whether the pills helped them or not. The results of their responses are shown in Table 7-31. Assuming that all patients told the truth, test the hypothesis that there is no difference between sleeping pills and sugar pills at a significance level of 0.05.

Table 7-31

	Slept Well	Did Not Sleep Well
Took Sleeping Pills	44	10
Took Sugar Pills	81	35

7.113. On a particular proposal of national importance, Democrats and Republicans cast votes as indicated in Table 7-32. At a level of significance of (a) 0.01, (b) 0.05, test the hypothesis that there is no difference between the two parties insofar as this proposal is concerned.

Table 7-32

	In Favor	Opposed	Undecided
Democrats	85	78	37
Republicans	118	61	25

7.114. Table 7-33 shows the relation between the performances of students in mathematics and physics. Test the hypothesis that performance in physics is independent of performance in mathematics, using (a) 0.05, (b) 0.01 significance level.

Table 7-33

		MATHEMATICS		
		High Grades	Medium Grades	Low Grades
PHYSICS	High Grades	56	71	12
	Medium Grades	47	163	38
	Low Grades	14	42	85

7.115. The results of a survey made to determine whether the age of a driver 21 years of age or older has any effect on the number of automobile accidents in which he is involved (including all minor accidents) are indicated in Table 7-34. At a level of significance of (a) 0.05, (b) 0.01, test the hypothesis that the number of accidents is independent of the age of the driver. What possible sources of difficulty in sampling techniques, as well as other considerations, could affect your conclusions?

Table 7-34

		AGE OF DRIVER				
		21–30	31–40	41–50	51–60	61–70
NUMBER OF ACCIDENTS	0	748	821	786	720	672
	1	74	60	51	66	50
	2	31	25	22	16	15
	More than 2	9	10	6	5	7

Coefficient of contingency

7.116. Table 7-35 shows the relationship between hair and eye color of a sample of 200 women. (a) Find the coefficient of contingency without and with Yates' correction. (b) Compare the result of (a) with the maximum coefficient of contingency.

Table 7-35

		HAIR COLOR	
		Blonde	Not Blonde
EYE COLOR	Blue	49	25
	Not Blue	30	96

7.117. Find the coefficient of contingency for the data of (a) Problem 7.109, (b) Problem 7.111 without and with Yates' correction.

7.118. Find the coefficient of contingency for the data of Problem 7.114.

Miscellaneous problems

7.119. Two urns, A and B, contain equal numbers of marbles, but the proportions of red and white marbles in each of the urns is unknown. A sample of 50 marbles selected with replacement from each of the urns revealed 32 red marbles from A and 23 red marbles from B. Using a significance level of 0.05, test the hypothesis that (a) the two urns have equal proportions of marbles and (b) A has a greater proportion of red marbles than B.

7.120. Referring to Problem 7.54, find the least number of questions a student must answer correctly before the instructor is sure at a significance level of (a) 0.05, (b) 0.01, (c) 0.001, (d) 0.06 that the student is not merely guessing. Discuss the results.

7.121. A coin that is tossed 8 times comes up heads 7 times. Can we reject the hypothesis that the coin is fair at a significance level of (a) 0.05? (b) 0.10? (c) 0.01? Use a two-tailed test.

7.122. The percentage of A's given in a physics course at a certain university over a long period of time was 10%. During one particular term there were 40 A's in a group of 300 students. Test the significance of this result at a level of (a) 0.05, (b) 0.01.

7.123. Using brand A gasoline, the mean number of miles per gallon traveled by 5 similar automobiles under identical conditions was 22.6 with a standard deviation of 0.48. Using brand B, the mean number was 21.4 with a standard deviation of 0.54. Choosing a significance level of 0.05, investigate whether brand A is really better than brand B in providing more mileage to the gallon.

7.124. In Problem 7.123 is there greater variability in miles per gallon using brand B than there is using brand A? Explain.

ANSWERS TO SUPPLEMENTARY PROBLEMS

7.63. (a) 0.2606.

7.64. (a) Accept the hypothesis if between 22 and 42 red marbles are drawn; reject it otherwise. (b) 0.99. (c) Accept the hypothesis if between 24 and 40 red marbles are drawn; reject it otherwise.

7.65. (a) (H_0: $p = 0.5$), (H_1: $p > 0.5$). (b) One-tailed test. (c) Reject H_0 if more than 39 red marbles are drawn, and accept it otherwise (or withhold decision). (d) Reject H_0 if more than 41 red marbles are drawn, and accept it otherwise (or withhold decision).

7.66. (a) We cannot reject the hypothesis at a 0.05 level.
(b) We can reject the hypothesis at a 0.05 level.

7.67. We cannot reject the hypothesis at a 0.01 level in either (a) or (b).

7.68. We can reject the claim at both levels using a one-tailed test.

7.69. Yes, at both levels, using a one-tailed test in each case.

7.70. The result is significant at a 0.05 level in both a one-tailed and two-tailed test.

7.71. The result is significant at a 0.01 level in a one-tailed test but not in a two-tailed test.

7.72. (a) Yes. (b) No.

7.73. A one-tailed test at both levels of significance shows that brand *B* is superior to brand *A*.

7.74. A one-tailed test shows that the difference is significant at a 0.05 level but not a 0.01 level.

7.75. A one-tailed test shows that the new fertilizer is superior at both levels of significance.

7.76. (a) A two-tailed test shows no difference in quality of performance at a 0.05 level.
(b) A one-tailed test shows that *B* is not performing better than *A* at a 0.05 level.

7.77. A two-tailed test shows that there is no evidence at either level that the mean lifetime has changed.

7.78. A one-tailed test indicates no decrease in the mean at either the 0.05 or 0.01 level.

7.79. A two-tailed test at both levels shows that the product does not meet specifications.

7.80. A one-tailed test at both levels shows that the mean copper content is higher than specifications require.

7.81. A one-tailed test shows that the process should not be introduced if the significance level adopted is 0.01 but should be introduced if the significance level adopted is 0.05.

7.82. Using a two-tailed test at a 0.05 level of significance, we would not conclude that there is a difference in acidity.

7.83. Using a one-tailed test at a 0.05 level of significance, we would conclude that the first group is not superior to the second.

7.84. The apparent increase in variability is not significant at either level.

7.85. The apparent decrease is significant at the 0.05 level but not at the 0.01 level.

7.86. We would conclude that the result is unusual at the 0.05 level but not at the 0.01 level.

7.87. We cannot conclude that the first variance is greater than the second at either level.

7.88. We can conclude that the first variance is greater than the second at both levels.

7.89. No. **7.90.** (a) 0.3112 (b) 0.0118 (c) 0 (d) 0 (e) 0.0118.

7.92. (a) 8.64 $\pm$ 0.96 oz (b) 8.64 $\pm$ 0.83 oz (c) 8.64 $\pm$ 0.63 oz **7.93.** (a) 6 (b) 4

7.94. $f(x) = {}_4C_x(0.32)^x(0.68)^{4-x}$; expected frequencies are 32, 60, 43, 13, and 2, respectively.

7.95. Expected frequencies are 1.7, 5.5, 12.0, 15.9, 13.7, 7.6, 2.7, and 0.6, respectively.

7.96. Expected frequencies are 1.1, 4.0, 11.1, 23.9, 39.5, 50.2, 49.0, 36.6, 21.1, 9.4, 3.1, and 1.0, respectively.

7.97. Expected frequencies are 41.7, 53.4, 34.2, 14.6, and 4.7, respectively.

7.98. $f(x) = \dfrac{(0.61)^x e^{-0.61}}{x!}$; expected frequencies are 108.7, 66.3, 20.2, 4.1, and 0.7, respectively.

7.99. The hypothesis cannot be rejected at either level.

7.100. The conclusion is the same as before.

7.101. The new instructor is not following the grade pattern of the others. (The fact that the grades happen to be better than average *may* be due to better teaching ability or lower standards or both.)

7.102. There is no reason to reject the hypothesis that the coins are fair.

7.103. There is no reason to reject the hypothesis at either level.

7.104. (a) 10, 60, 50, respectively (b) The hypothesis that the results are the same as those expected cannot be rejected at a 0.05 level of significance.

7.105. The difference is significant at the 0.05 level. **7.106.** (a) The fit is good. (b) No.

7.107. (a) The fit is "too good." (b) The fit is poor at the 0.05 level.

7.108. (a) The fit is very poor at the 0.05 level. Since the binomial distribution gives a good fit of the data, this is consistent with Problem 7.109. (b) The fit is good but not "too good."

7.109. The hypothesis can be rejected at the 0.05 level but not at the 0.01 level.

7.110. Same conclusion. **7.111.** The hypothesis cannot be rejected at either level.

7.112. The hypothesis cannot be rejected at the 0.05 level.

7.113. The hypothesis can be rejected at both levels.

7.114. The hypothesis can be rejected at both levels.

7.115. The hypothesis cannot be rejected at either level.

7.116. (a) 0.3863, 0.3779 (with Yates' correction)

7.117. (a) 0.2205, 0.1985 (corrected) (b) 0.0872, 0.0738 (corrected) **7.118.** 0.4651

7.119. (a) A two-tailed test at a 0.05 level fails to reject the hypothesis of equal proportions.
(b) A one-tailed test at a 0.05 level indicates that *A* has a greater proportion of red marbles than *B*.

7.120. (a) 9 (b) 10 (c) 10 (d) 8 **7.121.** (a) No. (b) Yes. (c) No.

7.122. Using a one-tailed test, the result is significant at the 0.05 level but is not significant at the 0.01 level.

7.123. We can conclude that brand *A* is better than brand *B* at the 0.05 level.

7.124. Not at the 0.05 level.

Curve Fitting, Regression, and Correlation

Curve Fitting

Very often in practice a relationship is found to exist between two (or more) variables, and one wishes to express this relationship in mathematical form by determining an equation connecting the variables.

A first step is the collection of data showing corresponding values of the variables. For example, suppose x and y denote, respectively, the height and weight of an adult male. Then a sample of n individuals would reveal the heights $x_1, x_2, \ldots, x_n$ and the corresponding weights $y_1, y_2, \ldots, y_n$.

A next step is to plot the points $(x_1, y_1), (x_2, y_2), \ldots, (x_n, y_n)$ on a rectangular coordinate system. The resulting set of points is sometimes called a *scatter diagram*.

From the scatter diagram it is often possible to visualize a smooth curve approximating the data. Such a curve is called an *approximating curve*. In Fig. 8-1, for example, the data appear to be approximated well by a straight line, and we say that a *linear relationship* exists between the variables. In Fig. 8-2, however, although a relationship exists between the variables, it is not a linear relationship and so we call it a *nonlinear relationship*. In Fig. 8-3 there appears to be no relationship between the variables.

The general problem of finding equations of approximating curves that fit given sets of data is called *curve fitting*. In practice the type of equation is often suggested from the scatter diagram. For Fig. 8-1 we could use a straight line

$$y = a + bx \tag{1}$$

while for Fig. 8-2 we could try a *parabola* or *quadratic curve*:

$$y = a + bx + cx^2 \tag{2}$$

Sometimes it helps to plot scatter diagrams in terms of *transformed variables*. For example, if $\log y$ vs. x leads to a straight line, we would try $\log y = a + bx$ as an equation for the approximating curve.

Regression

One of the main purposes of curve fitting is to estimate one of the variables (the *dependent variable*) from the other (the *independent variable*). The process of estimation is often referred to as *regression*. If y is to be estimated from x by means of some equation, we call the equation a *regression equation of y on x* and the corresponding curve a *regression curve of y on x*.

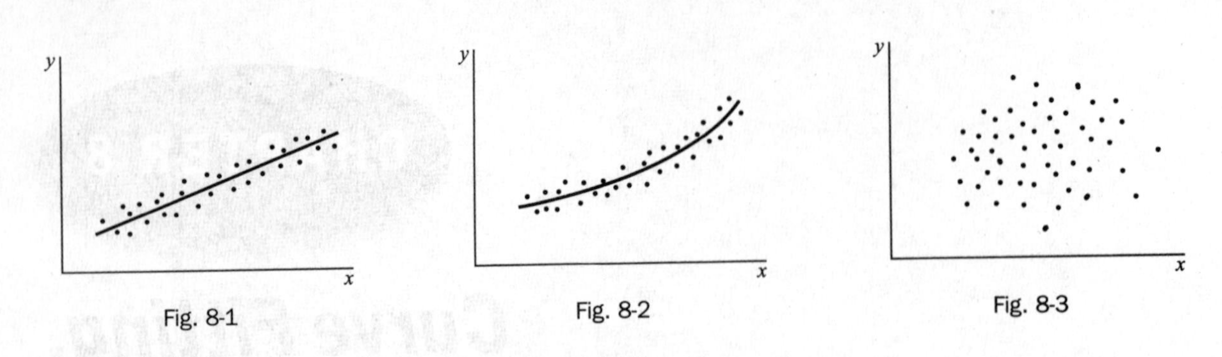

Fig. 8-1 Fig. 8-2 Fig. 8-3

The Method of Least Squares

Generally, more than one curve of a given type will appear to fit a set of data. To avoid individual judgment in constructing lines, parabolas, or other approximating curves, it is necessary to agree on a definition of a "best-fitting line," "best-fitting parabola," etc.

To motivate a possible definition, consider Fig. 8-4 in which the data points are $(x_1, y_1), \ldots, (x_n, y_n)$. For a given value of x, say, x_1, there will be a difference between the value y_1 and the corresponding value as determined from the curve C. We denote this difference by d_1, which is sometimes referred to as a *deviation*, *error*, or *residual* and may be positive, negative, or zero. Similarly, corresponding to the values $x_2, \ldots, x_n$, we obtain the deviations $d_2, \ldots, d_n$.

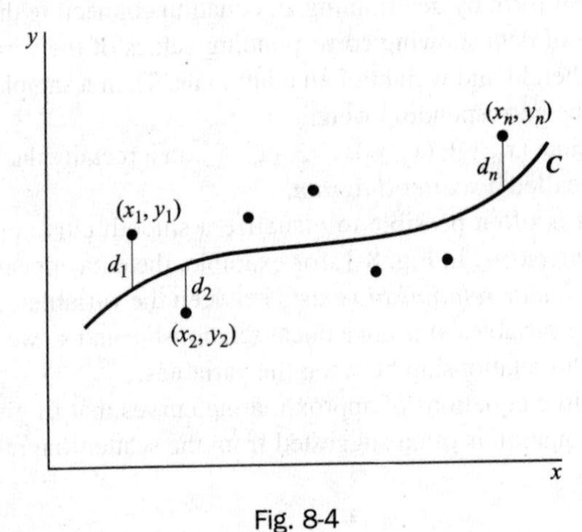

Fig. 8-4

A measure of the goodness of fit of the curve C to the set of data is provided by the quantity $d_1^2 + d_2^2 + \cdots + d_n^2$. If this is small, the fit is good, if it is large, the fit is bad. We therefore make the following definition.

Definition Of all curves in a given family of curves approximating a set of n data points, a curve having the property that

$$d_1^2 + d_2^2 + \cdots + d_n^2 = \text{a minimum}$$

is called a *best-fitting curve* in the family.

A curve having this property is said to fit the data in the *least-squares sense* and is called a *least-squares regression curve*, or simply a *least-squares curve*. A line having this property is called a *least-squares line*; a parabola with this property is called a *least-squares parabola*, etc.

It is customary to employ the above definition when x is the independent variable and y is the dependent variable. If x is the dependent variable, the definition is modified by considering horizontal instead of vertical deviations, which amounts to interchanging the x and y axes. These two definitions lead in general to two different least-squares curves. Unless otherwise specified, we shall consider y as the dependent and x as the independent variable.

It is possible to define another least-squares curve by considering perpendicular distances from the data points to the curve instead of either vertical or horizontal distances. However, this is not used very often.

The Least-Squares Line

By using the above definition, we can show (see Problem 8.3) that the least-squares line approximating the set of points $(x_1, y_1), \ldots, (x_n, y_n)$ has the equation

$$y = a + bx \tag{3}$$

where the constants a and b are determined by solving simultaneously the equations

$$\begin{aligned} \sum y &= an + b\sum x \\ \sum xy &= a\sum x + b\sum x^2 \end{aligned} \tag{4}$$

which are called the *normal equations* for the least-squares line. Note that we have for brevity used $\sum y$, $\sum xy$ instead of $\sum_{j=1}^{n} y_j$, $\sum_{j=1}^{n} x_j y_j$. The normal equations (4) are easily remembered by observing that the first equation can be obtained formally by summing on both sides of (3), while the second equation is obtained formally by first multiplying both sides of (3) by x and then summing. Of course, this is not a derivation of the normal equations but only a means for remembering them.

The values of a and b obtained from (4) are given by

$$a = \frac{\left(\sum y\right)\left(\sum x^2\right) - \left(\sum x\right)\left(\sum xy\right)}{n\sum x^2 - \left(\sum x\right)^2} \qquad b = \frac{n\sum xy - \left(\sum x\right)\left(\sum y\right)}{n\sum x^2 - \left(\sum x\right)^2} \tag{5}$$

The result for b in (5) can also be written

$$b = \frac{\sum (x - \bar{x})(y - \bar{y})}{\sum (x - \bar{x})^2} \tag{6}$$

Here, as usual, a bar indicates *mean*, e.g., $\bar{x} = (\sum x)/n$. Division of both sides of the first normal equation in (4) by n yields

$$\bar{y} = a + b\bar{x} \tag{7}$$

If desired, we can first find b from (5) or (6) and then use (7) to find $a = \bar{y} - b\bar{x}$. This is equivalent to writing the least-squares line as

$$y - \bar{y} = b(x - \bar{x}) \qquad \text{or} \qquad y - \bar{y} = \frac{\sum (x - \bar{x})(y - \bar{y})}{\sum (x - \bar{x})^2}(x - \bar{x}) \tag{8}$$

The result (8) shows that the constant b, which is the *slope* of the line (3), is the fundamental constant in determining the line. From (8) it is also seen that the least-squares line passes through the point $(\bar{x}, \bar{y})$, which is called the *centroid* or *center of gravity* of the data.

The slope b of the regression line is independent of the origin of coordinates. This means that if we make the transformation (often called a *translation of axes*) given by

$$x = x' + h \qquad y = y' + k \tag{9}$$

where h and k are any constants, then b is also given by

$$b = \frac{n\sum x'y' - \left(\sum x'\right)\left(\sum y'\right)}{n\sum x'^2 - \left(\sum x'\right)^2} = \frac{\sum (x' - \bar{x}')(y' - \bar{y}')}{\sum (x' - \bar{x}')^2} \tag{10}$$

where x, y have simply been replaced by x', y' [for this reason we say that b is *invariant under the transformation* (9)]. It should be noted, however, that a, which determines the intercept on the x axis, does depend on the origin (and so is not invariant).

In the particular case where $h = \bar{x}, k = \bar{y}$, (10) simplifies to

$$b = \frac{\sum x'y'}{\sum x'^2} \tag{11}$$

The results (10) or (11) are often useful in simplifying the labor involved in obtaining the least-squares line.

The above remarks also hold for the regression line of x on y. The results are formally obtained by simply interchanging x and y. For example, the least-squares regression line of x on y is

$$x - \bar{x} = \frac{\sum(x - \bar{x})(y - \bar{y})}{\sum(y - \bar{y})^2}(y - \bar{y}) \tag{12}$$

It should be noted that in general (12) is not the same line as (8).

The Least-Squares Line in Terms of Sample Variances and Covariance

The sample variances and covariance of x and y are given by

$$s_x^2 = \frac{\sum(x - \bar{x})^2}{n}, \quad s_y^2 = \frac{\sum(y - \bar{y})^2}{n}, \quad s_{xy} = \frac{\sum(x - \bar{x})(y - \bar{y})}{n} \tag{13}$$

In terms of these, the least-squares regression lines of y on x and of x on y can be written, respectively, as

$$y - \bar{y} = \frac{s_{xy}}{s_x^2}(x - \bar{x}) \quad \text{and} \quad x - \bar{x} = \frac{s_{xy}}{s_y^2}(y - \bar{y}) \tag{14}$$

if we formally define the *sample correlation coefficient* by [compare (54), page 82]

$$r = \frac{s_{xy}}{s_x s_y} \tag{15}$$

then (14) can be written

$$\frac{y - \bar{y}}{s_y} = r\left(\frac{x - \bar{x}}{s_x}\right) \quad \text{and} \quad \frac{x - \bar{x}}{s_x} = r\left(\frac{y - \bar{y}}{s_y}\right) \tag{16}$$

In view of the fact that $(x - \bar{x})/s_x$ and $(y - \bar{y})/s_y$ are standardized sample values or standard scores, the results in (16) provide a very simple way of remembering the regression lines. It is clear that the two lines in (16) are different unless $r = \pm 1$, in which case all sample points lie on a line [this will be shown in (26)] and there is *perfect linear correlation and regression*.

It is also of interest to note that if the two regression lines (16) are written as $y = a + bx$, $x = c + dy$, respectively, then

$$bd = r^2 \tag{17}$$

Up to now we have not considered the precise significance of the correlation coefficient but have only defined it formally in terms of the variances and covariance. On page 270, the significance will be given.

The Least-Squares Parabola

The above ideas are easily extended. For example, the *least-squares parabola* that fits a set of sample points is given by

$$y = a + bx + cx^2 \tag{18}$$

where a, b, c are determined from the *normal equations*

$$\begin{aligned} \sum y &= na + b\sum x + c\sum x^2 \\ \sum xy &= a\sum x + b\sum x^2 + c\sum x^3 \\ \sum x^2y &= a\sum x^2 + b\sum x^3 + c\sum x^4 \end{aligned} \tag{19}$$

These are obtained formally by summing both sides of (18) after multiplying successively by 1, x and x^2, respectively.

Multiple Regression

The above ideas can also be generalized to more variables. For example, if we feel that there is a linear relationship between a dependent variable z and two independent variables x and y, then we would seek an equation connecting the variables that has the form

$$z = a + bx + cy \tag{20}$$

This is called a *regression equation of z on x and y*. If x is the dependent variable, a similar equation would be called a *regression equation of x on y and z*.

Because (20) represents a plane in a three-dimensional rectangular coordinate system, it is often called a *regression plane*. To find the least-squares regression plane, we determine a, b, c in (20) so that

$$\begin{aligned} \sum z &= na + b\sum x + c\sum y \\ \sum xz &= a\sum x + b\sum x^2 + c\sum xy \\ \sum yz &= a\sum y + b\sum xy + c\sum y^2 \end{aligned} \tag{21}$$

These equations, called the *normal equations* corresponding to (20), are obtained as a result of applying a definition similar to that on page 266. Note that they can be obtained formally from (20) on multiplying by 1, x, y, respectively, and summing.

Generalizations to more variables, involving linear or nonlinear equations leading to *regression surfaces* in three- or higher-dimensional spaces, are easily made.

Standard Error of Estimate

If we let y_{est} denote the estimated value of y for a given value of x, as obtained from the regression curve of y on x, then a measure of the scatter about the regression curve is supplied by the quantity

$$s_{y.x} = \sqrt{\frac{\sum (y - y_{\text{est}})^2}{n}} \tag{22}$$

which is called the *standard error of estimate of y on x*. Since $\sum(y - y_{\text{est}})^2 = \sum d^2$, as used in the Definition on page 266, we see that out of all possible regression curves the least-squares curve has the smallest standard error of estimate.

In the case of a regression line $y_{\text{est}} = a + bx$, with a and b given by (4), we have

$$s_{y.x}^2 = \frac{\sum y^2 - a\sum y - b\sum xy}{n} \tag{23}$$

or

$$s_{y.x}^2 = \frac{\sum (y - \bar{y})^2 - b\sum (x - \bar{x})(y - \bar{y})}{n} \tag{24}$$

We can also express $s_{y.x}^2$ for the least-squares line in terms of the variance and correlation coefficient as

$$s_{y.x}^2 = s_y^2(1 - r^2) \tag{25}$$

from which it incidentally follows as a corollary that $r^2 \leq 1$, i.e., $-1 \leq r \leq 1$.

The standard error of estimate has properties analogous to those of standard deviation. For example, if we construct pairs of lines parallel to the regression line of y on x at respective vertical distances $s_{y.x}$, and $2s_{y.x}$, and $3s_{y.x}$ from it, we should find if n is large enough that there would be included between these pairs of lines about 68%, 95%, and 99.7% of the sample points, respectively. See Problem 8.23.

Just as there is an unbiased estimate of population variance given by $\hat{s}^2 = ns^2/(n - 1)$, so there is an unbiased estimate of the square of the standard error of estimate. This is given by $\hat{s}^2_{y.x} = ns^2_{y.x}/(n - 2)$. For this reason some statisticians prefer to give (22) with $n - 2$ instead of n in the denominator.

The above remarks are easily modified for the regression line of x on y (in which case the standard error of estimate is denoted by $s_{x.y}$) or for nonlinear or multiple regression.

The Linear Correlation Coefficient

Up to now we have defined the correlation coefficient formally by (15) but have not examined its significance. In attempting to do this, let us note that from (25) and the definitions of $s_{y.x}$ and s_y, we have

$$r^2 = 1 - \frac{\sum(y - y_{est})^2}{\sum(y - \bar{y})^2} \tag{26}$$

Now we can show that (see Problem 8.24)

$$\sum(y - \bar{y})^2 = \sum(y - y_{est})^2 + \sum(y_{est} - \bar{y})^2 \tag{27}$$

The quantity on the left of (27) is called the *total variation*. The first sum on the right of (27) is then called the *unexplained variation*, while the second sum is called the *explained variation*. This terminology arises because the deviations $y - y_{est}$ behave in a random or unpredictable manner while the deviations $y_{est} - \bar{y}$ are explained by the least-squares regression line and so tend to follow a definite pattern. It follows from (26) and (27) that

$$r^2 = \frac{\sum(y_{est} - \bar{y})^2}{\sum(y - \bar{y})^2} = \frac{\text{explained variation}}{\text{total variation}} \tag{28}$$

Therefore, r^2 can be interpreted as the fraction of the total variation that is explained by the least-squares regression line. In other words, r measures *how well* the least-squares regression line fits the sample data. If the total variation is *all* explained by the regression line, i.e., if $r^2 = 1$ or $r = \pm 1$, we say that there is *perfect linear correlation* (and in such case also *perfect linear regression*). On the other hand, if the total variation is all unexplained, then the explained variation is zero and so $r = 0$. In practice the quantity r^2, sometimes called the *coefficient of determination*, lies between 0 and 1.

The correlation coefficient can be computed from either of the results

$$r = \frac{s_{xy}}{s_x s_y} = \frac{\sum(x - \bar{x})(y - \bar{y})}{\sqrt{\sum(x - \bar{x})^2}\sqrt{\sum(y - \bar{y})^2}} \tag{29}$$

or

$$r^2 = \frac{\text{explained variation}}{\text{total variation}} = \frac{\sum(y_{est} - \bar{y})^2}{\sum(y - \bar{y})^2} \tag{30}$$

which for linear regression are equivalent. The formula (29) is often referred to as the *product-moment formula* for linear correlation.

Formulas equivalent to those above, which are often used in practice, are

$$r = \frac{n\sum xy - \left(\sum x\right)\left(\sum y\right)}{\sqrt{\left[n\sum x^2 - \left(\sum x\right)^2\right]\left[n\sum y^2 - \left(\sum y\right)^2\right]}} \tag{31}$$

and

$$r = \frac{\overline{xy} - \bar{x}\bar{y}}{\sqrt{(\overline{x^2} - \bar{x}^2)(\overline{y^2} - \bar{y}^2)}} \tag{32}$$

If we use the transformation (9), page 267, we find

$$r = \frac{n\sum x'y' - \left(\sum x'\right)\left(\sum y'\right)}{\sqrt{\left[n\sum x'^2 - \left(\sum x'\right)^2\right]\left[n\sum y'^2 - \left(\sum y'\right)^2\right]}} \tag{33}$$

which shows that r is invariant under a translation of axes. In particular, if $h = \bar{x}$, $k = \bar{y}$, (33) becomes

$$r = \frac{\sum x'y'}{\sqrt{\left(\sum x'^2\right)\left(\sum y'^2\right)}} \tag{34}$$

which is often useful in computation.

The linear correlation coefficient may be positive or negative. If r is positive, y tends to *increase* with x (the slope of the least-squares line is positive) while if r is negative, y tends to *decrease* with x (the slope is negative). The sign is *automatically* taken into account if we use the result (29), (31), (32), (33), or (34). However, if we use (30) to obtain r, we must apply the proper sign.

Generalized Correlation Coefficient

The definition (29) [or any of the equivalent forms (31) through (34)] for the correlation coefficient involves only sample values x, y. Consequently, it yields the same number for all forms of regression curves and is useless as a measure of fit, except in the case of linear regression, where it happens to coincide with (30). However, the latter definition, i.e.,

$$r^2 = \frac{\text{explained variation}}{\text{total variation}} = \frac{\sum(y_{\text{est}} - \bar{y})^2}{\sum(y - \bar{y})^2} \tag{35}$$

does reflect the form of the regression curve (via the y_{est}) and so is suitable as the definition of a *generalized correlation coefficient r*. We use (35) to obtain nonlinear correlation coefficients (which measure how well a *nonlinear regression curve* fits the data) or, by appropriate generalization, *multiple correlation coefficients*. The connection (25) between the correlation coefficient and the standard error of estimate holds as well for nonlinear correlation.

Since a correlation coefficient merely measures how well a given regression curve (or surface) fits sample data, it is clearly senseless to use a linear correlation coefficient where the data are nonlinear. Suppose, however, that one does apply (29) to nonlinear data and obtains a value that is numerically considerably less than 1. Then the conclusion to be drawn is not that there is *little correlation* (a conclusion sometimes reached by those unfamiliar with the fundamentals of correlation theory) but that there is *little linear* correlation. There may in fact be a *large* nonlinear correlation.

Rank Correlation

Instead of using precise sample values, or when precision is unattainable, the data may be ranked in order of size, importance, etc., using the numbers 1, 2, . . . , n. If two corresponding sets of values x and y are ranked in such manner, the *coefficient of rank correlation*, denoted by r_{rank}, or briefly r, is given by (see Problem 8.36)

$$r_{\text{rank}} = 1 - \frac{6\sum d^2}{n(n^2 - 1)} \tag{36}$$

where d = differences between ranks of corresponding x and y

n = number of pairs of values (x, y) in the data

The quantity r_{rank} in (36) is known as *Spearman's rank correlation coefficient*.

Probability Interpretation of Regression

A scatter diagram, such as that in Fig. 8-1, is a graphical representation of data points for a particular sample. By choosing a different sample, or enlarging the original one, a somewhat different scatter diagram would in general be obtained. Each scatter diagram would then result in a different regression line or curve, although we would expect that these would not differ significantly from each other if the samples are drawn from the same population.

From the concept of curve fitting in samples, we are led to curve fitting for the population from which samples are drawn. The scatter of points around a regression line or curve indicates that for a particular value of x, there are actually various values of y *distributed* about the line or curve. This idea of distribution leads us naturally to the realization that there is a connection between curve fitting and probability.

The connection is supplied by introducing the random variables X and Y, which can take on the various sample values x and y, respectively. For example, X and Y may represent heights and weights of adult males in a population from which samples are drawn. It is then assumed that X and Y have a joint probability function or density function, $f(x, y)$, according to whether they are considered discrete or continuous.

Given the joint density function or probability function, $f(x, y)$, of two random variables X and Y, it is natural from the above remarks to ask whether there is a function $g(X)$ such that

$$E\{[Y - g(X)]^2\} = \text{a minimum} \tag{37}$$

A curve with equation $y = g(x)$ having property (37) is called a *least-squares regression curve of Y on X*. We have the following theorem:

Theorem 8-1 If X and Y are random variables having joint density function or probability function $f(x, y)$, then there exists a least-squares regression curve of Y on X, having property (37), given by

$$y = g(x) = E(Y \mid X = x) \tag{38}$$

provided that X and Y each have a variance that is finite.

Note that $E(Y \mid X = x)$ is the conditional expectation of Y given $X = x$, as defined on page 82.

Similar remarks can be made for a *least-squares regression curve of X on Y*. In that case, (37) is replaced by

$$E\{[X - h(Y)]^2\} = \text{a minimum}$$

and (38) is replaced by $x = h(y) = E(X \mid Y = y)$. The two regression curves $y = g(x)$ and $x = h(y)$ are different in general.

An interesting case arises when the joint distribution is the bivariate normal distribution given by (49), page 117. We then have the following theorem:

Theorem 8-2 If X and Y are random variables having the bivariate normal distribution, then the least-squares regression curve of Y on X is a *regression line* given by

$$\frac{y - \mu_Y}{\sigma_Y} = \rho\left(\frac{x - \mu_X}{\sigma_X}\right) \tag{39}$$

where

$$\rho = \frac{\sigma_{XY}}{\sigma_X \sigma_Y} \tag{40}$$

represents the population correlation coefficient.

We can also write (39) as

$$y - \mu_Y = \beta(x - \mu_X) \tag{41}$$

where

$$\beta = \frac{\sigma_{XY}}{\sigma_X^2} \tag{42}$$

Similar remarks can be made for the least-squares regression curve of X on Y, which also turns out to be a line [given by (39) with X and Y, x and y, interchanged]. These results should be compared with corresponding ones on page 268.

In case $f(x, y)$ is not known, we can still use the criterion (37) to obtain approximating regression curves for the population. For example, if we assume $g(x) = \alpha + \beta x$, we obtain the least-squares regression line (39), where α, β are given in terms of the (unknown) parameters $\mu_X, \mu_Y, \sigma_X, \sigma_Y, \rho$. Similarly if $g(x) = \alpha + \beta x + \gamma x^2$, we can obtain a least-squares regression parabola, etc. See Problem 8.39.

In general, all of the remarks made on pages 266 to 271 for samples are easily extended to the population. For example, the standard error of estimate in the case of the population is given in terms of the variance and correlation coefficient by

$$\sigma_{Y.X}^2 = \sigma_Y^2(1 - \rho^2) \tag{43}$$

which should be compared with (25), page 269.

Probability Interpretation of Correlation

From the above remarks it is clear that a population correlation coefficient should provide a measure of how well a given population regression curve fits the population data. All the remarks previously made for correlation in a sample apply as well to the population. For example, if $g(x)$ is determined by (37), then

$$E[(Y - \bar{Y})^2] = E[(Y - Y_{\text{est}})^2] + E[(Y_{\text{est}} - \bar{Y})^2] \tag{44}$$

where $Y_{\text{est}} = g(X)$ and $\bar{Y} = E(Y)$. The three quantities in (44) are called the *total, unexplained*, and *explained variations*, respectively. This leads to the definition of the *population correlation coefficient* ρ, where

$$\rho^2 = \frac{\text{explained variation}}{\text{total variation}} = \frac{E[(Y_{\text{est}} - \bar{Y})^2]}{E[(Y - \bar{Y})^2]} \tag{45}$$

For the linear case this reduces to (40). Results similar to (31) through (34) can also be written for the case of a population and linear regression. The result (45) is also used to define ρ, in the nonlinear case.

Sampling Theory of Regression

The regression equation $y = a + bx$ is obtained on the basis of sample data. We are often interested in the corresponding regression equation $y = \alpha + \beta x$ for the population from which the sample was drawn. The following are some tests concerning a normal population. To keep the notation simple, we shall follow the common convention of indicating values of sampling random variables rather than the random variables themselves.

1. **TEST OF HYPOTHESIS $\beta = b$.** To test the hypothesis that the regression coefficient β is equal to some specified value b, we use the fact that the statistic

$$t = \frac{\beta - b}{s_{y.x}/s_x}\sqrt{n - 2} \tag{46}$$

has Student's distribution with $n - 2$ degrees of freedom. This can also be used to find confidence intervals for population regression coefficients from sample values. See Problems 8.43 and 8.44.

2. **TEST OF HYPOTHESES FOR PREDICTED VALUES.** Let y_0 denote the predicted value of y corresponding to $x = x_0$ as estimated from the sample regression equation, i.e., $y_0 = a + bx_0$. Let y_p denote the predicted value of y corresponding to $x = x_0$ for the population. Then the statistic

$$t = \frac{(y_0 - y_p)\sqrt{n - 2}}{s_{y.x}\sqrt{n + 1 + [n(x_0 - \bar{x})^2/s_x^2]}} \tag{47}$$

has Student's distribution with $n - 2$ degrees of freedom. From this, confidence limits for predicted population values can be found. See Problem 8.45.

3. **TEST OF HYPOTHESES FOR PREDICTED MEAN VALUES.** Let y_0 denote the predicted value of y corresponding to $x = x_0$ as estimated from the sample regression equation, i.e., $y_0 = a + bx_0$. Let $\bar{y}_p$

denote the predicted *mean value* of y corresponding to $x = x_0$ for the population [i.e., $\bar{y}_p = E(Y|X = x_0)$]. Then the statistic

$$t = \frac{(y_0 - \bar{y}_p)\sqrt{n-2}}{s_{y.x}\sqrt{1 + [(x_0 - \bar{x})^2/s_x^2]}} \tag{48}$$

has Student's distribution with $n - 2$ degrees of freedom. From this, confidence limits for predicted mean population values can be found. See Problem 8.46.

Sampling Theory of Correlation

We often have to estimate the population correlation coefficient ρ from the sampling correlation coefficient r or to test hypotheses concerning ρ. For this purpose we must know the sampling distribution of r. In case $\rho = 0$, this distribution is symmetric and a statistic having Student's distribution can be used. For $\rho \neq 0$, the distribution is skewed. In that case a transformation due to Fisher produces a statistic which is approximately normally distributed. The following tests summarize the procedures involved.

1. **TEST OF HYPOTHESIS $\rho = 0$.** Here we use the fact that the statistic

$$t = \frac{r\sqrt{n-2}}{\sqrt{1-r^2}} \tag{49}$$

has Student's distribution with $n - 2$ degrees of freedom. See Problems 8.47 and 8.48.

2. **TEST OF HYPOTHESIS $\rho = \rho_0 \neq 0$.** Here we use the fact that the statistic

$$Z = \frac{1}{2}\ln\left(\frac{1+r}{1-r}\right) = 1.1513\log_{10}\left(\frac{1+r}{1-r}\right) \tag{50}$$

is approximately normally distributed with mean and standard deviation given by

$$\mu_z = \frac{1}{2}\ln\left(\frac{1+\rho_0}{1-\rho_0}\right) = 1.1513\log_{10}\left(\frac{1+\rho_0}{1-\rho_0}\right), \qquad \sigma_Z = \frac{1}{\sqrt{n-3}} \tag{51}$$

These facts can also be used to find confidence limits for correlation coefficients. See Problems 8.49 and 8.50. The transformation (50) is called *Fisher's Z transformation*.

3. **SIGNIFICANCE OF A DIFFERENCE BETWEEN CORRELATION COEFFICIENTS.** To determine whether two correlation coefficients r_1 and r_2, drawn from samples of sizes n_1 and n_2, respectively, differ significantly from each other, we compute Z_1 and Z_2 corresponding to r_1 and r_2 using (50). We then use the fact that the test statistic

$$z = \frac{Z_1 - Z_2 - \mu_{Z_1 - Z_2}}{\sigma_{Z_1 - Z_2}} \tag{52}$$

where $\mu_{Z_1 - Z_2} = \mu_{Z_1} - \mu_{Z_2}, \qquad \sigma_{Z_1 - Z_2} = \sqrt{\sigma_{Z_1}^2 + \sigma_{Z_2}^2} = \sqrt{\frac{1}{n_1 - 3} + \frac{1}{n_2 - 3}}$ (53)

is normally distributed. See Problem 8.51.

Correlation and Dependence

Whenever two random variables X and Y have a nonzero correlation coefficient ρ, we know (Theorem 3-15, page 81) that they are *dependent* in the probability sense (i.e., their joint distribution does not factor into their marginal distributions). Furthermore, when $\rho \neq 0$, we can use an equation of the form (39) to *predict* the value of Y from the value of X.

It is important to realize that "correlation" and "dependence" in the above sense do not necessarily imply a direct causal interdependence of X and Y. This is shown in the following examples.

EXAMPLE 8.1 Let X and Y be random variables representing heights and weights of individuals. Here there is a direct interdependence between X and Y.

EXAMPLE 8.2 If X represents teachers' salaries over the years while Y represents the amount of crime, the correlation coefficient may be different from zero and we may be able to find a regression equation predicting one variable from the other. But we would hardly be willing to say that there is a direct interdependence between X and Y.

SOLVED PROBLEMS

The least-squares line

8.1. A straight line passes through the points (x_1, y_1) and (x_2, y_2). Show that the equation of the line is

$$y - y_1 = \left(\frac{y_2 - y_1}{x_2 - x_1}\right)(x - x_1)$$

The equation of a line is $y = a + bx$. Then since (x_1, y_1) and (x_2, y_2) are points on the line, we have

$$y_1 = a + bx_1, \qquad y_2 = a + bx_2$$

Therefore,

$$(1) \qquad y - y_1 = (a + bx) - (a + bx_1) = b(x - x_1)$$

$$(2) \qquad y_2 - y_1 = (a + bx_2) - (a + bx_1) = b(x_2 - x_1)$$

Obtaining $b = (y_2 - y_1)/(x_2 - x_1)$ from (2) and substituting in (1), the required result follows.

The graph of the line PQ is shown in Fig. 8-5. The constant $b = (y_2 - y_1)/(x_2 - x_1)$ is the slope of the line.

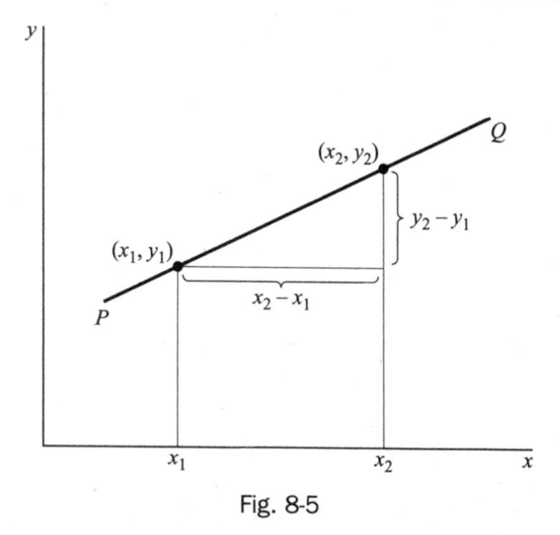

Fig. 8-5

8.2. (a) Construct a straight line that approximates the data of Table 8-1. (b) Find an equation for this line.

Table 8-1

x	1	3	4	6	8	9	11	14
y	1	2	4	4	5	7	8	9

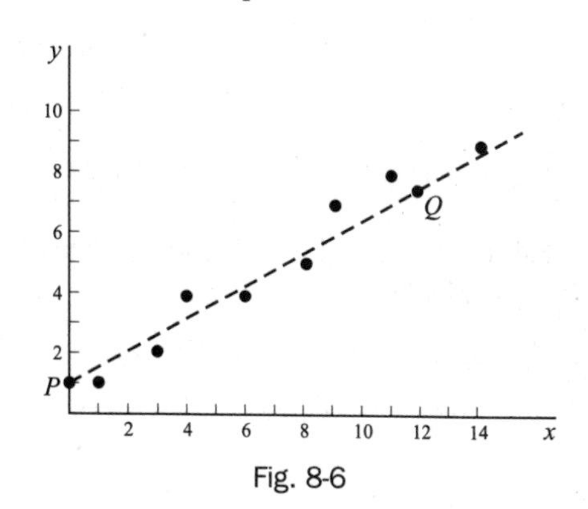

Fig. 8-6

(a) Plot the points (1, 1), (3, 2), (4, 4), (6, 4), (8, 5), (9, 7), (11, 8), and (14, 9) on a rectangular coordinate system as shown in Fig. 8-6.

A straight line approximating the data is drawn *freehand* in the figure. For a method eliminating the need for individual judgment, see Problem 8.4, which uses the method of least squares.

(b) To obtain the equation of the line constructed in (a), choose any two points on the line, such as P and Q. The coordinates of these points as read from the graph are approximately (0, 1) and (12, 7.5). Then from Problem 8.1,

$$y - 1 = \frac{7.5 - 1}{12 - 0}(x - 0)$$

or $y - 1 = 0.542x$ or $y = 1 + 0.542x$.

8.3. Derive the normal equations (4), page 267, for the least-squares line.

Refer to Fig. 8-7. The values of y on the least-squares line corresponding to $x_1, x_2, \ldots, x_n$ are

$$a + bx_1, \qquad a + bx_2, \qquad \ldots, \qquad a + bx_n$$

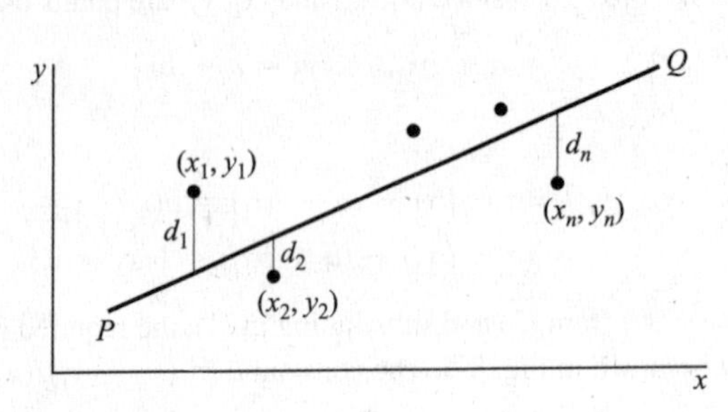

Fig. 8-7

The corresponding vertical deviations are

$$d_1 = a + bx_1 - y_1, \qquad d_2 = a + bx_2 - y_2, \qquad \ldots, \qquad d_n = a + bx_n - y_n$$

Then the sum of the squares of the deviations is

$$d_1^2 + d_2^2 + \cdots + d_n^2 = (a + bx_1 - y_1)^2 + (a + bx_2 - y_2)^2 + \cdots + (a + bx_n - y_n)^2$$

or

$$\sum d^2 = \sum (a + bx - y)^2$$

This is a function of a and b, i.e., $F(a, b) = \sum(a + bx - y)^2$. A necessary condition for this to be a minimum (or a maximum) is that $\partial F/\partial a = 0$, $\partial F/\partial b = 0$. Since

$$\frac{\partial F}{\partial a} = \sum \frac{\partial}{\partial a}(a + bx - y)^2 = \sum 2(a + bx - y)$$

$$\frac{\partial F}{\partial b} = \sum \frac{\partial}{\partial b}(a + bx - y)^2 = \sum 2x(a + bx - y)$$

we obtain

$$\sum (a + bx - y) = 0 \qquad \sum x(a + bx - y) = 0$$

i.e.,

$$\sum y = an + b\sum x \qquad \sum xy = a\sum x + b\sum x^2$$

as required. It can be shown that these actually yield a minimum.

8.4. Fit a least-squares line to the data of Problem 8.2 using (a) x as independent variable, (b) x as dependent variable.

(a) The equation of the line is $y = a + bx$. The normal equations are

$$\sum y = an + b\sum x$$

$$\sum xy = a\sum x + b\sum x^2$$

The work involved in computing the sums can be arranged as in Table 8-2. Although the last column is not needed for this part of the problem, it has been added to the table for use in part (b).

Since there are 8 pairs of values of x and y, $n = 8$ and the normal equations become

$$8a + 56b = 40$$
$$56a + 524b = 364$$

Solving simultaneously, $a = \frac{6}{11}$ or 0.545, $b = \frac{7}{11}$ or 0.636; and the required least-squares line is $y = \frac{6}{11} + \frac{7}{11}x$ or $y = 0.545 + 0.636x$. Note that this is not the line obtained in Problem 8.2 using the freehand method.

Table 8-2

x	y	x^2	xy	y^2
1	1	1	1	1
3	2	9	6	4
4	4	16	16	16
6	4	36	24	16
8	5	64	40	25
9	7	81	63	49
11	8	121	88	64
14	9	196	126	81
$\sum x = 56$	$\sum y = 40$	$\sum x^2 = 524$	$\sum xy = 364$	$\sum y^2 = 256$

Another method

$$a = \frac{\left(\sum y\right)\left(\sum x^2\right) - \left(\sum x\right)\left(\sum xy\right)}{n\sum x^2 - \left(\sum x\right)^2} = \frac{(40)(524) - (56)(364)}{(8)(524) - (56)^2} = \frac{6}{11} \quad \text{or} \quad 0.545$$

$$b = \frac{n\sum xy - \left(\sum x\right)\left(\sum y\right)}{n\sum x^2 - \left(\sum x\right)^2} = \frac{(8)(364) - (56)(40)}{(8)(524) - (56)^2} = \frac{7}{11} \quad \text{or} \quad 0.636$$

(b) If x is considered as the dependent variable and y as the independent variable, the equation of the least-squares line is $x = c + dy$ and the normal equations are

$$\sum x = cn + d\sum y$$
$$\sum xy = c\sum y + d\sum y^2$$

Then using Table 8-2, the normal equations become

$$8c + 40d = 56$$
$$40c + 256d = 364$$

from which $c = -\frac{1}{2}$ or -0.50, $d = \frac{3}{2}$ or 1.50.

These values can also be obtained from

$$c = \frac{\left(\sum x\right)\left(\sum y^2\right) - \left(\sum y\right)\left(\sum xy\right)}{n\sum y^2 - \left(\sum y\right)^2} = \frac{(56)(256) - (40)(364)}{(8)(256) - (40)^2} = -0.50$$

$$d = \frac{n\sum xy - \left(\sum x\right)\left(\sum y\right)}{n\sum y^2 - \left(\sum y\right)^2} = \frac{(8)(364) - (56)(40)}{(8)(256) - (40)^2} = 1.50$$

Therefore, the required equation of the least-squares line is $x = -0.50 + 1.50y$.

Note that by solving this equation for y, we obtain $y = 0.333 + 0.667x$, which is not the same as the line obtained in part (a).

8.5. Graph the two lines obtained in Problem 8.4.

The graphs of the two lines, $y = 0.545 + 0.636x$ and $x = -0.500 + 1.50y$, are shown in Fig. 8-8. Note that the two lines in this case are practically coincident, which is an indication that the data are very well described by a linear relationship.

The line obtained in part (a) is often called the *regression line of y on x* and is used for estimating y for given values of x. The line obtained in part (b) is called the *regression line of x on y* and is used for estimating x for given values of y.

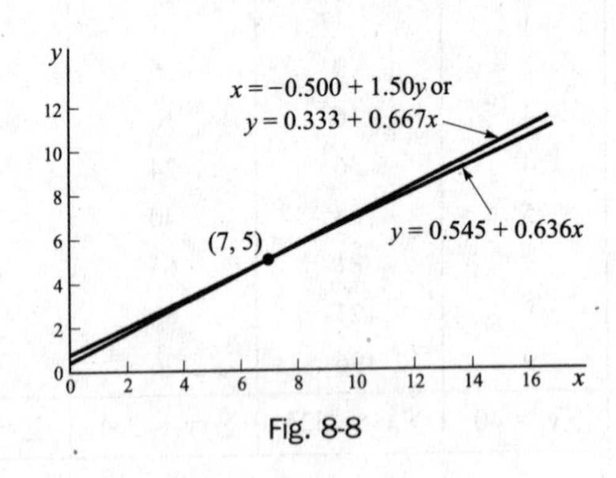

Fig. 8-8

8.6. (a) Show that the two least-squares lines obtained in Problem 8.4 intersect at point $(\bar{x}, \bar{y})$. (b) Estimate the value of y when $x = 12$. (c) Estimate the value of x when $y = 3$.

$$\bar{x} = \frac{\sum x}{n} = \frac{56}{8} = 7, \qquad \bar{y} = \frac{\sum y}{n} = \frac{40}{8} = 5$$

Then point $(\bar{x}, \bar{y})$, called the *centroid*, is (7, 5).

(a) Point (7, 5) lies on line $y = 0.545 + 0.636x$ or, more exactly, $y = \frac{6}{11} + \frac{7}{11}x$, since $5 = \frac{6}{11} + \frac{7}{11}(7)$.
Point (7, 5) lies on line $x = -\frac{1}{2} + \frac{3}{2}y$, since $7 = -\frac{1}{2} + \frac{3}{2}(5)$.

Another method

The equations of the two lines are $y = \frac{6}{11} + \frac{7}{11}x$ and $x = -\frac{1}{2} + \frac{3}{2}y$. Solving simultaneously, we find $x = 7, y = 5$. Therefore, the lines intersect in point (7, 5).

(b) Putting $x = 12$ into the regression line of y on x, $y = 0.545 + 0.636(12) = 8.2$.

(c) Putting $y = 3$ into the regression line of x on y, $x = -0.50 + 1.50(3) = 4.0$.

8.7. Prove that a least-squares line always passes through the point $(\bar{x}, \bar{y})$.

Case 1

x is the independent variable.

The equation of the least-squares line is (1) $y = a + bx$

A normal equation for the least-squares line is (2) $\sum y = an + b\sum x$

Dividing both sides of (2) by n gives (3) $\bar{y} = a + b\bar{x}$

Subtracting (3) from (1), the least-squares line can be written

$$y - \bar{y} = b(x - \bar{x}) \tag{4}$$

which shows that the line passes through the point $(\bar{x}, \bar{y})$.

Case 2

y is the independent variable.

Proceeding as in Case 1 with x and y interchanged and the constants a, b, replaced by c, d, respectively, we find that the least-squares line can be written

$$x - \bar{x} = d(y - \bar{y}) \tag{5}$$

which indicates that the line passes through the point $(\bar{x}, \bar{y})$.

Note that, in general, lines (4) and (5) are not coincident, but they intersect in $(\bar{x}, \bar{y})$.

8.8. Prove that the least-squares regression line of y on x can be written in the form (8), page 267.

We have from (4) of Problem 8.7, $y - \bar{y} = b(x - \bar{x})$. From the second equation in (5), page 267, we have

$$b = \frac{n\sum xy - \left(\sum x\right)\left(\sum y\right)}{n\sum x^2 - \left(\sum x\right)^2} \tag{1}$$

Now

$$\begin{aligned}
\sum(x - \bar{x})^2 &= \sum(x^2 - 2\bar{x}x + \bar{x}^2)\\
&= \sum x^2 - 2\bar{x}\sum x + \sum \bar{x}^2\\
&= \sum x^2 - 2n\bar{x}^2 + n\bar{x}^2\\
&= \sum x^2 - n\bar{x}^2\\
&= \sum x^2 - \frac{1}{n}\left(\sum x\right)^2\\
&= \frac{1}{n}\left[n\sum x^2 - \left(\sum x\right)^2\right]
\end{aligned}$$

Also

$$\begin{aligned}
\sum(x - \bar{x})(y - \bar{y}) &= \sum(xy - \bar{x}y - \bar{y}x + \bar{x}\bar{y})\\
&= \sum xy - \bar{x}\sum y - \bar{y}\sum x + \sum \bar{x}\bar{y}\\
&= \sum xy - n\bar{x}\bar{y} - n\bar{y}\bar{x} + n\bar{x}\bar{y}\\
&= \sum xy - n\bar{x}\bar{y}\\
&= \sum xy - \frac{\left(\sum x\right)\left(\sum y\right)}{n}\\
&= \frac{1}{n}\left[n\sum xy - \left(\sum x\right)\left(\sum y\right)\right]
\end{aligned}$$

Therefore, (1) becomes

$$b = \frac{\sum(x - \bar{x})(y - \bar{y})}{\sum(x - \bar{x})^2}$$

from which the result (8) is obtained. Proof of (12), page 268, follows on interchanging x and y.

8.9. Let $x = x' + h$, $y = y' + k$, where h and k are any constants. Prove that

$$b = \frac{n\sum xy - \left(\sum x\right)\left(\sum y\right)}{n\sum x^2 - \left(\sum x\right)^2} = \frac{n\sum x'y' - \left(\sum x'\right)\left(\sum y'\right)}{n\sum x'^2 - \left(\sum x'\right)^2}$$

From Problem 8.8 we have

$$b = \frac{n\sum xy - \left(\sum x\right)\left(\sum y\right)}{n\sum x^2 - \left(\sum x\right)^2} = \frac{\sum(x - \bar{x})(y - \bar{y})}{\sum(x - \bar{x})^2}$$

Now if $x = x' + h, y = y' + k$, we have

$$\bar{x} = \bar{x}' + h, \quad \bar{y} = \bar{x}' + k$$

Thus
$$\frac{\sum(x - \bar{x})(y - \bar{y})}{\sum(x - \bar{x})^2} = \frac{\sum(x' - \bar{x}')(y' - \bar{y}')}{\sum(x' - \bar{x}')^2}$$

$$= \frac{n\sum x'y' - \left(\sum x'\right)\left(\sum y'\right)}{n\sum x'^2 - \left(\sum x'\right)^2}$$

The result is useful in developing a shortcut for obtaining least-squares lines by subtracting suitable constants from the given values of x and y (see Problem 8.12).

8.10. If, in particular, $h = \bar{x}, k = \bar{y}$ in Problem 8.9, show that

$$b = \frac{\sum x'y'}{\sum x'^2}$$

This follows at once from Problem 8.9 since

$$\sum x' = \sum(x - \bar{x}) = \sum x - n\bar{x} = 0$$

and similarly $\sum y' = 0$.

8.11. Table 8-3 shows the respective heights x and y of a sample of 12 fathers and their oldest sons. (a) Construct a scatter diagram. (b) Find the least-squares regression line of y on x. (c) Find the least-squares regression line of x on y.

Table 8-3

Height x of Father (inches)	65	63	67	64	68	62	70	66	68	67	69	71
Height y of Son (inches)	68	66	68	65	69	66	68	65	71	67	68	70

(a) The scatter diagram is obtained by plotting the points (x, y) on a rectangular coordinate system as shown in Fig. 8-9.

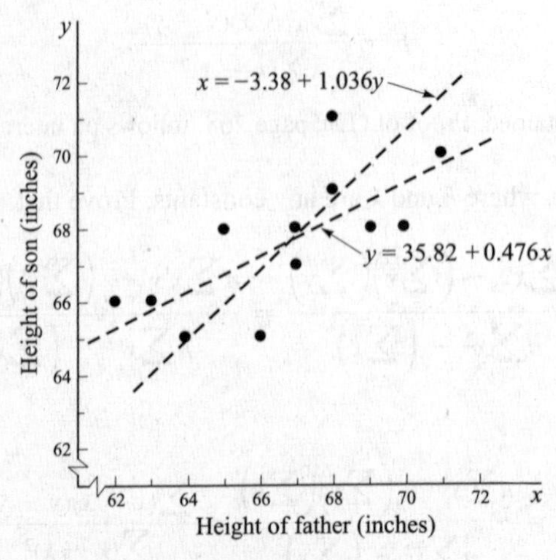

Fig. 8-9

(b) The regression line of y on x is given by $y = a + bx$, where a and b are obtained by solving the normal equations

$$\sum y = an + b\sum x$$
$$\sum xy = a\sum x + b\sum x^2$$

The sums are shown in Table 8-4, and so the normal equations become

$$12a + 800b = 811$$
$$800a + 53{,}418b = 54{,}107$$

from which we find $a = 35.82$ and $b = 0.476$, so that $y = 35.82 + 0.476x$. The graph of this equation is shown in Fig. 8-9.

Another method

$$a = \frac{\left(\sum y\right)\left(\sum x^2\right) - \left(\sum x\right)\left(\sum xy\right)}{n\sum x^2 - \left(\sum x\right)^2} = 35.82, \qquad b = \frac{n\sum xy - \left(\sum x\right)\left(\sum y\right)}{n\sum x^2 - \left(\sum x\right)^2} = 0.476$$

Table 8-4

x	y	x^2	xy	y^2
65	68	4225	4420	4624
63	66	3969	4158	4356
67	68	4489	4556	4624
64	65	4096	4160	4225
68	69	4624	4692	4761
62	66	3844	4092	4356
70	68	4900	4760	4624
66	65	4356	4290	4225
68	71	4624	4828	5041
67	67	4489	4489	4489
69	68	4761	4692	4624
71	70	5041	4970	4900
$\sum x = 800$	$\sum y = 811$	$\sum x^2 = 53{,}418$	$\sum = 54{,}107$	$\sum y^2 = 54{,}849$

(c) The regression line of x on y is given by $x = c + dy$, where c and d are obtained by solving the normal equations

$$\sum x = cn + d\sum y$$
$$\sum xy = c\sum y + d\sum y^2$$

Using the sums in Table 8-4, these become

$$12c + 811d = 800$$
$$811c + 54{,}849d = 54{,}107$$

from which we find $c = -3.38$ and $d = 1.036$, so that $x = -3.38 + 1.036y$. The graph of this equation is shown in Fig. 8-9.

Another method

$$c = \frac{\left(\sum x\right)\left(\sum y^2\right) - \left(\sum y\right)\left(\sum xy\right)}{n\sum y^2 - \left(\sum y\right)^2} = -3.38, \qquad d = \frac{n\sum xy - \left(\sum y\right)\left(\sum x\right)}{n\sum y^2 - \left(\sum y\right)^2} = 1.036$$

8.12. Work Problem 8.11 by using the method of Problem 8.9.

Subtract an appropriate value, say, 68, from x and y (the numbers subtracted from x and from y could be different). This leads to Table 8-5.

From the table we find

$$b = \frac{n\sum x'y' - \left(\sum x'\right)\left(\sum y'\right)}{n\sum x'^2 - \left(\sum x'\right)^2} = \frac{(12)(47) - (-16)(-5)}{(12)(106) - (16)^2} = 0.476$$

Also since $x' = x - 68$, $y' = y - 68$, we have $\bar{x}' = \bar{x} - 68$, $\bar{y}' = \bar{y} - 68$. Thus

$$\bar{x} = \bar{x}' + 68 = -\frac{16}{12} + 68 = 66.67, \qquad \bar{y} = \bar{y}' + 68 = -\frac{5}{12} + 68 = 67.58$$

The required regression equation of y on x is $y - \bar{y} = b(x - \bar{x})$, i.e.,

$$y - 67.58 = 0.476(x - 66.07) \quad \text{or} \quad y = 35.85 + 0.476x$$

in agreement with Problem 8.11, apart from rounding errors. In a similar manner we can obtain the regression equation of x on y.

Table 8-5

x'	y'	x'^2	$x'y'$	y'^2
−3	0	9	0	0
−5	−2	25	10	4
−1	0	1	0	0
−4	−3	16	12	9
0	1	0	0	1
−6	−2	36	12	4
2	0	4	0	0
−2	−3	4	6	9
0	3	0	0	9
−1	−1	1	1	1
1	0	1	0	0
3	2	9	6	4
$\sum x' = -16$	$\sum y' = -5$	$\sum x'^2 = 106$	$\sum x'y' = 47$	$\sum y'^2 = 41$

Nonlinear equations reducible to linear form

8.13. Table 8-6 gives experimental values of the pressure P of a given mass of gas corresponding to various values of the volume V. According to thermodynamic principles, a relationship having the form $PV^\gamma = C$, where γ and C are constants, should exist between the variables. (a) Find the values of γ and C. (b) Write the equation connecting P and V. (c) Estimate P when $V = 100.0$ in³.

Table 8-6

Volume V (in³)	54.3	61.8	72.4	88.7	118.6	194.0
Pressure P (lb/in²)	61.2	49.5	37.6	28.4	19.2	10.1

Since $PV^\gamma = C$, we have upon taking logarithms to base 10,

$$\log P + \gamma \log V = \log C \qquad \text{or} \qquad \log P = \log C - \gamma \log V$$

Setting $\log V = x$ and $\log P = y$, the last equation can be written

(1) $$y = a + bx$$

where $a = \log C$ and $b = -\gamma$.

Table 8-7 gives the values of x and y corresponding to the values of V and P in Table 8-6 and also indicates the calculations involved in computing the least-squares line (1).

Table 8-7

$x = \log V$	$y = \log P$	x^2	xy
1.7348	1.7868	3.0095	3.0997
1.7910	1.6946	3.2077	3.0350
1.8597	1.5752	3.4585	2.9294
1.9479	1.4533	3.7943	2.8309
2.0741	1.2833	4.3019	2.6617
2.2878	1.0043	5.2340	2.2976
$\sum x = 11.6953$	$\sum y = 8.7975$	$\sum x^2 = 23.0059$	$\sum xy = 16.8543$

The normal equations corresponding to the least-squares line (1) are

$$\sum y = an + b\sum x \qquad \sum xy = a\sum x + b\sum x^2$$

from which

$$a = \frac{\left(\sum y\right)\left(\sum x^2\right) - \left(\sum x\right)\left(\sum xy\right)}{n\sum x^2 - \left(\sum x\right)^2} = 4.20, \qquad b = \frac{n\sum xy - \left(\sum x\right)\left(\sum y\right)}{n\sum x^2 - \left(\sum x\right)^2} = -1.40$$

Then $y = 4.20 - 1.40x$.

(a) Since $a = 4.20 = \log C$ and $b = -1.40 = -\gamma$, $C = 1.60 \times 10^4$ and $\gamma = 1.40$.

(b) $PV^{1.40} = 16{,}000$.

(c) When $V = 100$, $x = \log V = 2$ and $y = \log P = 4.20 - 1.40(2) = 1.40$. Then $P = \text{antilog } 1.40 = 25.1 \text{ lb/in}^2$.

8.14. Solve Problem 8.13 by plotting the data on log-log graph paper.

For each pair of values of the pressure P and volume V in Table 8-6, we obtain a point that is plotted on the specially constructed *log-log graph paper* shown in Fig. 8-10.

A line (drawn freehand) approximating these points is also indicated. The resulting graph shows that there is a linear relationship between $\log P$ and $\log V$, which can be represented by the equation

$$\log P = a + b \log V \qquad \text{or} \qquad y = a + bx$$

The slope b, which is negative in this case, is given numerically by the ratio of the length of AB to the length of AC. Measurement in this case yields $b = -1.4$.

To obtain a, one point on the line is needed. For example, when $V = 100$, $P = 25$ from the graph. Then

$$a = \log P - b\log V = \log 25 + 1.4 \log 100 = 1.4 + (1.4)(2) = 4.2$$

so that

$$\log P + 1.4 \log V = 4.2, \qquad \log PV^{1.4} = 4.2, \qquad \text{and} \qquad PV^{1.4} = 16{,}000$$

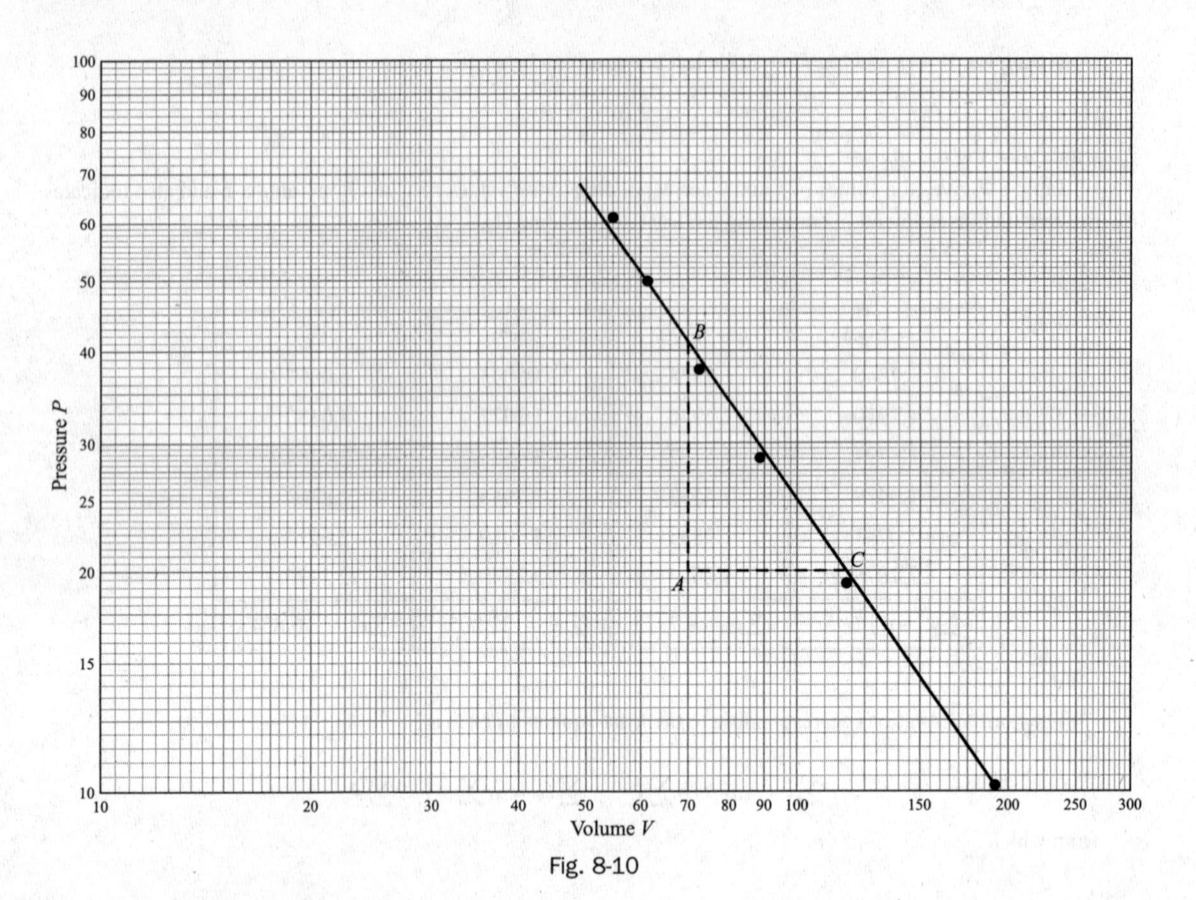

Fig. 8-10

The least-squares parabola

8.15. Derive the normal equations (19), page 269, for the least-squares parabola.

$$y = a + bx + cx^2$$

Let the sample points be $(x_1, y_1), (x_2, y_2), \ldots, (x_n, y_n)$. Then the values of y on the least-squares parabola corresponding to $x_1, x_2, \ldots, x_n$ are

$$a + bx_1 + cx_1^2, \qquad a + bx_2 + cx_2^2, \qquad \ldots, \qquad a + bx_n + cx_n^2$$

Therefore, the deviations from $y_1, y_2, \ldots, y_n$ are given by

$$d_1 = a + bx_1 + cx_1^2 - y_1, \qquad d_2 = a + bx_2 + cx_2^2 - y_2, \qquad \ldots, \qquad d_n = a + bx_n + cx_n^2 - y_n$$

and the sum of the squares of the deviations is given by

$$\sum d^2 = \sum (a + bx + cx^2 - y)^2$$

This is a function of a, b, and c, i.e.,

$$F(a,b,c) = \sum (a + bx + cx^2 - y)^2$$

To minimize this function, we must have

$$\frac{\partial F}{\partial a} = 0, \qquad \frac{\partial F}{\partial b} = 0, \qquad \frac{\partial F}{\partial c} = 0$$

Now

$$\frac{\partial F}{\partial a} = \sum \frac{\partial}{\partial a}(a + bx + cx^2 - y)^2 = \sum 2(a + bx + cx^2 - y)$$

$$\frac{\partial F}{\partial b} = \sum \frac{\partial}{\partial b}(a + bx + cx^2 - y)^2 = \sum 2x(a + bx + cx^2 - y)$$

$$\frac{\partial F}{\partial c} = \sum \frac{\partial}{\partial c}(a + bx + cx^2 - y)^2 = \sum 2x^2(a + bx + cx^2 - y)$$

Simplifying each of these summations and setting them equal to zero yields the equations (19), page 269.

8.16. Fit a least-squares parabola having the form $y = a + bx + cx^2$ to the data in Table 8-8.

Table 8-8

x	1.2	1.8	3.1	4.9	5.7	7.1	8.6	9.8
y	4.5	5.9	7.0	7.8	7.2	6.8	4.5	2.7

Then normal equations are

$$\begin{aligned}\sum y &= an + b\sum x + c\sum x^2\\ \sum xy &= a\sum x + b\sum x^2 + c\sum x^3\\ \sum x^2y &= a\sum x^2 + b\sum x^3 + c\sum x^4\end{aligned} \quad (1)$$

The work involved in computing the sums can be arranged as in Table 8-9.

Table 8-9

x	y	x^2	x^3	x^4	xy	x^2y
1.2	4.5	1.44	1.73	2.08	5.40	6.48
1.8	5.9	3.24	5.83	10.49	10.62	19.12
3.1	7.0	9.61	29.79	92.35	21.70	67.27
4.9	7.8	24.01	117.65	576.48	38.22	187.28
5.7	7.2	32.49	185.19	1055.58	41.04	233.93
7.1	6.8	50.41	357.91	2541.16	48.28	342.79
8.6	4.5	73.96	636.06	5470.12	38.70	332.82
9.8	2.7	96.04	941.19	9223.66	26.46	259.31
$\sum x =$ 42.2	$\sum y =$ 46.4	$\sum x^2 =$ 291.20	$\sum x^3 =$ 2275.35	$\sum x^4 =$ 18,971.92	$\sum xy =$ 230.42	$\sum x^2y =$ 1449.00

Then the normal equations (1) become, since $n = 8$,

$$\begin{aligned}8a + 42.2b + 291.20c &= 46.4\\ 42.2a + 291.20b + 2275.35c &= 230.42\\ 291.20a + 2275.35b + 18971.92c &= 1449.00\end{aligned} \quad (2)$$

Solving, $a = 2.588$, $b = 2.065$, $c = -0.2110$; hence the required least-squares parabola has the equation

$$y = 2.588 + 2.065x - 0.2110x^2$$

8.17. Use the least-squares parabola of Problem 8.16 to estimate the values of y from the given values of x.

For $x = 1.2$, $y_{est} = 2.588 + 2.065(1.2) - 0.2110(1.2)^2 = 4.762$. Similarly, other estimated values are obtained. The results are shown in Table 8-10 together with the actual values of y.

Table 8-10

y_{est}	4.762	5.621	6.962	7.640	7.503	6.613	4.741	2.561
y	4.5	5.9	7.0	7.8	7.2	6.8	4.5	2.7

Multiple regression

8.18. A variable z is to be estimated from variables x and y by means of a regression equation having the form $z = a + bx + cy$. Show that the least-squares regression equation is obtained by determining a, b, and c so that they satisfy (21), page 269.

Let the sample points be $(x_1, y_1, z_1), \ldots, (x_n, y_n, z_n)$. Then the values of z on the least-squares regression plane corresponding to $(x_1, y_1), \ldots, (x_n, y_n)$ are, respectively,

$$a + bx_1 + cy_1, \quad \ldots, \quad a + bx_n + cy_n$$

Therefore, the deviations from $z_1, \ldots, z_n$ are given by

$$d_1 = a + bx_1 + cy_1 - z_1, \quad \ldots, \quad d_n = a + bx_n + cy_n - z_n$$

and the sum of the squares of the deviations is given by

$$\sum d^2 = \sum (a + bx + cy - z)^2$$

Considering this as a function of a, b, c and setting the partial derivatives with respect to a, b, and c equal to zero, the required normal equations (21) on page 269, are obtained.

8.19. Table 8-11 shows the weights z to the nearest pound, heights x to the nearest inch, and ages y to the nearest year, of 12 boys, (a) Find the least-squares regression equation of z on x and y. (b) Determine the estimated values of z from the given values of x and y. (c) Estimate the weight of a boy who is 9 years old and 54 inches tall.

Table 8-11

Weight (z)	64	71	53	67	55	58	77	57	56	51	76	68
Height (x)	57	59	49	62	51	50	55	48	52	42	61	57
Age (y)	8	10	6	11	8	7	10	9	10	6	12	9

(a) The linear regression equation of z on x and y can be written

$$z = a + bx + cy$$

The normal equations (21), page 269, are given by

$$\begin{aligned} \sum z &= na + b\sum x + c\sum y \\ \sum xz &= a\sum x + b\sum x^2 + c\sum xy \\ \sum yz &= a\sum y + b\sum xy + c\sum y^2 \end{aligned} \tag{1}$$

The work involved in computing the sums can be arranged as in Table 8-12.

Table 8-12

z	x	y	z^2	x^2	y^2	xz	yx	xy
64	57	8	4096	3249	64	3648	512	456
71	59	10	5041	3481	100	4189	710	590
53	49	6	2809	2401	36	2597	318	294
67	62	11	4489	3844	121	4154	737	682
55	51	8	3025	2601	64	2805	440	408
58	50	7	3364	2500	49	2900	406	350
77	55	10	5929	3025	100	4235	770	550
57	48	9	3249	2304	81	2736	513	432
56	52	10	3136	2704	100	2912	560	520
51	42	6	2601	1764	36	2142	306	252
76	61	12	5776	3721	144	4636	912	732
68	57	9	4624	3249	81	3876	612	513
$\sum z =$ 753	$\sum x =$ 643	$\sum y =$ 106	$\sum z^2 =$ 48,139	$\sum x^2 =$ 34,843	$\sum y^2 =$ 976	$\sum xz =$ 40,830	$\sum yz =$ 6796	$\sum xy =$ 5779

Using this table, the normal equations (1) become

$$(2)\qquad \begin{aligned} 12a + 643b + 106c &= 753 \\ 643a + 34{,}843b + 5779c &= 40{,}830 \\ 106a + 5779b + 976c &= 6796 \end{aligned}$$

Solving, $a = 3.6512$, $b = 0.8546$, $c = 1.5063$, and the required regression equation is

$$(3)\qquad z = 3.65 + 0.855x + 1.506y$$

(b) Using the regression equation (3), we obtain the estimated values of z, denoted by z_{est}, by substituting the corresponding values of x and y. The results are given in Table 8-13 together with the sample values of z.

Table 8-13

z_{est}	64.414	69.136	54.564	73.206	59.286	56.925	65.717	58.229	63.153	48.582	73.857	65.920
z	64	71	53	67	55	58	77	57	56	51	76	68

(c) Putting $x = 54$ and $y = 9$ in (3), the estimated weight is $z_{\text{est}} = 63.356$, or about 63 lb.

Standard error of estimate

8.20. If the least-squares regression line of y on x is given by $y = a + bx$, prove that the standard error of estimate $s_{y.x}$ is given by

$$s_{y.x}^2 = \frac{\sum y^2 - a\sum y - b\sum xy}{n}$$

The values of y as estimated from the regression line are given by $y_{\text{est}} = a + bx$. Then

$$\begin{aligned} s_{y.x}^2 &= \frac{\sum (y - y_{\text{est}})^2}{n} = \frac{\sum (y - a - bx)^2}{n} \\ &= \frac{\sum y(y - a - bx) - a\sum (y - a - bx) - b\sum x(y - a - bx)}{n} \end{aligned}$$

But

$$\begin{aligned} \sum (y - a - bx) &= \sum y - an - b\sum x = 0 \\ \sum x(y - a - bx) &= \sum xy - a\sum x - b\sum x^2 = 0 \end{aligned}$$

since from the normal equations

$$\sum y = an + b\sum x \qquad \sum xy = a\sum x + b\sum x^2$$

Then

$$s_{y.x}^2 = \frac{\sum y(y - a - bx)}{n} = \frac{\sum y^2 - a\sum y - b\sum xy}{n}$$

This result can be extended to nonlinear regression equations.

8.21. Prove that the result in Problem 8.20 can be written

$$s_{y.x}^2 = \frac{\sum (y - \bar{y})^2 - b\sum (x - \bar{x})(y - \bar{y})}{n}$$

Method 1

Let $x = x' + \bar{x}$, $y = y' + \bar{y}$. Then from Problem 8.20

$$\begin{aligned} ns_{y.x}^2 &= \sum y^2 - a\sum y - b\sum xy \\ &= \sum (y' + \bar{y})^2 - a\sum (y' + \bar{y}) - b\sum (x' + \bar{x})(y' + \bar{y}) \\ &= \sum (y'^2 + 2y'\bar{y} + \bar{y}^2) - a\left(\sum y' + n\bar{y}\right) - b\sum (x'y' + \bar{x}y' + x'\bar{y} + \bar{x}\bar{y}) \end{aligned}$$

$$= \sum y'^2 + 2\bar{y}\sum y' + n\bar{y}^2 - an\bar{y} - b\sum x'y' - b\bar{x}\sum y' - b\bar{y}\sum x' - bn\bar{x}\bar{y}$$
$$= \sum y'^2 + n\bar{y}^2 - an\bar{y} - b\sum x'y' - bn\bar{x}\bar{y}$$
$$= \sum y'^2 - b\sum x'y' + n\bar{y}(\bar{y} - a - b\bar{x})$$
$$= \sum y'^2 - b\sum x'y'$$
$$= \sum(y - \bar{y})^2 - b\sum(x - \bar{x})(y - \bar{y})$$

where we have used the results $\sum x' = 0$, $\sum y' = 0$ and $\bar{y} = a + b\bar{x}$ (which follows on dividing both sides of the normal equation $\sum y = an + b\sum x$ by n). This proves the required result.

Method 2

We know that the regression line can be written as $y - \bar{y} = b(x - \bar{x})$, which corresponds to starting with $y = a + bx$ and then replacing a by zero, x by $x - \bar{x}$ and y by $y - \bar{y}$. When these replacements are made in Problem 8.20, the required result is obtained.

8.22. Compute the standard error of estimate, $s_{y.x}$, for the data of Problem 8.11.

From Problem 8.11(b) the regression line of y on x is $y = 35.82 + 0.476x$. In Table 8-14 are listed the actual values of y (from Table 8-3) and the estimated values of y, denoted by y_{est}, as obtained from the regression line. For example, corresponding to $x = 65$, we have $y_{\text{est}} = 35.82 + 0.476(65) = 66.76$.

Table 8-14

x	65	63	67	64	68	62	70	66	68	67	69	71
y	68	66	68	65	69	66	68	65	71	67	68	70
y_{est}	66.76	65.81	67.71	66.28	68.19	65.33	69.14	67.24	68.19	67.71	68.66	69.62
$y-y_{\text{est}}$	1.24	0.19	0.29	−1.28	0.81	0.67	−1.14	−2.24	2.81	−0.71	−0.66	0.38

Also listed are the values $y - y_{\text{est}}$, which are needed in computing $s_{y.x}$.

$$s_{y.x}^2 = \frac{\sum(y - y_{\text{est}})^2}{n} = \frac{(1.24)^2 + (0.19) + \cdots + (0.38)^2}{12} = 1.642$$

and $s_{y.x} = \sqrt{1.642} = 1.28$ inches.

8.23. (a) Construct two lines parallel to the regression line of Problem 8.11 and having vertical distance $s_{y.x}$ from it. (b) Determine the percentage of data points falling between these two lines.

(a) The regression line $y = 35.82 + 0.476x$ as obtained in Problem 8.11 is shown solid in Fig. 8-11. The two parallel lines, each having vertical distance $s_{y.x} = 1.28$ (see Problem 8.22) from it, are shown dashed in Fig. 8-11.

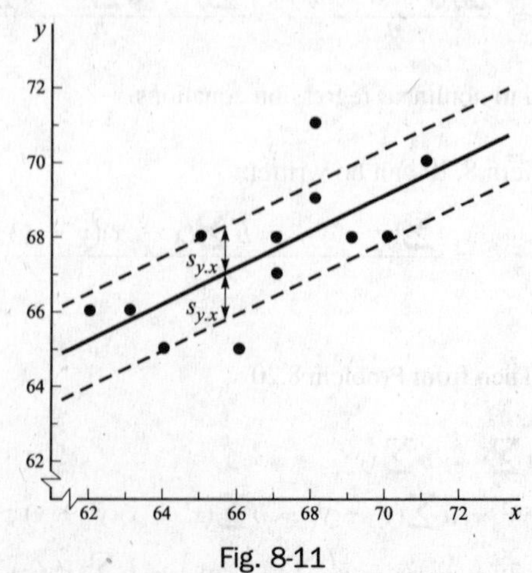

Fig. 8-11

(b) From the figure it is seen that of the 12 data points, 7 fall between the lines while 3 appear to lie on the lines. Further examination using the last line in Table 8-14 reveals that 2 of these 3 points lie between the lines. Then the required percentage is $9/12 = 75\%$.

Another method

From the last line in Table 8-14, $y - y_{\text{est}}$ lies between -1.28 and 1.28 (i.e., $\pm s_{y.x}$) for 9 points (x, y). Then the required percentage is $9/12 = 75\%$.

If the points are normally distributed about the regression line, theory predicts that about 68% of the points lie between the lines. This would have been more nearly the case if the sample size were large.

NOTE: A better estimate of the standard error of estimate of the population from which the sample heights were taken is given by $\hat{s}_{y.x} = \sqrt{n/(n-2)}s_{y.x} = \sqrt{12/10}(1.28) = 1.40$ inches.

The linear correlation coefficient

8.24. Prove that $\sum(y - \bar{y})^2 = \sum(y - y_{\text{est}})^2 + \sum(y_{\text{est}} - \bar{y})^2$.

Squaring both side of $y - \bar{y} = (y - y_{\text{est}}) + (y_{\text{est}} - \bar{y})$ and then summing, we have

$$\sum(y - \bar{y})^2 = \sum(y - y_{\text{est}})^2 + \sum(y_{\text{est}} - \bar{y})^2 + 2\sum(y - y_{\text{est}})(y_{\text{est}} - \bar{y})$$

The required result follows at once if we can show that the last sum is zero. In the case of linear regression this is so, since

$$\begin{aligned}\sum(y - y_{\text{est}})(y_{\text{est}} - \bar{y}) &= \sum(y - a - bx)(a + bx - \bar{y}) \\ &= a\sum(y - a - bx) + b\sum x(y - a - bx) - \bar{y}\sum(y - a - bx) \\ &= 0\end{aligned}$$

because of the normal equations $\sum(y - a - bx) = 0$, $\sum x(y - a - bx) = 0$.

The result can similarly be shown valid for nonlinear regression using a least-squares curve given by $y_{\text{est}} = a_0 + a_1x + a_2x^2 + \cdots + a_nx^n$.

8.25. Compute (a) the explained variation, (b) the unexplained variation, (c) the total variation for the data of Problem 8.11.

We have $\bar{y} = 67.58$ from Problem 8.12 (or from Table 8-4, since $\bar{y} = 811/12 = 67.58$). Using the values y_{est} from Table 8-14 we can construct Table 8-15.

Table 8-15

$y_{\text{est}} - \bar{y}$	-0.82	-1.77	0.13	-1.30	0.61	-2.25	1.56	-0.34	0.61	0.13	1.08	2.04

(a) Explained variation $= \sum(y_{\text{est}} - \bar{y})^2 = (-0.82)^2 + \cdots + (2.04)^2 = 19.22$.

(b) Unexplained variation $= \sum(y - y_{\text{est}})^2 = ns^2_{y.x} = 19.70$, from Problem 8.22.

(c) Total variation $= \sum(y - \bar{y})^2 = 19.22 + 19.70 = 38.92$, from Problem 8.24.

The results in (b) and (c) can also be obtained by direct calculation of the sum of squares.

8.26. Find (a) the coefficient of determination, (b) the coefficient of correlation for the data of Problem 8.11. Use the results of Problem 8.25.

(a) Coefficient of determination $= r^2 = \dfrac{\text{explained variation}}{\text{total variation}} = \dfrac{19.22}{38.92} = 0.4938$.

(b) Coefficient of correlation $= r = \pm\sqrt{0.4938} = \pm 0.7027$.

Since the variable y_{est} increases as x increases, the correlation is positive, and we therefore write $r = 0.7027$, or 0.70 to two significant figures.

8.27. Starting from the general result (30), page 270, for the correlation coefficient, derive the result (34), page 271 (the *product-moment formula*), in the case of linear regression.

The least-squares regression line of y on x can be written $y_{est} = a + bx$ or $y'_{est} = bx'$, where $b = \sum x'y'/\sum x'^2$, $x' = x - \bar{x}$, and $y'_{est} = y_{est} - \bar{y}$. Then, using $y' = y - \bar{y}$, we have

$$r^2 = \frac{\text{explained variation}}{\text{total variation}} = \frac{\sum (y_{est} - \bar{y})^2}{\sum (y - \bar{y})^2} = \frac{\sum y'^2_{est}}{\sum y'^2}$$

$$= \frac{\sum b^2 x'^2}{\sum y'^2} = \frac{b^2 \sum x'^2}{\sum y'^2} = \left(\frac{\sum x'y'}{\sum x'^2}\right)^2 \left(\frac{\sum x'^2}{\sum y'^2}\right) = \frac{\left(\sum x'y'\right)^2}{\sum x'^2 \sum y'^2}$$

and so

$$r = \pm \frac{\sum x'y'}{\sqrt{\sum x'^2 \sum y'^2}}$$

However, since $\sum x'y'$ is positive when y_{est} increases as x increases, but negative when y_{est} decreases as x increases, the expression for r automatically has the correct sign associated with it. Therefore, the required result follows.

8.28. By using the product-moment formula, obtain the linear correlation coefficient for the data of Problem 8.11.

The work involved in the computation can be organized as in Table 8-16. Then

$$r = \frac{\sum x'y'}{\sqrt{\left(\sum x'^2\right)\left(\sum y'^2\right)}} = \frac{40.34}{\sqrt{(84.68)(38.92)}} = 0.7027$$

agreeing with Problem 8.26(b).

Table 8-16

x	y	$x' = x - \bar{x}$	$y' = y - \bar{y}$	x'^2	$x'y'$	y'^2
65	68	−1.7	0.4	2.89	−0.68	0.16
63	66	−3.7	−1.6	13.69	5.92	2.56
67	68	0.3	0.4	0.09	0.12	0.16
64	65	−2.7	−2.6	7.29	7.02	6.76
68	69	1.3	1.4	1.69	1.82	1.96
62	66	−4.7	−1.6	22.09	7.52	2.56
70	68	3.3	0.4	10.89	1.32	0.16
66	65	−0.07	−2.6	0.49	1.82	6.76
68	71	1.3	3.4	1.69	4.42	11.56
67	67	0.3	−0.6	0.09	−0.18	0.36
69	68	2.3	0.4	5.29	0.92	0.16
71	70	4.3	2.4	18.49	10.32	5.76
$\sum x = 800$ $\bar{x} = 800/12 = 66.7$	$\sum y = 811$ $\bar{y} = 811/12 = 67.6$			$\sum x'^2 = 84.68$	$\sum x'y' = 40.34$	$\sum y'^2 = 38.92$

8.29. Prove the result (17), page 268.

The regression line of y on x is

$$y = a + bx \quad \text{where} \quad b = \frac{rs_y}{s_x}$$

Similarly, the regression line of x on y is

$$x = c + dy \quad \text{where} \quad d = \frac{rs_x}{s_y}$$

Then
$$bd = \left(\frac{rs_y}{s_x}\right)\left(\frac{rs_x}{s_y}\right) = r^2$$

8.30. Use the result of Problem 8.29 to find the linear correlation coefficient for the data of Problem 8.11.

From Problem 8.11(b) and 8.11(c), respectively,

$$b = \frac{484}{1016} = 0.476 \qquad d = \frac{484}{467} = 1.036$$

Then
$$r^2 = bd = \left(\frac{484}{1016}\right)\left(\frac{484}{467}\right) \quad \text{or} \quad r = 0.7027$$

agreeing with Problems 8.26(b) and 8.28.

8.31. Show that the linear correlation coefficient is given by

$$r = \frac{n\sum xy - \left(\sum x\right)\left(\sum y\right)}{\sqrt{\left[n\sum x^2 - \left(\sum x\right)^2\right]\left[n\sum y^2 - \left(\sum y\right)^2\right]}}$$

In Problem 8.27 it was shown that

$$(1) \qquad r = \frac{\sum x'y'}{\sqrt{\left(\sum x'^2\right)\left(\sum y'^2\right)}} = \frac{\sum(x - \bar{x})(y - \bar{y})}{\sqrt{\left[\sum(x - \bar{x})^2\right]\left[\sum(y - \bar{y})^2\right]}}$$

But
$$\begin{aligned}\sum(x - \bar{x})(y - \bar{y}) &= \sum(xy - \bar{x}y - x\bar{y} + \bar{x}\bar{y}) = \sum xy - \bar{x}\sum y - \bar{y}\sum x + n\bar{x}\bar{y} \\ &= \sum xy - n\bar{x}\bar{y} - n\bar{y}\bar{x} + n\bar{x}\bar{y} = \sum xy - n\bar{x}\bar{y} \\ &= \sum xy - \frac{\left(\sum x\right)\left(\sum y\right)}{n}\end{aligned}$$

since $\bar{x} = (\sum x)/n$ and $\bar{y} = (\sum y)/n$.

Similarly,
$$\begin{aligned}\sum(x - \bar{x})^2 &= \sum(x^2 - 2x\bar{x} + \bar{x}^2) = \sum x^2 - 2\bar{x}\sum x + n\bar{x}^2 \\ &= \sum x^2 - \frac{2\left(\sum x\right)^2}{n} + \frac{\left(\sum x\right)^2}{n} = \sum x^2 - \frac{\left(\sum x\right)^2}{n}\end{aligned}$$

and
$$\sum(y - \bar{y})^2 = \sum y^2 - \frac{\left(\sum y\right)^2}{n}$$

Then (1) becomes

$$r = \frac{\sum xy - \left(\sum x\right)\left(\sum y\right)/n}{\sqrt{\left[\sum x^2 - \left(\sum x\right)^2/n\right]\left[\sum y^2 - \left(\sum y\right)^2/n\right]}} = \frac{n\sum xy - \left(\sum x\right)\left(\sum y\right)}{\sqrt{\left[n\sum x^2 - \left(\sum x\right)^2\right]\left[n\sum y^2 - \left(\sum y\right)^2\right]}}$$

8.32. Use the formula of Problem 8.31 to obtain the linear correlation coefficient for the data of Problem 8.11.

From Table 8-4,

$$r = \frac{n\sum xy - \left(\sum x\right)\left(\sum y\right)}{\sqrt{\left[n\sum x^2 - \left(\sum x\right)^2\right]\left[n\sum y^2 - \left(\sum y\right)^2\right]}}$$

$$= \frac{(12)(54{,}107) - (800)(811)}{\sqrt{[(12)(53{,}418) - (800)^2][(12)(54{,}849) - (811)^2]}} = 0.7027$$

as in Problems 8.26(b), 8.28, and 8.30.

Generalized correlation coefficient

8.33. (a) Find the linear correlation coefficient between the variables x and y of Problem 8.16. (b) Find a nonlinear correlation coefficient between these variables, assuming the parabolic relationship obtained in Problem 8.16. (c) Explain the difference between the correlation coefficients obtained in (a) and (b). (d) What percentage of the total variation remains unexplained by the assumption of parabolic relationship between x and y?

(a) Using the calculations in Table 8-9 and the added fact that $\sum y^2 = 290.52$, we find

$$r = \frac{n\sum xy - \left(\sum x\right)\left(\sum y\right)}{\sqrt{\left[n\sum x^2 - \left(\sum x\right)^2\right]\left[n\sum y^2 - \left(\sum y\right)^2\right]}}$$

$$= \frac{(8)(230.42) - (42.2)(46.4)}{\sqrt{[(8)(291.20) - (42.2)^2][(8)(290.52) - (46.4)^2]}} = -0.3743$$

(b) From Table 8-9, $\bar{y} = (\sum y)/n = (46.4)/8 = 5.80$. Then

$$\text{Total variation} = \sum(y - \bar{y})^2 = 21.40$$

From Table 8-10,

$$\text{Explained variation} = \sum(y_{\text{est}} - \bar{y})^2 = 21.02$$

Therefore, $$r^2 = \frac{\text{explained variation}}{\text{total variation}} = \frac{21.02}{21.40} = 0.9822 \quad \text{and} \quad r = 0.9911$$

(c) The fact that part (a) shows a linear correlation coefficient of only -0.3743 indicates practically no *linear relationship* between x and y. However, there is a very good *nonlinear relationship* supplied by the parabola of Problem 8.16, as is indicated by the fact that the correlation coefficient in (b) is very nearly 1.

(d) $$\frac{\text{Unexplained variation}}{\text{Total variation}} = 1 - r^2 = 1 - 0.9822 = 0.0178$$

Therefore, 1.78% of the total variation remains unexplained. This could be due to random fluctuations or to an additional variable that has not been considered.

8.34. Find (a) s_y and (b) $s_{y.x}$ for the data of Problem 8.16.

(a) From Problem 8.33(b), $\Sigma(y - \bar{y})^2 = 21.40$. Then the standard deviation of y is

$$s_y = \sqrt{\frac{\sum(y - \bar{y})^2}{n}} = \sqrt{\frac{21.40}{8}} = 1.636 \text{ or } 1.64$$

(b) **First method**

Using (a) and Problem 8.33(b), the standard error of estimate of y on x is

$$s_{y.x} = s_y\sqrt{1 - r^2} = 1.636\sqrt{1 - (0.9911)^2} = 0.218 \text{ or } 0.22$$

Second method

Using Problem 8.33,

$$s_{y.x} = \sqrt{\frac{\sum(y - y_{\text{est}})^2}{n}} = \sqrt{\frac{\text{unexplained variation}}{n}} = \sqrt{\frac{21.40 - 21.02}{8}} = 0.218 \text{ or } 0.22$$

Third method

Using Problem 8.16 and the additional calculation $\sum y^2 = 290.52$, we have

$$s_{y.x} = \sqrt{\frac{\sum y^2 - a\sum y - b\sum xy - c\sum x^2y}{n}} = 0.218 \text{ or } 0.22$$

8.35. Explain how you would determine a multiple correlation coefficient for the variables in Problem 8.19.

Since z is determined from x and y, we are interested in the *multiple correlation coefficient of z on x and y*. To obtain this, we see from Problem 8.19 that

$$\begin{aligned}\text{Unexplained variation} &= \sum(z - z_{\text{est}})^2\\ &= (64 - 64.414)^2 + \cdots + (68 - 65.920)^2 = 258.88\\ \text{Total variation} &= \sum(z - \bar{z})^2 = \sum z^2 - n\bar{z}^2\\ &= 48{,}139 - 12(62.75)^2 = 888.25\\ \text{Explained variation} &= 888.25 - 258.88 = 629.37\end{aligned}$$

Then

$$\text{Multiple correlation coefficient of } z \text{ on } x \text{ and } y$$
$$= \sqrt{\frac{\text{explained variation}}{\text{total variation}}} = \sqrt{\frac{629.37}{888.25}} = 0.8418$$

It should be mentioned that if we were to consider the regression of x on y and z, the multiple correlation coefficient of x on y and z would in general be different from the above value.

Rank correlation

8.36. Derive Spearman's rank correlation formula (36), page 271.

Here we are considering nx values (e.g., weights) and n corresponding y values (e.g., heights). Let x_j be the rank given to the jth x value, and y_j the rank given to the jth y value. The ranks are the integers 1 through n. The mean of the x_j is then

$$\bar{x} = \frac{1 + 2 + \cdots + n}{n} = \frac{n(n + 1)/2}{n} = \frac{n + 1}{2}$$

while the variance is

$$\begin{aligned}s_x^2 = \overline{x^2} - \bar{x}^2 &= \frac{1^2 + 2^2 + \cdots + n^2}{n} - \left(\frac{n + 1}{2}\right)^2\\ &= \frac{n(n + 1)(2n + 1)/6}{n} - \left(\frac{n + 1}{2}\right)^2\\ &= \frac{n^2 - 1}{12}\end{aligned}$$

using the results 1 and 2 of Appendix A. Similarly, the mean $\bar{y}$ and variance s_y^2 are equal to $(n + 1)/2$ and $(n^2 - 1)/12$, respectively.

Now if $d_j = x_j - y_j$ are the deviations between the ranks, the variance of the deviations, s_d^2, is given in terms of s_x^2, s_y^2 and the correlation coefficient between ranks by

$$s_d^2 = s_x^2 + s_y^2 - 2r_{\text{rank}}s_xs_y$$

Then

(1) $$r_{\text{rank}} = \frac{s_x^2 + s_y^2 - s_d^2}{2s_x s_y}$$

Since $\bar{d} = 0, s_d^2 = (\sum d^2)/n$ and (1) becomes

(2) $$r_{\text{rank}} = \frac{(n^2 - 1)/12 + (n^2 - 1)/12 - \left(\sum d^2\right)/n}{(n^2 - 1)/6} = 1 - \frac{6\sum d^2}{n(n^2 - 1)}$$

8.37. Table 8-17 shows how 10 students were ranked according to their achievements in both the laboratory and lecture portions of a biology course. Find the coefficient of rank correlation.

Table 8-17

Laboratory	8	3	9	2	7	10	4	6	1	5
Lecture	9	5	10	1	8	7	3	4	2	6

The difference of ranks d in laboratory and lecture for each student is given in Table 8-18. Also given in the table are d^2 and $\sum d^2$.

Table 8-18

Difference of Ranks (d)	−1	−2	−1	1	−1	3	1	2	−1	−1	
d^2	1	4	1	1	1	9	1	4	1	1	$\sum d^2 = 24$

Then $$r_{\text{rank}} = 1 - \frac{6\sum d^2}{n(n^2 - 1)} = 1 - \frac{6(24)}{10(10^2 - 1)} = 0.8545$$

indicating that there is a marked relationship between achievements in laboratory and lecture.

8.38. Calculate the coefficient of rank correlation for the data of Problem 8.11, and compare your result with the correlation coefficient obtained by other methods.

Arranged in ascending order of magnitude, the fathers' heights are

(1) 62, 63, 64, 65, 66, 67, 67, 68, 68, 69, 70, 71

Since the 6th and 7th places in this array represent the same height (67 inches), we assign a *mean rank* 6.5 to both these places. Similarly, the 8th and 9th places are assigned the rank 8.5. Therefore, the fathers' heights are assigned the ranks

(2) 1, 2, 3, 4, 5, 6.5, 6.5, 8.5, 8.5, 10, 11, 12

Similarly, the sons' heights arranged in ascending order of magnitude are

(3) 65, 65, 66, 66, 67, 68, 68, 68, 68, 69, 70, 71

and since the 6th, 7th, 8th, and 9th places represent the same height (68 inches), we assign the *mean rank* 7.5 $(6 + 7 + 8 + 9)/4$ to these places. Therefore, the sons' heights are assigned the ranks

(4) 1.5, 1.5, 3.5, 3.5, 5, 7.5, 7.5, 7.5, 7.5, 10, 11, 12

Using the correspondences (1) and (2), (3) and (4), Table 8-3 becomes Table 8-19.

Table 8-19

Rank of Father	4	2	6.5	3	8.5	1	11	5	8.5	6.5	10	12
Rank of Son	7.5	3.5	7.5	1.5	10	3.5	7.5	1.5	12	5	7.5	11

The differences in ranks d, and the computations of d^2 and $\sum d^2$ are shown in Table 8-20.

Table 8-20

d	−3.5	−1.5	−1.0	1.5	−1.5	−2.5	3.5	3.5	−3.5	1.5	2.5	1.0	
d^2	12.25	2.25	1.00	2.25	2.25	6.25	12.25	12.25	12.25	2.25	6.25	1.00	$\sum d^2 = 72.50$

Then
$$r_{\text{rank}} = 1 - \frac{6\sum d^2}{n(n^2 - 1)} = 1 - \frac{6(72.50)}{12(12^2 - 1)} = 0.7465$$

which agrees well with the value $r = 0.7027$ obtained in Problem 8.26(b).

Probability interpretation of regression and correlation

8.39. Derive (39) from (37).

Assume that the regression equation is

$$y = E(Y \mid X = x) = \alpha + \beta x$$

For the least-squares regression line we must consider

$$\begin{aligned} E\{[Y - (\alpha + \beta X)]^2\} &= E\{[(Y - \mu_Y) - \beta(X - \mu_X) + (\mu_Y - \beta\mu_X - \alpha)]^2\} \\ &= E[(Y - \mu_Y)^2] + \beta^2 E[(X - \mu_X)^2] - 2\beta E[(X - \mu_X)(Y - \mu_Y)] + (\mu_Y - \beta\mu_X - \alpha)^2 \\ &= \sigma_Y^2 + \beta^2\sigma_X^2 - 2\beta\sigma_{XY} + (\mu_Y - \beta\mu_X - \alpha)^2 \end{aligned}$$

where we have used $E(X - \mu_X) = 0$, $E(Y - \mu_Y) = 0$.
Denoting the last expression by $F(\alpha, \beta)$, we have

$$\frac{\partial F}{\partial \alpha} = -2(\mu_Y - \beta\mu_X - \alpha), \qquad \frac{\partial F}{\partial \beta} = 2\beta\sigma_X^2 - 2\sigma_{XY} - 2\mu_X(\mu_Y - \beta\mu_X - \alpha)$$

Setting these equal to zero, which is a necessary condition for $F(\alpha, \beta)$ to be a minimum, we find

$$\mu_Y = \alpha + \beta\mu_X \qquad \beta\sigma_X^2 = \sigma_{XY}$$

Therefore, if $y = \alpha + \beta x$, then $y - \mu_Y = \beta(x - \mu_X)$ or

$$y - \mu_Y = \frac{\sigma_{XY}}{\sigma_X^2}(x - \mu_X)$$

or
$$\frac{y - \mu_Y}{\sigma_Y} = \rho\left(\frac{x - \mu_X}{\sigma_X}\right)$$

The similarity of the above proof for populations, using expectations, to the corresponding proof for samples, using summations, should be noted. In general, results for samples have analogous results for populations and conversely.

8.40. The joint density function of the random variables X and Y is

$$f(x, y) = \begin{cases} \frac{2}{3}(x + 2y) & 0 \le x \le 1, 0 \le y \le 1 \\ 0 & \text{otherwise} \end{cases}$$

Find the least-squares regression curve of (a) Y on X, (b) X on Y.

(a) The marginal density function of X is

$$f_1(x) = \int_0^1 \frac{2}{3}(x + 2y)\,dy = \frac{2}{3}(x + 1)$$

for $0 \le x \le 1$, and $f_1(x) = 0$ otherwise. Hence, for $0 \le x \le 1$, the conditional density of Y given X is

$$f_2(y \mid x) = \frac{f(x, y)}{f_1(x)} = \begin{cases} \dfrac{x + 2y}{x + 1} & 0 \le y \le 1 \\ 0 & y < 0 \text{ or } y > 1 \end{cases}$$

and the least-squares regression curve of Y on X is given by

$$y = E(Y \mid X = x) = \int_{-\infty}^{\infty} y f_2(y \mid x)\, dy$$

$$= \int_0^1 y\left(\frac{x + 2y}{x + 1}\right) dy = \frac{3x + 4}{6x + 6}$$

Neither $f_2(y \mid x)$ nor the least-squares regression curve is defined when $x < 0$ or $x > 1$.

(b) For $0 \le y \le 1$, the marginal density function of Y is

$$f_2(y) = \int_0^1 \frac{2}{3}(x + 2y)\, dx = \frac{1}{3}(1 + 4y)$$

Hence, for $0 \le y \le 1$, the conditional density of X given Y is

$$f_1(x \mid y) = \frac{f(x, y)}{f_2(y)} = \begin{cases} \dfrac{2x + 4y}{1 + 4y} & 0 \le x \le 1 \\ 0 & x < 0 \text{ or } x > 1 \end{cases}$$

and the least-squares regression curve of X on Y is given by

$$x = E(X \mid Y = y) = \int_{-\infty}^{\infty} x f_1(x \mid y)\, dx$$

$$= \int_0^1 x\left(\frac{2x + 4y}{1 + 4y}\right) dx = \frac{2 + 6y}{3 + 12y}$$

Neither $f_1(x \mid y)$ nor the least-squares regression curve is defined when $y < 0$ or $y > 1$.

Note that the two regression curves $y = (3x + 4)/(6x + 6)$ and $x = (2 + 6y)/(3 + 12y)$ are different.

8.41. Find (a) $\bar{X}$, (b) $\bar{Y}$, (c) σ_X^2, (d) σ_Y^2, (e) σ_{XY}, (f) ρ for the distribution in Problem 8.40.

(a) $$\bar{X} = \int_{x=0}^{1}\int_{y=0}^{1} x\left[\frac{2}{3}(x + 2y)\right] dx\, dy = \frac{5}{9}$$

(b) $$\bar{Y} = \int_{x=0}^{1}\int_{y=0}^{1} y\left[\frac{2}{3}(x + 2y)\right] dx\, dy = \frac{11}{18}$$

(c) $$\overline{X^2} = \int_{x=0}^{1}\int_{y=0}^{1} x^2\left[\frac{2}{3}(x + 2y)\right] dx\, dy = \frac{7}{18}$$

Then $$\sigma_X^2 = \overline{X^2} - \bar{X}^2 = \frac{7}{18} - \left(\frac{5}{9}\right)^2 = \frac{13}{162}$$

(d) $$\overline{Y^2} = \int_{x=0}^{1}\int_{y=0}^{1} y^2\left[\frac{2}{3}(x + 2y)\right] dx\, dy = \frac{4}{9}$$

Then $$\sigma_Y^2 = \overline{Y^2} - \bar{Y}^2 = \frac{4}{9} - \left(\frac{11}{18}\right)^2 = \frac{23}{324}$$

(e) $$\overline{XY} = \int_{x=0}^{1}\int_{y=0}^{1} xy\left[\frac{2}{3}(x + 2y)\right] dx\, dy = \frac{1}{3}$$

Then $$\sigma_{XY} = \overline{XY} - \bar{X}\bar{Y} = \frac{1}{3} - \left(\frac{5}{9}\right)\left(\frac{11}{18}\right) = -\frac{1}{162}$$

(f) $$\rho = \frac{\sigma_{XY}}{\sigma_X \sigma_Y} = \frac{-1/162}{\sqrt{13/162}\sqrt{23/324}} = -0.0818$$

Note that the linear correlation coefficient is small. This is to be expected from observation of the least-squares regression lines obtained in Problem 8.42.

8.42. Write the least-squares regression lines of (a) Y on X, (b) X on Y for Problem 8.40.

(a) The regression line of Y on X is $\dfrac{y - \bar{Y}}{\sigma_Y} = \rho\left(\dfrac{x - \bar{X}}{\sigma_X}\right)$ or

$$y - \bar{Y} = \frac{\sigma_{XY}}{\sigma_X^2}(x - \bar{X}) \quad \text{or} \quad y - \frac{11}{18} = \frac{-1/162}{13/162}\left(x - \frac{5}{9}\right)$$

(b) The regression line of X on Y is $\dfrac{x - \bar{X}}{\sigma_X} = \rho\left(\dfrac{y - \bar{Y}}{\sigma_Y}\right)$ or

$$x - \bar{X} = \frac{\sigma_{XY}}{\sigma_Y^2}(y - \bar{Y}) \quad \text{or} \quad x - \frac{5}{9} = \frac{-1/162}{23/324}\left(y - \frac{11}{18}\right)$$

Sampling theory of regression

8.43. In Problem 8.11 we found the regression equation of y on x to be $y = 35.82 + 0.476x$. Test the hypothesis at a 0.05 significance level that the regression coefficient of the population regression equation is as low as 0.180.

$$t = \frac{\beta - b}{s_{y.x}/s_x}\sqrt{n-2} = \frac{0.476 - 0.180}{1.28/2.66}\sqrt{12 - 2} = 1.95$$

since $s_{y.x} = 1.28$ (computed in Problem 8.22) and $s_x = \sqrt{\bar{x^2} - \bar{x}^2} = 2.66$ from Problem 8.11.

On the basis of a one-tailed test of Student's distribution at a 0.05 level, we would reject the hypothesis that the regression coefficient is as low as 0.180 if $t > t_{0.95} = 1.81$ for $12 - 2 = 10$ degrees of freedom. Therefore, we can reject the hypothesis.

8.44. Find 95% confidence limits for the regression coefficient of Problem 8.43.

$$\beta = b + \frac{t}{\sqrt{n-2}}\left(\frac{s_{y.x}}{s_x}\right)$$

Then 95% confidence limits for β (obtained by putting $t = \pm t_{0.975} = \pm 2.23$ for $12 - 2 = 10$ degrees of freedom) are given by

$$b \pm \frac{2.23}{\sqrt{12-2}}\left(\frac{s_{y.x}}{s_x}\right) = 0.476 \pm \frac{2.23}{\sqrt{10}}\left(\frac{1.28}{2.66}\right) = 0.476 \pm 0.340$$

i.e., we are 95% confident that β lies between 0.136 and 0.816.

8.45. In Problem 8.11, find 95% confidence limits for the heights of sons whose fathers' heights are (a) 65.0, (b) 70.0 inches.

Since $t_{0.975} = 2.23$ for $12 - 2 = 10$ degrees of freedom, the 95% confidence limits for y_p are

$$y_0 \pm \frac{2.23}{\sqrt{n-2}} s_{y.x}\sqrt{n + 1 + \frac{n(x_0 - \bar{x})^2}{s_x^2}}$$

where $y_0 = 35.82 + 0.476x_0$ (Problem 8.11), $s_{y.x} = 1.28$, $s_x = 2.66$ (Problem 8.43), and $n = 12$.

(a) If $x_0 = 65.0$, $y_0 = 66.76$ inches. Also, $(x_0 - \bar{x})^2 = (65.0 - 800/12)^2 = 2.78$. Then 95% confidence limits are

$$66.76 \pm \frac{2.23}{\sqrt{10}}(1.28)\sqrt{12 + 1 + \frac{12(2.78)}{(2.66)^2}} = 66.76 \pm 3.80 \text{ inches}$$

i.e., we can be about 95% confident that the sons' heights are between 63.0 and 70.6 inches.

(b) If $x_0 = 70.0$, $y_0 = 69.14$ inches. Also, $(x_0 - \bar{x})^2 = (70.0 - 800/12)^2 = 11.11$. Then the 95% confidence limits are computed to be 69.14 ± 5.09 inches, i.e., we can be about 95% confident that the sons' heights are between 64.1 and 74.2 inches.

Note that for large values of n, 95% confidence limits are given approximately by $y_0 \pm 1.96\, s_{y.x}$ or $y_0 \pm 2s_{y.x}$ provided that $x_0 - \bar{x}$ is not too large. This agrees with the approximate results mentioned on page 269. The methods of this problem hold regardless of the size of n or $x_0 - \bar{x}$, i.e., the sampling methods are exact for a normal population.

8.46. In Problem 8.11, find 95% confidence limits for the mean heights of sons whose fathers' heights are (a) 65.0, (b) 70.0 inches.

Since $t_{0.975} = 2.23$ for 10 degrees of freedom, the 95% confidence limits for $\bar{y}_p$ are

$$y_0 \pm \frac{2.23}{\sqrt{20}} s_{y.x}\sqrt{1 + \frac{(x_0 - \bar{x})^2}{s_x^2}}$$

where $y_0 = 35.82 + 0.476x_0$ (Problem 8.11), $s_{y.x} = 1.28$ (Problem 8.43).

(a) If $x_0 = 65.0$, we find [compare Problem 8.45(a)] the 95% confidence limits 66.76 ± 1.07 inches, i.e., we can be about 95% confident that the mean height of all sons whose fathers' heights are 65.0 inches will lie between 65.7 and 67.8 inches.

(b) If $x_0 = 70.0$, we find [compare Problem 8.45(b)] the 95% confidence limits 69.14 ± 1.45 inches, i.e., we can be about 95% confident that the mean height of all sons whose fathers' heights are 70.0 inches will lie between 67.7 and 70.6 inches.

Sampling theory of correlation

8.47. A correlation coefficient based on a sample of size 18 was computed to be 0.32. Can we conclude at a significance level of (a) 0.05, (b) 0.01 that the corresponding population correlation coefficient is significantly greater than zero?

We wish to decide between the hypotheses (H_0: $\rho = 0$) and (H_1: $\rho > 0$).

$$t = \frac{r\sqrt{n-2}}{\sqrt{1-r^2}} = \frac{0.32\sqrt{18-2}}{\sqrt{1-(0.32)^2}} = 1.35$$

(a) On the basis of a one-tailed test of Student's distribution at a 0.05 level, we would reject H_0 if $t > t_{0.95} = 1.75$ for $18 - 2 = 16$ degrees of freedom. Therefore, we cannot reject H_0 at a 0.05 level.

(b) Since we cannot reject H_0 at a 0.05 level, we certainly cannot reject it at a 0.01 level.

8.48. What is the minimum sample size necessary in order that we may conclude that a correlation coefficient of 0.32 is significantly greater than zero at a 0.05 level?

At a 0.05 level using a one-tailed test of Student's distribution, the minimum value of n must be such that

$$\frac{0.32\sqrt{n-2}}{\sqrt{1-(0.32)^2}} = t_{0.95} \quad \text{for } n - 2 \text{ degrees of freedom}$$

For $n = 26$, $\nu = 24$, $t_{0.95} = 1.71$, $t = 0.32\sqrt{24}/\sqrt{1-(0.32)^2} = 1.65$.

For $n = 27$, $\nu = 25$, $t_{0.95} = 1.71$, $t = 0.32\sqrt{25}/\sqrt{1-(0.32)^2} = 1.69$.

For $n = 28$, $\nu = 26$, $t_{0.95} = 1.71$, $t = 0.32\sqrt{26}/\sqrt{1-(0.32)^2} = 1.72$.

Then the minimum sample size is $n = 28$.

8.49. A correlation coefficient based on a sample of size 24 was computed to be $r = 0.75$. Can we reject the hypothesis that the population correlation coefficient is as small as (a) $\rho = 0.60$, (b) $\rho = 0.50$, at a 0.05 significance level?

(a) $$Z = 1.1513 \log\left(\frac{1 + 0.75}{1 - 0.75}\right) = 0.9730, \qquad \mu_Z = 1.1513 \log\left(\frac{1 + 0.60}{1 - 0.60}\right) = 0.6932,$$

$$\sigma_Z = \frac{1}{\sqrt{n - 3}} = \frac{1}{\sqrt{21}} = 0.2182$$

The standardized variable is then

$$z = \frac{Z - \mu_Z}{\sigma_Z} = \frac{0.9730 - 0.6932}{0.2182} = 1.28$$

At a 0.05 level of significance using a one-tailed test of the normal distribution, we would reject the hypothesis only if z were greater than 1.64. Therefore, we cannot reject the hypothesis that the population correlation coefficient is as small as 0.60.

(b) If $\rho = 0.50$, $\mu_Z = 1.1513 \log 3 = 0.5493$ and $z = (0.9730 - 0.5493)/0.2182 = 1.94$. Therefore, we can reject the hypothesis that the population correlation coefficient is as small as $\rho = 0.50$ at a 0.05 level of significance.

8.50. The correlation coefficient between physics and mathematics final grades for a group of 21 students was computed to be 0.80. Find 95% confidence limits for this coefficient.

Since $r = 0.80$ and $n = 21$, 95% confidence limits for μ_2 are given by

$$Z \pm 1.96\sigma_Z = 1.1513 \log\left(\frac{1 + r}{1 - r}\right) \pm 1.96\left(\frac{1}{\sqrt{n - 3}}\right) = 1.0986 \pm 0.4620$$

Then μ_Z has the 95% confidence interval 0.5366 to 1.5606.

If $$\mu_Z = 1.1513 \log\left(\frac{1 + \rho}{1 - \rho}\right) = 0.5366, \qquad \rho = 0.4904.$$

If $$\mu_Z = 1.1513 \log\left(\frac{1 + \rho}{1 - \rho}\right) = 1.5606, \qquad \rho = 0.9155.$$

Therefore, the 95% confidence limits for ρ are 0.49 and 0.92.

8.51. Two correlation coefficients obtained from samples of size $n_1 = 28$ and $n_2 = 35$ were computed to be $r_1 = 0.50$ and $r_2 = 0.30$, respectively. Is there a significant difference between the two coefficients at a 0.05 level?

$$Z_1 = 1.1513 \log\left(\frac{1 + r_1}{1 - r_1}\right) = 0.5493, \qquad Z_2 = 1.1513 \log\left(\frac{1 + r_2}{1 - r_2}\right) = 0.3095$$

and $$\sigma_{Z_1 - Z_2} = \sqrt{\frac{1}{n_1 - 3} + \frac{1}{n_2 - 3}} = 0.2669$$

We wish to decide between the hypotheses $(H_0: \mu_{Z_1} = \mu_{Z_2})$ and $(H_1: \mu_{Z_1} \neq \mu_{Z_2})$. Under the hypothesis H_0,

$$z = \frac{Z_1 - Z_2 - (\mu_{Z_1} - \mu_{Z_2})}{\sigma_{Z_1 - Z_2}} = \frac{0.5493 - 0.3095 - 0}{0.2669} = 0.8985$$

Using a two-tailed test of the normal distribution, we would reject H_0 only if $z > 1.96$ or $z < -1.96$. Therefore, we cannot reject H_0, and we conclude that the results are not significantly different at a 0.05 level.

Miscellaneous problems

8.52. Prove formula (25), page 269.

For the least-squares line we have, from Problems 8.20 and 8.21,

$$s_{y.x}^2 = \frac{\sum(y-\bar{y})^2}{n} - b\frac{\sum(x-\bar{x})(y-\bar{y})}{n}$$

But by definition,

$$\frac{\sum(y-\bar{y})^2}{n} = s_y^2 \qquad \frac{\sum(x-\bar{x})(y-\bar{y})}{n} = s_{xy}$$

and, by (6) on page 267,

$$b = \frac{\sum(x-\bar{x})(y-\bar{y})}{\sum(x-\bar{x})^2} = \frac{s_{xy}}{s_x^2}$$

Hence,
$$s_{y.x}^2 = s_y^2 - \frac{s_{xy}^2}{s_x^2} = s_y^2\left[1 - \left(\frac{s_{xy}}{s_x s_y}\right)^2\right] = s_y^2(1 - r^2)$$

An analogous formula holds for the population (see Problem 8.54).

8.53. Prove that $E[(Y - \bar{Y})^2] = E[(Y - Y_{\text{est}})^2] + [(Y_{\text{est}} - \bar{Y})^2]$ for the case of (a) a least-squares line, (b) a least-squares parabola.

We have
$$Y - \bar{Y} = (Y - Y_{\text{est}}) + (Y_{\text{est}} - \bar{Y})$$

Then
$$(Y - \bar{Y})^2 = (Y - Y_{\text{est}})^2 + (Y_{\text{est}} - \bar{Y})^2 + 2(Y - Y_{\text{est}})(Y_{\text{est}} - \bar{Y})$$

and so
$$E[(Y - \bar{Y})^2 = E[(Y - Y_{\text{est}})^2] + E[(Y_{\text{est}} - \bar{Y})^2] + 2E[(Y - Y_{\text{est}})(Y_{\text{est}} - \bar{Y})]$$

The required result follows if we can show that the last term is zero.

(a) For linear regression, $Y_{\text{est}} = \alpha + \beta X$. Then

$$\begin{aligned} E[(Y - Y_{\text{est}})(Y_{\text{est}} - \bar{Y})] &= E[(Y - \alpha - \beta X)(\alpha + \beta X - \bar{Y})] \\ &= (\alpha - \bar{Y})E(Y - \alpha - \beta X) + \beta E(XY - \alpha X - \beta X^2) \\ &= 0 \end{aligned}$$

because of the normal equations

$$E(Y - \alpha - \beta X) = 0, \qquad E(XY - \alpha X - \beta X^2) = 0$$

(Compare Problem 8.3.)

(b) For parabolic regression, $Y_{\text{est}} = \alpha + \beta X + \gamma X^2$. Then

$$\begin{aligned} E[(Y - Y_{\text{est}})(Y_{\text{est}} - \bar{Y})] &= E[(Y - \alpha - \beta X - \gamma X^2)(\alpha + \beta X + \gamma X^2 - \bar{Y})] \\ &= (\alpha - \bar{Y})E(Y - \alpha - \beta X - \gamma X^2) + \beta E[X(Y - \alpha - \beta X - \gamma X^2)] \\ &\quad + \gamma E[X^2(Y - \alpha - \beta X - \gamma X^2)] \\ &= 0 \end{aligned}$$

because of the normal equations

$$E(Y - \alpha - \beta X - \gamma X^2) = 0, \quad E[X(Y - \alpha - \beta X - \gamma X^2)] = 0, \quad E[X^2(Y - \alpha - \beta X - \gamma X^2)] = 0$$

Compare equations (19), page 269.

The result can be extended to higher-order least-squares curves.

8.54. Prove that $\sigma_{Y.X}^2 = \sigma_Y^2(1 - \rho^2)$ for least-squares regression.

By definition of the generalized correlation coefficient ρ, together with Problem 8.53, we have for either the linear or parabolic case

$$\rho^2 = \frac{E[(Y_{\text{est}} - \bar{Y})^2]}{E[(Y - \bar{Y})^2]} = 1 - \frac{E[(Y - Y_{\text{est}})^2]}{E[(Y - \bar{Y})^2]} = 1 - \frac{\sigma_{Y.X}^2}{\sigma_Y^2}$$

and the result follows at once.

The relation also holds for higher-order least-squares curves.

8.55. Show that for the case of linear regression the correlation coefficient as defined by (45) reduces to that defined by (40).

The square of the correlation coefficient, i.e., the *coefficient of determination*, as given by (45) is in the case of linear regression given by

(1)
$$\rho^2 = \frac{E[(Y_{\text{est}} - \bar{Y})^2]}{E[(Y - \bar{Y})^2]} = \frac{E[(\alpha + \beta X - \bar{Y})^2]}{\sigma_Y^2}$$

But since $\bar{Y} = \alpha + \beta\bar{X}$,

(2)
$$E[(\alpha + \beta X - \bar{Y})^2] = E[\beta^2(X - \bar{X})^2] = \beta^2 E[(X - \bar{X})^2]$$
$$= \frac{\sigma_{XY}^2}{\sigma_X^4}\sigma_X^2 = \frac{\sigma_{XY}^2}{\sigma_X^2}$$

Then (1) becomes

(3)
$$\rho^2 = \frac{\sigma_{XY}^2}{\sigma_X^2\sigma_Y^2} \quad \text{or} \quad \rho = \frac{\sigma_{XY}}{\sigma_X\sigma_Y}$$

as we were required to show. (The correct sign for ρ is included in σ_{XY}.)

8.56. Refer to Table 8-21. (a) Find a least-squares regression parabola fitting the data. (b) Compute the regression values (commonly called *trend values*) for the given years and compare with the actual values. (c) Estimate the population in 1945. (d) Estimate the population in 1960 and compare with the actual value, 179.3. (e) Estimate the population in 1840 and compare with the actual value, 17.1.

Table 8-21

Year	1850	1860	1870	1880	1890	1900	1910	1920	1930	1940	1950
U.S. Population (millions)	23.2	31.4	39.8	50.2	62.9	76.0	92.0	105.7	122.8	131.7	151.1

Source: Bureau of the Census.

(a) Let the variables x and y denote, respectively, the year and the population during that year. The equation of a least-squares parabola fitting the data is

(1)
$$y = a + bx + cx^2$$

where a, b, and c are found from the normal equations

(2)
$$\sum y = an + b\sum x + c\sum x^2$$
$$\sum xy = a\sum x + b\sum x^2 - c\sum x^3$$
$$\sum x^2y = a\sum x^2 + b\sum x^3 + c\sum x^4$$

It is convenient to locate the origin so that the middle year, 1900, corresponds to $x = 0$, and to choose a unit that makes the years 1910, 1920, 1930, 1940, 1950 and 1890, 1880, 1870, 1860, 1850 correspond to 1, 2, 3, 4, 5 and −1, −2, −3, −4, −5, respectively. With this choice $\sum x$ and $\sum x^3$ are zero and equations (2) are simplified.

The work involved in computation can be arranged as in Table 8-22. The normal equations (2) become

$$(3)\qquad \begin{aligned} 11a + 110c &= 886.8 \\ 110b &= 1429.8 \\ 110a + 1958c &= 9209.0 \end{aligned}$$

From the second equation in (3), $b = 13.00$; from the first and third equations, $a = 76.64$, $c = 0.3974$. Then the required equation is

$$(4)\qquad y = 76.64 + 13.00x + 0.3974x^2$$

where the origin, $x = 0$, is July 1, 1900, and the unit of x is 10 years.

(b) The trend values, obtained by substituting $x = -5, -4, -3, -2, -1, 0, 1, 2, 3, 4, 5$ in (4), are shown in Table 8-23 together with the actual values. It is seen that the agreement is good.

Table 8-22

Year	x	y	x^2	x^3	x^4	xy	x^2y
1850	−5	23.2	25	−125	625	−116.0	580.0
1860	−4	31.4	16	−64	256	−125.6	502.4
1870	−3	39.8	9	−27	81	−119.4	358.2
1880	−2	50.2	4	−8	16	−100.4	200.8
1890	−1	62.9	1	−1	1	−62.9	62.9
1900	0	76.0	0	0	0	0	0
1910	1	92.0	1	1	1	92.0	92.0
1920	2	105.7	4	8	16	211.4	422.8
1930	3	122.8	9	27	81	368.4	1105.2
1940	4	131.7	16	64	256	526.8	2107.2
1950	5	151.1	25	125	625	755.5	3777.5
	$\sum x = 0$	$\sum y =$ 886.8	$\sum x^2 =$ 110	$\sum x^3 = 0$	$\sum x^4 =$ 1958	$\sum xy =$ 1429.8	$\sum x^2y =$ 9209.0

Table 8-23

Year	$x = -5$ 1850	$x = -4$ 1860	$x = -3$ 1870	$x = -2$ 1880	$x = -1$ 1890	$x = 0$ 1900	$x = 1$ 1910	$x = 2$ 1920	$x = 3$ 1930	$x = 4$ 1940	$x = 5$ 1950
Trend Value	21.6	31.0	41.2	52.2	64.0	76.6	90.0	104.2	119.2	135.0	151.6
Actual Value	23.2	31.4	39.8	50.2	62.9	76.0	92.0	105.7	122.8	131.7	151.1

(c) 1945 corresponds to $x = 4.5$, for which $y = 76.64 + 13.00(4.5) + 0.3974(4.5)^2 = 143.2$.

(d) 1960 corresponds to $x = 6$, for which $y = 76.64 + 13.00(6) + 0.3974(6)^2 = 168.9$. This does not agree too well with the actual value, 179.3.

(e) 1840 corresponds to $x = -6$, for which $y = 76.64 + 13.00(-6) + 0.3974(-6)^2 = 12.9$. This does not agree with the actual value, 17.1.

This example illustrates the fact that a relationship which is found to be satisfactory for a range of values need not be satisfactory for an extended range of values.

8.57. The average prices of stocks and bonds listed on the New York Stock Exchange during the years 1950 through 1959 are given in Table 8-24. (a) Find the correlation coefficient, (b) interpret the results.

Table 8-24

Year	1950	1951	1952	1953	1954	1955	1956	1957	1958	1959
Average Price of Stocks (dollars)	35.22	39.87	41.85	43.23	40.06	53.29	54.14	49.12	40.71	55.15
Average Price of Bond (dollars)	102.43	100.93	97.43	97.81	98.32	100.07	97.08	91.59	94.85	94.65

Source: New York Stock Exchange.

(a) Denoting by x and y the average prices of stocks and bonds, the calculation of the correlation coefficient can be organized as in Table 8-25. Note that the year is used only to specify the corresponding values of x and y.

Table 8-25

x	y	$x' = x - \bar{x}$	$y' = y - \bar{y}$	x'^2	$x'y'$	y'^2
35.22	102.43	−10.04	4.91	100.80	−49.30	24.11
39.87	100.93	−5.39	3.41	29.05	−18.38	11.63
41.85	97.43	−3.41	−0.09	11.63	0.31	0.01
43.23	97.81	−2.03	0.29	4.12	−0.59	0.08
40.06	98.32	−5.20	0.80	27.04	−4.16	0.64
53.29	100.07	8.03	2.55	64.48	20.48	6.50
54.14	97.08	8.88	−0.44	78.85	−3.91	0.19
49.12	91.59	3.86	−5.93	14.90	−22.89	35.16
40.71	94.85	−4.55	−2.67	20.70	12.15	7.13
55.15	94.65	9.89	−2.87	97.81	−28.38	8.24
$\sum x =$ 452.64 $\bar{x} =$ 45.26	$\sum x =$ 975.16 $\bar{y} =$ 97.52			$\sum x'^2 =$ 449.38	$\sum x'y' =$ −94.67	$\sum y'^2 =$ 93.69

Then by the product-moment formula,

$$r = \frac{\sum x'y'}{\sqrt{\left(\sum x'^2\right)\left(\sum y'^2\right)}} = \frac{-94.67}{\sqrt{(449.38)(93.69)}} = -0.4614$$

(b) We conclude that there is some negative correlation between stock and bond prices (i.e., a tendency for stock prices to go down when bond prices go up, and vice versa), although this relationship is not marked.

Another method

Table 8-26 shows the ranks of the average prices of stocks and bonds for the years 1950 through 1959 in order of increasing prices. Also shown in the table are the differences in rank d and $\sum d^2$.

Table 8-26

Year	1950	1951	1952	1953	1954	1955	1956	1957	1958	1959	
Stock Prices in Order of Rank	1	2	5	6	3	8	9	7	4	10	
Bond Prices in Order of Rank	10	9	5	6	7	8	4	1	3	2	
Differences in Rank (d)	−9	−7	0	0	−4	0	5	6	1	8	
d^2	81	49	0	0	16	0	25	36	1	64	$\sum d^2 = 272$

Then $$r_{\text{rank}} = 1 - \frac{6\sum d^2}{n(n^2 - 1)} = 1 - \frac{6(272)}{10(10^2 - 1)} = -0.6485$$

This compares favorably with the result of the first method.

8.58. Table 8-27 shows the frequency distributions of the final grades of 100 students in mathematics and physics. With reference to this table determine (a) the number of students who received grades 70 through 79 in mathematics and 80 through 89 in physics, (b) the percentage of students with mathematics grades below 70, (c) the number of students who received a grade of 70 or more in physics and less than 80 in mathematics, (d) the percentage of students who passed at least one of the subjects assuming 60 to be the minimum passing grade.

Table 8-27
MATHEMATICS GRADES

PHYSICS GRADES	40–49	50–59	60–69	70–79	80–89	90–99	TOTALS
90–99				2	4	4	10
80–89			1	4	6	5	16
70–79			5	10	8	1	24
60–69	1	4	9	5	2		21
50–59	3	6	6	2			17
40–49	3	5	4				12
TOTALS	7	15	25	23	20	10	100

(a) Proceed down the column headed 70–79 (mathematics grade) to the row marked 80–89 (physics grade). The entry 4 gives the required number of students.

(b) Total number of students with mathematics grades below 70

$$= \text{(number with grades 40–49)} + \text{(number with grades 50–59)} + \text{(number with grades 60–69)}$$
$$= 7 + 15 + 25 = 47$$

Percentage of students with mathematics grades below 70 $= 47/100 = 47\%$.

(c) The required number of students in the total of the entries in Table 8-28, which represents part of Table 8-27.

Required number of students $= 1 + 5 + 2 + 4 + 10 = 22$.

Table 8-28

MATHEMATICS GRADES

PHYSICS GRADES	60–69	70–79
90–99		2
80–89	1	4
70–79	5	10

Table 8-29

MATHEMATICS GRADES

PHYSICS GRADES	40–49	50–59
50–59	3	6
40–49	3	5

(d) Referring to Table 8-29, which is taken from Table 8-27, it is seen that the number of students with grades below 60 in both mathematics and physics is $3 + 3 + 6 + 5 = 17$. Then the number of students with grades 60 or over in either physics or mathematics or both is $100 - 17 = 83$, and the required percentage is $83/100 = 83\%$.

Table 8-27 is sometimes called a *bivariate frequency table* or *bivariate frequency distribution*. Each square in the table is called a *cell* and corresponds to a pair of classes or class intervals. The number indicated in the cell is called the *cell frequency*. For example, in part (a) the number 4 is the frequency of the cell corresponding to the pair of class intervals 70–79 in mathematics and 80–89 in physics.

The totals indicated in the last row and last column are called *marginal totals* or *marginal frequencies*. They correspond, respectively, to the class frequencies of the separate frequency distributions of mathematics and physics grades.

8.59. Show how to modify the formula of Problem 8.31 for the case of data grouped as in Table 8-27.

For grouped data, we can consider the various values of the variables x and y as coinciding with the class marks, while f_x and f_y are the corresponding class frequencies or marginal frequencies indicated in the last row and column of the bivariate frequency table. If we let f represent the various cell frequencies corresponding to the pairs of class marks (x, y), then we can replace the formula of Problem 8.31 by

$$r = \frac{n\sum fxy - \left(\sum f_x x\right)\left(\sum f_y y\right)}{\sqrt{\left[n\sum f_x x^2 - \left(\sum f_x x\right)^2\right]\left[n\sum f_y y^2 - \left(\sum f_y y\right)^2\right]}} \tag{1}$$

If we let $x = x_0 + c_x u_x$ and $y = y_0 + c_y u_y$, where c_x and c_y are the class interval widths (assumed constant) and x_0 and y_0 are arbitrary class marks corresponding to the variables, the above formula becomes

$$r = \frac{n\sum fu_x u_y - \left(\sum f_x u_x\right)\left(\sum f_y u_y\right)}{\sqrt{\left[n\sum f_x u_x^2 - \left(\sum f_x u_x\right)^2\right]\left[n\sum f_y u_y^2 - \left(\sum f_y u_y\right)^2\right]}} \tag{2}$$

This is the *coding method* used in Chapter 5 as a short method for computing means, standard deviations, and higher moments.

8.60. Find the coefficient of linear correlation of the mathematics and physics grades of Problem 8.58.

We use formula (2) of Problem 8.59. The work can be arranged as in Table 8-30, which is called a *correlation table.*

Table 8-30

		Mathematics Grades, x									
	x	44.5	54.5	64.5	74.5	84.5	94.5	f_y	$f_y u_y$	$f_y u_y^2$	Sum of corner numbers in each row
Physics Grades, y: y	u_y \ u_x	-2	-1	0	1	2	3				
94.5	2				2 [4]	4 [16]	4 [24]	10	20	40	44
84.5	1			1 [0]	4 [4]	6 [12]	5 [15]	16	16	16	31
74.5	0			5 [0]	10 [0]	8 [0]	1 [0]	24	0	0	0
64.5	-1	1 [2]	4 [4]	9 [0]	5 [-5]	2 [-4]		21	-21	21	-3
54.5	-2	3 [12]	6 [12]	6 [0]	2 [-4]			17	-34	68	20
44.5	-3	3 [18]	5 [15]	4 [0]				12	-36	108	33
f_x		7	15	25	23	20	10	$\sum f_x = \sum f_y = n = 100$	$\sum f_y u_y = -55$	$\sum f_y u_y^2 = 253$	$\sum f u_x u_y = 125$
$f_x u_x$		-14	-15	0	23	40	30	$\sum f_x u_x = 64$			
$f_x u_x^2$		28	15	0	23	80	90	$\sum f_x u_x^2 = 236$			
Sum of corner numbers in each column		32	31	0	-1	24	39	$\sum f u_x u_y = 125$	Check		

The number in the corner of each cell represents the product $fu_x u_y$, where f is the cell frequency. The sum of these corner numbers in each row is indicated in the corresponding row of the last column. The sum of these corner numbers in each column is indicated in the corresponding column of the last row. The final totals of the last row and last column are equal and represent $\sum fu_x u_y$.

From Table 8-30 we have

$$r = \frac{n\sum fu_x u_y - \left(\sum f_x u_x\right)\left(\sum f_y u_y\right)}{\sqrt{\left[n\sum f_x u_x^2 - \left(\sum f_x u_x\right)^2\right]\left[n\sum f_y u_y^2 - \left(\sum f_y u_y\right)^2\right]}}$$

$$= \frac{(100)(125) - (64)(-55)}{\sqrt{[(100)(236) - (64)^2][(100)(253) - (-55)^2]}} = \frac{16{,}020}{\sqrt{(19{,}504)(22{,}275)}} = 0.7686$$

8.61. Use the correlation table of Problem 8.60 to compute (a) s_x, (b) s_y, (c) s_{xy}, and verify the formula $r = s_{xy}/s_x s_y$.

(a) $$s_x = c_x\sqrt{\frac{\sum f_x u_x^2}{n} - \left(\frac{\sum f_x u_x}{n}\right)^2} = 10\sqrt{\frac{236}{100} - \left(\frac{64}{100}\right)^2} = 13.966$$

(b) $$s_y = c_y\sqrt{\frac{\sum f_y u_y^2}{n} - \left(\frac{\sum f_y u_y}{n}\right)^2} = 10\sqrt{\frac{253}{100} - \left(\frac{-55}{100}\right)^2} = 14.925$$

(c) $$s_{xy} = c_x c_y\left[\frac{\sum fu_x u_y}{n} - \left(\frac{\sum f_x u_x}{n}\right)\left(\frac{\sum f_y u_y}{n}\right)\right] = (10)(10)\left[\frac{125}{100} - \left(\frac{64}{100}\right)\left(\frac{-55}{100}\right)\right] = 160.20$$

Therefore, the standard deviations of mathematics grades and physics grades are 14.0 and 14.9, respectively, while their covariance is 160.2. We have

$$\frac{s_{xy}}{s_x s_y} = \frac{160.20}{(13.966)(14.925)} = 0.7686$$

agreeing with r as found in Problem 8.60.

8.62. Write the equations of the regression lines of (a) y on x, (b) x on y for the data of Problem 8.60.

From Table 8-30 we have

$$\bar{x} = x_0 + c_x \frac{\sum f_x u_x}{n} = 64.5 + \frac{(10)(64)}{100} = 70.9$$

$$\bar{y} = y_0 + c_y \frac{\sum f_y u_y}{n} = 74.5 + \frac{(10)(-55)}{100} = 69.0$$

From the results of Problem 8.61, $s_x = 13.966$, $s_y = 14.925$ and $r = 0.7686$. We now use (16), page 268, to obtain the equations of the regression lines.

(a) $$y - \bar{y} = \frac{rs_y}{s_x}(x - \bar{x}), \quad y - 69.0 = \frac{(0.7686)(14.925)}{13.966}(x - 70.9),$$

or $$y - 69.0 = 0.821(x - 70.9)$$

(b) $$x - \bar{x} = \frac{rs_x}{s_y}(y - \bar{y}), \quad y - 70.0 = \frac{(0.7686)(13.966)}{14.925}(y - 69.0),$$

or $$x - 70.9 = 0.719(y - 69.0)$$

8.63. Compute the standard errors of estimate (a) $s_{y.x}$, (b) $s_{y.x}$ for the data of Problem 8.60. Use the results of Problem 8.61.

(a) $$s_{y.x} = s_y\sqrt{1 - r^2} = 14.925\sqrt{1 - (0.7686)^2} = 9.548$$

(b) $$s_{x.y} = s_x\sqrt{1 - r^2} = 13.966\sqrt{1 - (0.7686)^2} = 8.934$$

SUPPLEMENTARY PROBLEMS

The least-squares line

8.64. Fit a least-squares line to the data in Table 8-31 using (a) x as the independent variable, (b) x as the dependent variable. Graph the data and the least-squares lines using the same set of coordinate axes.

Table 8-31

x	3	5	6	8	9	11
y	2	3	4	6	5	8

8.65. For the data of Problem 8.64, find (a) the values of y when $x = 5$ and $x = 12$, (b) the value of x when $y = 7$.

8.66. Table 8-32 shows the final grades in algebra and physics obtained by 10 students selected at random from a large group of students. (a) Graph the data. (b) Find the least-squares line fitting the data, using x as the independent variable. (c) Find the least-squares line fitting the data, using y as independent variable. (d) If a student receives a grade of 75 in algebra, what is her expected grade in physics? (e) If a student receives a grade of 95 in physics, what is her expected grade in algebra?

Table 8-32

Algebra (x)	75	80	93	65	87	71	98	68	84	77
Physics (y)	82	78	86	72	91	80	95	72	89	74

8.67. Refer to Table 8-33. (a) Construct a scatter diagram. (b) Find the least-squares regression line of y on x. (c) Find the least-squares regression line of x on y. (d) Graph the two regression lines of (b) and (c) on the scatter diagram of (a).

Table 8-33

Grade on First Quiz (x)	6	5	8	8	7	6	10	4	9	7
Grade on Second Quiz (y)	8	7	7	10	5	8	10	6	8	6

Least-squares regression curves

8.68. Fit a least-squares parabola, $y = a + bx + cx^2$, to the data in Table 8-34.

Table 8-34

x	0	1	2	3	4	5	6
y	2.4	2.1	3.2	5.6	9.3	14.6	21.9

8.69. Table 8-35 gives the stopping distance d (feet) of an automobile traveling at speed v (miles per hour) at the instant danger is sighted. (a) Graph d against v. (b) Fit a least-squares parabola of the form $d = a + bv + cv^2$ to the data. (c) Estimate d when $v = 45$ miles per hour and 80 miles per hour.

Table 8-35

Speed, v (miles per hour)	20	30	40	50	60	70
Stopping Distance, d (feet)	54	90	138	206	292	396

8.70. The number y of bacteria per unit volume present in a culture after x hours is given in Table 8-36. (a) Graph the data on semilogarithmic graph paper, with the logarithmic scale used for y and the arithmetic scale for x. (b) Fit a least-squares curve having the form $y = ab^x$ to the data, and explain why this particular equation should yield good results, (c) Compare the values of y obtained from this equation with the actual values. (d) Estimate the value of y when $x = 7$.

Table 8-36

Number of Hours (x)	0	1	2	3	4	5	6
Number of Bacteria per Unit Volume (y)	32	47	65	92	132	190	275

Multiple regression

8.71. Table 8-37 shows the corresponding values of three variables x, y, and z. (a) Find the linear least-squares regression equation of z on x and y. (b) Estimate z when $x = 10$ and $y = 6$.

Table 8-37

x	3	5	6	8	12	14
y	16	10	7	4	3	2
z	90	72	54	42	30	12

Standard error of estimate and linear correlation coefficient

8.72. Find (a) $s_{y.x}$, (b) $s_{x.y}$ for the data in Problem 8.67.

8.73. Compute (a) the total variation in y, (b) the unexplained variation in y, (c) the explained variation in y for the data of Problem 8.67.

8.74. Use the results of Problem 8.73 to find the correlation coefficient between the two sets of quiz grades of Problem 8.67.

8.75. Find the covariance for the data of Problem 8.67 (a) directly, (b) by using the formula $s_{xy} = rs_x s_y$ and the result of Problem 8.74.

8.76. Table 8-38 shows the ages x and systolic blood pressures y of 12 women. (a) Find the correlation coefficient between x and y. (b) Determine the least-squares regression line of y on x. (c) Estimate the blood pressure of a woman whose age is 45 years.

Table 8-38

Age (x)	56	42	72	36	63	47	55	49	38	42	68	60
Blood Pressure (y)	147	125	160	118	149	128	150	145	115	140	152	155

8.77. Find the correlation coefficients for the data of (a) Problem 8.64, (b) Problem 8.66.

8.78. The correlation coefficient between two variables x and y is, $r = 0.60$. If $s_x = 1.50$, $s_y = 2.00$, $\bar{x} = 10$ and $\bar{y} = 20$, find the equations of the regression lines of (a) y on x, (b) x on y.

8.79. Compute (a) $s_{y.x}$, (b) $s_{x.y}$ for the data of Problem 8.78.

8.80. If $s_{y.x} = 3$ and $s_y = 5$, find r.

8.81. If the correlation coefficient between x and y is 0.50, what percentage of the total variation remains unexplained by the regression equation?

8.82. (a) Compute the correlation coefficient between the corresponding values of x and y given in Table 8-39. (b) Multiply each x value in the table by 2 and add 6. Multiply each y value in the table by 3 and subtract 15. Find the correlation coefficient between the two new sets of values, explaining why you do or do not obtain the same result as in part (a).

Table 8-39

x	2	4	5	6	8	11
y	18	12	10	8	7	5

Generalized correlation coefficient

8.83. Find the standard error of estimate of z on x and y for the data of Problem 8.71.

8.84. Compute the coefficient of multiple correlation for the data of Problem 8.71.

Rank correlation

8.85. Two judges in a contest, who were asked to rank 8 candidates, A, B, C, D, E, F, G, and H, in order of their preference, submitted the choice shown in Table 8-40. Find the coefficient of rank correlation and decide how well the judges agreed in their choices.

Table 8-40

Candidate	A	B	C	D	E	F	G	H
First Judge	5	2	8	1	4	6	3	7
Second Judge	4	5	7	3	2	8	1	6

8.86. Find the coefficient of rank correlation for the data of (a) Problem 8.67, (b) Problem 8.76.

8.87. Find the coefficient of rank correlation for the data of Problem 8.82.

Sampling theory of regression

8.88. On the basis of a sample of size 27 a regression equation y on x was found to be $y = 25.0 + 2.00x$. If $s_{y.x} = 1.50$, $s_x = 3.00$, and $\bar{x} = 7.50$, find (a) 95%, (b) 99%, confidence limits for the regression coefficient.

8.89. In Problem 8.88 test the hypothesis that the population regression coefficient is (a) as low as 1.70, (b) as high as 2.20, at a 0.01 level of significance.

8.90. In Problem 8.88 find (a) 95%, (b) 99%, confidence limits for y when $x = 6.00$.

8.91. In Problem 8.88 find (a) 95%, (b) 99%, confidence limits for the mean of all values of y corresponding to $x = 6.00$.

8.92. Referring to Problem 8.76, find 95% confidence limits for (a) the regression coefficient of y on x, (b) the blood pressures of all women who are 45 years old, (c) the mean of the blood pressures of all women who are 45 years old.

Sampling theory of correlation

8.93. A correlation coefficient based on a sample of size 27 was computed to be 0.40. Can we conclude at a significance level of (a) 0.05, (b) 0.01, that the corresponding population correlation coefficient is significantly greater than zero?

8.94. A correlation coefficient based on a sample of size 35 was computed to be 0.50. Can we reject the hypothesis that the population correlation coefficient is (a) as small as $\rho = 0.30$, (b) as large as $\rho = 0.70$, using a 0.05 significance level?

8.95. Find (a) 95%, (b) 99%, confidence limits for a correlation coefficient that is computed to be 0.60 from a sample of size 28.

8.96. Work Problem 8.95 if the sample size is 52.

8.97. Find 95% confidence limits for the correlation coefficient computed in Problem 8.76.

8.98. Two correlation coefficients obtained from samples of sizes 23 and 28 were computed to be 0.80 and 0.95, respectively. Can we conclude at a level of (a) 0.05, (b) 0.01, that there is a significant difference between the two coefficients?

Miscellaneous results

8.99. The sample least-squares regression lines for a set of data involving X and Y are given by $2x - 5y = 3$, $5x - 8y = 2$. Find the linear correlation coefficient.

8.100. Find the correlation coefficient between the heights and weights of 300 adult males in the United States as given in the Table 8-41.

Table 8-41
HEIGHTS x (inches)

WEIGHTS y (pounds)	59–62	63–66	67–70	71–74	75–78
90–109	2	1			
110–129	7	8	4	2	
130–149	5	15	22	7	1
150–169	2	12	63	19	5
170–189		7	28	32	12
190–209		2	10	20	7
210–229			1	4	2

8.101. (a) Find the least-squares regression line of y on x for the data of Problem 8.100. (b) Estimate the weights of two men whose heights are 64 and 72 inches, respectively.

8.102. Find (a) $s_{y.x}$, (b) $s_{x.y}$ for the data of Problem 8.100.

8.103. Find 95% confidence limits for the correlation coefficient computed in Problem 8.100.

8.104. Find the correlation coefficient between U.S. consumer price indexes and wholesale price indexes for all commodities as shown in Table 8-42. The base period 1947–1949 = 100.

Table 8-42

Year	1949	1950	1951	1952	1953	1954	1955	1956	1957	1958
Consumer Price Index	101.8	102.8	111.0	113.5	114.4	114.8	114.5	116.2	120.2	123.5
Wholesale Price Index	99.2	103.1	114.8	111.6	110.1	110.3	110.7	114.3	117.6	119.2

Source: Bureau of Labor Statistics.

8.105. Refer to Table 8-43. (a) Graph the data. (b) Find a least-squares line fitting the data and construct its graph. (c) Compute the trend values and compare with the actual values. (d) Predict the price index for medical care during 1958 and compare with the true value (144.4). (e) In what year can we expect the index of medical costs to be double that of 1947 through 1949, assuming present trends continue?

Table 8-43

Year	1950	1951	1952	1953	1954	1955	1956	1957
Consumer Price Index for Medical Care (1947–1949 = 100)	106.0	111.1	117.2	121.3	125.2	128.0	132.6	138.0

Source: Bureau of Labor Statistics.

8.106. Refer to Table 8-44. (a) Graph the data. (b) Find a least-squares parabola fitting the data. (c) Compute the trend values and compare with the actual values. (d) Explain why the equation obtained in (b) is not useful for extrapolation purposes.

Table 8-44

Year	1915	1920	1925	1930	1935	1940	1945	1950	1955
Birth Rate per 1000 Population	25.0	23.7	21.3	18.9	16.9	17.9	19.5	23.6	24.6

Source: Department of Health and Human Services.

ANSWERS TO SUPPLEMENTARY PROBLEMS

8.64. (a) $y = -\frac{1}{3} + \frac{5}{7}x$ or $y = -0.333 + 0.714x$ (b) $x = 1 + \frac{9}{7}y$ or $x = 1.00 + 1.29y$

8.65. (a) 3.24, 8.24 (b) 10.00 **8.66.** (b) $y = 29.13 + 0.661x$ (c) $x = -14.39 + 1.15y$ (d) 79 (e) 95

8.67. (b) $y = 4.000 + 0.500x$ (c) $x = 2.408 + 0.612y$

8.68. $y = 5.51 + 3.20(x - 3) + 0.733(x - 3)^2$ or $y = 2.51 - 1.20x + 0.733x^2$

8.69. (b) $d = 41.77 - 1.096v + 0.08786v^2$ (c) 170 ft, 516 feet

8.70. (b) $y = 32.14(1.427)^x$ or $y = 32.14(10)^{0.1544x}$ or $y = 32.14e^{0.3556x}$ (d) 387

8.71. (a) $z = 61.40 - 3.65x + 2.54y$ (b) 40 **8.72.** (a) 1.304 (b) 1.443

8.73. (a) 24.50 (b) 17.00 (c) 7.50 **8.74.** 0.5533 **8.75.** 1.5

8.76. (a) 0.8961 (b) $y = 80.78 + 1.138x$ (c) 132

8.77. (a) 0.958 (b) 0.872 **8.78.** (a) $y = 0.8x + 12$ (b) $x = 0.45y + 1$

8.79. (a) 1.60 (b) 1.20 **8.80.** ± 0.80 **8.81.** 75% **8.82.** (a) -0.9203

8.83. 3.12 **8.84.** 0.9927 **8.85.** $r_{\text{rank}} = \frac{2}{3}$ **8.86.** (a) 0.5182 (b) 0.9318

8.87. -1.0000 **8.88.** (a) 2.00 ± 0.21 (b) 2.00 ± 0.28

8.89. (a) Using a one-tailed test, we can reject the hypothesis.
(b) Using a one-tailed test, we cannot reject the hypothesis.

8.90. (a) 37.0 ± 3.6 (b) 37.0 ± 4.9 **8.91.** (a) 37.0 ± 1.5 (b) 37.0 ± 2.1

8.92. (a) 1.138 ± 0.398 (b) 132.0 ± 19.2 (c) 132.0 ± 5.4

8.93. (a) Yes. (b) No. **8.94.** (a) No. (b) Yes.

8.95. (a) 0.2923 and 0.7951 (b) 0.1763 and 0.8361

8.96. (a) 0.3912 and 0.7500 (b) 0.3146 and 0.7861

8.97. 0.7096 and 0.9653 **8.98.** (a) yes (b) no **8.99.** 0.8

8.100. 0.5440 **8.101.** (a) $y = 4.44x - 142.22$ (b) 141.9 and 177.5 pounds

8.102. (a) 16.92 1b (b) 2.07 in **8.103.** 0.4961 and 0.7235 **8.104.** 0.9263

8.105. (b) $y = 122.42 + 2.19x$ if x-unit is $\frac{1}{2}$ year and origin is at Jan. 1, 1954; or $y = 107.1 + 4.38x$ if x-unit is 1 year and origin is at July 1, 1950
(d) 142.1 (e) 1971

8.106. (b) $y = 18.16 - 0.1083x + 0.4653x^2$, where y is the birth rate per 1000 population and x-unit is 5 years with origin at July 1, 1935

CHAPTER 9

Analysis of Variance

The Purpose of Analysis of Variance

In Chapter 7 we used sampling theory to test the significance of differences between two sampling means. We assumed that the two populations from which the samples were drawn had the same variance. In many situations there is a need to test the significance of differences among three or more sampling means, or equivalently to test the null hypothesis that the sample means are all equal.

EXAMPLE 9.1 Suppose that in an agricultural experiment, four different chemical treatments of soil produced mean wheat yields of 28, 22, 18, and 24 bushels per acre, respectively. Is there a significant difference in these means, or is the observed spread simply due to chance?

Problems such as these can be solved by using an important technique known as the *analysis of variance,* developed by Fisher. It makes use of the F distribution already considered in previous chapters.

One-Way Classification or One-Factor Experiments

In a *one-factor experiment* measurements or observations are obtained for a independent groups of samples, where the number of measurements in each group is b. We speak of *a treatments*, each of which has *b repetitions* or *replications*. In Example 9.1, $a = 4$.

The results of a one-factor experiment can be presented in a table having a rows and b columns (Table 9-1). Here x_{jk} denotes the measurement in the jth row and kth column, where $j = 1, 2, \ldots, a$ and $k = 1, 2, \ldots, b$. For example, x_{35} refers to the fifth measurement for the third treatment.

Table 9-1

Treatment 1	x_{11} x_{12} $\cdots$ x_{1b}	$\bar{x}_{1.}$
Treatment 2	x_{21} x_{22} $\cdots$ x_{2b}	$\bar{x}_{2.}$
$\vdots$	$\vdots$	
Treatment a	x_{a1} x_{a2} $\cdots$ x_{ab}	$\bar{x}_{a.}$

We shall denote by $\bar{x}_{j.}$ the mean of the measurements in the jth row. We have

$$\bar{x}_{j.} = \frac{1}{b}\sum_{k=1}^{b} x_{jk} \qquad j = 1, 2, \ldots, a \tag{1}$$

The dot in $\bar{x}_{j.}$ is used to show that the index k has been summed out. The values $\bar{x}_{j.}$ are called *group means, treatment means,* or *row means*. The *grand mean*, or *overall mean*, is the mean of all the measurements in all the groups and is denoted by $\bar{x}$, i.e.,

$$\bar{x} = \frac{1}{ab}\sum_{j,k} x_{jk} = \frac{1}{ab}\sum_{j=1}^{a}\sum_{k=1}^{b} x_{jk} \tag{2}$$

Total Variation. Variation Within Treatments. Variation Between Treatments

We define the *total variation*, denoted by v, as the sum of the squares of the deviations of each measurement from the grand mean $\bar{x}$, i.e.,

$$\text{Total variation} = v = \sum_{j,k}(x_{jk} - \bar{x})^2 \tag{3}$$

By writing the identity,

$$x_{jk} - \bar{x} = (x_{jk} - \bar{x}_{j.}) + (\bar{x}_{j.} - \bar{x}) \tag{4}$$

and then squaring and summing over j and k, we can show (see Problem 9.1) that

$$\sum_{j,k}(x_{jk} - \bar{x})^2 = \sum_{j,k}(x_{jk} - \bar{x}_{j.})^2 + \sum_{j,k}(\bar{x}_{j.} - \bar{x})^2 \tag{5}$$

or

$$\sum_{j,k}(x_{jk} - \bar{x})^2 = \sum_{j,k}(x_{jk} - \bar{x}_{j.})^2 + b\sum_{j}(\bar{x}_{j.} - \bar{x})^2 \tag{6}$$

We call the first summation on the right of (5) or (6) the *variation within treatments* (since it involves the squares of the deviations of x_{jk} from the treatment means $\bar{x}_{j.}$) and denote it by v_w. Therefore,

$$v_w = \sum_{j,k}(x_{jk} - \bar{x}_{j.})^2 \tag{7}$$

The second summation on the right of (5) or (6) is called the *variation between treatments* (since it involves the squares of the deviations of the various treatment means $\bar{x}_{j.}$ from the grand mean $\bar{x}$) and is denoted by v_b. Therefore,

$$v_b = \sum_{j,k}(\bar{x}_{j.} - \bar{x})^2 = b\sum_{j}(\bar{x}_{j.} - \bar{x})^2 \tag{8}$$

Equations (5) or (6) can thus be written

$$v = v_w + v_b \tag{9}$$

Shortcut Methods for Obtaining Variations

To minimize the labor in computing the above variations, the following forms are convenient:

$$v = \sum_{j,k} x_{jk}^2 - \frac{\tau^2}{ab} \tag{10}$$

$$v_b = \frac{1}{b}\sum_{j}\tau_{j.}^2 - \frac{\tau^2}{ab} \tag{11}$$

$$v_w = v - v_b \tag{12}$$

where τ is the total of all values x_{jk} and $\tau_{j.}$ is the total of all values in the jth treatment, i.e.,

$$\tau = \sum_{j,k} x_{jk} \qquad \tau_{j.} = \sum_{k} x_{jk} \tag{13}$$

In practice it is convenient to subtract some fixed value from all the data in the table; this has no effect on the final results.

Linear Mathematical Model for Analysis of Variance

We can consider each row of Table 9-1 as representing a random sample of size b from the population for that particular treatment. Therefore, for treatment j we have the independent, identically distributed random variables $X_{j1}, X_{j2}, \ldots, X_{jb}$, which, respectively, take on the values $x_{j1}, x_{j2}, \ldots, x_{jb}$. Each of the X_{jk} ($k = 1, 2, \ldots, b$)

can be expressed as the sum of its expected value and a "chance" or "error" term:

$$X_{jk} = \mu_j + \Delta_{jk} \tag{14}$$

The Δ_{jk} can be taken as independent (relative to j as well as to k), normally distributed random variables with mean zero and variance σ^2. This is equivalent to assuming that the X_{jk} ($j = 1, 2, \ldots, a$; $k = 1, 2, \ldots, b$) are mutually independent, normal variables with means μ_j and common variance σ^2.

Let us define the constant μ by

$$\mu = \frac{1}{a}\sum_j \mu_j$$

We can think of μ as the mean for a sort of grand population comprising all the treatment populations. Then (14) can be rewritten as (see Problem 9.18)

$$X_{jk} = \mu + \alpha_j + \Delta_{jk} \qquad \text{where} \qquad \sum_j \alpha_j = 0 \tag{15}$$

The constant α_j can be viewed as the special effect of the jth treatment.

The null hypothesis that all treatment means are equal is given by (H_0: $\alpha_j = 0$; $j = 1, 2, \ldots, a$) or equivalently by (H_0: $\mu_j = \mu$; $j = 1, 2, \ldots, a$). If H_0 is true, the treatment populations, which by assumption are normal, have a common mean as well as a common variance. Then there is just one treatment population, and all treatments are statistically identical.

Expected Values of the Variations

The between-treatments variation V_b, the within-treatments variation V_w, and the total variation V are random variables that, respectively, assume the values v_b, v_w, and v as defined in (8), (7), and (3). We can show (Problem 9.19) that

$$E(V_b) = (a - 1)\sigma^2 + b\sum_j \alpha_j^2 \tag{16}$$

$$E(V_w) = a(b - 1)\sigma^2 \tag{17}$$

$$E(V) = (ab - 1)\sigma^2 + b\sum \alpha_j^2 \tag{18}$$

From (17) it follows that

$$E\left[\frac{V_w}{a(b - 1)}\right] = \sigma^2 \tag{19}$$

so that

$$\hat{S}_w^2 = \frac{V_w}{a(b - 1)} \tag{20}$$

is always a best (unbiased) estimate of σ^2 regardless of whether H_0 is true or not. On the other hand, from (16) and (18) we see that only if H_0 is true will we have

$$E\left(\frac{V_b}{a - 1}\right) = \sigma^2 \qquad E\left(\frac{V}{ab - 1}\right) = \sigma^2 \tag{21}$$

so that only in such case will

$$\hat{S}_b^2 = \frac{V_b}{a - 1} \qquad \hat{S}^2 = \frac{V}{ab - 1} \tag{22}$$

provide unbiased estimates of σ^2. If H_0 is not true, however, then we have from (16)

$$E(\hat{S}_b^2) = \sigma^2 + \frac{b}{a-1}\sum_j \alpha_j^2 \tag{23}$$

Distributions of the Variations

Using Theorem 4-4, page 115, we can prove the following fundamental theorems concerning the distributions of the variations V_w, V_b, and V.

Theorem 9-1 V_w/σ^2 is chi-square distributed with $a(b-1)$ degrees of freedom.

Theorem 9-2 Under the null hypothesis H_0, V_b/σ^2 and V/σ^2 are chi-square distributed with $a-1$ and $ab-1$ degrees of freedom, respectively.

It is important to emphasize that Theorem 9-1 is valid whether or not we assume H_0, while Theorem 9-2 is valid only if H_0 is assumed.

The *F* Test for the Null Hypothesis of Equal Means

If the null hypothesis H_0 is not true, i.e., if the treatment means are not equal, we see from (23) that we can expect $\hat{S}_b^2$ to be greater than σ^2, with the effect becoming more pronounced as the discrepancy between means increases. On the other hand, from (19) and (20) we can expect $\hat{S}_w^2$ to be equal to σ^2 regardless of whether the means are equal or not. It follows that a good statistic for testing the hypothesis H_0 is provided by $\hat{S}_b^2/\hat{S}_w^2$. If this is significantly large, we can conclude that there is a significant difference between treatment means and thus reject H_0. Otherwise we can either accept H_0 or reserve judgment pending further analysis.

In order to use this statistic, we must know its distribution. This is provided in the following theorem, which is a consequence of Theorem 5-8, page 159.

Theorem 9-3 The statistic $F = \hat{S}_b^2/\hat{S}_w^2$ has the F distribution with $a-1$ and $a(b-1)$ degrees of freedom.

Theorem 9-3 enables us to test the null hypothesis at some specified significance level using a one-tailed test of the F distribution.

Analysis of Variance Tables

The calculations required for the above test are summarized in Table 9-2, which is called an *analysis of variance table*. In practice we would compute v and v_b using either the long method, (3) and (8), or the short method, (10) and (11), and then compute $v_w = v - v_b$. It should be noted that the degrees of freedom for the total variation, i.e., $ab-1$, is equal to the sum of the degrees of freedom for the between-treatments and within-treatments variations.

Table 9-2

Variation	Degrees of Freedom	Mean Square	F
Between Treatments, $v_b = b\sum_j(\bar{x}_{j.} - \bar{x})^2$	$a-1$	$\hat{s}_b^2 = \dfrac{v_b}{a-1}$	$\dfrac{\hat{s}_b^2}{\hat{s}_w^2}$ with $a-1$, $a(b-1)$ degrees of freedom
Within Treatments, $v_w = v - v_b$	$a(b-1)$	$\hat{s}_w^2 = \dfrac{v_w}{a(b-1)}$	
Total, $v = v_b + v_w = \sum_{j,k}(x_{jk} - \bar{x})^2$	$ab-1$		

Modifications for Unequal Numbers of Observations

In case the treatments 1, . . . , a have different numbers of observations equal to $n_1, \ldots, n_a$, respectively, the above results are easily modified. We therefore obtain

$$v = \sum_{j,k}(x_{jk} - \bar{x})^2 = \sum_{j,k} x_{jk}^2 - \frac{\tau^2}{n} \tag{24}$$

$$v_b = \sum_{j,k}(\bar{x}_{j.} - \bar{x})^2 = \sum_{j} n_j(\bar{x}_{j.} - \bar{x})^2 = \sum_{j} \frac{\tau_{j.}^2}{n_j} - \frac{\tau^2}{n} \tag{25}$$

$$v_w = v - v_b \tag{26}$$

where $\sum_{j,k}$ denotes the summation over k from 1 to n_j and then over j from 1 to a, $n = \sum_j n_j$ is the total number of observations in all treatments, τ is the sum of all observations, $\tau_{j.}$ is the sum of all values in the jth treatment, and $\sum_j$ is the sum from $j = 1$ to a.

The analysis of variance table for this case is given in Table 9-3.

Table 9-3

Variation	Degrees of Freedom	Mean Square	F
Between Treatments, $v_b = \sum_j n_j(\bar{x}_{j.} - \bar{x})^2$	$a - 1$	$\hat{s}_b^2 = \dfrac{v_b}{a-1}$	$\dfrac{\hat{s}_b^2}{\hat{s}_w^2}$ with $a - 1, n - a$ degrees of freedom
Within Treatments, $v_w = v - v_b$	$n - a$	$\hat{s}_w^2 = \dfrac{v_w}{n-a}$	
Total, $v = v_b + v_w = \sum_{j,k}(x_{jk} - \bar{x})^2$	$n - 1$		

Two-Way Classification or Two-Factor Experiments

The ideas of analysis of variance for one-way classification or one-factor experiments can be generalized. We illustrate the procedure for *two-way classification* or *two-factor experiments.*

EXAMPLE 9.2 Suppose that an agricultural experiment consists of examining the yields per acre of 4 different varieties of wheat, where each variety is grown on 5 different plots of land. Then a total of (4)(5) = 20 plots are needed. It is convenient in such case to combine plots into *blocks*, say, 4 plots to a block, with a different variety of wheat grown on each plot within a block. Therefore, 5 blocks would be required here.

In this case there are two classifications or factors, since there may be differences in yield per acre due to (i) the particular type of wheat grown or (ii) the particular block used (which may involve different soil fertility, etc.).

By analogy with the agricultural experiment of Example 9.2, we often refer to the two classifications or factors in an experiment as *treatments* and *blocks*, but of course we could simply refer to them as Factor 1 and Factor 2, etc.

Notation for Two-Factor Experiments

Assuming that we have a treatments and b blocks, we construct Table 9-4, where it is supposed that there is one experimental value (for example, yield per acre) corresponding to each treatment and block. For treatment j and block k we denote this value by x_{jk}. The mean of the entries in the jth row is denoted by $\bar{x}_{j.}$, where $j = 1, \ldots, a$, while the mean of the entries in the kth column is denoted by $\bar{x}_{.k}$, where $k = 1, \ldots, b$. The *overall*, or *grand, mean* is denoted by $\bar{x}$. In symbols,

$$\bar{x}_{j.} = \frac{1}{b}\sum_{k=1}^{b} x_{jk}, \qquad \bar{x}_{.k} = \frac{1}{a}\sum_{j=1}^{a} x_{jk}, \qquad \bar{x} = \frac{1}{ab}\sum_{j,k} x_{jk} \tag{27}$$

Table 9-4

Blocks

Treatments	1	2	...	b	
1	x_{11}	x_{12}	...	x_{1b}	$\bar{x}_{1.}$
2	x_{21}	x_{22}	...	x_{2b}	$\bar{x}_{2.}$
⋮	⋮	⋮	⋮		
a	x_{a1}	x_{a2}	...	x_{ab}	$\bar{x}_{a.}$
	$\bar{x}_{.1}$	$\bar{x}_{.2}$	...	$\bar{x}_{.b}$	

Variations for Two-Factor Experiments

As in the case of one-factor experiments, we can define variations for two-factor experiments. We first define the *total variation,* as in (3), to be

$$v = \sum_{j,k}(x_{jk} - \bar{x})^2 \tag{28}$$

By writing the identity

$$x_{jk} - \bar{x} = (x_{jk} - \bar{x}_{j.} - \bar{x}_{.k} + \bar{x}) + (\bar{x}_{j.} - \bar{x}) + (\bar{x}_{.k} - \bar{x}) \tag{29}$$

and then squaring and summing over j and k, we can show that

$$v = v_e + v_r + v_c \tag{30}$$

where v_e = variation due to error or chance $= \sum_{j,k}(x_{jk} - \bar{x}_{j.} - \bar{x}_{.k} + \bar{x})^2$

v_r = variation between rows (treatments) $= b\sum_{j=1}^{a}(\bar{x}_{j.} - \bar{x})^2$

v_c = variation between columns (blocks) $= a\sum_{k=1}^{b}(\bar{x}_{.k} - \bar{x})^2$

The variation due to error or chance is also known as the *residual variation.*

The following are short formulas for computation, analogous to (10), (11), and (12).

$$v = \sum_{j,k} x_{jk}^2 - \frac{\tau^2}{ab} \tag{31}$$

$$v_r = \frac{1}{b}\sum_{j=1}^{a} \tau_{j.}^2 - \frac{\tau^2}{ab} \tag{32}$$

$$v_c = \frac{1}{a}\sum_{k=1}^{b} \tau_{.k}^2 - \frac{\tau^2}{ab} \tag{33}$$

$$v_e = v - v_r - v_c \tag{34}$$

where $\tau_{j.}$ is the total of entries of the jth row, $\tau_{.k}$ is the total of entries in the kth column, and τ is the total of all entries.

Analysis of Variance for Two-Factor Experiments

For the mathematical model of two-factor experiments, let us assume that the random variables X_{jk} whose values are the x_{jk} can be written as

$$X_{jk} = \mu + \alpha_j + \beta_k + \Delta_{jk} \tag{35}$$

Here μ is the population grand mean, α_j is that part of X_{jk} due to the different treatments (sometimes called the *treatment effects*), β_k is that part of X_{jk} due to the different blocks (sometimes called the *block effects*), and Δ_{jk} is that part of X_{jk} due to chance or error. As before, we can take the Δ_{jk} as independent normally distributed random variables with mean zero and variance σ^2, so that the X_{jk} are also independent normally distributed variables with variance σ^2. Under suitable assumptions on the means of the X_{jk}, we have

$$\sum_j \alpha_j = 0 \qquad \sum_k \beta_k = 0 \tag{36}$$

which makes

$$\mu = \frac{1}{ab}\sum_{j,k} E(X_{jk})$$

Corresponding to the results (16) through (18), we can prove that

$$E(V_r) = (a-1)\sigma^2 + b\sum_j \alpha_j^2 \tag{37}$$

$$E(V_c) = (b-1)\sigma^2 + a\sum_k \beta_k^2 \tag{38}$$

$$E(V_e) = (a-1)(b-1)\sigma^2 \tag{39}$$

$$E(V) = (ab-1)\sigma^2 + b\sum_j \alpha_j^2 + a\sum_k \beta_k^2 \tag{40}$$

There are two null hypotheses that we would want to test:

$H_0^{(1)}$: All treatment (row) means are equal, i.e., $\alpha_j = 0, j = 1, \ldots, a$
$H_0^{(2)}$: All block (column) means are equal, i.e., $\beta_k = 0, k = 1, \ldots, b$

We see from (39) that, without regard to $H_0^{(1)}$ or $H_0^{(2)}$, a best (unbiased) estimate of σ^2 is provided by

$$\hat{S}_e^2 = \frac{V_e}{(a-1)(b-1)} \qquad \text{i.e.,} \quad E(\hat{S}_e^2) = \sigma^2 \tag{41}$$

Also, if the hypotheses $H_0^{(1)}$ and $H_0^{(2)}$ are true, then

$$\hat{S}_r^2 = \frac{V_r}{a-1}, \quad \hat{S}_c^2 = \frac{V_c}{b-1}, \quad \hat{S}^2 = \frac{V}{ab-1} \tag{42}$$

will be unbiased estimates of σ^2. If $H_0^{(1)}$ and $H_0^{(2)}$ are not true, however, we have from (37) and (38), respectively,

$$E(\hat{S}_r^2) = \sigma^2 + \frac{b}{a-1}\sum_j \alpha_j^2 \tag{43}$$

$$E(\hat{S}_c^2) = \sigma^2 + \frac{a}{b-1}\sum_k \beta_k^2 \tag{44}$$

The following theorems are similar to Theorems 9-1 and 9-2.

Theorem 9-4 V_e/σ^2 is chi-square distributed with $(a-1)(b-1)$ degrees of freedom, without regard to $H_0^{(1)}$ or $H_0^{(2)}$.

Theorem 9-5 Under the hypothesis $H_0^{(1)}$, V_r/σ^2 is chi-square distributed with $a - 1$ degrees of freedom. Under the hypothesis $H_0^{(2)}$, V_c/σ^2 is chi-square distributed with $b - 1$ degrees of freedom. Under both hypotheses $H_0^{(1)}$ and $H_0^{(2)}$, V/σ^2 is chi-square distributed with $ab - 1$ degrees of freedom.

To test the hypothesis $H_0^{(1)}$ it is natural to consider the statistic $\hat{S}_r^2/\hat{S}_e^2$ since we can see from (43) that $\hat{S}_r^2$ is expected to differ significantly from σ^2 if the row (treatment) means are significantly different. Similarly, to test the hypothesis $H_0^{(2)}$, we consider the statistic $\hat{S}_c^2/\hat{S}_e^2$. The distributions of $\hat{S}_r^2/\hat{S}_e^2$ and $\hat{S}_c^2/\hat{S}_e^2$ are given in the following analog to Theorem 9-3.

Theorem 9-6 Under the hypothesis $H_0^{(1)}$ the statistic $\hat{S}_r^2/\hat{S}_e^2$ has the F distribution with $a - 1$ and $(a - 1)(b - 1)$ degrees of freedom. Under the hypothesis $H_0^{(2)}$ the statistic $\hat{S}_c^2/\hat{S}_e^2$ has the F distribution with $b - 1$ and $(a - 1)(b - 1)$ degrees of freedom.

The theorem enables us to accept or reject $H_0^{(1)}$ and $H_0^{(2)}$ at specified significance levels. For convenience, as in the one-factor case, an analysis of variance table can be constructed as shown in Table 9-5.

Table 9-5

Variation	Degrees of Freedom	Mean Square	F
Between Treatments, $v_r = b\sum_j(\bar{x}_{j.} - \bar{x})^2$	$a - 1$	$\hat{s}_r^2 = \dfrac{v_r}{a-1}$	$\hat{s}_r^2/\hat{s}_e^2$ with $a - 1$, $(a-1)(b-1)$ degrees of freedom
Between Blocks, $v_c = a\sum_k(\bar{x}_{.k} - \bar{x})^2$	$b - 1$	$\hat{s}_c^2 = \dfrac{v_r}{b-1}$	$\hat{s}_c^2/\hat{s}_e^2$ with $b - 1$, $(a-1)(b-1)$ degrees of freedom
Residual or Random, $v_e = v - v_r - v_c$	$(a-1)(b-1)$	$\hat{s}_e^2 = \dfrac{v_e}{(a-1)(b-1)}$	
Total, $v = v_r + v_c + v_e = \sum_{j,k}(x_{jk} - \bar{x})^2$	$ab - 1$		

Two-Factor Experiments with Replication

In Table 9-4 there is only one entry corresponding to a given treatment and a given block. More information regarding the factors can often be obtained by repeating the experiment, a process called *replication.* In that case there will be more than one entry corresponding to a given treatment and a given block. We shall suppose that there are c entries for every position; appropriate changes can be made when the replication numbers are not all equal.

Because of replication an appropriate model must be used to replace that given by (35), page 320. To obtain this, we let X_{jkl} denote the random variable corresponding to the jth row or treatment, the kth column or block, and the lth repetition or replication. The model is then given by

$$X_{jkl} = \mu + \alpha_j + \beta_k + \gamma_{jk} + \Delta_{jkl}$$

where μ, α_j, β_k are defined as before, Δ_{jkl} are independent normally distributed random variables with mean zero and variance σ^2, while γ_{jk} denote row-column or treatment-block *interaction effects* (often simply called *interactions*). Corresponding to (36) we have

$$\sum_j \alpha_j = 0, \qquad \sum_k \beta_k = 0, \qquad \sum_j \gamma_{jk} = 0, \qquad \sum_k \gamma_{jk} = 0 \tag{45}$$

As before, the total variation v of all the data can be broken up into variations due to rows v_r, columns v_c, and random or residual error v_e:

$$v = v_r + v_c + v_i + v_e \tag{46}$$

where

$$v = \sum_{j,k,l}(x_{jkl} - \bar{x})^2 \tag{47}$$

$$v_r = bc\sum_{j=1}^{a}(\bar{x}_{j..} - \bar{x})^2 \tag{48}$$

$$v_c = ac\sum_{k=1}^{b}(\bar{x}_{.k.} - \bar{x})^2 \tag{49}$$

$$v_i = c\sum_{j,k}(\bar{x}_{jk.} - \bar{x}_{j..} - \bar{x}_{.k.} + \bar{x})^2 \tag{50}$$

$$v_e = \sum_{j,k,l}(x_{jkl} - \bar{x}_{jk.})^2 \tag{51}$$

In these results the dots in subscripts have meanings analogous to those given before (page 319). For example,

$$\bar{x}_{j..} = \frac{1}{bc}\sum_{k,l}x_{jkl} = \frac{1}{b}\sum_{k}\bar{x}_{jk.} \tag{52}$$

Using the appropriate number of degrees of freedom (df) for each source of variation, we can set up the analysis of variation table, Table 9-6.

Table 9-6

Variation	Degrees of Freedom	Mean Square	F
Between Treatments, v_r	$a - 1$	$\hat{s}_r^2 = \dfrac{v_r}{a-1}$	$\hat{s}_r^2/\hat{s}_e^2$ with $a - 1$, $ab(c-1)$ degrees of freedom
Between Blocks, v_c	$b - 1$	$\hat{s}_c^2 = \dfrac{v_c}{b-1}$	$\hat{s}_c^2/\hat{s}_e^2$ with $b - 1$, $ab(c-1)$ degrees of freedom
Interaction, v_i	$(a-1)(b-1)$	$\hat{s}_i^2 = \dfrac{v_i}{(a-1)(b-1)}$	$\hat{s}_i^2/\hat{s}_e^2$ with $(a-1)(b-1)$, $ab(c-1)$ degrees of freedom
Residual or Random, v_e	$ab(c-1)$	$\hat{s}_e^2 = \dfrac{v_e}{ab(c-1)}$	
Total, v	$abc - 1$		

The F ratios in the last column of Table 9-6 can be used to test the null hypotheses

$H_0^{(1)}$: All treatment (row) means are equal, i.e., $\alpha_j = 0$

$H_0^{(2)}$: All block (column) means are equal, i.e., $\beta_k = 0$

$H_0^{(3)}$: There are no interactions between treatments and blocks, i.e., $\gamma_{jk} = 0$

From a practical point of view we should first decide whether or not $H_0^{(3)}$ can be rejected at an appropriate level of significance using the F ratio $\hat{s}_i^2/\hat{s}_e^2$ of Table 9-6. Two possible cases then arise.

Case I $H_0^{(3)}$ *Cannot Be Rejected*: In this case we can conclude that the interactions are not too large. We can then test $H_0^{(1)}$ and $H_0^{(2)}$ by using the F ratios $\hat{s}_r^2/\hat{s}_e^2$ and $\hat{s}_c^2/\hat{s}_e^2$, respectively, as shown in Table 9-6. Some statisticians recommend pooling the variations in this case by taking the total $v_i + v_e$ and dividing it by the total corresponding degrees of freedom, $(a - 1)(b - 1) + ab(c - 1)$, and using this value to replace the denominator $\hat{s}_e^2$ in the F test.

Case II $H_0^{(3)}$ *Can Be Rejected*: In this case we can conclude that the interactions are significantly large. Differences in factors would then be of importance only if they were large compared with such interactions. For this reason many statisticians recommend that $H_0^{(1)}$ and $H_0^{(2)}$ be tested using the F ratios $\hat{s}_r^2/\hat{s}_i^2$ and $\hat{s}_c^2/\hat{s}_i^2$ rather than those given in Table 9-6. We shall use this alternate procedure also.

The analysis of variance with replication is most easily performed by first totaling replication values that correspond to particular treatments (rows) and blocks (columns). This produces a two-factor table with single entries, which can be analyzed as in Table 9-5. The procedure is illustrated in Problem 9.13.

Experimental Design

The techniques of analysis of variance discussed above are employed after the results of an experiment have been obtained. However, in order to gain as much information as possible, the details of an experiment must be carefully planned in advance. This is often referred to as the *design of the experiment*. In the following we give some important examples of experimental design.

1. **COMPLETE RANDOMIZATION.** Suppose that we have an agricultural experiment as in Example 9.1, page 314. To design such an experiment, we could divide the land into $4 \times 4 = 16$ plots (indicated in Fig. 9-1 by squares, although physically any shape can be used) and assign each treatment, indicated by A, B, C, D, to four blocks chosen completely at random. The purpose of the randomization is to eliminate various sources of error such as soil fertility.

D	A	C	C
B	D	B	A
D	C	B	D
A	B	C	A

Complete randomization

Fig. 9-1

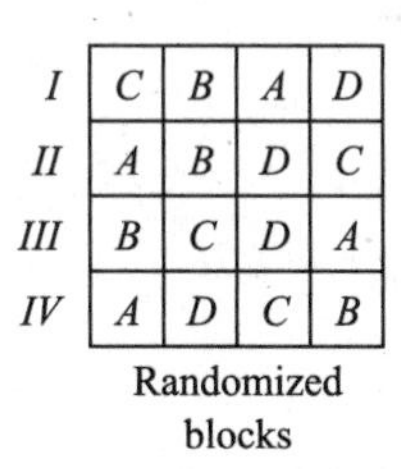

Randomized blocks

Fig. 9-2

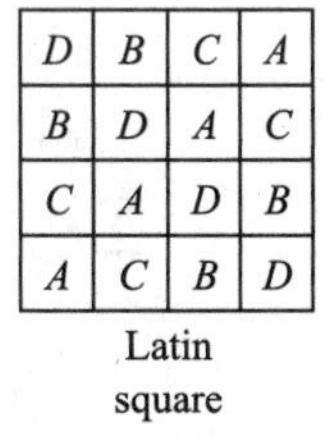

Latin square

Fig. 9-3

B_γ	A_β	D_δ	C_α
A_δ	B_α	C_γ	D_β
D_α	C_δ	B_β	A_γ
C_β	D_γ	A_α	B_δ

Graeco-Latin square

Fig. 9-4

2. **RANDOMIZED BLOCKS.** When, as in Example 9.2, it is necessary to have a complete set of treatments for each block, the treatments A, B, C, D are introduced in random order within each block *I*, *I*, *III*, *IV* (see Fig. 9-2) and for this reason the blocks are referred to as *randomized blocks*. This type of design is used when it is desired to control *one source of error or variability*, namely, the difference in blocks (rows in Fig. 9-2).

3. **LATIN SQUARES.** For some purposes it is necessary to control *two sources of error or variability* at the same time, such as the difference in rows and the difference in columns. In the experiment of Example 9.1, for instance, errors in different rows and columns could be due to changes in soil fertility in different parts of the land. In that case it is desirable that each treatment should occur once in each row and once in each column, as in Fig. 9-3. The arrangement is called a *Latin square* from the fact that Latin letters A, B, C, D are used.

4. GRAECO-LATIN SQUARES. If it is necessary to control *three* sources of error or variability, a *Graeco-Latin square* is used, as indicated in Fig. 9-4. Such a square is essentially two Latin squares superimposed on each other, with Latin letters A, B, C, D used for one square while Greek letters, $\alpha, \beta, \gamma, \delta$ are used for the other squares. The additional requirement that must be met is that each Latin letter must be used once and only once with each Greek letter. When this property is met the square is said to be *orthogonal*.

SOLVED PROBLEMS

One-way classification or one-factor experiments

9.1. Prove that

$$\sum_{j,k}(x_{jk} - \bar{x})^2 = \sum_{j,k}(x_{jk} - \bar{x}_{j.})^2 + \sum_{j,k}(\bar{x}_{j.} - \bar{x})^2$$

We have $x_{jk} - \bar{x} = (x_{jk} - \bar{x}_{j.}) + (\bar{x}_{j.} - \bar{x})$. Then squaring and summing over j and k, we find

$$\sum_{j,k}(x_{jk} - \bar{x})^2 = \sum_{j,k}(x_{jk} - \bar{x}_{j.})^2 + \sum_{j,k}(\bar{x}_{j.} - \bar{x})^2 + 2\sum_{j,k}(x_{jk} - \bar{x}_{j.})(\bar{x}_{j.} - \bar{x})$$

To prove the required result, we must show that the last summation is zero. In order to do this, we proceed as follows.

$$\sum_{j,k}(x_{jk} - \bar{x}_{j.})(\bar{x}_{j.} - \bar{x}) = \sum_{j=1}^{a}(\bar{x}_{j.} - \bar{x})\left[\sum_{k=1}^{b}(x_{jk} - \bar{x}_{j.})\right]$$

$$= \sum_{j=1}^{a}(\bar{x}_{j.} - \bar{x})\left[\left(\sum_{k=1}^{b}x_{jk}\right) - b\bar{x}_{j.}\right] = 0$$

since $\bar{x}_{j.} = \frac{1}{b}\sum_{k=1}^{b}x_{jk}$.

9.2. Verify that (a) $\tau = ab\bar{x}$, (b) $\tau_{j.} = b\bar{x}_{j.}$, (c) $\sum_j \tau_{j.} = ab\bar{x}$, using the notation on page 315.

(a)
$$\tau = \sum_{j,k}x_{jk} = ab\left(\frac{1}{ab}\sum_{j,k}x_{jk}\right) = ab\bar{x}$$

(b)
$$\tau_{j.} = \sum_{k}x_{jk} = b\left(\frac{1}{b}\sum_{k}x_{jk}\right) = b\bar{x}_{j.}$$

(c) Since $\tau_{j.} = \sum_k x_{jk}$, we have

$$\sum_{j}\tau_{j.} = \sum_{j}\sum_{k}x_{jk} = \tau = ab\bar{x}$$

by part (a).

9.3. Verify the shortcut formulas (10) through (12), page 315.

We have

$$v = \sum_{j,k}(x_{jk} - \bar{x})^2 = \sum_{j,k}\left(x_{jk}^2 - 2\bar{x}x_{jk} + \bar{x}^2\right)$$

$$= \sum_{j,k}x_{jk}^2 - 2\bar{x}\sum_{j,k}x_{jk} + ab\bar{x}^2$$

$$= \sum_{j,k}x_{jk}^2 - 2\bar{x}(ab\bar{x}) + ab\bar{x}^2$$

$$= \sum_{j,k}x_{jk}^2 - ab\bar{x}^2$$

$$= \sum_{j,k}x_{jk}^2 - \frac{\tau^2}{ab}$$

using Problem 9.2(a) in the third and last lines above. Similarly

$$\begin{aligned} v_b &= \sum_{j,k}(\bar{x}_{j.} - \bar{x})^2 = \sum_{j,k}(\bar{x}_{j.}^2 - 2\bar{x}\bar{x}_{j.} + \bar{x}^2) \\ &= \sum_{j,k}\bar{x}_{j.}^2 - 2\bar{x}\sum_{j,k}\bar{x}_{j.} + ab\bar{x}^2 \\ &= \sum_{j,k}\left(\frac{\tau_{j.}}{b}\right)^2 - 2\bar{x}\sum_{j,k}\frac{\tau_{j.}}{b} + ab\bar{x}^2 \\ &= \frac{1}{b^2}\sum_{j=1}^{a}\sum_{k=1}^{b}\tau_{j.}^2 - 2\bar{x}(ab\bar{x}) + ab\bar{x}^2 \\ &= \frac{1}{b}\sum_{j=1}^{a}\tau_{j.}^2 - ab\bar{x}^2 \\ &= \frac{1}{b}\sum_{j=1}^{a}\tau_{j.}^2 - \frac{\tau^2}{ab} \end{aligned}$$

using Problem 9.2(b) in the third line and Problem 9.2(a) in the last line.

Finally, (12) follows from the fact that $v = v_b + v_w$ or $v_w = v - v_b$.

9.4. Table 9-7 shows the yields in bushels per acre of a certain variety of wheat grown in a particular type of soil treated with chemicals A, B, or C. Find (a) the mean yields for the different treatments, (b) the grand mean for all treatments, (c) the total variation, (d) the variation between treatments, (e) the variation within treatments. Use the long method.

Table 9-7

A	48	49	50	49
B	47	49	48	48
C	49	51	50	50

Table 9-8

3	4	5	4
2	4	3	3
4	6	5	5

To simplify the arithmetic, we may subtract some suitable number, say, 45, from all the data without affecting the values of the variations. We then obtain the data of Table 9-8.

(a) The treatment (row) means for Table 9-8 are given, respectively, by

$$\bar{x}_{1.} = \frac{1}{4}(3 + 4 + 5 + 4) = 4, \qquad \bar{x}_{2.} = \frac{1}{4}(2 + 4 + 3 + 3) = 3, \qquad \bar{x}_{3.} = \frac{1}{4}(4 + 6 + 5 + 5) = 5$$

Therefore, the mean yields, obtained by adding 45 to these, are 49, 48, and 50 bushels per acre for A, B, and C, respectively.

(b)
$$\bar{x} = \frac{1}{12}(3 + 4 + 5 + 4 + 2 + 4 + 3 + 3 + 4 + 6 + 5 + 5) = 4$$

Therefore, the grand mean for the original set of data is $45 + 4 = 49$ bushels per acre.

(c)
$$\begin{aligned} \text{Total variation} = v &= \sum_{j,k}(x_{jk} - \bar{x})^2 \\ &= (3-4)^2 + (4-4)^2 + (5-4)^2 + (4-4)^2 \\ &\quad + (2-4)^2 + (4-4)^2 + (3-4)^2 + (3-4)^2 \\ &\quad + (4-4)^2 + (6-4)^2 + (5-4)^2 + (5-4)^2 \\ &= 14 \end{aligned}$$

(d)
$$\begin{aligned} \text{Variation between treatments} = v_b &= b\sum_{j}(\bar{x}_{j.} - \bar{x})^2 \\ &= 4[(4-4)^2 + (3-4)^2 + (5-4)^2] = 8 \end{aligned}$$

(e)
$$\text{Variation within treatments} = v_w = v - v_b = 14 - 8 = 6$$

Another method

$$v_w = \sum_{j,k}(x_{jk} - \bar{x}_{j.})^2$$
$$= (3 - 4)^2 + (4 - 4)^2 + (5 - 4)^2 + (4 - 4)^2$$
$$+ (2 - 3)^2 + (4 - 3)^2 + (3 - 3)^2 + (3 - 3)^2$$
$$+ (4 - 5)^2 + (6 - 5)^2 + (5 - 5)^2 + (5 - 5)^2$$
$$= 6$$

9.5. Referring to Problem 9.4, find an unbiased estimate of the population variance σ^2 from (a) the variation between treatments under the null hypothesis of equal treatment means, (b) the variation within treatments.

(a) $$\hat{s}_b^2 = \frac{v_b}{a - 1} = \frac{8}{3 - 1} = 4$$

(b) $$\hat{s}_w^2 = \frac{v_w}{a(b - 1)} = \frac{6}{3(4 - 1)} = \frac{2}{3}$$

9.6. Referring to Problem 9.4, can we reject the null hypothesis of equal means at (a) the 0.05 significance level? (b) the 0.01 significance level?

We have $$F = \frac{\hat{s}_b^2}{\hat{s}_w^2} = \frac{4}{2/3} = 6$$

with $a - 1 = 3 - 1 = 2$ and $a(b - 1) = 3(4 - 1) = 9$ degrees of freedom.

(a) Referring to Appendix F, with $\nu_1 = 2$ and $\nu_2 = 9$, we see that $F_{0.95} = 4.26$. Since $F = 6 > F_{0.95}$, we can reject the null hypothesis of equal means at the 0.05 level.

(b) Referring to Appendix F, with $\nu_1 = 2$ and $\nu_2 = 9$, we see that $F_{0.99} = 8.02$. Since $F = 6 < F_{0.99}$, we cannot reject the null hypothesis of equal means at the 0.01 level.

The analysis of variance table for Problems 9.4 through 9.6 is shown in Table 9-9.

Table 9-9

Variation	Degrees of Freedom	Mean Square	F
Between Treatments, $v_b = 8$	$a - 1 = 2$	$\hat{s}_b^2 = \frac{8}{2} = 4$	$F = \frac{\hat{s}_b^2}{\hat{s}_w^2} = \frac{4}{2/3}$ $= 6$
Within Treatments, $v_w = v - v_b$ $= 14 - 8 = 6$	$a(b - 1) = (3)(3) = 9$	$\hat{s}_w^2 = \frac{6}{9} = \frac{2}{3}$	with 2, 9 degrees of freedom
Total, $v = 14$	$ab - 1 = (3)(4) - 1$ $= 11$		

9.7. Use the shortcut formulas (10) through (12) to obtain the results of Problem 9.4.

(a) We have

$$\sum_{j,k} x_{jk}^2 = 9 + 16 + 25 + 16 + 4 + 16 + 9 + 9 + 16 + 36 + 25 + 25 = 206$$

Also

$$\tau = 3 + 4 + 5 + 4 + 2 + 4 + 3 + 3 + 4 + 6 + 5 + 5 = 48$$

Therefore,

$$v = \sum_{j,k} x_{jk}^2 - \frac{\tau^2}{ab}$$

$$= 206 - \frac{(48)^2}{(3)(4)} = 206 - 192 = 14$$

(b) The totals of the rows are

$$\tau_{1.} = 3 + 4 + 5 + 4 = 16$$
$$\tau_{2.} = 2 + 4 + 3 + 3 = 12$$
$$\tau_{3.} = 4 + 6 + 5 + 5 = 20$$

Also

$$\tau = 16 + 12 + 20 = 48$$

Then

$$v_b = \frac{1}{b}\sum_j \tau_{j.}^2 - \frac{\tau^2}{ab}$$

$$= \frac{1}{4}(16^2 + 12^2 + 20^2) - \frac{(48)^2}{(3)(4)} = 200 - 192 = 8$$

(c)

$$v_w = v - v_b = 14 - 8 = 6$$

It is convenient to arrange the data as in Table 9-10.

Table 9-10

					$\tau_{j.}$	$\tau_{j.}^2$
A	3	4	5	4	16	256
B	2	4	3	3	12	144
C	4	6	5	5	20	400
	$\sum_{j,k} x_{jk}^2 = 206$				$\tau = \sum_j \tau_{j.} = 48$	$\sum_j \tau_{j.}^2 = 800$

$$v = 206 - \frac{(48)^2}{(3)(4)} = 14$$

$$v_b = \frac{1}{4}(800) - \frac{(48)^2}{(3)(4)} = 8$$

The results agree with those obtained in Problem 9.4 and from this point the analysis proceeds as before.

9.8. A company wishes to purchase one of five different machines *A, B, C, D, E*. In an experiment designed to decide whether there is a difference in performance of the machines, five experienced operators each work on the machines for equal times. Table 9-11 shows the number of units produced. Test the hypothesis that there is no difference among the machines at the (a) 0.05, (b) 0.01 level of significance.

Table 9-11

A	68	72	75	42	53
B	72	52	63	55	48
C	60	82	65	77	75
D	48	61	57	64	50
E	64	65	70	68	53

Table 9-12

						$T_{j\cdot}$	$T_{j\cdot}^2$
A	8	12	15	−18	−7	10	100
B	12	−8	3	−5	−2	0	0
C	0	22	6	17	15	60	3600
D	−12	1	−3	4	−10	−20	400
E	4	5	10	8	−7	20	400
	$\sum x_{jk}^2 = 2356$					70	4500

Subtract a suitable number, say, 60, from all the data to obtain Table 9-12.

Then

$$v = 2356 - \frac{(70)^2}{(5)(4)} = 2356 - 245 = 2111$$

$$v_b = \frac{1}{5}(4500) - \frac{(70)^2}{(5)(4)} = 900 - 245 = 655$$

We now form Table 9-13.

Table 9-13

Variation	Degrees of Freedom	Mean Square	F
Between Treatments, $v_c = 655$	$a - 1 = 4$	$\hat{s}_b^2 = \frac{655}{4} = 163.75$	$F = \frac{\hat{s}_b^2}{\hat{s}_w^2} = 2.25$
Within Treatments, $v_w = 1456$	$a(b - 1) = 5(4) = 20$	$\hat{s}_w^2 = \frac{1456}{(5)(4)} = 72.8$	
Total, $v = 2111$	$ab - 1 = 24$		

For 4, 20 degrees of freedom we have $F_{0.95} = 2.87$. Therefore, we cannot reject the null hypothesis at a 0.05 level and therefore certainly cannot reject it at a 0.01 level.

Modifications for unequal numbers of observations

9.9. Table 9-14 shows the lifetimes in hours of samples from three different types of television tubes manufactured by a company. Using the long method, test at (a) the 0.05, (b) the 0.01 significance level whether there is a difference in the three types.

Table 9-14

Sample 1	407	411	409		
Sample 2	404	406	408	405	402
Sample 3	410	408	406	408	

Table 9-15

						Total	Mean
Sample 1	7	11	9			27	9
Sample 2	4	6	8	5	2	25	5
Sample 3	10	8	6	8		32	8
	$\bar{x}$ = grand mean = $\frac{84}{12} = 7$						

It is convenient to subtract a suitable number, say, 400, obtaining Table 9-15.

In this table we have indicated the row totals, the sample or group means, and the grand mean. We then have

$$v = \sum_{j,k}(x_{jk} - \bar{x})^2 = (7 - 7)^2 + (11 - 7)^2 + \cdots + (8 - 7)^2 = 72$$

$$v_b = \sum_{j,k}(\bar{x}_{j.} - \bar{x})^2 = \sum_j n_j(\bar{x}_{j.} - \bar{x})^2$$

$$= 3(9 - 7)^2 + 5(7 - 5)^2 + 4(8 - 7)^2 = 36$$

$$v_w = v - v_b = 72 - 36 = 36$$

We can also obtain v_w directly by observing that it is equal to

$$(7 - 9)^2 + (11 - 9)^2 + (9 - 9)^2 + (4 - 5)^2 + (6 - 5)^2 + (8 - 5)^2 + (5 - 5)^2$$
$$+ (2 - 5)^2 + (10 - 8)^2 + (8 - 8)^2 + (6 - 8)^2 + (8 - 8)^2$$

The data can be summarized in the analysis of variance table, Table 9-16.

Table 9-16

Variation	Degrees of Freedom	Mean Square	F
$v_b = 36$	$a - 1 = 2$	$\hat{s}_b^2 = \frac{36}{2} = 18$	$\frac{\hat{s}_b^2}{\hat{s}_w^2} = \frac{18}{4}$
$v_w = 36$	$n - a = 9$	$\hat{s}_w^2 = \frac{36}{9} = 4$	$= 4.5$

Now for 2 and 9 degrees of freedom we find from Appendix F that $F_{0.95} = 4.26$, $F_{0.99} = 8.02$. Therefore, we can reject the hypothesis of equal means (i.e., there is no difference in the three types of tubes) at the 0.05 level but not at the 0.01 level.

9.10. Work Problem 9.9 by using the shortcut formulas included in (24), (25), and (26).

From Table 9-15,

$$n_1 = 3, \quad n_2 = 5, \quad n_3 = 4, \quad n = 12, \quad \tau_{1.} = 27, \quad \tau_{2.} = 25, \quad \tau_{3.} = 32, \quad \tau = 84$$

We therefore have

$$v = \sum_{j,k} x_{jk}^2 - \frac{\tau^2}{n} = 7^2 + 11^2 + \cdots + 6^2 + 8^2 - \frac{(84)^2}{12} = 72$$

$$v_b = \sum_j \frac{\tau_{j.}^2}{n_j} - \frac{\tau^2}{n} = \frac{(27)^2}{3} + \frac{(25)^2}{5} + \frac{(32)^2}{4} - \frac{(84)^2}{12} = 36$$

$$v_w = v - v_b = 36$$

Using these, the analysis of variance then proceeds as in Problem 9.9.

Two-way classification or two-factor experiments

9.11. Table 9-17 shows the yields per acre of four different plant crops grown on lots treated with three different types of fertilizer. Using the long method, test at the 0.01 level of significance whether (a) there is a significant difference in yield per acre due to fertilizers, (b) there is a significant difference in yield per acre due to crops.

Table 9-17

	Crop I	Crop II	Crop III	Crop IV
Fertilizer A	4.5	6.4	7.2	6.7
Fertilizer B	8.8	7.8	9.6	7.0
Fertilizer C	5.9	6.8	5.7	5.2

Compute the row totals and row means, as well as the column totals and column means and grand mean, as shown in Table 9-18.

Table 9-18

	Crop I	Crop II	Crop III	Crop IV	Row Totals	Row Means
Fertilizer A	4.5	6.4	7.2	6.7	24.8	6.2
Fertilizer B	8.8	7.8	9.6	7.0	33.2	8.3
Fertilizer C	5.9	6.8	5.7	5.2	23.6	5.9
Column Totals	19.2	21.0	22.5	18.9	Grand total = 81.6	
Column Means	6.4	7.0	7.5	6.3	Grand mean = 6.8	

$$v_r = \text{variation of row means from grand mean}$$
$$= 4[(6.2 - 6.8)^2 + (8.3 - 6.8)^2 + (5.9 - 6.8)^2] = 13.68$$
$$v_c = \text{variation of column means from grand mean}$$
$$= 3[(6.4 - 6.8)^2 + (7.0 - 6.8)^2 (7.5 - 6.8)^2 + (6.3 - 6.8)^2] = 2.82$$
$$v = \text{total variation}$$
$$= (4.5 - 6.8)^2 + (6.4 - 6.8)^2 + (7.2 - 6.8)^2 + (6.7 - 6.8)^2$$
$$+ (8.8 - 6.8)^2 + (7.8 - 6.8)^2 + (9.6 - 6.8)^2 + (7.0 - 6.8)^2$$
$$+ (5.9 - 6.8)^2 + (6.8 - 6.8)^2 + (5.7 - 6.8)^2 + (5.2 - 6.8)^2$$
$$= 23.08$$
$$v_e = \text{random variation} = v - v_r - v_c = 6.58$$

This leads to the analysis of variance in Table 9-19.

At the 0.05 level of significance with 2, 6 degrees of freedom, $F_{0.95} = 5.14$. Then, since $6.24 > 5.14$, we can reject the hypothesis that the row means are equal and conclude that at the 0.05 level there is a significant difference in yield due to fertilizers.

Since the F value corresponding to differences in column means is less than 1, we can conclude that there is no significant difference in yield due to crops.

Table 9-19

Variation	Degrees of Freedom	Mean Square	F
$v_r = 13.68$	2	$\hat{s}_r^2 = 6.84$	$F = \hat{s}_r^2/\hat{s}_e^2 = 6.24$ df: 2, 6
$v_c = 2.82$	3	$\hat{s}_c^2 = 0.94$	$F = \hat{s}_c^2/\hat{s}_e^2 = 0.86$ df: 3, 6
$v_e = 6.58$	6	$\hat{s}_e^2 = 1.097$	
$v = 23.08$	11		

9.12. Use the short computational formulas to obtain the results of Problem 9.11.

We have from Table 9-18:

$$\sum_{j,k} x_{jk}^2 = (4.5)^2 + (6.4)^2 + \cdots + (5.2)^2 = 577.96$$

$$\tau = 24.8 + 33.2 + 23.6 = 8.16$$

$$\sum \tau_{j.}^2 = (24.8)^2 + (33.2)^2 + (23.6)^2 = 2274.24$$

$$\sum \tau_{.k}^2 = (19.2)^2 + (21.0)^2 + (22.5)^2 + (18.9)^2 = 1673.10$$

Then

$$v = \sum_{j,k} x_{jk}^2 - \frac{\tau^2}{ab} = 577.96 - 554.88 = 23.08$$

$$v_r = \frac{1}{b}\sum \tau_{j.}^2 - \frac{\tau^2}{ab} = \frac{1}{4}(2274.24) - 554.88 = 13.68$$

$$v_c = \frac{1}{a}\sum \tau_{.k}^2 - \frac{\tau^2}{ab} = \frac{1}{3}(1673.10) - 554.88 = 2.82$$

$$v_e = v - v_r - v_c = 23.08 - 13.68 - 2.82 = 6.58$$

in agreement with Problem 9.11.

Two-factor experiments with replication

9.13. A manufacturer wishes to determine the effectiveness of four types of machines, A, B, C, D, in the production of bolts. To accomplish this, the number of defective bolts produced by each machine on the days of a given week are obtained for each of two shifts. The results are indicated in Table 9-20. Perform an analysis of variance to test at the 0.05 level of significance whether there is (a) a difference in machines, (b) a difference in shifts.

Table 9-20

	FIRST SHIFT					SECOND SHIFT				
	Mon	Tues	Wed	Thurs	Fri	Mon	Tues	Wed	Thurs	Fri
A	6	4	5	5	4	5	7	4	6	8
B	10	8	7	7	9	7	9	12	8	8
C	7	5	6	5	9	9	7	5	4	6
D	8	4	6	5	5	5	7	9	7	10

The data can be equivalently organized as in Table 9-21. In this table the two main factors, namely, *Machine* and *Shift*, are indicated. Note that for each machine two shifts have been indicated. The days of the week can be considered as replicates or repetitions of performance of each machine for the two shifts.

Table 9-21

FACTOR I	FACTOR II	REPLICATES					
Machine	Shift	Mon	Tues	Wed	Thurs	Fri	TOTALS
A	1	6	4	5	5	4	24
	2	5	7	4	6	8	30
B	1	10	8	7	7	9	41
	2	7	9	12	8	8	44
C	1	7	5	6	5	9	32
	2	9	7	5	4	6	31
D	1	8	4	6	5	5	28
	2	5	7	9	7	10	38
	TOTALS	57	51	54	47	59	268

The total variation for all data of Table 9-21 is

$$v = 6^2 + 4^2 + 5^2 + \cdots + 7^2 + 10^2 - \frac{(268)^2}{40} = 1946 - 1795.6 = 150.4$$

In order to consider the two main factors, *Machine* and *Shift*, we limit our attention to the total of replication values corresponding to each combination of factors. These are arranged in Table 9-22, which thus is a two-factor table with single entries.

Table 9-22

	First Shift	Second Shift	TOTALS
A	24	30	54
B	41	44	85
C	32	31	63
D	28	38	66
TOTALS	125	143	268

The total variation for Table 9-22, which we shall call the *subtotal variation* v_s, is given by

$$v_s = \frac{(24)^2}{5} + \frac{(41)^2}{5} + \frac{(32)^2}{5} + \frac{(28)^2}{5} + \frac{(30)^2}{5} + \frac{(44)^2}{5} + \frac{(31)^2}{5} + \frac{(38)^2}{5} - \frac{(268)^2}{40}$$
$$= 1861.2 - 1795.6 = 65.6$$

The variation between rows is given by

$$v_r = \frac{(54)^2}{10} + \frac{(85)^2}{10} + \frac{(63)^2}{10} + \frac{(66)^2}{10} - \frac{(268)^2}{40} = 1846.6 - 1795.6 = 51.0$$

The variation between columns is given by

$$v_c = \frac{(125)^2}{20} + \frac{(143)^2}{20} - \frac{(268)^2}{40} = 1803.7 - 1795.6 = 8.1$$

If we now subtract from the subtotal variation v_s the sum of the variations between rows and columns $(v_r + v_c)$, we obtain the variation due to *interaction* between rows and columns. This is given by

$$v_i = v_s - v_r - v_c = 65.6 - 51.0 - 8.1 = 6.5$$

Finally, the residual variation, which we can think of as the random or error variation v_e (provided that we believe that the various days of the week do not provide any important differences), is found by subtracting the sum of the row, column, and interaction variations (i.e., the subtotal variation) from the total variation v. This yields

$$v_e = v - (v_r + v_c + v_i) = v - v_s = 150.4 - 65.6 = 84.8$$

These variations are indicated in the analysis of variance, Table 9-23. The table also gives the number of degrees of freedom corresponding to each type of variation. Therefore, since there are 4 rows in Table 9-22, the variation due to rows has $4 - 1 = 3$ degrees of freedom, while the variation due to the 2 columns has $2 - 1 = 1$ degrees of freedom. To find the degrees of freedom due to interaction, we note that there are 8 entries in Table 9-22. Therefore, the total degrees of freedom is $8 - 1 = 7$. Since 3 of these are due to rows and 1 to columns, the remainder, $7 - (3 + 1) = 3$, is due to interaction. Since there are 40 entries in the original Table 9-21, the total degrees of freedom is $40 - 1 = 39$. Therefore, the degrees of freedom due to random or residual variation is $39 - 7 = 32$.

Table 9-23

Variation	Degrees of Freedom	Mean Square	F
Rows (Machines), $v_r = 51.0$	3	$\hat{s}_r^2 = 17.0$	$\frac{17.0}{2.65} = 6.42$
Column (Shifts), $v_c = 8.1$	1	$\hat{s}_c^2 = 8.1$	$\frac{8.1}{2.65} = 3.06$
Interaction, $v_i = 6.5$	3	$\hat{s}_i^2 = 2.167$	$\frac{2.167}{2.65} = 0.817$
Subtotal, $v_s = 65.6$	7		
Random or Residual, $v_e = 84.8$	32	$\hat{s}_e^2 = 2.65$	
Total, $v = 150.4$	39		

To proceed further, we must first determine if there is any significant interaction between the basic factors (i.e., rows and columns of Table 9-22). From Table 9-23 we see that for interaction $F = 0.817$, which shows that interaction is not significant, i.e., we cannot reject hypothesis $H_0^{(3)}$ of page 323. Following the rules on page 323, we see that the computed F for rows is 6.42. Since $F_{0.95} = 2.90$ for 3, 32 degrees of freedom we can reject the hypothesis $H_0^{(1)}$ that the rows have equal means. This is equivalent to saying that at the 0.05 level, we can conclude that the machines are not equally effective.

For 1, 32 degrees of freedom $F_{0.95} = 4.15$. Then since the computed F for columns is 3.06, we *cannot* reject the hypothesis $H_0^{(2)}$ that the columns have equal means. This is equivalent to saying that at the 0.05 level there is no significant difference between shifts.

If we choose to analyze the results by pooling the interaction and residual variations as recommended by some statisticians, we find for the pooled variation and pooled degrees of freedom (df) $v_i + v_e = 6.5 + 84.8 = 91.3$ and $3 + 32 = 35$, respectively, which lead to a pooled variance of $91.3/35 = 2.61$. Use of this value instead of 2.65 for the denominator of F in Table 9-23 does not affect the conclusions reached above.

9.14. Work Problem 9.13 if the 0.01 level is used.

At this level there is still no appreciable interaction, so we can proceed further.

Since $F_{0.99} = 4.47$ for 3, 32 df, and since the computed F for rows is 6.42, we can conclude that even at the 0.01 level the machines are not equally effective.

Since $F_{0.99} = 7.51$ for 1, 32 df, and since the computed F for columns is 3.06, we can conclude that at the 0.01 level there is no significant difference in shifts

Latin squares

9.15. A farmer wishes to test the effects of four different fertilizers, A, B, C, D, on the yield of wheat. In order to eliminate sources of error due to variability in soil fertility, he uses the fertilizers in a Latin square arrangement as indicated in Table 9-24, where the numbers indicate yields in bushels per unit area. Perform an analysis of variance to determine if there is a significant difference between the fertilizers at the (a) 0.05, (b) 0.01 levels of significance.

Table 9-24

A 18	*C* 21	*D* 25	*B* 11
D 22	*B* 12	*A* 15	*C* 19
B 15	*A* 20	*C* 23	*D* 24
C 22	*D* 21	*B* 10	*A* 17

Table 9-25

					TOTALS
	A 18	*C* 21	*D* 25	*B* 11	75
	D 22	*B* 12	*A* 15	*C* 19	68
	B 15	*A* 20	*C* 23	*D* 24	82
	C 22	*D* 21	*B* 10	*A* 17	70
TOTALS	77	74	73	71	295

Table 9-26

	A	*B*	*C*	*D*	
TOTAL	70	48	85	92	295

We first obtain totals for rows and columns as indicated in Table 9-25. We also obtain total yields for each of the fertilizers as shown in Table 9-26. The total variation and the variations for rows, columns, and treatments are then obtained as usual. We find

$$\text{Total variation} = v = (18)^2 + (21)^2 + (25)^2 + \cdots + (10)^2 + (17)^2 - \frac{(295)^2}{16}$$
$$= 5769 - 5439.06 = 329.94$$

$$\text{Variation between rows} = v_r = \frac{(75)^2}{4} + \frac{(68)^2}{4} + \frac{(82)^2}{4} + \frac{(70)^2}{4} - \frac{(295)^2}{16}$$
$$= 5468.25 - 5439.06 = 29.19$$

$$\text{Variation between columns} = v_c = \frac{(77)^2}{4} + \frac{(74)^2}{4} + \frac{(73)^2}{4} + \frac{(71)^2}{4} - \frac{(295)^2}{16}$$
$$= 5443.75 - 5439.06 = 4.69$$

$$\text{Variation between treatments} = v_t = \frac{(70)^2}{4} + \frac{(48)^2}{4} + \frac{(85)^2}{4} + \frac{(92)^2}{4} - \frac{(295)^2}{16}$$
$$= 5723.25 - 5439.06 = 284.19$$

The analysis of variance is now shown in Table 9-27.

(a) Since $F_{0.95,3,6} = 4.76$, we can reject at the 0.05 level the hypothesis that there are equal row means. It follows that at the 0.05 level there is a difference in the fertility of the soil from one row to another.

Since the F value for columns is less than 1, we conclude that there is no difference in soil fertility in the columns.

Since the F value for treatments is $47.9 > 4.76$, we can conclude that there is a difference between fertilizers.

Table 9-27

Variation	Degrees of Freedom	Mean Square	F
Rows, 29.19	3	9.73	4.92
Columns, 4.69	3	1.563	0.79
Treatments, 284.19	3	94.73	47.9
Residuals, 11.87	6	1.978	
Total, 329.94	15		

(b) Since $F_{0.99,3,6} = 9.78$, we can accept the hypothesis that there is no difference in soil fertility in the rows (or the columns) at a 0.01 level of significance. However, we must still conclude that there is a difference between fertilizers at the 0.01 level.

Graeco-Latin squares

9.16. It is of interest to determine if there is any difference in mileage per gallon between gasolines A, B, C, D. Design an experiment using four different drivers, four different cars, and four different roads.

Since the same number (four) of gasolines, drivers, cars, and roads are involved, we can use a *Graeco-Latin square*. Suppose that the different cars are represented by the rows and the different drivers by the columns, as indicated in Table 9-28. We now assign the different gasolines A, B, C, D to rows and columns at random, subject only to the requirement that each letter appear just once in each row and just once in each column. Therefore, each driver will have an opportunity to drive each car and use each type of gasoline (and no car will be driven twice with the same gasoline).

We now assign at random the four roads to be used, denoted by $\alpha, \beta, \gamma, \delta$, subjecting them to the same requirement imposed on the Latin letters. Therefore, each driver will have the opportunity to drive along each of the roads also. One possible arrangement is that given in Table 9-28.

Table 9-28

CARS \ DRIVERS	1	2	3	4
1	B_γ	A_β	D_δ	C_α
2	A_α	B_α	C_γ	D_β
3	D_α	C_δ	B_β	A_γ
4	C_β	D_γ	A_α	B_δ

9.17. Suppose that, in carrying out the experiment of Problem 9.16, the numbers of miles per gallon are as given in Table 9-29. Use analysis of variance to determine if there are any significant differences at the 0.05 level.

We first obtain row and column totals as shown in Table 9-30.

Table 9-29

CARS \ DRIVERS	1	2	3	4
1	B_γ 19	A_β 16	D_δ 16	C_α 14
2	A_δ 15	B_α 18	C_γ 11	D_β 15
3	D_α 14	C_δ 11	B_β 21	A_γ 16
4	C_β 16	D_γ 16	A_α 15	B_δ 23

Table 9-30

					TOTALS
	B_γ 19	A_β 16	D_δ 16	C_α 14	65
	A_δ 15	B_α 18	C_γ 11	D_β 15	59
	D_α 14	C_δ 11	B_β 21	A_γ 16	62
	C_β 16	D_γ 16	A_α 15	B_δ 23	70
TOTALS	64	61	63	68	256

Then we obtain totals for each Latin letter and for each Greek letter, as follows:

$$A \text{ total: } 15 + 16 + 15 + 16 = 62$$
$$B \text{ total: } 19 + 18 + 21 + 23 = 81$$
$$C \text{ total: } 16 + 11 + 11 + 14 = 52$$
$$D \text{ total: } 14 + 16 + 16 + 15 = 61$$
$$\alpha \text{ total: } 14 + 18 + 15 + 14 = 61$$
$$\beta \text{ total: } 16 + 16 + 21 + 15 = 68$$
$$\gamma \text{ total: } 19 + 16 + 11 + 16 = 62$$
$$\delta \text{ total: } 15 + 11 + 16 + 23 = 65$$

We now compute the variations corresponding to all of these, using the shortcut method.

$$\text{Rows: } \frac{(65)^2}{4} + \frac{(59)^2}{4} + \frac{(62)^2}{4} + \frac{(70)^2}{4} - \frac{(256)^2}{16} = 4112.50 - 4096 = 16.50$$

$$\text{Columns: } \frac{(64)^2}{4} + \frac{(61)^2}{4} + \frac{(63)^2}{4} + \frac{(68)^2}{4} - \frac{(256)^2}{16} = 4102.50 - 4096 = 6.50$$

$$\text{Gasolines } (A, B, C, D)\text{: } \frac{(62)^2}{4} + \frac{(81)^2}{4} + \frac{(52)^2}{4} + \frac{(61)^2}{4} - \frac{(256)^2}{16} = 4207.50 - 4096 = 111.50$$

$$\text{Roads } (\alpha, \beta, \gamma, \delta)\text{: } \frac{(61)^2}{4} + \frac{(68)^2}{4} + \frac{(62)^2}{4} + \frac{(65)^2}{4} - \frac{(256)^2}{16} = 4103.50 - 4096 = 7.50$$

The total variation is

$$(19)^2 + (16)^2 + (16)^2 + \cdots + (15)^2 + (23)^2 - \frac{(256)}{16} = 4244 - 4096 = 148.00$$

so that the variation due to error is

$$148.00 - 16.50 - 6.50 - 111.50 - 7.50 = 6.00$$

The results are shown in the analysis of variance, Table 9-31. The total number of degrees of freedom is $n^2 - 1$ for an $n \times n$ square. Rows, columns, Latin letters, and Greek letters each have $n - 1$ degrees of freedom. Therefore, the degrees of freedom for error is $n^2 - 1 - 4(n - 1) = (n - 1)(n - 3)$. In our case $n = 4$.

Table 9-31

Variation	Degrees of Freedom	Mean Square	F
Rows (Cars), 16.50	3	5.500	$\frac{5.500}{2.000} = 2.75$
Columns (Drivers), 6.50	3	2.167	$\frac{2.167}{2.000} = 1.08$
Gasolines (A, B, C, D), 111.50	3	37.167	$\frac{37.167}{2.000} = 18.6$
Roads ($\alpha, \beta, \gamma, \delta$), 7.50	3	2.500	$\frac{2.500}{2.000} = 12.5$
Error, 6.00	3	2.000	
Total, 148.00	15		

We have $F_{0.95,3,3} = 9.28$ and $F_{0.99,3,3} = 29.5$. Therefore, we can reject the hypothesis that the gasolines are the same at the 0.05 level but not at the 0.1 level.

Miscellaneous problems

9.18. Prove that $\sum \alpha_j = 0$ [(15), page 316].

The treatment population means are given by $\mu_j = \mu + \alpha_j$. Hence,

$$\sum_{j=1}^{a} \mu_j = \sum_{j=1}^{a} \mu + \sum_{j=1}^{a} \alpha_j = a\mu + \sum_{j=1}^{a} \alpha_j = \sum_{j=1}^{a} \mu_j + \sum_{j=1}^{a} \alpha_j$$

where we have used the definition $\mu = (\sum \mu_j)/a$. It follows that $\sum \alpha_j = 0$.

9.19. Derive (a) equation (17), (b) equation (16), on page 316.

(a) By definition we have

$$\begin{aligned} V_w &= \sum_{j,k} (X_{jk} - \bar{X}_{j\cdot})^2 \\ &= b \sum_{j=1}^{a} \left[\frac{1}{b} \sum_{k=1}^{b} (X_{jk} - \bar{X}_{j\cdot})^2 \right] \\ &= b \sum_{j=1}^{a} S_j^2 \end{aligned}$$

where S_j^2 is the sample variance for the jth treatment, as defined by (15), Chapter 5. Then, since the sample size is b,

$$\begin{aligned} E(V_w) &= b \sum_{j=1}^{a} E(S_j^2) \\ &= b \sum_{j=1}^{a} \left(\frac{b-1}{b} \sigma^2 \right) \\ &= a(b-1)\sigma^2 \end{aligned}$$

using (16) of Chapter 5.

(b) By definition,

$$\begin{aligned} V_b &= b \sum_{j=1}^{a} (\bar{X}_{j.} - \bar{X})^2 \\ &= b \sum_{j=1}^{a} \bar{X}_{j.}^2 - 2b\bar{X} \sum_{j=1}^{a} \bar{X}_{j.} + ab\bar{X}^2 \\ &= b \sum_{j=1}^{a} \bar{X}_{j.}^2 - ab\bar{X}^2 \end{aligned}$$

since

$$\bar{X} = \frac{\sum_j \bar{X}_{j.}^2}{a}$$

Then, omitting the summation index, we have

$$E(V_b) = b \sum E(\bar{X}_{j.}^2) - abE(\bar{X})^2 \tag{1}$$

Now for any random variable U, $E(U^2) = \text{Var}(U) + [E(U)]^2$. Therefore,

(2) $$E(\bar{X}_{j.}^2) = \text{Var}(\bar{X}_{j.}) + [E(\bar{X}_{j.})]^2$$

(3) $$E(\bar{X}^2) = \text{Var}(\bar{X}) + [E(\bar{X})]^2$$

But since the treatment populations are normal, with means μ_j and common variance σ^2, we have from Theorem 5-4, page 156:

(4) $$\text{Var}(\bar{X}_{j.}) = \frac{\sigma^2}{b}$$

(5) $$\text{Var}(\bar{X}) = \frac{\sigma^2}{ab}$$

(6) $$E(\bar{X}_{j.}) = \mu_j = \mu + \alpha_j$$

(7) $$E(\bar{X}) = \mu$$

Using the results (2) through (7), plus the result of Problem 9.18, in (1) we have

$$\begin{aligned} E(V_b) &= b\sum\left[\frac{\sigma^2}{b} + (\mu + \alpha_j)^2\right] - ab\left[\frac{\sigma^2}{ab} + \mu^2\right] \\ &= a\sigma^2 + b\sum(\mu + \alpha_j)^2 - \sigma^2 - ab\mu^2 \\ &= (a - 1)\sigma^2 + ab\mu^2 + 2b\mu\sum\alpha_j + b\sum\alpha_j^2 - ab\mu^2 \\ &= (a - 1)\sigma^2 + b\sum\alpha_j^2 \end{aligned}$$

9.20. Prove Theorem 9-1, page 317.

As shown in Problem 9.19(a),

$$V_w = b\sum_{j=1}^{a} S_j^2 \quad \text{or} \quad \frac{V_w}{\sigma^2} = \sum_{j=1}^{a}\frac{bS_j^2}{\sigma^2}$$

where S_j^2 is the sample variance for samples of size b drawn from the population of treatment j. By Theorem 5-6, page 158, bS_j^2/σ^2 has a chi-square distribution with $b - 1$ degrees of freedom. Then, since the variances S_j^2 are independent, we conclude from Theorem 4-4, page 121, that V_w/σ^2 is chi square distributed with $a(b - 1)$ degrees of freedom.

9.21. In Problem 9.13 we assumed that there were no significant differences in replications, i.e., the different days of the week. Can we support this conclusion at a (a) 0.05, (b) 0.01 significance level?

If there is any variation due to the replications, it is included in what was called the "residual" or "random, error," $v_e = 84.8$, in Table 9-23. To find the variation due to replication, we use the column totals in Table 9-21, obtaining

$$\begin{aligned} v_{\text{rep}} &= \frac{(57)^2}{8} + \frac{(51)^2}{8} + \frac{(54)^2}{8} + \frac{(47)^2}{8} + \frac{(59)^2}{8} - \frac{(268)^2}{40} \\ &= 1807 - 1795.6 = 11.4 \end{aligned}$$

Since there are 5 replications, the number of degrees of freedom associated with this variation is $5 - 1 = 4$. The residual variation after subtracting variation due to replication is $v'_e = 84.8 - 11.4 = 73.4$. The other variations are the same as in Table 9-23. The final analysis of variance table, taking into account replications, is Table 9-32.

From the table we see that the computed F for replication is 1.09. But since $F_{0.95} = 2.71$ for 4, 28 degrees of freedom, we can conclude that there is no significant variation at the 0.05 level (and therefore at the 0.01 level) due to replications, i.e., the days of the week are not significant. The conclusions concerning *Machines* and *Shifts* are the same as those obtained in Problem 9.13.

Table 9-32

Variation	Degrees of Freedom	Mean Square	F
Rows (Machines), $v_r = 51.0$	3	17.0	$\frac{17.0}{2.621} = 6.49$
Columns (Shifts), $v_c = 8.1$	1	8.1	$\frac{8.1}{2.621} = 3.05$
Replications (Days of Week), $v_{\text{rep}} = 11.4$	4	2.85	$\frac{2.85}{2.621} = 1.09$
Interaction, $v_i = 6.5$	3	2.167	$\frac{2.167}{2.621} = 0.827$
Random or Residual, $v'_e = 73.4$	28	2.621	
Total, $v = 150.4$	39		

9.22. Describe how analysis of variance techniques can be used for three-way classification or three-factor experiments (with single entries). Display the analysis of variance table to be used in such case.

We assume that classification is made into A groups, denoted by $A_1, \ldots, A_a$; B groups, denoted by $B_1, \ldots, B_b$; and C groups, denoted by $C_1, \ldots, C_c$. The value which is in A_j, B_k, and C_l is denoted by x_{jkl}. The value $\bar{x}_{jk.}$, for example, denotes the mean of values in the C class when A_j and B_k are kept fixed. Similar meanings are given to $\bar{x}_{j.l}$ and $\bar{x}_{.kl}$. The value $\bar{x}_{j..}$ is the mean of values for the B and C classes when A_j is fixed. Finally $\bar{x}$ denotes the grand mean.

There will be a *total variation* given by

$$v = \sum_{j,k,l}(x_{jkl} - \bar{x})^2 \tag{1}$$

which can be broken into seven variations, as indicated in Table 9-33. These variations are between classes of the same type and between classes of different types (*interactions*). The interaction between all classes is as before called the *residual*, or *random*, *variation*.

The seven variations into which (1) can be broken are given by

$$v = v_A + v_B + v_C + v_{AB} + v_{BC} + v_{CA} + v_{ABC}$$

where

$$v_A = bc\sum_j(\bar{x}_{j..} - \bar{x})^2, \qquad v_B = ca\sum_k(\bar{x}_{.k.} - \bar{x})^2, \qquad v_C = ab\sum_l(\bar{x}_{..l} - \bar{x})^2$$

$$v_{AB} = c\sum_{j,k}(\bar{x}_{jk.} - \bar{x}_{j..} - \bar{x}_{.k.} + \bar{x})^2$$

$$v_{BC} = a\sum_{k,l}(\bar{x}_{.kl} - \bar{x}_{.k} - \bar{x}_{..l} + \bar{x})^2$$

$$v_{CA} = b\sum_{j,l}(\bar{x}_{j.l} - \bar{x}_{..l} - \bar{x}_{j..} + \bar{x})^2$$

$$v_{ABC} = \sum_{j,k,l}(x_{jkl} - \bar{x}_{jk.} - \bar{x}_{j.l} - \bar{x}_{.kl} + \bar{x}_{j..} + \bar{x}_{.k.} + \bar{x}_{..l} - \bar{x})^2$$

Table 9-33

Variation	Degrees of Freedom	Mean Square	F
v_A (Between A Groups)	$a-1$	$\hat{s}_A^2 = \dfrac{v_A}{a-1}$	$\hat{s}_A^2/\hat{s}_{ABC}^2$ $a-1,$ $(a-1)(b-1)(c-1)$ df
v_B (Between B Groups)	$b-1$	$\hat{s}_B^2 = \dfrac{v_B}{b-1}$	$\hat{s}_B^2/\hat{s}_{ABC}^2$ $b-1,$ $(a-1)(b-1)(c-1)$ df
v_c (Between C Groups)	$c-1$	$\hat{s}_C^2 = \dfrac{v_C}{c-1}$	$\hat{s}_C^2/\hat{s}_{ABC}^2$ $c-1,$ $(a-1)(b-1)(c-1)$ df
v_{AB} (Between A and B Groups)	$(a-1)(b-1)$	$\hat{s}_{AB}^2 = \dfrac{v_{AB}}{(a-1)(b-1)}$	$\hat{s}_{AB}^2/\hat{s}_{ABC}^2$ $(a-1)(b-1),$ $(a-1)(b-1)(c-1)$ df
v_{BC} (Between B and C Groups)	$(b-1)(c-1)$	$\hat{s}_{BC}^2 = \dfrac{v_{BC}}{(b-1)(c-1)}$	$\hat{s}_{BC}^2/\hat{s}_{ABC}^2$ $(b-1)(c-1),$ $(a-1)(b-1)(c-1)$ df
v_{CA} (Between C and A Groups)	$(c-1)(a-1)$	$\hat{s}_{CA}^2 = \dfrac{v_{CA}}{(c-1)(a-1)}$	$\hat{s}_{CA}^2/\hat{s}_{ABC}^2$ $(c-1)(a-1),$ $(a-1)(b-1)(c-1)$ df
v_{ABC} (Between A, B, and C Groups)	$(a-1)(b-1)(c-1)$	$\hat{s}_{ABC}^2 = \dfrac{v_{ABC}}{(a-1)(b-1)(c-1)}$	
v (Total)	$abc-1$		

SUPPLEMENTARY PROBLEMS

One-way classification or one-factor experiments

9.23. An experiment is performed to determine the yields of 5 different varieties of wheat, A, B, C, D, E. Four plots of land are assigned to each variety, and the yields (in bushels per acre) are as shown in Table 9-34. Assuming the plots to be of similar fertility and that varieties are assigned at random to plots, determine if there is a significant difference in yields at levels of significance (a) 0.05, (b) 0.01.

Table 9-34

A	20	12	15	19
B	17	14	12	15
C	23	16	18	14
D	15	17	20	12
E	21	14	17	18

Table 9-35

A	33	38	36	40	31	35
B	32	40	42	38	30	34
C	31	37	35	33	34	30
D	29	34	32	30	33	31

9.24. A company wishes to test 4 different types of tires, A, B, C, D. The lifetimes of the tires, as determined from their treads, are given (in thousands of miles) in Table 9-35, where each type has been tried on 6 similar automobiles assigned at random to tires. Test at the (a) 0.05, (b) 0.01 levels whether there is a difference in tires.

9.25. A teacher wishes to test three different teaching methods, I, II, III. To do this, three groups of 5 students each are chosen at random, and each group is taught by a different method. The same examination is then given to all the students, and the grades in Table 9-36 are obtained. Determine at (a) the 0.05, (b) the 0.01 level whether there is a significant difference in the teaching methods.

Table 9-36

Method I	75	62	71	58	73
Method II	81	85	68	92	90
Method III	73	79	60	75	81

Modifications for unequal numbers of observations

9.26. Table 9-37 gives the numbers of miles to the gallon obtained by similar automobiles using 5 different brands of gasoline. Test at the (a) 0.05, (b) 0.01 level of significance whether there is any significant difference in brands.

Table 9-37

Brand A	12	15	14	11	15
Brand B	14	12	15		
Brand C	11	12	10	14	
Brand D	15	18	16	17	14
Brand E	10	12	14	12	

Table 9-38

Mathematics	72	80	83	75	
Science	81	74	77		
English	88	82	90	87	80
Economics	74	71	77	70	

9.27. During one semester a student received grades in various subjects as shown in Table 9-38. Test at the (a) 0.05, (b) 0.01 levels whether there is any significant difference in his grades in these subjects.

Two-way classification or two-factor experiments

9.28. Articles manufactured by a company are produced by 3 operators using 3 different machines. The manufacturer wishes to determine whether there is a difference (a) between operators, (b) between machines. An experiment is performed to determine the number of articles per day produced by each operator using each machine; the results are given in Table 9-39. Provide the desired information using a level of significance of 0.05.

Table 9-39

	Operator 1	Operator 2	Operator 3
Machine A	23	27	24
Machine B	34	30	28
Machine C	28	25	27

9.29. Work Problem 9.28 using a 0.01 level of significance.

9.30. Seeds of 4 different types of corn are planted in 5 blocks. Each block is divided into 4 plots, which are then randomly assigned to the 4 types. Test at a 0.05 level whether the yields in bushels per acre, as shown in Table 9-40, vary significantly with (a) soil differences (i.e., the 5 blocks), (b) differences in type of corn.

Table 9-40

		TYPES OF CORN			
		I	*II*	*III*	*IV*
	A	12	15	10	14
	B	15	19	12	11
BLOCKS	*C*	14	18	15	12
	D	11	16	12	16
	E	16	17	11	14

9.31. Work Problem 9.30 using a 0.01 level of significance.

9.32. Suppose that in Problem 9.24 the first observation for each type of tire is made using one particular kind of automobile, the second observation using a second particular kind, and so on. Test at the 0.05 level if there is a difference in (a) the types of tires, (b) the kinds of automobiles.

9.33. Work Problem 9.32 using a 0.01 level of significance.

9.34. Suppose that in Problem 9.25 the first entry for each teaching method corresponds to a student at one particular school, the second to a student at another school, and so on. Test the hypothesis at the 0.05 level that there is a difference in (a) teaching methods, (b) schools.

9.35. An experiment is performed to test whether color of hair and heights of adult female students in the United States have any bearing on scholastic achievement. The results are given in Table 9-41, where the numbers indicate individuals in the top 10% of those graduating. Analyze the experiment at a 0.05 level.

Table 9-41

	Redhead	Blonde	Brunette
Tall	75	78	80
Medium	81	76	79
Short	73	75	77

9.36. Work Problem 9.35 at a 0.01 level.

Two-factor experiments with replication

9.37. Suppose that the experiment of Problem 9.23 was carried out in the southern part of the United States and that the columns of Table 9-34 now indicate 4 different types of fertilizer, while a similar experiment performed in the western part yields the results in Table 9-42. Test at the 0.05 level whether there is a difference in (a) fertilizers, (b) locations.

Table 9-42

A	16	18	20	23
B	15	17	16	19
C	21	19	18	21
D	18	22	21	23
E	17	18	24	20

9.38. Work Problem 9.37 using a 0.01 level.

9.39. Table 9-43 gives the number of articles produced by 4 different operators working on two different types of machines, *I* and *II*, on different days of the week. Determine at the 0.05 level whether there are significant differences in (a) the operators, (b) the machines.

Table 9-43

	Machine *I*					Machine *II*				
	Mon	Tues	Wed	Thurs	Fri	Mon	Tues	Wed	Thurs	Fri
Operator *A*	15	18	17	20	12	14	16	18	17	15
Operator *B*	12	16	14	18	11	11	15	12	16	12
Operator *C*	14	17	18	16	13	12	14	16	14	11
Operator *D*	19	16	21	23	18	17	15	18	20	17

Latin square

9.40. An experiment is performed to test the effect on corn yield of 4 different fertilizer treatments, *A*, *B*, *C*, *D*, and of soil variations in two perpendicular directions. The Latin square Table 9-44 is obtained, where the numbers indicate corn yield per unit area. Test at a 0.01 level the hypothesis that there is no difference in (a) fertilizers, (b) soil variations.

Table 9-44

C 8	*A* 10	*D* 12	*B* 11
A 14	*C* 12	*B* 11	*D* 15
D 10	*B* 14	*C* 16	*A* 10
B 7	*D* 16	*A* 14	*C* 12

9.41. Work Problem 9.40 using a 0.05 level.

9.42. Referring to Problem 9.35 suppose that we introduce an additional factor giving the section *E*, *M*, or *W* of the United States in which a student was born, as shown in Table 9-45. Determine at a 0.05 level whether there is a significant difference in scholastic achievement of female students due to differences in (a) height, (b) hair color, (c) birthplace.

Table 9-45

E 75	*W* 78	*M* 80
M 81	*E* 76	*W* 79
W 73	*M* 75	*E* 77

Graeco-Latin squares

9.43. In order to produce a superior type of chicken feed, 4 different quantities of each of two chemicals are added to the basic ingredients. The different quantities of the first chemical are indicated by A, B, C, D while those of the second chemical are indicated by $\alpha, \beta, \gamma, \delta$. The feed is given to baby chicks arranged in groups according to 4 different initial weights, W_1, W_2, W_3, W_4, and 4 different species, S_1, S_2, S_3, S_4. The increases in weight per unit time are given in the Graeco-Latin square of Table 9-46. Perform an analysis of variance of the experiment at a 0.05 level of significance, stating any conclusions that can be drawn.

Table 9-46

	W	W_2	W_3	W_4
S_1	C_γ 8	B_β 6	A_α 5	D_δ 6
S_2	A_δ 4	D_α 3	C_β 7	B_γ 3
S_3	D_β 5	A_γ 6	B_δ 5	C_α 6
S_4	B_α 6	C_δ 10	D_γ 10	A_β 8

9.44. Four different types of cables, T_1, T_2, T_3, T_4, are manufactured by each of 4 companies, C_1, C_2, C_3, C_4. Four operators, A, B, C, D, using four different machines, $\alpha, \beta, \gamma, \delta$, measure the cable strengths. The average strengths obtained are given in the Graeco-Latin square of Table 9-47. Perform an analysis of variance at the 0.05 level, stating any conclusions that can be drawn.

Table 9-47

	C_1	C_2	C_3	C_4
T_1	A_β 164	B_γ 181	C_α 193	D_δ 160
T_2	C_δ 171	D_α 162	A_γ 183	B_β 145
T_3	D_γ 198	C_β 221	B_δ 207	A_α 188
T_4	B_α 157	A_δ 172	D_β 166	C_γ 136

Miscellaneous problems

9.45. Table 9-48 gives data on the accumulated rust on iron treated with chemical A, B, or C, respectively. Determine at the (a) 0.05, (b) 0.01 level whether there is a significant difference in the treatments.

Table 9-48

A	3	5	4	4
B	4	2	3	3
C	6	4	5	5

Table 9-49

Tall	110	105	118	112	90	
Short	95	103	115	107		
Medium	108	112	93	104	96	102

9.46. An experiment measures the IQs of adult male students of tall, short, and medium stature. The results are indicated in Table 9-49. Determine at the (a) 0.05, (b) 0.01 level whether there is any significant difference in the IQ scores relative to height differences.

9.47. An examination is given to determine whether veterans or nonveterans of different IQs performed better. The scores obtained are shown in Table 9-50. Determine at the 0.05 level whether there is a difference in scores due to differences in (a) veteran status, (b) IQ.

Table 9-50

	High IQ	Medium IQ	Low IQ
Veteran	90	81	74
Nonveteran	85	78	70

9.48. Work Problem 9.47 using a 0.01 level.

9.49. Table 9-51 shows test scores for a sample of college students from different parts of the country having different IQs. Analyze the table at a 0.05 level of significance and state your conclusions.

Table 9-51

	High	Medium	Low
East	88	80	72
West	84	78	75
South	86	82	70
North & Central	80	75	79

9.50. Work Problem 9.49 at a 0.01 level.

9.51. Suppose that the results in Table 9-48 of Problem 9.48 hold for the northeastern part of the United States, while corresponding results for the western part are given in Table 9-52. Determine at the 0.05 level whether there are differences due to (a) chemicals, (b) location.

Table 9-52

A	5	4	6	3
B	3	4	2	3
C	5	7	4	6

9.52. Referring to Problems 9.23 and 9.37, suppose that an additional experiment performed in the northeastern part of the United States produced the results in Table 9-53. Test at the 0.05 level whether there is a difference in (a) fertilizers, (b) the three locations.

Table 9-53

A	17	14	18	12
B	20	10	20	15
C	18	15	16	17
D	12	11	14	11
E	15	12	19	14

9.53. Work Problem 9.52 using a 0.01 level.

9.54. Perform an analysis of variance on the Latin square of Table 9-54 at a 0.05 level and state conclusions.

Table 9-54

	FACTOR 1		
FACTOR 2	B 16	C 21	A 15
	A 18	B 23	C 14
	C 15	A 18	B 12

9.55. Perform an analysis of variance on the Graeco-Latin square of Table 9-55 at a 0.05 level, and state conclusions.

Table 9-55

	FACTOR 1			
FACTOR 2	A_γ 6	B_β 12	C_δ 4	D_α 18
	B_δ 3	A_α 8	D_γ 15	C_β 14
	D_β 15	C_γ 20	B_α 9	A_δ 5
	C_α 16	D_δ 6	A_β 17	B_γ 7

ANSWERS TO SUPPLEMENTARY PROBLEMS

9.23. There is a significant difference in yield at both levels.

9.24. There is no significant difference in tires at either level.

9.25. There is a significant difference in teaching methods at the 0.05 level but not the 0.01 level.

9.26. There is a significant difference in brands at the 0.05 level but not the 0.01 level.

9.27. There is a significant difference in his grades at both levels.

9.28. There is no significant difference in operators or machines.

9.29. There is no significant difference in operators or machines.

9.30. There is a significant difference in types of corn but not in soils at the 0.05 level.

9.31. There is no significant difference in type of corn or soils at the 0.01 level.

9.32. There is a significant difference in both tires and automobiles at the 0.05 level.

9.33. There is no significant difference in either tires or automobiles at the 0.01 level.

9.34. There is a significant difference in teaching methods but no significant difference in schools at the 0.05 level.

9.35. There is no significant difference in either hair color or height.

9.36. Same answer as Problem 9.35.

9.37. There is a significant difference in locations at the 0.05 level but not in fertilizers.

9.38. There is no significant difference in locations of fertilizers at the 0.01 level.

9.39. There is a significant difference in operators but not in machines.

9.40. There is no significant difference in either fertilizers or soils.

9.41. Same answer as Problem 9.40.

9.42. There is no significant difference in scholastic achievement due to differences in height, hair color, or birthplace.

9.43. There are significant differences in species and quantities of the first chemical but no other significant differences.

9.44. There are significant differences in types of cables but no significant differences in cable strengths due to operators, machines, or companies.

9.45. There is no significant difference in treatments at either level.

9.46. There is no significant difference in IQ scores at either level.

9.47. There are significant differences in examination scores due to both veteran status and IQ at the 0.05 level.

9.48. At the 0.01 level the differences in examination scores due to veteran status are not significant, but those due to IQ are significant.

9.49. There are no significant differences in test scores of students from different parts of the country, but there are significant differences in test scores due to IQ.

9.50. Same answer as Problem 9.49.

9.51. There is a significant difference due to chemicals or locations at the 0.05 level.

9.52. There are significant differences due to locations but not to fertilizers.

9.53. There are no significant differences due to locations or fertilizers.

9.54. There are no significant differences due to factor 1, factor 2, or treatments A, B, C.

9.55. There are no significant differences due to factors or treatments.

Nonparametric Tests

Introduction

Most tests of hypotheses and significance (or decision rules) considered in previous chapters require various assumptions about the distribution of the population from which the samples are drawn. For example, in Chapter 5 the population distributions often are required to be normal or nearly normal.

Situations arise in practice in which such assumptions may not be justified or in which there is doubt that they apply, as in the case where a population may be highly skewed. Because of this, statisticians have devised various tests and methods that are independent of population distributions and associated parameters. These are called *nonparametric tests*.

Nonparametric tests can be used as shortcut replacements for more complicated tests. They are especially valuable in dealing with nonnumerical data, such as arise when consumers rank cereals or other products in order of preference.

The Sign Test

Consider Table 10-1, which shows the numbers of defective bolts produced by two different types of machines (I and II) on 12 consecutive days and which assumes that the machines have the same total output per day. We wish to test the hypothesis H_0 that there is no difference between the machines: that the observed differences between the machines in terms of the numbers of defective bolts they produce are merely the result of chance, which is to say that the samples come from the same population.

A simple nonparametric test in the case of such paired samples is provided by the *sign test*. This test consists of taking the difference between the numbers of defective bolts for each day and writing only the *sign* of the difference; for instance, for day 1 we have 47–71, which is negative. In this way we obtain from Table 10-1 the sequence of signs

$$- \quad - \quad + \quad - \quad - \quad - \quad + \quad - \quad + \quad - \quad - \quad - \tag{1}$$

(i.e., 3 pluses and 9 minuses). Now if it is just as likely to get a $+$ as a $-$, we would expect to get 6 of each. The test of H_0 is thus equivalent to that of whether a coin is fair if 12 tosses result in 3 heads ($+$) and 9 tails ($-$). This involves the binomial distribution of Chapter 4. Problem 10.1 shows that by using a two-tailed test of this distribution at the 0.05 significance level, we cannot reject H_0; that is, there is no difference between the machines at this level.

Remark 1 If on some day the machines produced the same number of defective bolts, a difference of *zero* would appear in sequence (1). In that case we can omit these sample values and use 11 instead of 12 observations.

Remark 2 A normal approximation to the binomial distribution, using a correction for continuity, can also be used (see Problem 10.2).

Table 10-1

Day	1	2	3	4	5	6	7	8	9	10	11	12
Machine I	47	56	54	49	36	48	51	38	61	49	56	52
Machine II	71	63	45	64	50	55	42	46	53	57	75	60

Although the sign test is particularly useful for paired samples, as in Table 10-1, it can also be used for problems involving single samples (see Problems 10.3 and 10.4).

The Mann–Whitney *U* Test

Consider Table 10-2, which shows the strengths of cables made from two different alloys, I and II. In this table we have two samples: 8 cables of alloy I and 10 cables of alloy II. We would like to decide whether or not there is a difference between the samples or, equivalently, whether or not they come from the same population. Although this problem can be worked by using the t test of Chapter 7, a nonparametric test called the *Mann–Whitney U test*, or briefly the *U test*, is useful. This test consists of the following steps:

Table 10-2

Alloy I				Alloy II				
18.3	16.4	22.7	17.8	12.6	14.1	20.5	10.7	15.9
18.9	25.3	16.1	24.2	19.6	12.9	15.2	11.8	14.7

Step 1. Combine all sample values in an array from the smallest to the largest, and assign ranks (in this case from 1 to 18) to all these values. If two or more sample values are identical (i.e., there are *tie scores*, or briefly *ties*), the sample values are each assigned a rank equal to the *mean* of the ranks that would otherwise be assigned. If the entry 18.9 in Table 10-2 were 18.3, two identical values 18.3 would occupy ranks 12 and 13 in the array so that the rank assigned to each would be $\frac{1}{2}(12 + 13) = 12.5$.

Step 2. Find the sum of the ranks for each of the samples. Denote these sums by R_1 and R_2, where N_1 and N_2 are the respective sample sizes. For convenience, choose N_1 as the smaller size if they are unequal, so that $N_1 \leq N_2$. A significant difference between the rank sums R_1 and R_2 implies a significant difference between the samples.

Step 3. To test the difference between the rank sums, use the statistic

$$U = N_1N_2 + \frac{N_1(N_1 + 1)}{2} - R_1 \tag{2}$$

corresponding to sample 1. The sampling distribution of U is symmetrical and has a mean and variance given, respectively, by the formulas

$$\mu_U = \frac{N_1N_2}{2} \qquad \sigma_U^2 = \frac{N_1N_2(N_1 + N_2 + 1)}{12} \tag{3}$$

If N_1 and N_2 are both at least equal to 8, it turns out that the distribution of U is nearly normal, so that

$$Z = \frac{U - \mu_U}{\sigma_U} \tag{4}$$

is normally distributed with mean 0 and variance 1. Using Appendix C, we can then decide whether the samples are significantly different. Problem 10.5 shows that there is a significant difference between the cables at the 0.05 level.

Remark 3 A value corresponding to sample 2 is given by the statistic

$$U = N_1N_2 + \frac{N_2(N_2 + 1)}{2} - R_2 \tag{5}$$

and has the same sampling distribution as statistic (2), with the mean and variance of formulas (3). Statistic (5) is related to statistic (2), for if U_1 and U_2 are the values corresponding to statistics (2) and (5), respectively, then we have the result

$$U_1 + U_2 = N_1N_2 \tag{6}$$

We also have

$$R_1 + R_2 = \frac{N(N + 1)}{2} \tag{7}$$

where $N = N_1 + N_2$. Result (7) can provide a check for calculations.

Remark 4 The statistic U in equation (2) is the total number of times that sample 1 values precede sample 2 values when all sample values are arranged in increasing order of magnitude. This provides an alternative *counting method* for finding U.

The Kruskal–Wallis *H* Test

The U test is a nonparametric test for deciding whether or not two samples come from the same population. A generalization of this for k samples is provided by the *Kruskal–Wallis H test*, or briefly the *H test*.

This test may be described thus: Suppose that we have k samples of sizes $N_1, N_2, \ldots, N_k$, with the total size of all samples taken together being given by $N = N_1 + N_2 + \cdots + N_k$. Suppose further that the data from all the samples taken together are ranked and that the sums of the ranks for the k samples are $R_1, R_2, \ldots, R_k$, respectively. If we define the statistic

$$H = \frac{12}{N(N + 1)} \sum_{j=1}^{k} \frac{R_j^2}{N_j} - 3(N + 1) \tag{8}$$

then it can be shown that the sampling distribution of H is very nearly a *chi-square distribution* with $k - 1$ degrees of freedom, provided that $N_1, N_2, \ldots, N_k$ are all at least 5.

The H test provides a nonparametric method in the *analysis of variance* for one-way classification, or one-factor experiments, and generalizations can be made.

The *H* Test Corrected for Ties

In case there are too many ties among the observations in the sample data, the value of H given by statistic (8) is smaller than it should be. The corrected value of H, denoted by H_c, is obtained by dividing the value given in statistic (8) by the correction factor

$$1 - \frac{\sum(T^3 - T)}{N^3 - N} \tag{9}$$

where T is the number of ties corresponding to each observation and where the sum is taken over all the observations. If there are no ties, then $T = 0$ and factor (9) reduces to 1, so that no correction is needed. In practice, the correction is usually negligible (i.e., it is not enough to warrant a change in the decision).

The Runs Test for Randomness

Although the word *random* has been used many times in this book (such as in "random sampling" and "tossing a coin at random"), no previous chapter has given any test for randomness. A non-parametric test for randomness is provided by the *theory of runs*.

To understand what a run is, consider a sequence made up of two symbols, a and b, such as

$$a \;\; a \;|\; b \;\; b \;\; b \;|\; a \;|\; b \;\; b \;|\; a \;\; a \;\; a \;\; a \;\; a \;|\; b \;\; b \;\; b \;|\; a \;\; a \;\; a \;\; a \;| \tag{10}$$

In tossing a coin, for example, a could represent heads and b could represent tails. Or in sampling the bolts produced by a machine, a could represent defective and b could represent nondefective.

A *run* is defined as a set of identical (or related) symbols contained between two different symbols or no symbol (such as at the beginning or end of the sequence). Proceeding from left to right in sequence (10), the first run, indicated by a vertical bar, consists of two a's; similarly, the second run consists of three b's, the third run consists of one a, etc. There are seven runs in all.

It seems clear that some relationship exists between randomness and the number of runs. Thus for the sequence

$$a \mid b \mid a \mid b \mid a \mid b \mid a \mid b \mid a \mid b \mid a \mid b \tag{11}$$

there is a *cyclic pattern*, in which we go from a to b, back to a again, etc., which we could hardly believe to be random. In that case we have *too many* runs (in fact, we have the maximum number possible for the given number of a's and b's).

On the other hand, for the sequence

$$a \quad a \quad a \quad a \quad a \quad a \mid b \quad b \quad b \quad b \mid a \quad a \quad a \quad a \quad a \mid b \quad b \quad b \mid \tag{12}$$

there seems to be a *trend pattern*, in which the a's and b's are grouped (or clustered) together. In such case there are *too few* runs, and we could not consider the sequence to be random.

Thus a sequence would be considered nonrandom if there are either too many or too few runs, and random otherwise. To quantify this idea, suppose that we form all possible sequences consisting of N_1 a's and N_2 b's, for a total of N symbols in all ($N_1 + N_2 = N$). The collection of all these sequences provides us with a sampling distribution. Each sequence has an associated number of runs, denoted by V. In this way we are led to the sampling distribution of the statistic V. It can be shown that this sampling distribution has a mean and variance given, respectively, by the formulas

$$\mu_V = \frac{2N_1N_2}{N_1 + N_2} + 1 \qquad \sigma_V^2 = \frac{2N_1N_2(2N_1N_2 - N_1 - N_2)}{(N_1 + N_2)^2(N_1 + N_2 - 1)} \tag{13}$$

By using formulas (13), we can test the hypothesis of randomness at appropriate levels of significance. It turns out that if both N_1 and N_2 are at least equal to 8, then the sampling distribution of V is very nearly a normal distribution. Thus

$$Z = \frac{V - \mu_V}{\sigma_V} \tag{14}$$

is normally distributed with mean 0 and variance 1, and thus Appendix C can be used.

Further Applications of the Runs Test

The following are other applications of the runs test to statistical problems:

1. **ABOVE- AND BELOW-MEDIAN TEST FOR RANDOMNESS OF NUMERICAL DATA.** To determine whether numerical data (such as collected in a sample) are random, first place the data in the *same order* in which they were collected. Then find the median of the data and replace each entry with the letter a or b according to whether its value is *above* or *below* the median. If a value is the same as the median, omit it from the sample. The sample is random or not according to whether the sequence of a's and b's is random or not. (See Problem 10.20.)

2. **DIFFERENCES IN POPULATIONS FROM WHICH SAMPLES ARE DRAWN.** Suppose that two samples of sizes m and n are denoted by $a_1, a_2, \ldots, a_m$ and $b_1, b_2, \ldots, b_n$, respectively. To decide whether the samples do or do not come from the same population, first arrange all $m + n$ sample values in a sequence of increasing values. If some values are the same, they should be ordered by a random process (such as by using random numbers). If the resulting sequence is random, we can conclude that the samples are not really different and thus come from the same population; if the sequence is not random, no such conclusion can be drawn. This test can provide an alternative to the Mann–Whitney U test. (See Problem 10.21.)

Spearman's Rank Correlation

Nonparametric methods can also be used to measure the correlation of two variables, X and Y. Instead of using precise values of the variables, or when such precision is unavailable, the data may be ranked from 1 to N in order of size, importance, etc. If X and Y are ranked in such a manner, the *coefficient of rank correlation*, or *Spearman's formula for rank correlation* (as it is often called), is given by

$$r_S = 1 - \frac{6\sum D^2}{N(N^2 - 1)} \tag{15}$$

where D denotes the differences between the ranks of corresponding values of X and Y, and where N is the number of pairs of values (X, Y) in the data.

SOLVED PROBLEMS

The sign test

10.1. Referring to Table 10-1, test the hypothesis H_0 that there is no difference between machines I and II against the alternative hypothesis H_1 that there is a difference at the 0.05 significance level.

Figure 10-1 is a graph of the binomial distribution (and a normal approximation to it) that gives the probabilities of x heads in 12 tosses of a fair coin, where $x = 0, 1, 2, \ldots, 12$. From Chapter 4 the probability of x heads is

$$\Pr\{x\} = \binom{12}{x}\left(\frac{1}{2}\right)^x\left(\frac{1}{2}\right)^{12-x} = \binom{12}{x}\left(\frac{1}{2}\right)^{12}$$

whereby $\Pr\{0\} = 0.00024$, $\Pr\{1\} = 0.00293$, $\Pr\{2\} = 0.01611$, and $\Pr\{3\} = 0.05371$.

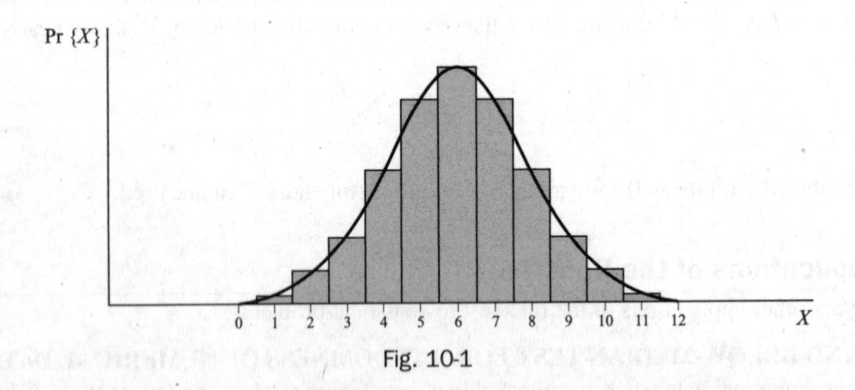

Fig. 10-1

Since H_1 is the hypothesis that there is a *difference* between the machines, rather than the hypothesis that machine I is *better* than machine II, we use a two-tailed test. For the 0.05 significance level, each tail has the associated probability $\frac{1}{2}(0.05) = 0.025$. We now add the probabilities in the left-hand tail until the sum exceeds 0.025. Thus

$$\Pr\{0, 1, \text{ or } 2 \text{ heads}\} = 0.00024 + 0.00293 + 0.01611 = 0.01928$$

$$\Pr\{0, 1, 2, \text{ or } 3 \text{ heads}\} = 0.00024 + 0.00293 + 0.01611 + 0.05371 = 0.07299$$

Since 0.025 is greater than 0.01928 but less than 0.07299, we can reject hypothesis H_0 if the number of heads is 2 or less (or, by symmetry, if the number of heads is 10 or more); however, the number of heads [the + signs in sequence (1) of this chapter] is 3. Thus we cannot reject H_0 at the 0.05 level and must conclude that there is no difference between the machines at this level.

10.2. Work Problem 10.1 by using a normal approximation to the binomial distribution.

For a normal approximation to the binomial distribution, we use the fact that the z score corresponding to the number of heads is

$$Z = \frac{X - \mu}{\sigma} = \frac{X - Np}{\sqrt{Npq}}.$$

Because the variable X for the binomial distribution is discrete while that for a normal distribution is continuous, we make a *correction for continuity* (for example, 3 heads are really a value between 2.5 and 3.5 heads). This amounts to decreasing X by 0.5 if $X > Np$ and to increasing X by 0.5 if $X < Np$. Now $N = 12$, $\mu = Np = (12)(0.5) = 6$, and $\sigma = \sqrt{Npq} = \sqrt{(12)(0.5)(0.5)} = 1.73$, so that

$$z = \frac{(3 + 0.5) - 6}{1.73} = -1.45$$

Since this is greater than -1.96 (the value of z for which the area in the left-hand tail is 0.025), we arrive at the same conclusion in Problem 10.1.

Note that $\Pr\{Z \leq -1.45\} = 0.0735$, which agrees very well with the $\Pr\{X \leq 3 \text{ heads}\} = 0.07299$ of Problem 10.1.

10.3. The PQR Company claims that the lifetime of a type of battery that it manufactures is more than 250 hours. A consumer advocate wishing to determine whether the claim is justified measures the lifetimes of 24 of the company's batteries; the results are listed in Table 10-3. Assuming the sample to be random, determine whether the company's claim is justified at the 0.05 significance level.

Table 10-3

271	230	198	275	282	225	284	219
253	216	262	288	236	291	253	224
264	295	211	252	294	243	272	268

Let H_0 be the hypothesis that the company's batteries have a lifetime equal to 250 hours, and let H_1 be the hypothesis that they have a lifetime greater than 250 hours. To test H_0 against H_1, we can use the sign test. To do this, we subtract 250 from each entry in Table 10-3 and record the signs of the differences, as shown in Table 10-4. We see that there are 15 plus signs and 9 minus signs.

Table 10-4

+	−	−	+	+	−	+	−
+	−	+	+	−	+	+	−
+	+	−	+	+	−	+	+

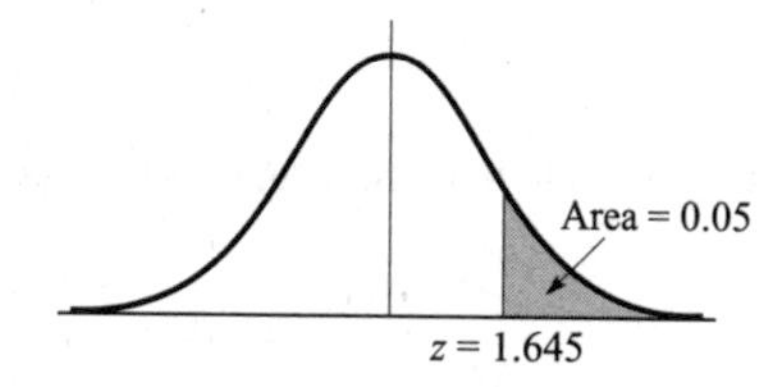

Fig. 10-2

Using a one-tailed test at the 0.05 significance level, we would reject H_0 if the z score were greater than 1.645 (Fig. 10-2). Since the z score, using a correction for continuity, is

$$z = \frac{(15 - 0.5) - (24)(0.5)}{\sqrt{(24)(0.5)(0.5)}} = 1.02$$

the company's claim cannot be justified at the 0.05 level.

10.4. A sample of 40 grades from a statewide examination is shown in Table 10-5. Test the hypothesis at the 0.05 significance level that the median grade for all participants is (a) 66, (b) 75.

Table 10-5

71	67	55	64	82	66	74	58	79	61
78	46	84	93	72	54	78	86	48	52
67	95	70	43	70	73	57	64	60	83
73	40	78	70	64	86	76	62	95	66

(a) Subtracting 66 from all the entries of Table 10-5 and retaining only the associated signs gives us Table 10-6, in which we see that there are 23 pluses, 15 minuses, and 2 zeros. Discarding the 2 zeros, our sample consists of 38 signs: 23 pluses and 15 minuses. Using a two-tailed test of the normal distribution with probabilities $\frac{1}{2}(0.05) = 0.025$ in each tail (Fig. 10-3), we adopt the following decision rule:

Accept the hypothesis if $-1.96 \leq z \leq 1.96$.
Reject the hypothesis otherwise.

Table 10-6

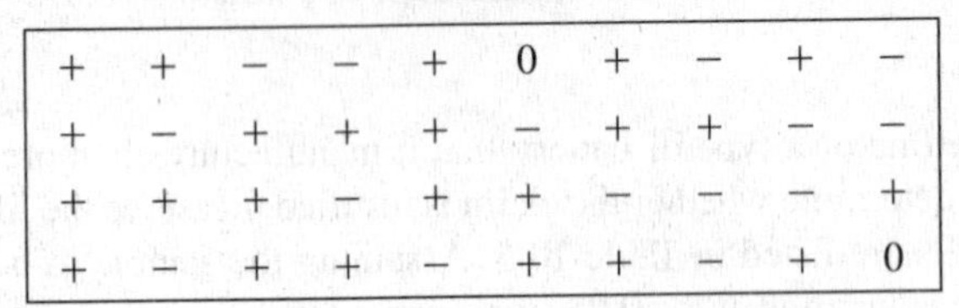

+	+	−	−	+	0	+	−	+	−
+	−	+	+	+	−	+	+	−	−
+	+	+	−	+	+	−	−	−	+
+	−	+	+	−	+	+	−	+	0

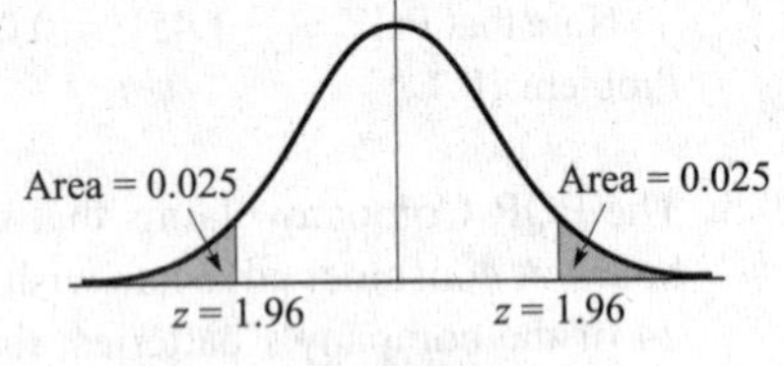

Fig. 10-3

Since
$$Z = \frac{X - Np}{\sqrt{Npq}} = \frac{(23 - 0.5) - (38)(0.5)}{\sqrt{(38)(0.5)(0.5)}} = 1.14$$

we accept the hypothesis that the median is 66 at the 0.05 level.

Note that we could also have used 15, the number of minus signs. In this case

$$z = \frac{(15 + 0.5) - (38)(0.5)}{\sqrt{(38)(0.5)(0.5)}} = -1.14$$

with the same conclusion.

(b) Subtracting 75 from all the entries in Table 10-5 gives us Table 10-7, in which there are 13 pluses and 27 minuses. Since

$$z = \frac{(13 + 0.5) - (40)(0.5)}{\sqrt{(40)(0.5)(0.5)}} = -2.06$$

we reject the hypothesis that the median is 75 at the 0.05 level.

Table 10-7

−	−	−	−	+	−	−	−	+	−
+	−	+	+	−	−	+	+	−	−
−	+	−	−	−	−	−	−	−	+
−	−	+	−	−	+	+	−	+	−

Using this method, we can arrive at a 95% confidence interval for the median grade on the examination. (see Problem 10.30.)

The Mann–Whitney *U* test

10.5. Referring to Table 10-2, determine whether there is a difference at the 0.05 significance level between cables made of alloy I and alloy II.

We organize the work in accordance with steps 1, 2, and 3 (described earlier in this chapter):

Step 1. Combining all 18 sample values in an array from the smallest to the largest gives us the first line of Table 10-8. These values are numbered 1 to 18 in the second line, which gives us the ranks.

Step 2. To find the sum of the ranks for each sample, rewrite Table 10-2 by using the associated ranks from Table 10-8; this gives us Table 10-9. The sum of the ranks is 106 for alloy I and 65 for alloy II.

Table 10-8

10.7	11.8	12.6	12.9	14.1	14.7	15.2	15.9	16.1	16.4	17.8	18.3	18.9	19.6	20.5	22.7	24.2	25.3
1	2	3	4	5	6	7	8	9	10	11	12	13	14	15	16	17	18

Step 3. Since the alloy I sample has the smaller size, $N_1 = 8$ and $N_2 = 10$. The corresponding sums of the ranks are $R_1 = 106$ and $R_2 = 65$. Then

$$U = N_1N_2 + \frac{N_1(N_1 + 1)}{2} - R_1 = (8)(10) + \frac{(8)(9)}{2} - 106 = 10$$

$$\mu_U = \frac{N_1N_2}{2} = \frac{(8)(10)}{2} = 40 \qquad \sigma_U^2 = \frac{N_1N_2(N_1 + N_2 + 1)}{12} = \frac{(8)(10)(19)}{12} = 126.67$$

Thus $\sigma_U = 11.25$ and

$$Z = \frac{U - \mu_U}{\sigma_U} = \frac{10 - 40}{11.25} = -2.67$$

Table 10-9

Alloy I		Alloy II	
Cable Strength	Rank	Cable Strength	Rank
18.3	12	12.6	3
16.4	10	14.1	5
22.7	16	20.5	15
17.8	11	10.7	1
18.9	13	15.9	8
25.3	18	19.6	14
16.1	9	12.9	4
24.2	17	15.2	7
Sum	106	11.8	2
		14.7	6
		Sum	65

Since the hypothesis H_0 that we are testing is whether there is *no* difference between the alloys, a two-tailed test is required. For the 0.05 significance level, we have the decision rule:

Accept H_0 if $-1.96 \le z \le 1.96$.
Reject H_0 otherwise.

Because $z = -2.67$, we reject H_0 and conclude that there is a difference between the alloys at the 0.05 level.

10.6. Verify results (6) and (7) of this chapter for the data of Problem 10.5.

(a) Since samples 1 and 2 yield values for U given by

$$U_1 = N_1N_2 + \frac{N_1(N_1 + 1)}{2} - R_1 = (8)(10) + \frac{(8)(9)}{2} - 106 = 10$$

$$U_2 = N_1N_2 + \frac{N_2(N_2 + 1)}{2} - R_2 = (8)(10) + \frac{(10)(11)}{2} - 65 = 70$$

we have $U_1 + U_2 = 10 + 70 = 80$, and $N_1N_2 = 106 + 65 = 171$ and

$$\frac{N(N+1)}{2} = \frac{(N_1 + N_2)(N_1 + N_2 + 1)}{2} = \frac{(18)(19)}{2} = 171$$

10.7. Work Problem 10.5 by using the statistic U for the alloy II sample.

For the alloy II sample,

$$U = N_1N_2 + \frac{N_2(N_2+1)}{2} - R_2 = (8)(10) + \frac{(10)(11)}{2} - 65 = 70$$

so that

$$z = \frac{U - \mu_U}{\sigma_U} = \frac{70 - 40}{11.25} = 2.67$$

This value of z is the *negative* of the z in Problem 10.5, and the right-hand tail of the normal distribution is used instead of the left-hand tail. Since this value of z also lies outside $-1.96 \le z \le 1.96$, the conclusion is the same as that for Problem 10.5.

10.8. A professor has two classes in psychology: a morning class of 9 students, and an afternoon class of 12 students. On a final examination scheduled at the same time for all students, the classes received the grades shown in Table 10-10. Can one conclude at the 0.05 significance level that the morning class performed worse than the afternoon class?

Table 10-10

Morning class	73	87	79	75	82	66	95	75	70			
Afternoon class	86	81	84	88	90	85	84	92	83	91	53	84

Step 1. Table 10-11 shows the array of grades and ranks. Note that the rank for the two grades of 75 is $\frac{1}{2}(5 + 6) = 5.5$, while the rank for the three grades of 84 is $\frac{1}{3}(11 + 12 + 13) = 12$.

Step 2. Rewriting Table 10-10 in terms of ranks gives us Table 10-12.

Check: $R_1 = 73$, $R_2 = 158$, and $N = N_1 + N_2 = 9 + 12 = 21$; thus $R_1 + R_2 = 73 + 158 = 231$ and

$$\frac{N(N+1)}{2} = \frac{(21)(22)}{2} = 231 = R_1 + R_2$$

Table 10-11

53	66	70	73	75	75	79	81	82	83	84	84	84	85	86	87	88	90	91	92	95
1	2	3	4	5.5		7	8	9	10		12		14	15	16	17	18	19	20	21

Table 10-12

													Sum of Ranks
Morning class	4	16	7	5.5	9	2	21	5.5	3				73
Afternoon class	15	8	12	17	18	14	12	20	10	19	1	12	158

Step 3.

$$U = N_1N_2 + \frac{N_1(N_1+1)}{2} - R_1 = (9)(12) + \frac{(9)(10)}{2} - 73 = 80$$

$$\mu_U = \frac{N_1N_2}{2} = \frac{(9)(12)}{2} = 54 \qquad \sigma_U^2 = \frac{N_1N_2(N_1+N_2+1)}{12} = \frac{(9)(12)(22)}{12} = 198$$

Therefore,
$$Z = \frac{U - \mu_U}{\sigma_U} = \frac{80-54}{14.07} = -1.85$$

Since we wish to test the hypothesis H_1 that the morning class performs worse than the afternoon class against the hypothesis H_0 that there is no difference at the 0.05 level, a one-tailed test is needed. Referring to Fig. 10-2, which applies here, we have the decision rule:

Accept H_0 if $z \leq 1.645$.
Reject H_0 if $z > 1.645$.

Since the actual value of $z = 1.85 > 1.645$, we reject H_0 and conclude that the morning class performed worse than the afternoon class at the 0.05 level. This conclusion cannot be reached, however, for the 0.01 level (see Problem 10.33).

10.9. Find U for the data of Table 10-13 by using (a) formula (2) of this chapter, (b) the counting method (as described in Remark 4 of this chapter).

(a) Arranging the data from both samples in an array in increasing order of magnitude and assigning ranks from 1 to 5 gives us Table 10-14. Replacing the data of Table 10-13 with the corresponding ranks gives us Table 10-15, from which the sums of the ranks are $R_1 = 5$ and $R_2 = 10$. Since $N_1 = 2$ and $N_2 = 3$, the value of U for sample 1 is

$$U = N_1N_2 + \frac{N_1(N_1+1)}{2} - R_1 = (2)(3) + \frac{(2)(3)}{2} - 5 = 4$$

The value U for sample 2 can be found similarly to be $U = 2$.

Table 10-13

Sample 1	22	10	
Sample 2	17	25	14

Table 10-14

Data	10	14	17	22	25
Rank	1	2	3	4	5

Table 10-15

				Sum of Ranks
Sample 1	4	1		5
Sample 2	3	5	2	10

(b) Let us replace the sample values in Table 10-14 with I or II, depending on whether the value belongs to sample 1 or 2. Then the first line of Table 10-14 becomes

Data	I	II	II	I	II

From this we see that

Number of sample 1 values preceding first sample 2 value = 1
Number of sample 1 values preceding second sample 2 value = 1
Number of sample 1 values preceding third sample 2 value = 2
Total = 4

Thus the value of U corresponding to the first sample is 4.

Similarly, we have

$$\begin{aligned} \text{Number of sample 2 values preceding first sample 1 value} &= 0 \\ \text{Number of sample 2 values preceding second sample 1 value} &= \underline{2} \\ \text{Total} &= 2 \end{aligned}$$

Thus the value of U corresponding to the second sample is 2.

Note that since $N_1 = 2$ and $N_2 = 3$, these values satisfy $U_1 + U_2 = N_1N_2$; that is, $4 + 2 = (2)(3) = 6$.

10.10. A population consists of the values 7, 12, and 15. Two samples are drawn without replacement from this population; sample 1, consisting of one value, and sample 2, consisting of two values. (Between them, the two samples exhaust the population.)

(a) Find the sampling distribution of U.

(b) Find the mean and variance of the distribution in part (a).

(c) Verify the results found in part (b) by using formulas (3) of this chapter.

(a) We choose sampling without replacement to avoid ties—which would occur if, for example, the value 12 were to appear in both samples.

There are $3 \cdot 2 = 6$ possibilities for choosing the samples, as shown in Table 10-16. It should be noted that we could just as easily use ranks 1, 2, and 3 instead of 7, 12, and 15. The value U in Table 10-16 is that found for sample 1, but if U for sample 2 were used, the distribution would be the same.

Table 10-16

Sample 1	Sample 2		U
7	12	15	2
7	15	12	2
12	7	15	1
12	15	7	1
15	7	12	0
15	12	7	0

(b) The mean and variance found from Table 10-15 are given by

$$\mu_U = \frac{2 + 2 + 1 + 1 + 0 + 0}{6} = 1$$

$$\sigma_U^2 = \frac{(2-1)^2 + (2-1)^2 + (1-1)^2 + (1-1)^2 + (0-1)^2 + (0-1)^2}{6} = \frac{2}{3}$$

(c) By formulas (3),

$$\mu_U = \frac{N_1N_2}{2} = \frac{(1)(2)}{2} = 1$$

$$\sigma_U^2 = \frac{N_1N_2(N_1 + N_2 + 1)}{12} = \frac{(1)(2)(1 + 2 + 1)}{12} = \frac{2}{3}$$

showing agreement with part (a).

10.11. (a) Find the sampling distribution of U in Problem 10.9 and graph it.

(b) Obtain the mean and variance of U directly from the results of part (a).

(c) Verify part (b) by using formulas (3) of this chapter.

(a) In this case there are $5 \cdot 4 \cdot 3 \cdot 2 = 120$ possibilities for choosing values for the two samples and the method of Problem 10.9 is too laborious. To simplify the procedure, let us concentrate on the smaller sample (of size $N_1 = 2$) and the possible sums of the ranks, R_1. The sum of the ranks for sample 1 is the *smallest* when the sample consists of the two lowest-ranking numbers (1, 2): then $R_1 = 1 + 2 = 3$. Similarly, the sum of the ranks for sample 1 is the *largest* when the sample consists of the two highest-ranking numbers (4, 5); then $R_1 = 4 + 5 = 9$. Thus R_1 varies from 3 to 9.

Column 1 of Table 10-17 lists these values of R_1 (from 3 to 9), and column 2 shows the corresponding sample 1 values, whose sum is R_1. Column 3 gives the frequency (or number) of samples with sum R_1; for example, there are $f = 2$ samples with $R_1 = 5$. Since $N_1 = 2$ and $N_2 = 3$, we have

$$U = N_1N_2 + \frac{N_1(N_1 + 1)}{2} - R_1 = (2)(3) + \frac{(2)(3)}{2} - R_1 = 9 - R_1$$

The probability that $U = R_1$ (i.e., $\Pr\{U = R_1\}$) is shown in column 5 of Table 10-17 and is obtained by finding the relative frequency. The relative frequency is found by dividing each frequency f by the sum of all the frequencies, or 10; for example, $\Pr\{U = 5\} = \frac{2}{10} = 0.2$.

Table 10-17

R_1	Sample 1 Values	f	U	$\Pr\{U = R_1\}$
3	(1, 2)	1	6	0.1
4	(1, 3)	1	5	0.1
5	(1, 4), (2, 3)	2	4	0.2
6	(1, 5), (2, 4)	2	3	0.2
7	(2, 5), (3, 4)	2	2	0.2
8	(3, 5)	1	1	0.1
9	(4, 5)	1	0	0.1

(b) From columns 3 and 4 of Table 10-17 we have

$$\mu_U = \bar{U} = \frac{\sum fU}{\sum f} = \frac{(1)(6) + (1)(5) + (2)(4) + (2)(3) + (2)(2) + (1)(1) + (1)(0)}{1 + 1 + 2 + 2 + 2 + 1 + 1} = 3$$

$$\sigma_U^2 = \frac{\sum f(U - \bar{U})^2}{\sum f}$$

$$= \frac{(1)(6-3)^2 + (1)(5-3)^2 + (2)(4-3)^2 + (2)(3-3)^2 + (2)(2-3)^2 + (1)(1-3)^2 + (1)(0-3)^2}{10}$$

$$= 3$$

Another method

$$\sigma_U^2 = \overline{U^2} - \bar{U}^2 = \frac{(1)(6)^2 + (1)(5)^2 + (2)(4)^2 + (2)(3)^2 + (2)(2)^2 + (1)(1)^2 + (1)(0)^2}{10} - (3)^2 = 3$$

(c) By formulas (3), using $N_1 = 2$ and $N_2 = 3$, we have

$$\mu_U = \frac{N_1N_2}{2} = \frac{(2)(3)}{2} = 3 \qquad \sigma_U^2 = \frac{N_1N_2(N_1 + N_2 + 1)}{12} = \frac{(2)(3)(6)}{12} = 3$$

10.12. If N numbers in a set are ranked from 1 to N, prove that the sum of the ranks is $[N(N + 1)]/2$.

Let R be the sum of the ranks. Then we have

$$R = 1 + 2 + 3 + \cdots + (N - 1) + N \tag{16}$$

$$R = N + (N - 1) + (N - 2) + \cdots + 2 + 1 \tag{17}$$

where the sum in equation (17) is obtained by writing the sum in (16) backward. Adding equations (16) and (17) gives

$$2R = (N + 1) + (N + 1) + (N + 1) + \cdots + (N + 1) + (N + 1) = N(N + 1)$$

since $(N + 1)$ occurs N times in the sum; thus $R = [N(N + 1)]/2$. This can also be obtained by using a result from elementary algebra on arithmetic progressions and series.

10.13. If R_1 and R_2 are the respective sums of the ranks for samples 1 and 2 in the U test, prove that $R_1 + R_2 = [N(N + 1)]/2$.

We assume that there are no ties in the sample data. Then R_1 must be the sum of some of the ranks (numbers) in the set 1, 2, 3, . . . , N, while R_2 must be the sum of the remaining ranks in the set. Thus the sum $R_1 + R_2$ must be the sum of all the ranks in the set; that is, $R_1 + R_2 = 1 + 2 + 3 + \cdots + N = [N(N + 1)]/2$ by Problem 10.12.

The Kruskal–Wallis *H* test

10.14. A company wishes to purchase one of five different machines: *A, B, C, D*, or *E*. In an experiment designed to determine whether there is a performance difference between the machines, five experienced operators each work on the machines for equal times. Table 10-18 shows the number of units produced by each machine. Test the hypothesis that there is no difference between the machines at the (a) 0.05, (b) 0.01 significance levels.

Table 10-18

A	68	72	77	42	53
B	72	53	63	53	48
C	60	82	64	75	72
D	48	61	57	64	50
E	64	65	70	68	53

Table 10-19

						Sum of Ranks
A	17.5	21	24	1	6.5	70
B	21	6.5	12	6.5	2.5	48.5
C	10	25	14	23	21	93
D	2.5	11	9	14	4	40.5
E	14	16	19	17.5	6.5	73

Since there are five samples (*A, B, C, D*, and *E*), $k = 5$. And since each sample consists of five values, we have $N_1 = N_2 = N_3 = N_4 = N_5 = 5$, and $N = N_1 + N_2 + N_3 + N_4 + N_5 = 25$. By arranging all the values in increasing order of magnitude and assigning appropriate ranks to the ties, we replace Table 10-18 with Table 10-19, the right-hand column of which shows the sum of the ranks. We see from Table 10-19 that $R_1 = 70$, $R_2 = 48.5$, $R_3 = 93$, $R_4 = 40.5$, and $R_5 = 73$. Thus

$$H = \frac{12}{N(N + 1)} \sum_{j=1}^{k} \frac{R_j^2}{N_j} - 3(N + 1)$$

$$= \frac{12}{(25)(26)} \left[\frac{(70)^2}{5} + \frac{(48.5)^2}{5} + \frac{(93)^2}{5} + \frac{(40.5)^2}{5} + \frac{(73)^2}{5} \right] - 3(26) = 6.44$$

For $k - 1 = 4$ degrees of freedom at the 0.05 significance level, from Appendix E we have $\chi^2_{0.95} = 9.49$. Since $6.44 < 9.49$, we cannot reject the hypothesis of no difference between the machines at the 0.05 level and therefore certainly cannot reject it at the 0.01 level. In other words, we can accept the hypothesis (or reserve judgment) that there is no difference between the machines at both levels.

Note that we have already worked this problem by using analysis of variance (see Problem 9.8) and have arrived at the same conclusion.

10.15. Work Problem 10.14 if a correction for ties is made.

Table 10-20 shows the number of ties corresponding to each of the tied observations. For example, 48 occurs two times, whereby $T = 2$, and 53 occurs four times, whereby $T = 4$. By calculating $T_3 - T$ for each of these values of T and adding, we find that $\sum(T^3 - T) = 6 + 60 + 24 + 6 + 24 = 120$, as shown in Table 10-20. Then, since $N = 25$, the correction factor is

$$1 - \frac{\sum(T^3 - T)}{N^3 - N} = 1 - \frac{120}{(25)^3 - 25} = 0.9923$$

Table 10-20

Observation	48	53	64	68	72	
Number of ties (T)	2	4	3	2	3	
$T^3 - T$	6	60	24	6	24	$\sum(T^3 - T) = 120$

and the corrected value of H is

$$H_c = \frac{6.44}{0.9923} = 6.49$$

This correction is not sufficient to change the decision made in Problem 10.14.

10.16. Three samples are chosen at random from a population. Arranging the data according to rank gives us Table 10-21. Determine whether there is any difference between the samples at the (a) 0.05, (b) 0.01 significance levels.

Table 10-21

Sample 1	7	4	6	10	
Sample 2	11	9	12		
Sample 3	5	1	3	8	2

Here $k = 3$, $N_1 = 4$, $N_2 = 3$, $N_3 = 5$, $N = N_1 + N_2 + N_3 = 12$, $R_1 = 7 + 4 + 6 + 10 = 27$, $R_2 = 11 + 9 + 12 = 32$, and $R_3 = 5 + 1 + 3 + 8 + 2 = 19$. Thus

$$H = \frac{12}{N(N+1)} \sum_{j=1}^{k} \frac{R_j^2}{N_j} - 3(N+1) = \frac{12}{(12)(13)} \left[\frac{(27)^2}{4} + \frac{(32)^2}{3} + \frac{(19)^2}{5} \right] - 3(13) = 6.83$$

(a) For $k - 1 = 3 - 1 = 2$ degrees of freedom, $\chi^2_{0.95} = 5.99$. Thus, since $6.83 > 5.99$, we can conclude that there is a significant difference between the samples at the 0.05 level.

(b) For 2 degrees of freedom, $\chi^2_{0.95} = 9.21$. Thus, since $6.83 < 9.21$, we cannot conclude that there is a difference between the samples at the 0.01 level.

The runs test for randomness

10.17. In 30 tosses of a coin, the following sequence of heads (H) and tails (T) is obtained:

H T T H T H H H T H H T T H T

H T H H T H T T H T H H T H T

(a) Determine the number of runs, V.

(b) Test at the 0.05 significance level whether the sequence is random.

(a) Using a vertical bar to indicate a run, we see from

H | T T | H | T | H H H | T | H H | T T | H | T |

H | T | H H | T | H | T T | H | T | H H | T | H | T |

that the number of runs is $V = 22$.

(b) There are $N_1 = 16$ heads and $N_2 = 14$ tails in the given sample of tosses, and from part (a), the number of runs is $V = 22$. Thus from formulas (13) of this chapter we have

$$\mu_V = \frac{2(16)(14)}{16 + 14} + 1 = 15.93 \qquad \sigma_V^2 = \frac{2(16)(14)[2(16)(14) - 16 - 14]}{(16 + 14)^2(16 + 14 - 1)} = 7.175$$

or $\sigma_V = 2.679$. The z score corresponding to $V = 22$ runs is therefore

$$Z = \frac{V - \mu_V}{\sigma_V} = \frac{22 - 15.93}{2.679} = 2.27$$

Now for a two-tailed test at the 0.05 significance level, we would accept the hypothesis H_0 of randomness if $-1.96 \leq z \leq 1.96$ and would reject it otherwise (see Fig. 10-4). Since the calculated value of z is $2.27 > 1.96$, we conclude that the tosses are not random at the 0.05 level. The test shows that there are *too many* runs, indicating a *cyclic pattern* in the tosses.

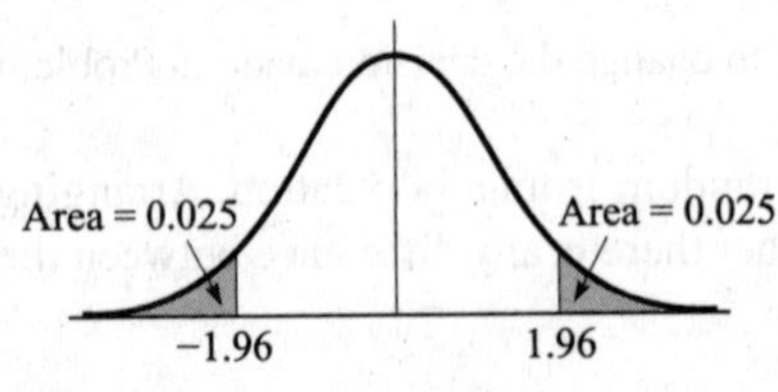

Fig. 10-4

If a correction for continuity is used, the above z score is replaced by

$$z = \frac{(22 - 0.5) - 15.93}{2.679} = 2.08$$

and the same conclusion is reached.

10.18. A sample of 48 tools produced by a machine shows the following sequence of good (G) and defective (D) tools:

G G G G G G D D G G G G G G G G

G G D D D D G G G G G G D G G G

G G G G G G D D G G G G G D G G

Test the randomness of the sequence at the 0.05 significance level.

The numbers of D's and G's are $N_1 = 10$ and $N_2 = 38$, respectively, and the number of runs is $V = 11$. Thus the mean and variance are given by

$$\mu_V = \frac{2(10)(38)}{10 + 38} + 1 = 16.83 \qquad \sigma_V^2 = \frac{2(10)(38)[2(10)(38) - 10 - 38]}{(10 + 38)^2(10 + 38 - 1)} = 4.997$$

so that $\sigma_V = 2.235$.

For a two-tailed test at the 0.05 level, we would accept the hypothesis H_0 of randomness if $-1.96 \leq z \leq 1.96$ (see Fig. 10-4) and would reject if otherwise. Since the z score corresponding to $V = 11$ is

$$Z = \frac{V - \mu_V}{\sigma_V} = \frac{11 - 16.83}{2.235} = -2.61$$

and $-2.61 < -1.96$, we can reject H_0 at the 0.05 level.

The test shows that there are *too few* runs, indicating a clustering (or bunching) of defective tools. In other words, there seems to be a *trend pattern* in the production of defective tools. Further examination of the production process is warranted.

10.19. (a) Form all possible sequences consisting of three a's and two b's, and give the numbers of runs, V, corresponding to each sequence.

(b) Obtain the sampling distribution of V.

(c) Obtain the probability distribution of V.

(a) The number of possible sequences consisting of three a's and two b's is

$$\binom{5}{2} = \frac{5!}{2!3!} = 10$$

These sequences are shown in Table 10-22, along with the number of runs corresponding to each sequence.

Table 10-22

Sequence	Runs (V)
a a a b b	2
a a b a b	4
a a b b a	3
a b a b a	5
a b b a a	3
a b a a b	4
b b a a a	2
b a b a a	4
b a a a b	3
b a a b a	4

Table 10-23

V	f
2	2
3	3
4	4
5	1

(b) The sampling distribution of V is given in Table 10-23 (obtained from Table 10-21), where V denotes the number of runs and f denotes the frequency. For example, Table 10-23 shows that there is one 5, four 4s, etc.

(c) The probability distribution of V is obtained from Table 10-23 by dividing each frequency by the total frequency $2 + 3 + 4 + 1 = 10$. For example, $\Pr\{V = 5\} = \frac{1}{10} = 0.1$.

10.20. Find (a) the mean, (b) the variance of the number of runs in Problem 10.19 directly from the results obtained there.

(a) From Table 10-22 we have

$$\mu_V = \frac{2 + 4 + 3 + 5 + 3 + 4 + 2 + 4 + 3 + 4}{10} = \frac{17}{5}$$

Another method

From Table 10-22 the grouped-data method gives

$$\mu_V = \frac{\sum fV}{\sum f} = \frac{(2)(2) + (3)(3) + (4)(4) + (1)(5)}{2 + 3 + 4 + 1} = \frac{17}{5}$$

(b) Using the grouped-data method for computing the variance, from Table 10-23 we have

$$\sigma_V^2 = \frac{\sum f(V - \bar{V})^2}{\sum f} = \frac{1}{10}\left[(2)\left(2 - \frac{17}{5}\right)^2 + (3)\left(3 - \frac{17}{5}\right)^2 + (4)\left(4 - \frac{17}{5}\right)^2 + (1)\left(5 - \frac{17}{2}\right)^2\right] = \frac{21}{25}$$

Another method

As in Chapter 5, the variance is given by

$$\sigma_V^2 = \overline{V^2} - \bar{V}^2 = \frac{(2)(2)^2 + (3)(3)^2 + (4)(4)^2 + (1)(5)^2}{10} - \left(\frac{17}{5}\right)^2 = \frac{21}{25}$$

10.21. Work Problem 10.20 by using formulas (13) of this chapter.

Since there are three a's and two b's, we have $N_1 = 3$ and $N_2 = 2$. Thus

(a) $$\mu_V = \frac{2N_1N_2}{N_1 + N_2} + 1 = \frac{2(3)(2)}{3 + 2} + 1 = \frac{17}{5}$$

(b) $$\sigma_V^2 = \frac{2N_1N_2(2N_1N_2 - N_1 - N_2)}{(N_1 + N_2)^2(N_1 + N_2 - 1)} = \frac{2(3)(2)[2(3)(2) - 3 - 2]}{(3 + 2)^2(3 + 2 - 1)} = \frac{21}{25}$$

Further applications of the runs test

10.22. Referring to Problem 10.3, and assuming a significance level of 0.05, determine whether the sample lifetimes of the batteries produced by the PQR Company are random.

Table 10-24 shows the batteries' lifetimes in increasing order of magnitude. Since there are 24 entries in the table, the median is obtained from the middle two entries, 253 and 262, as $\frac{1}{2}(253 + 262) = 257.5$. Rewriting the data of Table 10-23 by using an a if the entry is above the median and a b if it is below the median, we obtain Table 10-25, in which we have 12 a's, 12 b's, and 15 runs. Thus $N_1 = 12$, $N_2 = 12$, $N = 24$, $V = 15$, and we have

$$\mu_V = \frac{2N_1N_2}{N_1 + N_2} + 1 = \frac{2(12)(12)}{12 + 12} + 1 = 13 \qquad \sigma_V^2 = \frac{2(12)(12)(264)}{(24)^2(23)} = 5.739$$

so that $$Z = \frac{V - \mu_V}{\sigma_V} = \frac{15 - 13}{2.396} = 0.835$$

Using a two-tailed test at the 0.05 significance level, we would accept the hypothesis of randomness if $-1.96 \le z \le 1.96$. Since 0.835 falls within this range, we conclude that the sample is random.

Table 10-24

198	211	216	219	224	225	230	236
243	252	253	253	262	264	268	271
272	275	282	284	288	291	294	295

Table 10-25

a	b	b	a	a	b	a	b
b	b	a	a	b	a	b	b
a	a	b	b	a	b	a	a

10.23. Work Problem 10.5 by using the runs test for randomness.

The arrangement of all values from both samples already appears in line 1 of Table 10-8. Using the symbols a and b for the data from samples I and II, respectively, the arrangement becomes

$$b \; b \; b \; b \; b \; b \; b \; b \; a \; a \; a \; a \; a \; b \; b \; a \; a \; a$$

Since there are four runs, we have $V = 4$, $N_1 = 8$, and $N_2 = 10$. Then

$$\mu_V = \frac{2N_1N_2}{N_1 + N_2} + 1 = \frac{2(8)(10)}{18} + 1 = 9.889$$

$$\sigma_V^2 = \frac{2N_1N_2(2N_1N_2 - N_1 - N_2)}{(N_1 + N_2)^2(N_1 + N_2 - 1)} = \frac{2(8)(10)(142)}{(18)^2(17)} = 4.125$$

so that $$Z = \frac{V - \mu_V}{\sigma_V} = \frac{4 - 9.889}{2.031} = -2.90$$

If H_0 is the hypothesis that there is no difference between the alloys, it is also the hypothesis that the above sequence is random. We would accept this hypothesis if $-1.96 \leq z \leq 1.96$ and would reject it otherwise. Since $= -2.90$ lies outside this interval, we reject H_0 and reach the same conclusion as for Problem 10.5.

Note that if a correction is made for continuity,

$$Z = \frac{V - \mu_V}{\sigma_V} = \frac{(4 + 0.5) - 9.889}{2.031} = -2.65$$

and we reach the same conclusion.

Rank correlation

10.24. Table 10-26 shows how 10 students, arranged in alphabetical order, were ranked according to their achievements in both the laboratory and lecture sections of a biology course. Find the coefficient of rank correlation.

Table 10-26

Laboratory	8	3	9	2	7	10	4	6	1	5
Lecture	9	5	10	1	8	7	3	4	2	6

The difference in ranks, D, in the laboratory and lecture sections for each student is given in Table 10-27, which also gives D^2 and $\sum D^2$. Thus

$$r_s = 1 - \frac{6\sum D^2}{N(N^2 - 1)} = 1 - \frac{6(24)}{10(10^2 - 1)} = 0.8545$$

indicating that there is a marked relationship between the achievements in the course's laboratory and lecture sections.

Table 10-27

Difference of ranks (D)	−1	−2	−1	1	−1	3	1	2	−1	−1	
D^2	1	4	1	1	1	9	1	4	1	1	$\sum D^2 = 24$

10.25. Table 10-28 shows the heights of a sample of 12 fathers and their oldest adult sons. Find the coefficient of rank correlation.

Table 10-28

Height of father (inches)	65	63	67	64	68	62	70	66	68	67	69	71
Height of son (inches)	68	66	68	65	69	66	68	65	71	67	68	70

Arranged in ascending order of magnitude, the fathers' heights are

$$62 \quad 63 \quad 64 \quad 65 \quad 66 \quad 67 \quad 67 \quad 68 \quad 68 \quad 69 \quad 71 \tag{18}$$

Since the sixth and seventh places in this array represent the same height (67 inches), we assign a mean rank $\frac{1}{2}(6 + 7) = 6.5$ to these places. Similarly, the eighth and ninth places are assigned the rank $\frac{1}{2}(8 + 9) = 8.5$. Thus the fathers' heights are assigned the ranks

$$1 \quad 2 \quad 3 \quad 4 \quad 5 \quad 6.5 \quad 6.5 \quad 8.5 \quad 8.5 \quad 10 \quad 11 \quad 12 \tag{19}$$

Similarly, arranged in ascending order of magnitude, the sons' heights are

$$65 \quad 65 \quad 66 \quad 66 \quad 67 \quad 68 \quad 68 \quad 68 \quad 68 \quad 69 \quad 70 \quad 71 \tag{20}$$

and since the sixth, seventh, eighth, and ninth places represent the same height (68 inches), we assign the mean rank $\frac{1}{4}(6 + 7 + 8 + 9) = 7.5$ to these places. Thus the sons' heights are assigned the ranks

$$1.5 \quad 1.5 \quad 3.5 \quad 3.5 \quad 5 \quad 7.5 \quad 7.5 \quad 7.5 \quad 7.5 \quad 10 \quad 11 \quad 12 \tag{21}$$

Using the correspondences (18) and (19), and (20) and (21), we can replace Table 10-28 with Table 10-29. Table 10-30 shows the difference in ranks, D, and the computations of D^2 and $\sum D^2$, whereby

$$r_s = 1 - \frac{6\sum D^2}{N(N^2 - 1)} = 1 - \frac{6(72.50)}{12(12^2 - 1)} = 0.7465$$

The result agrees well with the correlation coefficient obtained by other methods (see Problems 8.26, 8.28, 8.30, and 8.32).

Table 10-29

Rank of father	4	2	6.5	3	8.5	1	11	5	8.5	6.5	10	12
Rank of son	7.5	3.5	7.5	1.5	10	3.5	7.5	1.5	12	5	7.5	11

Table 10-30

D	−3.5	−1.5	−1.0	1.5	−1.5	−2.5	3.5	3.5	−3.5	1.5	2.5	1.0	
D^2	12.25	2.25	1.00	2.25	2.25	6.25	12.25	12.25	12.25	2.25	6.25	1.00	$\sum D^2 = 72.50$

SUPPLEMENTARY PROBLEMS

The sign test

10.26. A company claims that if its product is added to an automobile's gasoline tank, the mileage per gallon will improve. To test the claim, 15 different automobiles are chosen and the mileage per gallon with and without the additive is measured; the results are shown in Table 10-31. Assuming that the driving conditions are the same, determine whether there is a difference due to the additive at significance levels of (a) 0.05, (b) 0.01.

Table 10-31

With additive	34.7	28.3	19.6	25.1	15.7	24.5	28.7	23.5	27.7	32.1	29.6	22.4	25.7	28.1	24.3
Without additive	31.4	27.2	20.4	24.6	14.9	22.3	26.8	24.1	26.2	31.4	28.8	23.1	24.0	27.3	22.9

10.27. Can one conclude at the 0.05 significance level that the mileage per gallon achieved in Problem 10.26 is *better* with the additive than without it?

10.28. A weight-loss club advertises that a special program that it has designed will produce a weight loss of at least 6% in 1 month if followed precisely. To test the club's claim, 36 adults undertake the program. Of these, 25 realize the desired loss, 6 gain weight, and the rest remain essentially unchanged. Determine at the 0.05 significance level whether the program is effective.

10.29. A training manager claims that by giving a special course to company sales personnel, the company's annual sales will increase. To test this claim, the course is given to 24 people. Of these 24, the sales of 16 increase, those of 6 decrease, and those of 2 remain unchanged. Test at the 0.05 significance level the hypothesis that the course increased the company's sales.

10.30. The MW Soda Company sets up "taste tests" in 27 locations around the country in order to determine the public's relative preference for two brands of cola, *A* and *B*. In eight locations brand *A* is preferred over brand *B*, in 17 locations brand *B* is preferred over brand *A*, and in the remaining locations there is indifference. Can one conclude at the 0.05 significance level that brand *B* is preferred over brand *A*?

10.31. The breaking strengths of a random sample of 25 ropes made by a manufacturer are given in Table 10-32. On the basis of this sample, test at the 0.05 significance level the manufacturer's claim that the breaking strength of a rope is (a) 25, (b) 30, (c) 35, (d) 40.

Table 10-32

41	28	35	38	23
37	32	24	46	30
25	36	22	41	37
43	27	34	27	36
42	33	28	31	24

10.32. Show how to obtain 95% confidence limits for the data in Problem 10.4.

10.33. Make up and solve a problem involving the sign test.

The Mann–Whitney *U* test

10.34. Instructors *A* and *B* both teach a first course in chemistry at XYZ University. On a common final examination, their students received the grades shown in Table 10-33. Test at the 0.05 significance level the hypothesis that there is no difference between the two instructors' grades.

Table 10-33

A	88	75	92	71	63	84	55	64	82	96				
B	72	65	84	53	76	80	51	60	57	85	94	87	73	61

10.35. Referring to Problem 10.34, can one conclude at the 0.01 significance level that the students' grades in the morning class are worse than those in the afternoon class?

10.36. A farmer wishes to determine whether there is a difference in yields between two different varieties of wheat, I and II. Table 10-34 shows the production of wheat per unit area using the two varieties. Can the farmer conclude at significance levels of (a) 0.05, (b) 0.01 that a difference exists?

Table 10-34

Wheat I	15.9	15.3	16.4	14.9	15.3	16.0	14.6	15.3	14.5	16.6	16.0
Wheat II	16.4	16.8	17.1	16.9	18.0	15.6	18.1	17.2	15.4		

10.37. Can the farmer of Problem 10.36 conclude at the 0.05 level that wheat II produces a larger yield than wheat I?

10.38. A company wishes to determine whether there is a difference between two brands of gasoline, *A* and *B*. Table 10-35 shows the distances traveled per gallon for each brand. Can we conclude at the 0.05 significance level (a) that there is a difference between the brands, (b) that brand *B* is better than brand *A*?

Table 10-35

A	30.4	28.7	29.2	32.5	31.7	29.5	30.8	31.1	30.7	31.8
B	33.5	29.8	30.1	31.4	33.8	30.9	31.3	29.6	32.8	33.0

10.39. Can the U test be used to determine whether there is a difference between machines I and II of Table 10-1?

10.40. Make up and solve a problem using the U test.

10.41. Find U for the data of Table 10-36, using (a) the formula method, (b) the counting method.

Table 10-36

Sample 1	15	25
Sample 2	20	32

Table 10-37

Sample 1	40	27	30	56
Sample 2	10	35		

10.42. Work Problem 10.41 for the data of Table 10-37.

10.43. A population consists of the values 2, 5, 9, and 12. Two samples are drawn from this population, the first consisting of one of these values and the second consisting of the other three values.

(a) Obtain the sampling distribution of U and its graph.

(b) Obtain the mean and variance of this distribution, both directly and by formula.

10.44. Prove that $U_1 + U_2 = N_1N_2$.

10.45. Prove that $R_1 + R_2 = [N(N + 1)]/2$ for the case where the number of ties is (a) 1, (b) 2, (c) any number.

10.46. If $N_1 = 14$, $N_2 = 12$, and $R_1 = 105$, find (a) R_2, (b) U_1, (c) U_2.

10.47. If $N_1 = 10$, $N_2 = 16$, and $U_2 = 60$, find (a) R_1, (b) R_2, (c) U_1.

10.48. What is the largest number of the values N_1, N_2, R_1, R_2, U_1, and U_2 that can be determined from the remaining ones? Prove your answer.

The Kruskal–Wallis *H* test

10.49. An experiment is performed to determine the yields of five different varieties of wheat: *A*, *B*, *C*, *D*, and *E*. Four plots of land are assigned to each variety. The yields (in bushels per acre) are shown in Table 10-38. Assuming that the plots have similar fertility and that the varieties are assigned to the plots at random, determine whether there is a significant difference between the yields at the (a) 0.05, (b) 0.01 levels.

Table 10-38

A	20	12	15	19
B	17	14	12	15
C	23	16	18	14
D	15	17	20	12
E	21	14	17	18

Table 10-39

A	33	38	36	40	31	35
B	32	40	42	38	30	34
C	31	37	35	33	34	30
D	27	33	32	29	31	28

10.50. A company wishes to test four different types of tires: *A*, *B*, *C*, and *D*. The lifetimes of the tires, as determined from their treads, are given (in thousands of miles) in Table 10-39; each type has been tried on six similar automobiles assigned to the tires at random. Determine whether there is a significant difference between the tires at the (a) 0.05, (b) 0.01 levels.

10.51. A teacher wishes to test three different teaching methods: I, II, and III. To do this, the teacher chooses at random three groups of five students each and teaches each group by a different method. The same examination is then given to all the students, and the grades in Table 10-40 are obtained. Determine at the (a) 0.05, (b) 0.01 significance levels whether there is a difference between the teaching methods.

Table 10-40

Method I	78	62	71	58	73
Method II	76	85	77	90	87
Method III	74	79	60	75	80

Table 10-41

Mathematics	72	80	83	75	
Science	81	74	77		
English	88	82	90	87	80
Economics	74	71	77	70	

10.52. During one semester a student received in various subjects the grades shown in Table 10-41. Test at the (a) 0.05, (b) 0.01 significance levels whether there is a difference between the grades in these subjects.

10.53. Using the *H* test, work (a) Problem 9.14, (b) Problem 9.23, (c) Problem 9.24.

10.54. Using the *H* test, work (a) Problem 9.25, (b) Problem 9.26, (c) Problem 9.27.

The runs test for randomness

10.55. Determine the number of runs, *V*, for each of these sequences:

(a) A B A B B A A A B B A B

(b) H H T H H H T T T T H H T H H T H T

10.56. Twenty-five individuals were sampled as to whether they liked or did not like a product (indicated by Y and N, respectively). The resulting sample is shown by the following sequence:

Y Y N N N N Y Y Y N Y N N Y N N N N N Y Y Y Y N N

(a) Determine the number of runs, *V*.

(b) Test at the 0.05 significance level whether the responses are random.

10.57. Use the runs test on sequences (10) and (11) in this chapter, and state any conclusions about randomness.

10.58. (a) Form all possible sequences consisting of two *a*'s and one *b*, and give the number of runs, *V*, corresponding to each sequence.

(b) Obtain the sampling distribution of *V*.

(c) Obtain the probability distribution of *V*.

10.59. In Problem 10.58, find the mean and variance of *V* (a) directly from the sampling distribution, (b) by formula.

10.60. Work Problems 10.58 and 10.59 for the cases in which there are (a) two *a*'s and two *b*'s, (b) one *a* and three *b*'s, (c) one *a* and four *b*'s.

10.61. Work Problems 10.58 and 10.59 for the cases in which there are (a) two a's and four b's, (b) three a's and three b's.

Further applications of the runs test

10.62. Assuming a significance level of 0.05, determine whether the sample of 40 grades in Table 10-5 is random.

10.63. The closing prices of a stock on 25 successive days are given in Table 10-42. Determine at the 0.05 significance level whether the prices are random.

Table 10-42

10.375	11.125	10.875	10.625	11.500
11.625	11.250	11.375	10.750	11.000
10.875	10.750	11.500	11.250	12.125
11.875	11.375	11.875	11.125	11.750
11.375	12.125	11.750	11.500	12.250

10.64. The first digits of $\sqrt{2}$ are 1.41421 35623 73095 0488 · · ·. What conclusions can you draw concerning the randomness of the digits?

10.65. What conclusions can you draw concerning the randomness of the following digits?

(a) $\sqrt{3}$ = 1.73205 08075 68877 2935 · · ·

(b) π = 3.14159 26535 89793 2643 · · ·

10.66. Work Problem 10.30 by using the runs test for randomness.

10.67. Work Problem 10.32 by using the runs test for randomness.

10.68. Work Problem 10.34 by using the runs test for randomness.

Rank correlation

10.69. In a contest, two judges were asked to rank eight candidates (numbered 1 through 8) in order of preference. The judges submitted the choices shown in Table 10-43.

(a) Find the coefficient of rank correlation.

(b) Decide how closely the judges agreed in their choices.

Table 10-43

First judge	5	2	8	1	4	6	3	7
Second judge	4	5	7	3	2	8	1	6

10.70. The rank correlation coefficient is derived by using the ranked data in the product-moment formula of Chapter 8. Illustrate this by using both methods to work a problem.

10.71. Can the rank correlation coefficient be found for grouped data? Explain this, and illustrate your answer with an example.

ANSWERS TO SUPPLEMENTARY PROBLEMS

10.26. There is a difference at the 0.05 level, but not at the 0.01 level. **10.27.** Yes.

10.28. The program is effective at the 0.05 level.

10.29. We can reject the hypothesis of increased sales at the 0.05 level. **10.30.** No.

10.31. (a) Reject. (b) Accept. (c) Accept. (d) Reject.

10.34. There is no difference at the 0.05 level. **10.35.** No.

10.36. (a) Yes. (b) Yes. **10.37.** Yes. **10.38.** (a) Yes. (b) Yes.

10.41. 3 **10.42.** 6 **10.49.** There is no significant difference at either level.

10.50. The difference is significant at the 0.05 level, but not at the 0.01 level.

10.51. The difference is significant at the 0.05 level, but not at the 0.01 level.

10.52. There is a significant difference between the grades at both levels.

10.55. (a) 8. (b) 10. **10.56.** (a) 10. (b) The responses are random at the 0.05 level.

10.62. The sample is not random at the 0.05 level. There are *too many* runs, indicating a cyclic pattern.

10.63. The sample is not random at the 0.05 level. There are *too few* runs, indicating a trend pattern.

10.64. The digits are random at the 0.05 level.

10.65. (a) The digits are random at the 0.05 level. (b) The digits are random at the 0.05 level.

10.69. (a) 0.67. (b) The judges did not agree too closely in their choices.

Bayesian Methods

Subjective Probability

The statistical methods developed thus far in this book are based entirely on the classical and frequency approaches to probability (see page 5). Bayesian methods, on the other hand, rely also on a third—the so-called subjective or personal—view of probability.

Central to Bayesian methods is the process of assigning probabilities to parameters, hypotheses, and models and updating these probabilities on the basis of observed data. For example, Bayesians do not treat the mean θ of a normal population as an unknown constant; they regard it as the realized value of a random variable, say Θ, with a probability density function over the real line. Similarly, the hypothesis that a coin is fair may be assigned a probability of 0.3 of being true, reflecting our degree of belief in the coin being fair.

In the Bayesian approach, the property of randomness thus appertains to hypotheses, models, and fixed quantities such as parameters as well as to variable and observable quantities such as conventional random variables. Probabilities that describe the extent of our knowledge and ignorance of such nonvariable entities are usually referred to as *subjective probabilities* and are usually determined using one's intuition and past experience, prior to and independently of any current or future observations. In this book, we shall not discuss the controversial yet pivotal issue of the meaning and measurement of subjective probabilities. Rather, our focus will be on how prior probabilities are utilized in the Bayesian treatment of some of the statistical problems covered earlier.

EXAMPLE 11.1 Statements involving classical probabilities: (a) the chances of rolling a 3 or a 5 with a fair die are one in three; (b) the probability of picking a red chip out of a box containing two red and three green chips is two in five. Examples of the frequency approach to probability: (a) based on official statistics, the chances are practically zero that specific person in the U.S. will die from food poisoning next year; (b) I toss a coin 100 times and estimate the probability of a head coming up to be $37/100 = 0.37$. Statements involving subjective probabilities: (a) he is 80% sure that he will get an A in this course; (b) I believe the chances are only 1 in 10 that there is life on Mars; (c) the mean of this Poisson distribution is equally likely to be 1, 1.5, or 2.

Prior and Posterior Distributions

The following example is helpful for introducing some of the common terminology of Bayesian statistics.

EXAMPLE 11.2 A box contains two fair coins and a biased coin with probability for heads $P(H) = 0.2$. A coin is chosen at random from the box and tossed three times. If two heads and a tail are obtained, what is the probability of the event F, that the chosen coin is fair, and what is the probability of the event B, that the coin is biased?

Let D denote the event (data) that two heads and a tail are obtained in three tosses. The conditional probability $P(D|F)$ of observing the data under the hypothesis that a fair coin is tossed is a binomial probability and may be obtained from (1), (see Chapter 4). The conditional probability $P(D|B)$ of observing D when a biased coin is tossed may be obtained similarly. Bayes' theorem (page 8) then gives us

$$P(F|D) = \frac{P(D|F)P(F)}{P(D|F)P(F) + P(D|B)P(B)} = \frac{[3(0.5)^3]\cdot\left(\frac{2}{3}\right)}{[3(0.5)^3]\cdot\left(\frac{2}{3}\right) + [3(0.2)^2(0.8)]\cdot\left(\frac{1}{3}\right)} = \frac{250}{282} \approx 0.89$$

Also, $P(B|D) = 1 - P(F|D) \approx 0.11$.

In the Bayesian context, the unconditional probability $P(F)$ in the preceding example is usually referred to as the *prior* probability (*before* any observations are collected) of the hypothesis F, that a fair coin was tossed, and the conditional probability $P(F \mid D)$ is called and the *posterior* probability of the hypothesis F (*after* the fact that D was observed). Analogously, $P(B)$, and $P(B \mid D)$ are the respective prior and posterior probabilities that the biased coin was tossed. The prior probabilities used here are classical probabilities.

The following example involves a simple modification of Example 11.2 that necessitates an extension of the concept of randomness and brings into play the notion of a subjective probability.

EXAMPLE 11.3 A box contains an unknown number of fair coins and biased coins (with $P(H) = 0.2$ each). A coin is chosen at random from the box and tossed three times. If two heads and a tail are obtained, what is the probability that the chosen coin is biased?

In Example 11.2, the prior probability $P(F)$ for choosing a fair coin could be determined using combinatorial reasoning. Since the proportion of fair coins in the box is now unknown, we cannot access $P(F)$ as a classical probability without resorting to repeated independent drawings from the box and approximating it as a frequency ratio. We cannot therefore apply Bayes' theorem to determine the posterior probability for F.

Bayesians, nonetheless, would provide a solution to this by first positing that the unknown prior probability $P(F)$ is a random quantity, say Θ, by virtue of our uncertainty as to its exact value and then reasoning that it is possible to arrive at a probability or density function $\pi(\theta)$ for Θ that reflects our degree of belief in various propositions concerning $P(F)$. For example, one could argue that in the absence of any evidence to the contrary *before the coin is tossed*, it is reasonable to assume that the box contains an equal number of fair and biased coins. Since $P(H) = 0.2$ for a biased coin and 0.5 for a fair coin, the unknown parameter Θ then would have the subjective prior probability function shown in Table 11-1.

Table 11-1

θ	0.2	0.5
$\pi(\theta)$	1/2	1/2

Prior distributions that give equal weight to all possible values of a parameter are examples of *diffuse*, *vague*, or *noninformative* priors which are often recommended when virtually no prior information about the parameter is available. When a parameter can take on any value in a finite interval, the diffuse prior would usually be the uniform density on that interval. We will also encounter situations where uniform prior densities over the entire real line are used; such densities will be called *improper* since the total area under them is infinite.

Starting from the prior probability function in Table 11-1, the posterior probability function for Θ after observing D (two heads and a tail in three tosses), $\pi(\theta \mid D)$, may be obtained using Bayes' theorem as in Example 11.2, and is given in Table 11-2 (see Problem 11.3).

Table 11-2

θ	0.2	0.5
$\pi(\theta \mid D)$	32/157	125/157

It is convenient at this point to introduce some notation that is particularly helpful for presenting Bayesian methods. Suppose that X is a random variable with probability or density function $f(x)$ that depends on an unknown parameter θ. We assume that our uncertainty as to the value of θ may be represented by the probability or density function $\pi(\theta)$ of a random variable Θ. The function $f(x)$ may then be thought of as the conditional probability or density function of X given $\Theta = \theta$; we shall therefore denote $f(x)$ by $f(x \mid \theta)$ throughout this chapter. Also, we shall denote the joint probability or density function of X and Θ by $f(x; \theta) = f(x \mid \theta)\,\pi(\theta)$ and the posterior (or conditional) probability or density function of Θ given $X = x$ by $\pi(\theta \mid x)$. If $x_1, x_2, \ldots, x_n$ is a random sample of values of X, then the joint density function of the sample (also known as the *likelihood function*, see (19), Chapter 6) will be written using the vector notation $\boldsymbol{x} = (x_1, x_2, \ldots, x_n)$ as $f(\boldsymbol{x}|\theta) = f(x_1|\theta) \cdot f(x_2|\theta) \cdots f(x_n|\theta)$; similarly, the posterior probability or density function of θ given the sample will be denoted by $\pi(\theta|\boldsymbol{x})$.

The following version of Bayes' theorem for random variables is a direct consequence of (*26*) and (*43*), Chapter 2:

$$\pi(\theta|\boldsymbol{x}) = \frac{f(\boldsymbol{x};\theta)}{f(\boldsymbol{x})} = \frac{f(\boldsymbol{x}|\theta)\pi(\theta)}{\int_\Theta f(\boldsymbol{x}|\theta)\pi(\theta)\,d\theta} \tag{1}$$

where the integral is over the range of values of Θ and is replaced with a sum if Θ is discrete.

In our applications of Bayes' theorem, we seldom have to perform the integration (or summation) appearing in the denominator of (1) since its value is independent of θ. We can therefore write (1) in the form

$$\pi(\theta|\boldsymbol{x}) \propto f(\boldsymbol{x}|\theta)\pi(\theta) \tag{2}$$

meaning that $\pi(\theta|\boldsymbol{x}) = C \cdot f(\boldsymbol{x}|\theta)\pi(\theta)$, where C is a proportionality constant that is free of θ. Once the functional form of the posterior density is known, the "normalizing" constant C can be determined so as to make $\pi(\theta|\boldsymbol{x})$ a probability density function. (See Example 11.4.)

Remark 1 The convention of using upper case letters for random variables is often ignored in Bayesian presentations when dealing with parameters, and we shall follow this practice in the sequel. For instance, in the next example, we use λ to denote both the random parameter (rather than Λ) and its possible values.

EXAMPLE 11.4 The random variable X has a Poisson distribution with an unknown parameter λ. It has been determined that λ has the subjective prior probability function given in Table 11-3. A random sample of size 3 yields the X-values 2, 0, and 3. We wish to find the posterior distribution of λ.

Table 11-3

λ	0.5	1.0	1.5
$\pi(\lambda)$	1/2	1/3	1/6

The likelihood of the data is $f(\boldsymbol{x}|\lambda) = e^{-3\lambda}\dfrac{\lambda^{x_1+x_2+x_3}}{x_1!x_2!x_3!}$. From (1) and (2), we have the posterior density

$$\pi(\lambda|\boldsymbol{x}) = \frac{\dfrac{e^{-3\lambda}\lambda^{x_1+x_2+x_3}\pi(\lambda)}{x_1!x_2!x_3!}}{\dfrac{1}{x_1!x_2!x_3!}\sum_\lambda e^{-3\lambda}\lambda^{x_1+x_2+x_3}\pi(\lambda)} \propto e^{-3\lambda}\lambda^5\pi(\lambda) \qquad \lambda = 0.5, 1, 1.5$$

The constant of proportionality in the preceding is simply the reciprocal of the sum $\sum_\lambda e^{-3\lambda}\lambda^5\pi(\lambda)$ over the three possible values of λ. By substituting $\lambda = 0.5, 1.0, 1.5$, respectively, and $\pi(\lambda)$ from Table 11-3 into the preceding sum, and then normalizing so that the sum of the probabilities $\pi(\lambda|x)$ is equal to 1, we obtain the values in Table 11-4.

Table 11-4

λ	0.5	1.0	1.5
$\pi(\lambda\|x)$	0.10	0.49	0.41

EXAMPLE 11.5 The random variable X has a binomial distribution with probability function given by

$$f(x|\theta) = \binom{n}{x}\theta^x(1-\theta)^{n-x} \qquad x = 1, 2, \ldots, n$$

Suppose that nothing is known about the parameter θ so that a uniform (vague) prior distribution on the interval [0, 1] is chosen for θ. If a sample of size 4 yielded 3 successes, then the posterior probability density function of θ may be obtained using (2):

$$\pi(\theta\,|\,x) = \frac{f(x\,|\,\theta)\pi(\theta)}{\int_\Theta f(x\,|\,\theta)\pi(\theta)\,d\theta} = \frac{\binom{4}{3}\theta^3(1-\theta)\cdot 1}{\int_0^1 \binom{4}{3}\theta^3(1-\theta)\,d\theta} \propto \theta^3(1-\theta)$$

The last expression may be recognized as a beta density (see (34), Chapter 4) with $\alpha = 4$ and $\beta = 2$. Since the normalizing constant here should be $\frac{1}{B(4,2)} = \frac{5!}{3!1!}$ (see Appendix A), we deduce that the constant of proportionality is 20 and $\pi(\theta\,|\,x) = 20\theta^3(1-\theta)$, $0 < \theta < 1$. The graphs of the prior (uniform) and posterior densities are shown in Fig. 11-1. The mean and variance are, respectively, 0.5 and $1/12 \approx 0.08$ for the prior density whereas they are $2/3 \approx 0.67$ and $8/252 \approx 0.03$ for the posterior density. The shift to the right and the increased concentration about the mean as we move from the prior to the posterior density are evident in Fig. 11-1.

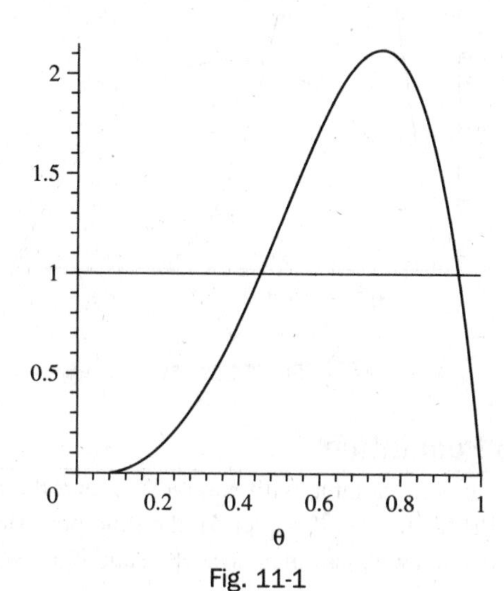

Fig. 11-1

Sampling From a Binomial Population

The result obtained in Example 11.5 may be generalized in a straightforward manner. Suppose that X has a binomial distribution with parameters n and θ (see (1), Chapter 4) and that the prior probability distribution of θ is beta with density function (see (34), Chapter 4):

$$\pi(\theta) = \frac{\theta^{\alpha-1}(1-\theta)^{\beta-1}}{B(\alpha,\beta)} \qquad 0 < \theta < 1 \qquad (\alpha, \beta > 0) \tag{3}$$

where $B(\alpha, \beta)$ is the beta function (see Appendix A). (Note that if $\alpha = \beta = 1$, then $\pi(\theta)$ is the uniform density on [0, 1]—the situation discussed in Example 11.5.) Then the posterior density $\pi(\theta\,|\,x)$ corresponding to any observed value x is given by

$$\pi(\theta\,|\,x) = \frac{f(x\,|\,\theta)\pi(\theta)}{\int_\Theta f(x\,|\,\theta)\pi(\theta)\,dp} \propto \theta^x(1-\theta)^{n-x}\cdot\theta^{\alpha-1}(1-\theta)^{\beta-1} = \frac{\theta^{x+\alpha-1}(1-\theta)^{n-x+\beta-1}}{B(x+\alpha, n-x+\beta)} \qquad 0 < \theta < 1 \tag{4}$$

This may be recognized as a beta density with parameters $x + \alpha$ and $n - x + \beta$. We thus have the following:

Theorem 11-1 If X is a binomial random variable with parameters n and θ, and the prior density of θ is beta with parameters α and β, then the posterior density of θ after observing $X = x$ is beta with parameters $x + \alpha$ and $n - x + \beta$.

EXAMPLE 11.6 Suppose that X is binomial with parameters $n = 10$ and unknown θ and that $\pi(\theta)$ is beta with parameters $\alpha = \beta = 2$. If an observation on X yielded $x = 2$, then the posterior density $\pi(\theta \mid x)$ may be determined as follows.

From Theorem 11-1 we see that $\pi(\theta \mid x)$ is beta with parameters 4 and 10. The prior (symmetric about 0.5) and posterior densities are shown in Fig. 11-2. It is clear that the effect of the observation on the prior density of θ is to shift its mean from 0.5 down to $4/14 \approx 0.29$ and to shrink the variance from 0.05 down to 0.014 (see (36), Chapter 4).

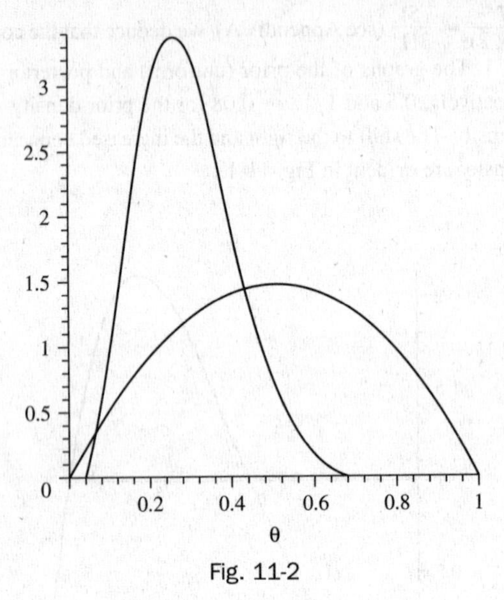

Fig. 11-2

Sampling From a Poisson Population

Theorem 11-2 If X is a Poisson random variable with parameter λ and the prior density of λ is gamma with parameters α and β (as in (31), Chapter 4), then the posterior density of λ, given the sample $x_1, x_2, \ldots, x_n$, is gamma with parameters $n\bar{x} + \alpha$ and $\beta/(1 + n\beta)$, where $\bar{x}$ is the sample mean.

If $x_1, x_2, \ldots, x_n$ is a sample of n observations on X, then the likelihood of $x = (x_1, x_2, \ldots, x_n)$ may be written as $f(\mathbf{x} \mid \theta) = e^{-n\lambda} \dfrac{\lambda^{n\bar{x}}}{x_1!x_2!\cdots x_n!}$. We are given the prior density of λ:

$$\pi(\lambda) = \frac{\lambda^{\alpha-1}e^{-\lambda/\beta}}{\beta^{\alpha}\Gamma(\alpha)} \qquad \lambda > 0 \tag{5}$$

It follows that the posterior density of λ is

$$\pi(\lambda \mid \mathbf{x}) = \frac{f(\mathbf{x} \mid \lambda)\pi(\lambda)}{\displaystyle\int_{\Lambda} f(\mathbf{x} \mid \lambda)\pi(\lambda)\,d\lambda} \propto \frac{e^{-n\lambda}\lambda^{n\bar{x}} \cdot \lambda^{\alpha-1}e^{-\lambda/\beta}}{\displaystyle\int_0^{\infty} e^{-\lambda}\lambda^{n\bar{x}} \cdot \lambda^{\alpha-1}e^{-\lambda/\beta}\,d\lambda} = \frac{(1 + n\beta)^{n\bar{x}+\alpha}\lambda^{(n\bar{x}+\alpha)-1}e^{-\lambda(n\beta+1)/\beta}}{\beta^{n\bar{x}+\alpha}\Gamma(n\bar{x} + \alpha)} \qquad \lambda > 0 \tag{6}$$

The last expression may be recognized as a gamma density, thus proving Theorem 11-2.

EXAMPLE 11.7 The number of defects in a 1000-foot spool of yarn manufactured by a machine has a Poisson distribution with unknown mean λ. The prior distribution of λ is gamma with parameters $\alpha = 3$ and $\beta = 1$. A total of eight defects were found in a sample of five spools that were examined. The posterior distribution of λ is gamma with parameters $\alpha = 11$ and $\beta = 1/6 \approx 0.17$. The prior mean and variance are both 3 while the posterior mean and variance are respectively 1.87 and 0.32. The two densities are shown in Fig. 11-3.

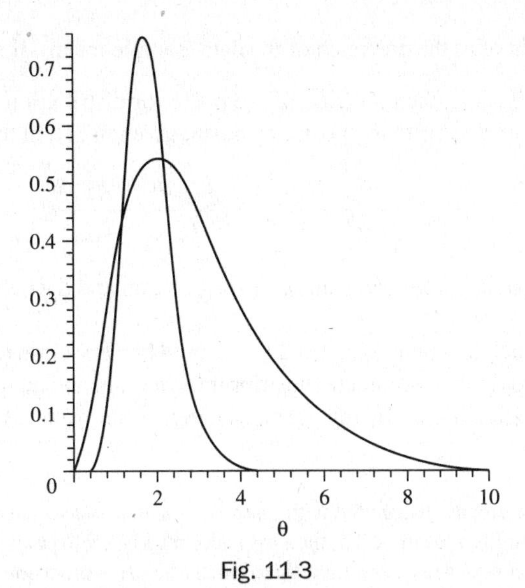

Fig. 11-3

Sampling From a Normal Population with Known Variance

Theorem 11-3 Suppose that a random sample of size n is drawn from a normal distribution with unknown mean θ and known variance σ^2. Also suppose that the prior distribution of θ is normal with mean μ and variance v^2. Then the posterior distribution of θ is also normal, with mean μ_{post} and variance v^2_{post} given by

$$\mu_{\text{post}} = \frac{\sigma^2\mu + nv^2\bar{x}}{\sigma^2 + nv^2} \qquad v^2_{\text{post}} = \frac{\sigma^2 v^2}{\sigma^2 + nv^2} \tag{7}$$

The likelihood of the observations is given by

$$f(\boldsymbol{x}\,|\,\theta) = \frac{1}{(2\pi)^{n/2}\sigma^n}\exp\left\{-\frac{1}{2\sigma^2}\sum_{i=1}^{n}(x_i - \theta)^2\right\}$$

We know from Problem 5.20 (see Method 2) that $\sum(x_i - \theta)^2 = \sum(x_i - \bar{x})^2 + n(\bar{x} - \theta)^2$. Using this, and ignoring multiplicative constants not involving θ, we can write the likelihood as

$$f(\boldsymbol{x}\,|\,\theta) \propto \exp\left\{-\frac{n}{2\sigma^2}(\theta - \bar{x})^2\right\}$$

Using (2) and the fact that $\pi(\theta) = \dfrac{1}{v\sqrt{2\pi}}\exp\left\{-\dfrac{1}{2v^2}(\theta - \mu)^2\right\}$, we get the posterior density of θ as

$$\pi(\theta\,|\,\boldsymbol{x}) \propto \exp\left\{-\frac{1}{2}\left[\frac{n}{\sigma^2}(\theta - \bar{x})^2 + \frac{1}{v^2}(\theta - \mu)^2\right]\right\}$$

Completing the square in the expression in brackets, we get

$$\pi(\theta\,|\,\boldsymbol{x}) \propto \exp\left\{-\frac{[\theta - (\bar{x}v^2 + \mu\sigma^2/n)/(v^2 + \sigma^2/n)]^2}{[2(\sigma^2/n)v^2]/[v^2 + (\sigma^2/n)]}\right\} \qquad -\infty < \theta < \infty$$

This proves that the posterior density of θ is normal with mean and variance given in (7).

A comparison of the prior and posterior variances of θ in Theorem 11-3 brings out some important facts. It is convenient to do the comparison in terms of the reciprocal of the variance, which is known as the *precision* of the distribution or random variable. Clearly, the smaller the variance of a distribution, the larger would be its precision. Precision is thus a measure of how concentrated a random variable is or how well we know it. In Theorem 11-3, if we denote the precision of the prior and posterior distributions of θ, respectively, by ξ_{prior} and ξ_{post}, then we have

$$\xi_{\text{prior}} = \frac{1}{v^2} \quad \text{and} \quad \xi_{\text{post}} = \frac{\sigma^2 + nv^2}{\sigma^2 v^2} = \frac{1}{v^2} + \frac{n}{\sigma^2} \tag{8}$$

The quantity $\frac{n}{\sigma^2}$ may be thought of as the precision of the data (sample mean). If we denote this by ξ_{data}, we have the result $\xi_{\text{post}} = \xi_{\text{prior}} + \xi_{\text{data}}$. That is, the precision of the posterior distribution is the sum of the precisions of the prior and of the data. We can also write the posterior mean, given in (7), in the form

$$\mu_{\text{post}} = \frac{\sigma^2\mu + nv^2\bar{x}}{\sigma^2 + nv^2} = \frac{\xi_{\text{prior}}\mu + \xi_{\text{data}}\bar{x}}{\xi_{\text{prior}} + \xi_{\text{data}}} \tag{9}$$

This says that the posterior mean is a weighted sum of the prior mean and the data, with weights proportional to the respective precisions.

Now suppose that ξ_{prior} is much less than ξ_{data}. Then ξ_{post} would be very close to ξ_{data}, and μ_{post} would be close to $\bar{x}$. In other words, the data would then dominate the prior information, and the posterior distribution would essentially be proportional to the likelihood. In any event, as can be seen from (8) and (9), the data would dominate the prior for very large n.

EXAMPLE 11.8 Suppose X is normally distributed with unknown mean θ and variance 4 and that $\pi(\theta)$ is standard normal. If a sample of size $n = 10$ yields a mean of 0.5, then by Theorem 11-3, $\pi(\theta \mid \mathbf{x})$ is normal with mean 0.36 and variance 0.29. The posterior precision (=3.5) is more than three times the prior precision (=1), which is evident from the densities shown in Fig. 11-4. The precision of the data is $10/4 = 2.5$, which is reasonably larger than the prior precision of 1; and this is reflected in the posterior mean of 0.36 being closer to $\bar{x} = 0.5$ than to the prior mean, 0.

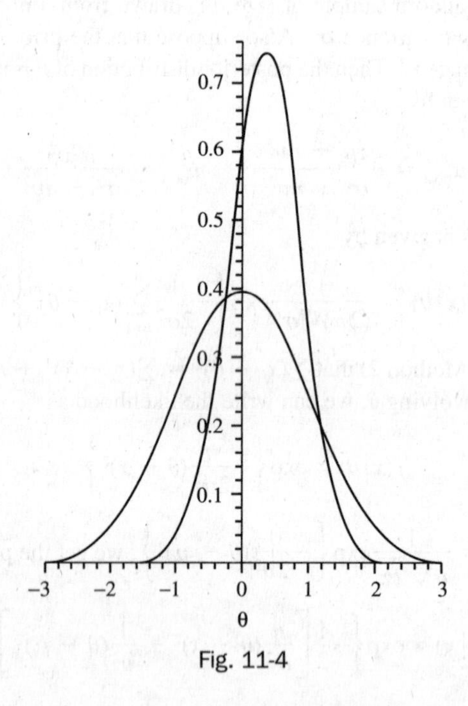

Fig. 11-4

Improper Prior Distributions

The prior probability density functions $\pi(\theta)$ we have seen until now are all *proper* in the sense that (i) $\pi(\theta) \geq 0$ and (ii) $\int_{-\infty}^{\infty} \pi(\theta)\, d\theta = 1$ (see page 37). Prior densities that satisfy the first condition but violate the second due to the integral being divergent have, however, been employed within the Bayesian framework and are referred to as *improper priors*. They often arise as natural choices for representing vague prior information about parameters with infinite range.

For example, when sampling from a normal population with known mean, say 0, but unknown variance θ, we may assume that the prior density for θ is given by $\pi(\theta) = \frac{1}{\theta}$, $\theta > 0$. Given a sample of observations

$\boldsymbol{x} = (x_1, x_2, \ldots, x_n)$, if we overlook the fact that the prior is improper and apply formula (1), we get the posterior density

$$\pi(\theta \mid \boldsymbol{x}) \propto \frac{1}{\theta^{n/2}} \exp\left\{\frac{-\sum_i x_i^2}{2\theta}\right\} \cdot \frac{1}{\theta} = \theta^{-\frac{n}{2}-1} \exp\left\{\frac{-\sum_i x_i^2}{2\theta}\right\} \qquad \theta > 0 \tag{10}$$

This is a proper density, known as an inverse gamma, with parameters $\alpha = n/2$ and $\beta = \sum_i x_i^2/2$ (see Problem 11.99). We have thus arrived at a proper posterior density starting with an improper prior. Indeed, this will be true in all of the situations with improper priors that we encounter here, although this is not always the case.

EXAMPLE 11.9 Suppose that X is binomial with known n and unknown success probability θ. The prior density for θ given by $\pi(\theta) = \dfrac{1}{\theta(1-\theta)}$, $0 < \theta < 1$ is improper and is known as *Haldane's prior*. Let us overlook the fact that $\pi(\theta)$ is improper, and proceed formally to derive the posterior density $\pi(\theta \mid x)$ corresponding to an observed value x of X:

$$\pi(\theta \mid x) = \frac{f(x \mid \theta)\pi(\theta)}{\int_\Theta f(x \mid \theta)\pi(\theta)d\theta} \propto \frac{\theta^x(1-\theta)^{n-x}}{\theta(1-\theta)} = \frac{\theta^{x-1}(1-\theta)^{n-x-1}}{B(x, n-x)} \qquad 0 < \theta < 1$$

We see that the posterior is a proper beta density with parameters x and $n - x$.

EXAMPLE 11.10 Suppose X is normally distributed with unknown mean θ and known variance σ^2. An improper prior distribution for θ in this case is given by $\pi(\theta) = 1$, $-\infty < \theta < \infty$. This density may be thought of as representing prior ignorance in that intervals of the same length have the same weight regardless of their location on the real line. Given the observation vector $\boldsymbol{x} = (x_1, x_2, \ldots, x_n)$, the posterior distribution of θ under this prior is given by

$$\pi(\theta \mid \boldsymbol{x}) \propto f(\boldsymbol{x} \mid \theta)\pi(\theta) \propto \exp\left\{-\frac{\sum_i (x_i - \theta)^2}{2\sigma^2}\right\} \cdot 1 \propto \exp\left\{-\frac{n(\theta - \bar{x})^2}{2\sigma^2}\right\} \qquad -\infty < \theta < \infty$$

which is normal with mean $\bar{x}$ and variance σ^2/n.

Conjugate Prior Distributions

Note that Theorems 11-1, 11-2, and 11-3 share an important characteristic in that the prior and posterior densities in each belong to the same family of distributions. Whenever this happens, we say that the family of prior distributions used is *conjugate* (or *closed*) with respect to the population density $f(x \mid \theta)$. Thus the beta family is conjugate with respect to the binomial distribution (Theorem 11-1), the gamma family is conjugate with respect to the Poisson distribution (Theorem 11-2), and the normal family is conjugate with respect to the normal distribution with known variance (Theorem 11-3).

Since $\pi(\theta \mid \boldsymbol{x}, y) \propto f(y \mid \theta)\pi(\theta \mid \boldsymbol{x})$ whenever $\boldsymbol{x}$ and $\boldsymbol{y}$ are two independent samples from $f(x \mid \theta)$, conjugate families make it easier to update prior densities in a sequential manner by just changing the parameters of the family (see Example 11.11). Conjugate families are thus desirable in Bayesian analysis and they exist for most of the commonly encountered distributions. In practice, however, prior distributions are to be chosen on the basis of how well they represent one's prior knowledge and beliefs rather than on mathematical convenience. If, however, a conjugate prior distribution closely approximates an appropriate but otherwise unwieldy prior distribution, then the former naturally is a prudent choice.

We now show that the gamma family is conjugate with respect to the exponential distribution. Suppose that X has the exponential density, $f(x \mid \theta) = \theta e^{-\theta x}$, $x > 0$, with unknown θ, and that the prior density of θ is gamma with parameters α and β. The posterior density of θ is then given by

$$\pi(\theta \mid \boldsymbol{x}) \propto f(\boldsymbol{x} \mid \theta) \cdot \pi(\theta) = \frac{\theta^n e^{-\theta n \bar{x}} \theta^{\alpha-1} e^{-\theta/\beta}}{\beta^\alpha \Gamma(\alpha)} = \frac{(1 + n\beta\bar{x})^{n+\alpha}\theta^{n+\alpha-1}e^{-\theta\left(\frac{1}{\beta}+n\bar{x}\right)}}{\beta^{n+\alpha}\Gamma(n+\alpha)} \qquad \theta > 0 \tag{11}$$

This establishes the following theorem.

Theorem 11-4 If X has the exponential density, $f(x|\theta) = \theta e^{-\theta x}$, $x > 0$, with unknown θ and the prior density of θ is gamma with parameters α and β, then the posterior density of θ is gamma with parameters $\alpha + n$ and $\beta/(1 + n\beta\bar{x})$.

EXAMPLE 11.11 In Example 11.6, suppose an additional, independent observation on the same binomial population yields the sample value $y = 3$. The posterior density $\pi(\theta|x, y)$ may then be found either (a) directly from the prior density $\pi(\theta)$ given in Example 11.6 or (b) using the posterior density $\pi(\theta|x)$ derived there.

(a) We assume that the prior density is beta with parameters $\alpha = 2$ and $\beta = 2$ and that a sample value of 5 is observed on a binomial random variable with $n = 20$. Theorem 11-1 then gives us the posterior beta density with parameters $\alpha = 2 + 5 = 7$ and $\beta = 15 + 2 = 17$.

(b) We assume that the prior density is the posterior density obtained in Example 11.6, namely a beta with parameters $\alpha = 4$ and $\beta = 10$, and that a sample value of 3 is observed on a binomial random variable with $n = 10$. Theorem 11-1 gives a posterior beta density with parameters $\alpha = 4 + 3 = 7$ and $\beta = 7 + 10 = 17$.

EXAMPLE 11.12 A random sample of size n is drawn from a geometric distribution with parameter θ (see page 123): $f(x; \theta) = \theta(1 - \theta)^{x-1}$, $x = 1, 2, \ldots$ Suppose that the prior density of θ is beta with parameters α and β. Then the posterior density of θ is

$$\pi(\theta|\mathbf{x}) = \frac{f(\mathbf{x}|\theta)\pi(\theta)}{\int_{\Theta} f(\mathbf{x}|\theta)\pi(\theta)dp} \propto \frac{\theta^n(1-\theta)^{n(\bar{x}-1)} \cdot \theta^{\alpha-1}(1-\theta)^{\beta-1}}{\int_0^1 \theta^n(1-\theta)^{n(\bar{x}-1)}\theta^{\alpha-1}(1-\theta)^{\beta-1}d\theta} = \frac{\theta^{\alpha+n-1}(1-\theta)^{\beta+n\bar{x}-n-1}}{B(\alpha+n, \beta+n\bar{x}-n)} \qquad 0 < \theta < 1$$

which is also a beta, with parameters $\alpha + n$ and $\beta + n\bar{x} - n$, where $\bar{x}$ is the sample mean. In other words, the beta family is conjugate with respect to the geometric distribution.

Bayesian Point Estimation

A central tenet in Bayesian statistics is that everything one needs to know about an unknown parameter is to be found in its posterior distribution. Accordingly, Bayesian point estimation of a parameter essentially amounts to finding appropriate single-number summaries of the posterior distribution of the parameter. We shall now present some summary measures employed for this purpose and their relative merits as to how well they represent the parameter.

EXAMPLE 11.13 We saw in Example 11.5 that when sampling from a binomial distribution with a uniform prior, the posterior density of θ is beta with parameters $\alpha = 4$ and $\beta = 2$. The graph of this density is shown in Fig. 11-1. A natural candidate for single-number summary status here would be the mean of the posterior density. We know from (36), Chapter 4 that the posterior mean is given by $\alpha/(\alpha + \beta) = 2/3$.

The median and mode (see page 83) of the posterior density are two other possible choices as point estimates for θ. The mode is given by (see (37), Chapter 4) $(\alpha - 1)/(\alpha + \beta - 2) = 3/4$. Note that the mode coincides with the maximum likelihood estimate (see pages 198–199) of θ, namely the sample proportion of successes. As a corollary to Theorem 11-5, we see that this is true in general of the binomial distribution with a uniform prior.

The median in this case is not attractive from a practical standpoint since it has to be numerically determined due to the lack of a closed form expression for the median of a beta distribution. Nevertheless, as we shall see later, the median in general is an optimal summary measure in a certain sense.

The following theorem generalizes some of the results from Example 11.13.

Theorem 11-5 If X is a binomial random variable with parameters n and θ and the prior density of θ is beta with parameters α and β, then the respective estimates of θ provided by the posterior mean and mode are $\mu_{\text{post}} = (x + \alpha)/(n + \alpha + \beta)$ and $\gamma_{\text{post}} = (x + \alpha - 1)/(n + \alpha + \beta - 2)$.

Remark 2 A special case of this theorem, when α and β both equal 1, is of some interest. The posterior mean estimate of θ is then $(x + 1)/(n + 2)$. Accordingly, if all n trials result in successes (i.e., if $x = n$), then the probability that the next trial will also be a success is given by $(n + 1)/(n + 2)$. This result has a venerable history and is known as *Laplace's law of succession*.

When $\alpha = \beta = 1$ in Theorem 11-5, the posterior mode estimate γ_{post} of θ reduces to the maximum likelihood estimate x/n. This was also pointed out in Example 11.13 but the result is obviously not true for general α and β. But, regardless of the values of α and β, when the sample size is large enough, both μ_{post} and γ_{post} will be close to the sample proportion x/n. Furthermore, for all n, μ_{post} is a convex combination of the prior mean of θ and the sample proportion. (See Problem 11.38.)

EXAMPLE 11.14 Suppose that a random sample of size n is drawn from a normal distribution with unknown mean θ and variance 1. Also suppose that the prior distribution of θ is normal with mean 0 and variance 1. From Theorem 11-3, we see that the posterior distribution of θ is normal with mean $n\bar{x}/(1 + n)$.

Clearly, the posterior mean, median, and mode are identical here and would therefore lead to the same point estimate, $n\bar{x}/(1 + n)$, of θ. It was shown in Problem 6.25, page 206, that the maximum likelihood estimate for θ in this case is the sample mean $\bar{x}$, which is known to be unbiased (Theorem 5-1). On the other hand, the Bayesian estimates derived here are biased, although they are asymptotically unbiased.

A general result along these lines follows easily from Theorem 11-3 and is as follows.

Theorem 11-6 Suppose that a random sample of size n is drawn from a normal distribution with unknown mean θ and known variance σ^2. Also suppose that the prior distribution of θ is normal with mean μ and variance v^2. Then the posterior mean, median, and mode all provide the same estimate of θ, namely $(\sigma^2\mu + nv^2\bar{x})/(\sigma^2 + nv^2)$, where $\bar{x}$ is the sample mean.

As we saw in the binomial case, the posterior mean estimate μ_{post} just obtained lies between the prior mean μ and the maximum likelihood estimate $\bar{x}$ of θ. This may be seen by writing μ_{post} in the form $[\sigma^2/(\sigma^2 + nv^2)] \cdot \mu + [nv^2/(\sigma^2 + nv^2)] \cdot \bar{x}$, as a convex combination of the two. We can also see from this expression that for large n, μ_{post} will be close to $\bar{x}$ and will not be appreciably influenced by the prior mean μ.

An optimality property of μ_{post} as an estimate of θ directly follows from Theorem 3-6. Indeed, we can prove a more general result along these lines using this theorem. Suppose we are interested in estimating a function of θ, say $\tau(\theta)$. For any set of observations x from $f(x|\theta)$, if we define the statistic $T(x)$ as the posterior expectation of $\tau(\theta)$, namely

$$T(x) = E(\tau(\theta)|x) = \int_{-\infty}^{\infty} \tau(\theta)\pi(\theta|x)\,d\theta$$

then it follows from Theorem 3-6 that

$$E[(\tau(\theta) - a(x))^2|x] = \int_{-\infty}^{\infty} (\tau(\theta) - a(x))^2\pi(\theta|x)\,d\theta$$

is a minimum when $a(x) = T(x)$. In other words, $T(x)$ satisfies the property

$$E[(\tau(\theta) - T(x))^2|x] = \min_a E[(\tau(\theta) - a(x))^2|x] \text{ for each } x \tag{12}$$

since $T(x)$ is the mean of $\tau(\theta)$ with respect to the posterior density $\pi(\theta|x)$.

In the general theory of Bayesian estimation, we typically start with a loss function $L(\tau(\theta), a)$ that measures the distance between the parameter and an estimate. We then seek an estimator, say $\delta^*(X)$, with the property that

$$E[L(\tau(\theta), \delta^*(x))|x] = \min_a E[L(\tau(\theta), a(x))|x] \text{ for each value } x \text{ of } X \tag{13}$$

where the expectation is over the parameter space endowed with the posterior density. An estimator satisfying equation (13) is called a *Bayes estimator of $\tau(\theta)$ with respect to the loss function* $L(\tau(\theta), a)$. The following theorem then is just a restatement of (12):

Theorem 11-7 The mean of $\tau(\theta)$ with respect to the posterior distribution $\pi(\theta|X)$ is the Bayes estimator of $\tau(\theta)$ for the squared error loss function $L(\theta, a) = (\theta - a)^2$.

Another common loss function is the absolute error loss function $L(\theta, a) = |\theta - a|$. It is shown in Problem 11.100 that the median of the posterior density is the Bayes estimator for this loss function.

Theorem 11-8 The median of $\tau(\theta)$ with respect to the posterior distribution $\pi(\theta|X)$ is the Bayes estimator of $\tau(\theta)$ for the absolute error loss function $L(\theta, a) = |\theta - a|$.

When $\tau(\theta) = \theta$, these two theorems reduce to the optimality results mentioned earlier for the posterior mean and median as estimates of θ.

EXAMPLE 11.15 Suppose X is a binomial random variable with parameters n and θ, and the prior density of θ is beta with parameters $\alpha = \beta = 1$. Theorems 11-7 and 11-8 may then be used to obtain the Bayes estimates of $\theta(1 - \theta)$ for the (a) squared error and (b) absolute error loss functions.

(a) We obtain the posterior mean of $\theta(1 - \theta)$ from Theorem 11-1. We have

$$E(\theta(1-\theta)|x) = E(\theta|x) - E(\theta^2|x) = \frac{x+1}{n+2} - \left[\frac{(x+1)(x+2)}{(n+2)(n+3)}\right] = \frac{(x+1)(n-x+1)}{(n+2)(n+3)}$$

(b) The median of the posterior distribution of $\theta(1 - \theta)$ may be obtained numerically from the posterior distribution of θ using computer software. To show the work involved, let us assume that $n = 10$ and $x = 4$. The posterior distribution of θ is beta with parameters 5 and 7. The median of $\theta(1 - \theta)$, say m, satisfies the condition $P(\theta(1 - \theta) \leq m) = 0.5$, which is equivalent to the requirement that $P\left(\theta \leq \frac{1}{2} - \frac{\sqrt{1-4m}}{2}\right) + P\left(\theta \geq \frac{1}{2} + \frac{\sqrt{1-4m}}{2}\right) = 0.5$ under the beta distribution with parameters 5 and 7. The solution is $m = 0.247$. (The posterior mean of $\theta(1 - \theta)$ in this case is 0.224.)

Bayesian Interval Estimation

Given the posterior density function $\pi(\theta|x)$ for a parameter θ, any θ − interval $[\theta_L, \theta_U]$ with the property

$$\int_{\theta_L}^{\theta_U} \pi(\theta|x)\,d\theta = 1 - \alpha \tag{14}$$

is called a Bayesian $(1 - \alpha) \times 100\%$ *credibility interval* for θ. Of the various possible intervals that satisfy this property, two deserve special mention: the equal tail area interval and the highest posterior density (*HPD*) interval.

The equal tail area $(1 - \alpha) \times 100\%$ interval has the property that the area under the posterior density to the left of θ_L equals the area to the right of θ_U:

$$\int_{-\infty}^{\theta_L} \pi(\theta|x)\,d\theta = \int_{\theta_U}^{\infty} \pi(\theta|x)\,d\theta = (1 - \alpha)/2$$

The requirement for the *HPD* interval is that, in addition to (14), we have $\pi(\theta|x) \geq \pi(\theta'|x)$ if $\theta \in [\theta_L, \theta_U]$ and $\theta' \notin [\theta_L, \theta_U]$. Clearly if $\pi(\theta|x)$ does not have a unique mode, then the set of θ − values satisfying the last condition may not be an interval. To avoid this possibility, we shall assume here that the posterior density is unimodal. It follows directly from this assumption that $\pi(\theta_L|x) = \pi(\theta_U|x)$ and that for any α the *HPD* interval is the shortest of all possible $(1 - \alpha) \times 100\%$ credibility intervals. But equal tail area intervals are much easier to construct from the readily available percentiles of most common distributions. The two intervals coincide when the posterior density is symmetric and unimodal.

EXAMPLE 11.16 Suppose that a random sample of size 9 from a normal distribution with unknown mean θ and variance 1 yields a sample mean of 2.5. Also suppose that the prior distribution of θ is normal with mean 0 and variance 1. From Theorem 11-3, we see that the posterior distribution of θ is normal with mean 2.25 and variance 0.1. A 95% equal tail credibility interval for θ is given by $[\theta_L, \theta_U]$ with θ_L and θ_U equal, respectively, to the 2.5th percentile and 97.5th percentile of the normal density with mean 2.25 and variance 0.1. From Appendix C, we then have $\theta_L \approx 2.25 - (2.36 \times 0.32) = 1.49$ and $\theta_U \approx 2.25 + (2.36 \times 0.32) = 3.01$. The 95% Bayesian equal tail credibility interval (*and* the *HPD* interval, because of the symmetry of the normal density) is thus given by [1.49, 3.01].

EXAMPLE 11.17 In Problem 6.6, we obtained traditional confidence intervals for a normal mean θ based on a sample of size $n = 200$ assuming that the population standard deviation was $\sigma = 0.042$. The 95% confidence interval for the population mean came out to be [0.82, 0.83]. It is instructive now to obtain the actual posterior probability for this interval obtained assuming normal prior distribution for θ with mean $\mu = 1$ and standard deviation $\upsilon = 0.05$.

From Theorem 11-3, we see that the posterior density has mean $\mu_{\text{post}} \approx 0.825$ and standard deviation $\upsilon_{\text{post}} \approx 0.003$. The area under this density over the interval [0.82, 0.83] is 0.9449.

A basic conceptual difference between conventional confidence intervals and Bayesian credibility intervals should be pointed out. The confidence statement associated with a 100 α% confidence interval for a parameter θ is the probability statement $P_X(L(X) \le \theta \le U(X)) = \alpha$ in the sample space of observations, with the frequency interpretation that in repeated sampling the random interval $[L(X), U(X)]$ will enclose the constant θ 100 α% of the times. But, given a random sample $\boldsymbol{x} = (x_1, x_2, \ldots, x_n)$ of observations on X, the statement $P(L(\boldsymbol{x}) \le \theta \le U(\boldsymbol{x})) = \alpha$ (in words, "we are 100 α% sure that θ lies between $L(\boldsymbol{x})$ and $U(\boldsymbol{x})$") is devoid of any sense simply because θ, $L(\boldsymbol{x})$, and $U(\boldsymbol{x})$ are all constants.

The credibility statement associated with a Bayesian 100 α% credibility interval is the probability statement $P_\Theta(L(\boldsymbol{x}) \le \theta \le U(\boldsymbol{x})) = \alpha$ in the parameter space endowed with the probability density $\pi(\theta|\boldsymbol{x})$. Although this statement may not have a frequency interpretation, it nonetheless is a valid and useful summary description of the distribution of the parameter to the effect that the interval $[L(\boldsymbol{x}), U(\boldsymbol{x})]$ carries a probability of α under the posterior density $\pi(\theta|\boldsymbol{x})$.

Bayesian Hypothesis Tests

Suppose we wish to test the null hypothesis $H_0 : \theta \le \theta_0$ against the alternative hypothesis $H_1 : \theta > \theta_0$. Then a reasonable rule for rejecting H_0 in favor of H_1 could be based on the posterior probability of the null hypothesis given the data,

$$P(H_0|\boldsymbol{x}) = \int_{-\infty}^{\theta_0} \pi(\theta|\boldsymbol{x})\,d\theta \tag{15}$$

For instance, we could specify an $\alpha > 0$ and decide to reject H_0 whenever $\boldsymbol{x}$ is such that $P(H_0|\boldsymbol{x}) \le \alpha$. A test based on this rejection criterion is known as a *Bayes* α *test.*

Remark 3 The Bayesian posterior probability of the null hypothesis shown in (15) is quite different from the P value of a test (see page 215) although the two are frequently confused for each other, and the latter is often loosely referred to as the probability of the null hypothesis.

We now show an optimality property enjoyed by Bayes α tests. We saw in Chapter 7 that the quantities of primary interest in assessing the performance of a test are the probabilities of Type I error and Type II error for each θ. If C is the critical region for a test, then these two probabilities are given by

$$P_{\text{I}}(\theta) = \begin{cases} \int_C f(\boldsymbol{x}, \theta)\,d\boldsymbol{x}, & \theta \le \theta_0 \\ 0 & \theta > \theta_0 \end{cases} \quad \text{and} \quad P_{\text{II}}(\theta) = \begin{cases} \int_{C'} f(\boldsymbol{x}, \theta)\,d\boldsymbol{x}, & \theta > \theta_0 \\ 0 & \theta \le \theta_0 \end{cases}$$

For any specified α, the following weighted mean of these two probabilities is known as the *Bayes risk* of the test.

$$r(C) = (1 - \alpha) \int_{-\infty}^{\theta_0}\int_C \pi(\theta|\boldsymbol{x}) P_{\text{I}}(\theta)\,d\boldsymbol{x}\,d\theta + \alpha \int_{\theta_0}^{\infty}\int_{C'} \pi(\theta|\boldsymbol{x}) P_{\text{II}}(\theta)\,d\boldsymbol{x}\,d\theta \tag{16}$$

For each fixed $\boldsymbol{x}$, the quantity on the right may be written as

$$(1 - \alpha)P(\theta \le \theta_0|\boldsymbol{x})I_C(\boldsymbol{x}) + \alpha P(\theta > \theta_0|\boldsymbol{x})I_{C'}(\boldsymbol{x}) = (1 - \alpha)P(\theta \le \theta_0|\boldsymbol{x})I_C(\boldsymbol{x}) + \alpha P(\theta > \theta_0|\boldsymbol{x})(1 - I_C(\boldsymbol{x}))$$

$$= \alpha P(\theta > \theta_0|\boldsymbol{x}) + [(1 - \alpha)P(\theta \le \theta_0|\boldsymbol{x})I_C(\boldsymbol{x}) - \alpha P(\theta > \theta_0|\boldsymbol{x})I_C(\boldsymbol{x})]$$

where $I_E(\boldsymbol{x})$ denotes the indicator function of the set E. The term inside brackets is minimized when the critical region C is defined so that

$$I_C(\boldsymbol{x}) = \begin{cases} 1 & \text{if } (1 - \alpha)P(\theta \le \theta_0|\boldsymbol{x}) \le \alpha P(\theta > \theta_0|\boldsymbol{x}) \\ 0 & \text{otherwise} \end{cases}$$

This shows that $r(C)$ is minimized when C consists of those data points $\boldsymbol{x}$ for which $P(\theta \le \theta_0|\boldsymbol{x}) \le \alpha$.

We have thus established that the *Bayes α test* minimizes the Bayes risk defined by (16). In general, we have the following theorem.

Theorem 11-9 For any subset Θ_0 of the parameter space, among all tests of the null hypothesis $H_0: \theta \in \Theta_0$ against the alternative $H_1: \theta \in \Theta_0'$, the *Bayes α test*, which rejects H_0 if $P(\theta \in \Theta_0|\boldsymbol{x}) \le \alpha$ minimizes the Bayes risk defined by

$$r(C) = (1-\alpha)\iint_{\Theta_0 C} \pi(\theta|\boldsymbol{x})P_{\mathrm{I}}(\theta)\,dx\,d\theta + \alpha\iint_{\Theta_0' C'} \pi(\theta|\boldsymbol{x})P_{\mathrm{II}}(\theta)\,dx\,d\theta$$

EXAMPLE 11.18 Suppose that the reaction time (in seconds) of an individual to certain stimuli is known to be normally distributed with unknown mean θ but a known standard deviation of 0.30 sec. The prior density of θ is normal with $\mu = 0.4$ sec and $v^2 = 0.13$. A sample of 20 observations yielded a mean reaction time of 0.35 sec. We wish to test the null hypothesis $H_0: \theta \le 0.3$ against the alternative $H_1: \theta > 0.3$ using a Bayes 0.05 test.

By Theorem 11-3, the posterior density is normal with mean 0.352 and variance 0.004. The posterior probability of H_0 is therefore given by $P(\theta \le 0.3) = P\left(Z \le \frac{0.3 - 0.352}{0.063}\right) \approx 0.20$. Since this probability is greater than 0.05, we cannot reject H_0.

EXAMPLE 11.19 X is a Bernoulli random variable with success probability θ, which is known to be either 0.3 or 0.6. It is desired to test the null hypothesis $H_0: \theta = 0.3$ against the alternative $H_1: \theta = 0.6$ using a Bayes 0.05 test assuming the vague prior probability distribution for $\theta: P(\theta = 0.3) = P(\theta = 0.6) = 0.5$. A sample of 30 trials on X yields 16 successes. To check the rejection criterion of the Bayes 0.05 test, we need the posterior probability of the null hypothesis:

$$P(\theta = 0.3|x = 16) = \frac{P(x = 16|\theta = 0.3) \cdot P(\theta = 0.3)}{P(x = 16|\theta = 0.3) \cdot P(\theta = 0.3) + P(x = 16|\theta = 0.6) \cdot P(\theta = 0.6)}$$

$$= \frac{(0.0042)(0.5)}{(0.0042)(0.5) + (0.1101)(0.5)} \approx 0.037$$

Since this probability is less than 0.05, we reject the null hypothesis.

Bayes Factors

When the prior distribution involved is proper, Bayesian statistical inference can be formulated in the language of odds (see page 5) using what are known as *Bayes factors*. Bayes factors may be regarded as the Bayesian analogues of likelihood ratios on which most of the classical tests in Chapter 7 are based.

Consider the hypothesis testing problem discussed in the previous section. We are interested in testing the null hypothesis $H_0: \theta \in \Theta_0$ against the alternative hypothesis $H_1: \theta \in \Theta_0'$. The quantities

$$\frac{P(H_0)}{P(H_1)} = \frac{\int_{\Theta_0} \pi(\theta)\,d\theta}{\int_{\Theta_0'} \pi(\theta)\,d\theta} \quad \text{and} \quad \frac{P(H_0|\boldsymbol{x})}{P(H_1|\boldsymbol{x})} = \frac{\int_{\Theta_0} \pi(\theta|\boldsymbol{x})\,d\theta}{\int_{\Theta_0'} \pi(\theta|\boldsymbol{x})\,d\theta} \tag{17}$$

are known respectively as the *prior and posterior odds ratios* of H_0 relative to H_1. The *Bayes factor* (*BF* for short) is defined as the posterior odds ratio over the prior odds ratio. Using the fact that $\pi(\theta|\boldsymbol{x}) \propto f(\boldsymbol{x}|\theta)\pi(\theta)$, we can write the Bayes factor in the following form:

$$BF = \frac{\text{Posterior odds ratio}}{\text{Prior odds ratio}} = \left(\frac{P(H_0|\boldsymbol{x})}{P(H_1|\boldsymbol{x})}\right)\Big/\left(\frac{P(H_0)}{P(H_1)}\right) = \frac{\frac{1}{P(H_0)}\int_{\Theta_0} f(\boldsymbol{x}|\theta)\pi(\theta)\,d\theta}{\frac{1}{P(H_1)}\int_{\Theta_0'} f(\boldsymbol{x}|\theta)\pi(\theta)\,d\theta} \tag{18}$$

The Bayes factor is thus the ratio of the marginals (or the averages) of the likelihood under the two hypotheses. It can also be seen from (18) that when the hypotheses are both simple, say $H_0: \theta = \theta_0$ and $H_1: \theta = \theta_1$, the Bayes factor becomes the familiar likelihood ratio of classical inference: $BF = \dfrac{f(\boldsymbol{x}\,|\,\theta_0)}{f(\boldsymbol{x}\,|\,\theta_1)}$.

EXAMPLE 11.20 In Example 11.18, let us calculate the Bayes factor for the null hypothesis $H_0: \theta \le 0.3$ against the alternative, $H_1: \theta > 0.3$, using (18). We need $P(H_0) = P(\theta \le 0.3)$, where θ is a normal random variable with mean 0.4 and variance 0.13. This equals $P\left(Z \le \dfrac{0.3 - 0.4}{0.36}\right) \approx 0.39$. The posterior probability of the null hypothesis, available from Example 11.18, is $P(H_0|\boldsymbol{x}) \approx 0.20$. The Bayes factor is $\left(\dfrac{P(H_0|\boldsymbol{x})}{P(H_1|\boldsymbol{x})}\right)\Big/\left(\dfrac{P(H_0)}{P(H_1)}\right) = \left(\dfrac{0.20}{0.80}\right)\Big/\left(\dfrac{0.39}{0.61}\right) \approx 0.39$.

EXAMPLE 11.21 A box contains a fair coin and two biased coins (each with $P(\text{"heads"}) = 0.2$). A coin is randomly chosen from the box and tossed 10 times. If 4 heads are obtained, what is the Bayes factor for the null hypothesis H_0 that the chosen coin is fair relative to the alternative H_1 that it is biased? The prior probabilities are $P(H_0) = 1/3$ and $P(H_1) = 2/3$, so the prior odds ratio is 0.5. The posterior probabilities are $P(H_0|\boldsymbol{x}) = \dfrac{(0.5)^{10}}{(0.5)^{10} + 2(0.2)^4(0.8)^6} \approx 0.54$ and $P(H_1|\boldsymbol{x}) = \dfrac{(0.2)^4(0.8)^6}{(0.5)^{10} + (0.2)^4(0.8)^6} \approx 0.46$, so the posterior odds ratio is $0.54/0.56 \approx 1.16$. The Bayes factor is therefore $1.16/0.5 \approx 3.32$. We can also get the same result directly as the ratio of the likelihoods under the two hypotheses $P(\boldsymbol{x}|H_0) = \dbinom{10}{4}(0.5)^{10}$ and $P(\boldsymbol{x}|H_1) = \dbinom{10}{4}(0.2)^4(0.8)^6$.

It can be seen from (18) that the Bayes factor quantifies the strength of evidence afforded by the data for or against the null hypothesis relative to the alternative hypothesis. Generally speaking, we could say that if the Bayes factor is larger than 1, the observed data adds confirmation to the null hypothesis and if it is less than 1, then the data disconfirms the null hypothesis. Furthermore, the larger the Bayes factor, the stronger the evidence in favor of the null hypothesis. The calibration of the Bayes factor to reflect the actual strength of evidence for or against the null hypothesis is a topic that will not be discussed here. We can, however, prove the following theorem:

Theorem 11-10 The Bayes α test is equivalent to the test that rejects the null hypothesis if

$$BF \le \frac{\alpha[1 - P(H_0)]}{(1 - \alpha)P(H_0)}.$$

To see this, note that the rejection criterion of a Bayes α test, namely $P(H_0|\boldsymbol{x}) \le \alpha$, is equivalent to the condition $\dfrac{P(H_0|\boldsymbol{x})}{P(H_1|\boldsymbol{x})} \le \dfrac{\alpha}{1 - \alpha}$ and that this inequality is equivalent to the condition $BF \le \dfrac{\alpha[1 - P(H_0)]}{(1 - \alpha)P(H_0)}$.

Remark 4 An *ad hoc* rule sometimes used is to reject the null hypothesis if $BF < 1$. It can be shown that this is equivalent to the Bayes α test with $\alpha = P(H_0)$: Reject H_0 if $P(H_0|\boldsymbol{x}) \le P(H_0)$.

EXAMPLE 11.22 Let us determine the rejection criterion in terms of the Bayes factor for the test used in Example 11.19. We have $\alpha = 0.05$, and $P(H_0) = P(\theta = 0.3) = 0.5$. Therefore, by Theorem 11-10, the test criterion is to reject H_0 if $BF \le \dfrac{(0.05)(0.5)}{(0.95)(0.5)} \approx 0.053$. The Bayes factor corresponding to 16 successes out of 30 trials is $\left(\dfrac{P(H_0|\boldsymbol{x})}{P(H_1|\boldsymbol{x})}\right)\Big/\left(\dfrac{P(H_0)}{P(H_1)}\right) = \left(\dfrac{0.037}{1 - 0.037}\right)\Big/\left(\dfrac{0.5}{0.5}\right) \approx 0.038$. Since this is less than 0.053, we reject the null hypothesis.

EXAMPLE 11.23 In Example 11.18, suppose we wish to employ the decision rule to reject H_0 if the Bayes factor is less than 1. We already know that the probability of the null hypothesis under the posterior density of θ is 0.20. The posterior odds for H_0 are therefore $\dfrac{0.20}{0.80} = \dfrac{1}{4}$. The prior probability for H_0 is given by $P(\theta \le 0.3) = P\left(Z \le \dfrac{0.3 - 0.4}{0.36}\right) \approx 0.39$, so the prior odds for H_0 are $\dfrac{0.39}{0.61}$. The Bayes factor for H_0 is $(1/4)/(39/61) \approx 0.39 < 1$. Our decision therefore is to reject H_0.

Bayesian Predictive Distributions

The Bayesian framework makes it possible to obtain the conditional distribution of future observations on the basis of a currently available prior or posterior distribution for the population parameter. These are known as *predictive distributions* and the basic process involved in their derivation is straightforward marginalization of the joint distribution of the future observations and the parameter (see pages 40–41).

Suppose that n Bernoulli trials with unknown success probability θ result in x successes and that the prior density of θ is beta with parameters α and β. If m further trials are contemplated on the same Bernoulli population, what can we say about the number of successes obtained? We know from Theorem 11-1 that the posterior distribution of θ, given x, is beta with parameters $x + \alpha$ and $n - x + \beta$. If $f(y|\theta)$ is the probability function of the number Y of successes in the m future trials, the joint density of Y and θ is

$$f(y, \theta|x) = f(y|\theta)\pi(\theta|x) = \binom{m}{y}\theta^y(1-\theta)^{m-y} \cdot \frac{\theta^{x+\alpha-1}(1-\theta)^{n-x+\beta-1}}{B(x+\alpha, n-x+\beta)}$$

$$= \binom{m}{y}\frac{\theta^{x+y+\alpha-1}(1-\theta)^{m+n-x-y+\beta-1}}{B(x+\alpha, n-x+\beta)} \qquad 0 < \theta < 1, y = 0, 1, \ldots, m$$

The predictive probability function of Y, denoted by $f^*(y)$, is the marginal density of Y obtained from the above joint density by integrating out θ:

$$f^*(y) = \int_0^1 \binom{m}{y}\frac{\theta^{x+y+\alpha-1}(1-\theta)^{n+m-x-y+\beta-1}}{B(x+\alpha, n-x+\beta)}d\theta \tag{19}$$

$$= \binom{m}{y}\frac{B(x+y+\alpha, m+n-x-y+\beta)}{B(x+\alpha, n-x+\beta)} \qquad y = 0, 1, \ldots, m \tag{20}$$

We thus have the following theorem.

Theorem 11-11 If n Bernoulli trials with unknown success probability θ result in x successes, and the prior density of θ is beta with parameters α and β, then the predictive density of the number of successes Y in m future trials on the same Bernoulli population is given by (20).

Remark 5 It is evident from (19) that $f^*(y)$ may also be regarded as the expectation, $E_\Theta(f(y|\theta))$, of the probability function of Y with respect to the posterior density $\pi(\theta|x)$ of θ.

EXAMPLE 11.24 Suppose that 7 successes were obtained in 10 Bernoulli trials with success probability θ. An independent set of 8 more Bernoulli trials with the same success probability is being contemplated. What could be said about the number of future successes if θ has a uniform prior density in the interval [0, 1]?

The predictive distribution of the number of future successes may be obtained from (20) with $\alpha = \beta = 1$, $n = 10$, $m = 8$, and $x = 7$:

$$f^*(y) = \binom{8}{y}\frac{B(y+8, 12-y)}{B(8, 4)} \qquad y = 0, 1, \ldots, 8$$

Table 11-5 summarizes the numerical results.

Table 11-5

y	0	1	2	3	4	5	6	7	8
$f^*(y)$	0.002	0.012	0.040	0.089	0.153	0.210	0.227	0.182	0.085

Remark 6 In an earlier remark, following Theorem 11-5, on *Laplace's law of succession*, it was pointed out that if all n trials of a binomial experiment resulted in successes, then the probability that a future trial will also result in success may be estimated by the posterior mean of the success parameter θ, namely $(n + 1)/(n + 2)$. The same result can be obtained as a special case of (20) with $\alpha = \beta = 1$, $m = 1$, and $x = n$. The predictive distribution of a future observation turns out to be binomial with success probability $(n + 1)/(n + 2)$. The two approaches, however, do not lead to the same results beyond $n = 1$. For instance, if we take the posterior mean $(n + 1)/(n + 2)$ as the success probability for the m future trials, then the probability that all of them are successes would be $[(n + 1)/(n + 2)]^m$, but (20) gives us $(n + 1)/(m + n + 1)$.

EXAMPLE 11.25 In Example 11.24, suppose that we are interested in predicting the outcome of the first 10 Bernoulli trials before they are performed. Determine the predictive distribution of the number of successes, say X, in the 10 trials, again assuming that θ has a uniform prior density in the interval [0, 1].

The joint distribution of X and θ is given by

$$f(x;\theta) = \binom{10}{x}\theta^x(1-\theta)^{10-x}\cdot 1 \qquad 0<\theta<1 \qquad x = 0, 1, \ldots, 10$$

The marginal density of x may be obtained from this by integrating out θ:

$$f^*(x) = \int_0^1 \binom{10}{x}\theta^x(1-\theta)^{10-x}d\theta = \binom{10}{x}B(x+1, 11-x) = \frac{1}{11} \qquad y = 0, 1, \ldots, 10$$

Remark 7 The predictive distributions obtained in Examples 11.24 and 11.25 are different in that they are based, respectively, on a posterior and a prior distribution of the parameter. A distinction between *prior* predictive distributions and *posterior* predictive distributions is sometimes made to indicate the nature of the parameter distribution used.

Predictive distributions for future normal samples may be derived analogously. We saw in Theorem 11-3 that if we have a sample of size n from a normal distribution with unknown mean θ and known variance σ^2 and if θ is normal with mean μ and variance v^2, then the posterior distribution of θ is also normal, with mean μ_{post} and variance v^2_{post} given by (7). Suppose that another observation, say Y, is made on the original population. We now show that the predictive distribution of Y is normal with mean μ_{post} and variance $\sigma^2 + v^2_{\text{post}}$.

The predictive density $f^*(y)$ of Y is given by

$$f^*(y) = f(y|\mathbf{x}) = \int f(y,\theta|\mathbf{x})d\theta = \int f(y|\theta)\cdot\pi(\theta|\mathbf{x})d\theta$$

$$\propto \int_{-\infty}^{\infty} e^{-\frac{1}{2\sigma^2}(y-\theta)^2} e^{-\frac{1}{2v^2_{\text{post}}}(\theta-\mu_{\text{post}})^2}d\theta$$

After some simplification we get

$$f^*(y) = \int_{-\infty}^{\infty} e^{-\frac{1}{2(\sigma^2 v^2_{\text{post}})/(\sigma^2+v^2_{\text{post}})}\left[\theta - \frac{(v^2_{\text{post}}y+\sigma^2\mu_{\text{post}})}{(\sigma^2+v^2_{\text{post}})}\right]^2} \times e^{-\frac{1}{2(\sigma^2 v^2_{\text{post}})/(\sigma^2+v^2_{\text{post}})}\left[-\left(\frac{v^2_{\text{post}}y+\sigma^2\mu_{\text{post}}}{\sigma^2+v^2_{\text{post}}}\right)^2 + \frac{v^2_{\text{post}}y+\sigma^2\mu_{\text{post}}}{\sigma^2+v^2_{\text{post}}}\right]}d\theta$$

The exponent in the second factor may be further simplified to yield

$$f^*(y) \propto \int_{-\infty}^{\infty} e^{-\frac{1}{2(\sigma^2 v^2_{\text{post}})/(\sigma^2+v^2_{\text{post}})}\left[\theta - \frac{(v^2_{\text{post}}y+\sigma^2\mu_{\text{post}})}{\sigma^2+v^2_{\text{post}}}\right]^2} \times e^{-\frac{1}{2(\sigma^2+v^2_{\text{post}})}(y-\mu_{\text{post}})^2}d\theta$$

The second factor here is free from θ. The first factor is a normal density in θ and integrates out to an expression free from θ and y. Therefore, the preceding integral becomes

$$e^{-\frac{1}{2(\sigma^2+v^2_{\text{post}})}(y-\mu_{\text{post}})^2}$$

This may be recognized as a normal density with mean μ_{post} and variance $\sigma^2 + v^2_{\text{post}}$. We thus see that predictive density of the future observation Y is normal with mean equal to the posterior mean of θ and variance equal to the sum of the population variance and the posterior variance of θ.

The following theorem is a straightforward generalization of this result (see Problem 11.96).

Theorem 11-12 Suppose that a random sample of size n is drawn from a normal distribution with unknown mean θ and known variance σ^2 and that the prior distribution of θ is normal with mean μ and variance v^2. If a second independent sample of size m is drawn from the same population, then the predictive distribution of the sample mean is normal with mean μ_{post} and variance $\left(\frac{\sigma^2}{m} + v^2_{\text{post}}\right)$, where $\mu_{\text{post}} = \frac{\sigma^2\mu + nv^2\bar{x}}{\sigma^2 + nv^2}$, $v^2_{\text{post}} = \frac{\sigma^2 v^2}{\sigma^2 + nv^2}$, and $\bar{x}$ is the mean of the first sample, of size n.

EXAMPLE 11.26 The shipping weight of packages handled by a company is normally distributed with mean θ lb and variance 8. If the first 25 packages handled on a given day have an average weight of 15 lb what are the chances the next 25 packages to be handled will have an average in excess of 16 lb? Assume that θ has a prior normal distribution with mean 12 and variance 9.

From Theorem 11-12, the mean and variance of the predictive density of the average weight of the future sample are given by 14.90 and 0.31. The probability we need is $P(\bar{Y} > 16) = P(Z > 1.98) = 0.0234$, so the chances are about 2% that the future average weight would exceed 16 lbs.

Point and interval summaries of the predictive density may be obtained as for the posterior density of a parameter, and they serve similar purposes. For example, given the predictive density function $f^*(\bar{y})$ of the sample mean $\bar{Y}$ of a future sample from a population, the expectation, median, or mode of $f^*(\bar{y})$ may be used as a *predictive point estimate* of $\bar{Y}$. Also, intervals $[\bar{y}_L, \bar{y}_U]$ satisfying the property

$$\int_{\bar{y}_L}^{\bar{y}_U} f^*(\bar{y})\,d\bar{y} = 1 - \alpha \tag{21}$$

may be used as Bayesian $(1 - \alpha) \times 100\%$ *predictive intervals* for $\bar{Y}$; and equal tail area and *HPD* predictive intervals may be defined as in the case of credibility intervals for a parameter.

EXAMPLE 11.27 In Example 11.24, find the predictive (a) mean, (b) median, and (c) mode of the number of future successes.

(a) The predictive distribution of Y is given in Table 11-5. The predictive mean number of future successes is the expectation of Y, which is 5.34.

(b) The predictive median is between 5 and 6, and we may take it to be 5.5.

(c) The predictive mode is 6.

EXAMPLE 11.28 In Example 11.26, find a 95% equal tail area predictive interval for the average weight of the 25 future packages.

The predictive distribution is normal with mean 14.90 and variance 0.31. The 95% equal tail interval is given by $14.90 \pm (1.96 \times 0.56) = [13.8, 16.0]$.

SOLVED PROBLEMS

Subjective probability

11.1. Identify the type of probability used: (a) The probability that my daughter will attend college is 0.9. (b) The chances of getting three heads out of three tosses of a fair coin are 1 in 8. (c) I am 40% sure it will rain on the 4th of July this year because it did in 12 out of the past 30 years. (d) We are 70% sure that the variance of this distribution does not exceed 3.5. (e) Some economists believe there is a better than even chance that the economy will go into recession next year. (f) The chances are just 2% that she will miss both of her free throws. (g) I am 90% sure this coin is not fair. (h) The probability that all three children are boys in a three-child family is about 0.11. (i) The odds are 3 to 1 the Badgers will not make it to the Super Bowl this year. (j) You have one chance in a million of winning this lottery. (k) You have a better than even chance of finding a store that carries this item.

(a), (d), (e), (g), (i): subjective; (b), (j), (k): classical; (c), (f), (h): frequency

Prior and posterior distributions

11.2. A box contains two fair coins and two biased coins (each with $P(\text{"heads"}) = 0.3$). A coin is chosen at random from the box and tossed four times. If two heads and two tails are obtained, find the posterior probabilities of the event F that the chosen coin is fair and the event B that the coin is biased.

Let D denote the event (data) that two heads and two tails are obtained in four tosses. We then have, from Bayes theorem,

$$P(F|D) = \frac{P(D|F)P(F)}{P(D|F)P(F) + P(D|B)P(B)} = \frac{\left[\binom{4}{2}(0.5)^4\right] \cdot \left(\frac{1}{2}\right)}{\left[\binom{4}{2}(0.5)^4\right] \cdot \left(\frac{1}{2}\right) + \left[\binom{4}{2}(0.3)^2(0.7)^2\right] \cdot \left(\frac{1}{2}\right)} = \frac{625}{1066} \approx 0.59$$

$$P(B|D) = 1 - P(F|D) = 441/1066 \approx 0.41$$

11.3. Verify the posterior probability values given in Table 11-2.

$$P(\theta = 0.2|D) = \frac{P(D|\theta = 0.2)P(\theta = 0.2)}{P(D|\theta = 0.2)P(\theta = 0.2) + P(D|\theta = 0.5)P(\theta = 0.5)} = \frac{[3(0.2)^2(0.8)] \cdot \left(\frac{1}{2}\right)}{[3(0.2)^2(0.8)] \cdot \left(\frac{1}{2}\right) + [3(0.5)^3] \cdot \left(\frac{1}{2}\right)}$$

$$= \frac{32}{157} \approx 0.20$$

$$P(\theta = 0.5|D) = 1 - P(\theta = 0.2|D) = \frac{125}{157} \approx 0.80$$

11.4. The random variable X has a Poisson distribution with an unknown parameter λ. It has been determined that λ has the subjective prior probability function given in Table 11-6. A random sample of size 2 yields the X-values 2 and 0. Find the posterior distribution of λ.

Table 11-6

λ	0.5	1.0	1.5
$\pi(\lambda)$	1/2	1/3	1/6

The likelihood of the data is $f(\mathbf{x}|\lambda) = e^{-2\lambda}\dfrac{\lambda^{x_1+x_2}}{x_1!x_2!}$. The posterior density is (up to factors free from λ)

$$\pi(\lambda|\mathbf{x}) \propto e^{-2\lambda}\lambda^{x_1+x_2}\pi(\lambda) \propto e^{-2\lambda}\lambda^2\pi(\lambda) \quad \text{for} \quad \lambda = 0.5, 1, 1.5$$

The results are summarized in Table 11.7.

Table 11-7

λ	0.5	1.0	1.5
$\pi(\lambda|\mathbf{x})$	0.42	0.41	0.17

11.5. In a lot of n bolts produced by a machine, an unknown number ρ are defective. Assume that ρ has a prior binomial distribution with parameter p. Find the posterior distribution of ρ if a bolt chosen at random from the lot is (a) defective; (b) not defective.

(a) We are given the prior probability function $\pi(\rho) = \binom{n}{\rho}p^{\rho}(1-p)^{n-\rho}, \rho = 0, 1, \ldots, n$. The posterior probability function of ρ, given the event D = "defective," is $\pi(\rho|D) \propto \frac{\rho}{n} \cdot \binom{n}{\rho}p^{\rho}(1-p)^{n-\rho}, \rho = 0,1, \ldots, n$

$$= \binom{n-1}{\rho-1}p^{\rho}(1-p)^{n-\rho}, \rho = 1, \ldots, n$$

Since $\sum_{\rho=1}^{n}\binom{n-1}{\rho-1}p^{\rho-1}(1-p)^{n-\rho} = 1$, the constant of proportionality in the preceding probability function must be $\frac{1}{p}$. Therefore, $\pi(\rho|D) = \binom{n-1}{\rho-1}p^{\rho-1}(1-p)^{n-\rho}, \rho = 1, \ldots, n$.

(b) $\pi(\rho|D') \propto \frac{n-\rho}{n} \cdot \binom{n}{\rho} p^{\rho}(1-p)^{n-\rho}, \rho = 0, 1, \ldots, n-1$

$= \binom{n-1}{\rho} p^{\rho}(1-p)^{n-\rho}, \rho = 0, \ldots, n-1$

Since $\sum_{\rho=0}^{n-1} \binom{n-1}{\rho} p^{\rho}(1-p)^{n-1-\rho} = 1$, the constant of proportionality in the preceding probability function must be $\frac{1}{1-p}$. Therefore, $\pi(\rho|D') = \binom{n-1}{\rho} p^{\rho}(1-p)^{n-1-\rho}, \rho = 0, \ldots, n-1$.

11.6. X is a binomial random variable with known n and unknown success probability θ. Find the posterior density of θ assuming a prior density $\pi(\theta)$ equal to (a) $2\theta, 0 < \theta < 1$; (b) $3\theta^2, 0 < \theta < 1$; (c) $4\theta^3, 0 < \theta < 1$.

(a) $\pi(\theta|\boldsymbol{x}) \propto \theta^x(1-\theta)^{n-x} \cdot \theta = \theta^{x+1}(1-\theta)^{n-x}, 0 < \theta < 1$.

Since this is a beta density with parameters $x + 2$ and $n - x + 1$, the normalizing constant is $1/B(x + 2, n - x + 1)$ and we get $\pi(\theta|\boldsymbol{x}) = \frac{1}{B(x+2, n-x+1)}\theta^{x+1}(1-\theta)^{n-x}, 0 < \theta < 1$.

(b) The posterior is the beta density: $\pi(\theta|\boldsymbol{x}) = \frac{1}{B(x+3, n-x+1)}\theta^{x+2}(1-\theta)^{n-x}, 0 < \theta < 1$.

(c) The posterior is the beta density: $\pi(\theta|\boldsymbol{x}) = \frac{1}{B(x+4, n-x+1)}\theta^{x+3}(1-\theta)^{n-x}, 0 < \theta < 1$.

11.7. A random sample $\boldsymbol{x} = (x_1, x_2, \ldots, x_n)$ of size n is taken from a population with density function $f(x|\theta) = 3\theta x^2 e^{-\theta x^3}, x > 0$. θ has a prior gamma density with parameters α and β. Find the posterior density of θ.

$\pi(\theta|\boldsymbol{x}) \propto \theta^n e^{-\theta\sum x^3} \cdot \theta^{\alpha-1}e^{-\theta/\beta} \propto \theta^{n+\alpha-1}e^{-\theta\left(\sum x^3 + \frac{1}{\beta}\right)}$. This may be recognized as a gamma density with parameters $n + \alpha$ and $\frac{\beta}{1 + \beta\sum x^3}$. The normalizing constant should therefore be $\left(\frac{1+\beta\sum x^3}{\beta}\right)^{n+\alpha} \cdot \frac{1}{\Gamma(n+\alpha)}$ and the posterior density is $\pi(\theta|\boldsymbol{x}) = \frac{1}{\Gamma(n+\alpha)}\left(\frac{1+\beta\sum x^3}{\beta}\right)^{n+\alpha}\theta^{n+\alpha-1}e^{-\theta\left(\sum x^3+\frac{1}{\beta}\right)}, \theta > 0$.

11.8. X is normal with mean 0 and unknown precision ξ which has prior gamma density with parameters α and β. Find the posterior distribution of ξ based on a random sample $\boldsymbol{x} = (x_1, x_2, \ldots, x_n)$ from X.

$\pi(\xi|\boldsymbol{x}) \propto \xi^{n/2}e^{-\frac{\xi}{2}\sum x^2} \cdot \xi^{\alpha-1}e^{-\xi/\beta} \propto \xi^{\frac{n}{2}+\alpha-1}e^{-\xi\left(\frac{\sum x^2}{2}+\frac{1}{\beta}\right)}, \xi > 0$

Therefore, ξ has a gamma distribution with parameters $\frac{n}{2} + \alpha$ and $\frac{2\beta}{\beta\sum x^2 + 2}$.

Sampling from a binomial population

11.9. A poll of 100 voters chosen at random from all voters in a given district indicated that 55% of them were in favor of a particular candidate. Suppose that prior to the poll we believe that the true proportion θ of voters in that district favoring that candidate has a uniform density over the interval [0, 1]. Find the posterior density of θ.

Applying Theorem 11-1 with $n = 100$ and $x = 55$, the posterior density of θ is beta with parameters $\alpha = 56$ and $\beta = 46$.

11.10. In 40 tosses of a coin, 24 heads were obtained. Find the posterior distribution of the proportion θ of heads that would be obtained in an unlimited number of tosses of the coin. Use a uniform prior for θ.

By Theorem 11-1, the posterior density of θ is beta with $\alpha = 25$ and $\beta = 17$.

11.11. A poll to predict the fate of a forthcoming referendum found that 480 out of 1000 people surveyed were in favor of the referendum. What are the chances that the referendum would be lost?

Assume a vague prior distribution (uniform on [0, 1]) for the proportion θ of people in the population who favor the referendum. The posterior distribution of θ, given the poll result, is beta with parameters 481, 521. We need the probability that $\theta < 0.5$. Computer software gives 0.90 for this probability, so we can be 90% sure that the referendum would lose.

11.12. In the previous problem, suppose an additional 1000 people were surveyed and 530 were found to be in favor of the referendum. What can we conclude now?

We take the prior now to be beta with parameters 481 and 521. The posterior becomes beta with parameters 1011 and 991. The probability for $\theta < 0.5$ is 0.33. This means there is only a 33% chance now of the referendum losing.

Sampling from a Poisson population

11.13. The number of accidents during a six-month period at an intersection has a Poisson distribution with mean λ. It is believed that λ has a gamma prior density with parameters $\alpha = 2$ and $\beta = 5$. If eight accidents were observed during the first six months of the year, find the (a) posterior density, (b) posterior mean, and (c) posterior variance.

(a) We know from Theorem 11-2 that the posterior density is gamma with parameters $n\bar{x} + \alpha = 10$ and $\beta/(1 + n\beta) = 5/6$.

(b) From (32), Chapter 4, the posterior mean $= 50/6 \approx 8.33$ and (c) the posterior variance $= 250/36 \approx 6.94$.

11.14. The number of defects in a 1000-foot spool of yarn manufactured by a machine has a Poisson distribution with unknown mean λ. The prior distribution of λ is gamma with parameters $\alpha = 2$ and $\beta = 1$. A total of 23 defects were found in a sample of 10 spools that were examined. Determine the posterior density of λ.

By Theorem 11-2, the posterior density is gamma with parameters $n\bar{x} + \alpha = 23 + 2 = 25$ and $\beta/(1 + n\beta) = 1/11 \approx 0.091$.

Sampling from a normal population

11.15. A sample of 100 measurements of the diameter of a sphere gave a mean $\bar{x} = 4.38$ inch. Based on prior experience, we know that the diameter is normally distributed with unknown mean θ and variance 0.36. Determine the posterior density of θ assuming a normal prior density with mean 4.5 inch and variance 0.4.

From Theorem 11-3, we see that the posterior density is normal with mean 4.381 and variance 0.004.

11.16. The reaction time of an individual to certain stimuli is known to be normally distributed with unknown mean θ but a known standard deviation of 0.35 sec. A sample of 20 observations yielded a mean reaction time of 1.18 sec. Assume that the prior density of θ is normal with mean $\mu = 1$ sec and variance $v^2 = 0.13$. Find the posterior density of θ.

By Theorem 11-3, the posterior density is normal with mean 1.17 and variance 0.006.

11.17. A random sample of 25 observations is taken from a normal population with unknown mean θ and variance 16. The prior distribution of θ is standard normal. Find (a) the posterior mean and (b) its precision. (c) Find the precision of the maximum likelihood estimator.

(a) By Theorem 11-3, the posterior mean of θ is $\left(\dfrac{25}{25 + 16}\right)\bar{x} = \dfrac{25\bar{x}}{41}$.

(b) The precision of the estimate is the reciprocal of its variance. The variance of $\dfrac{25\bar{x}}{41} = \left(\dfrac{25}{41}\right)^2 \dfrac{16}{25} = 0.24$, so the precision is roughly 4.2.

(c) The maximum likelihood estimate of θ is $\bar{x}$. Its variance is $16/25$, so the precision is about 1.6.

11.17. X is normal with mean 0 and unknown precision ξ, which has prior gamma distribution with parameters α and β. Find the posterior distribution of ξ based on a random sample $\boldsymbol{x} = (x_1, x_2, \ldots, x_n)$ from X.

$$\pi(\xi \mid \boldsymbol{x}) \propto \xi^{n/2} e^{-\frac{\xi}{2}\sum x^2} \cdot \xi^{\alpha-1} e^{-\xi/\beta} = \xi^{\frac{n}{2}+\alpha-1} e^{-\xi\left(\sum x^2 + \frac{1}{\beta}\right)}$$

Therefore ξ is gamma distributed with parameters $\dfrac{n}{2} + \alpha$ and $\dfrac{\beta}{\beta \sum x^2 + 1}$.

Improper prior distributions

11.19. An improper prior density for a Poisson mean λ is defined by $\pi(\lambda) = 1, \lambda > 0$. Show that the posterior density in this case is gamma with parameters $n\bar{x} + 1$ and $\frac{1}{n}$.

Given the observation vector $\boldsymbol{x}$, the posterior density of λ is $\pi(\lambda | x) \propto e^{-n\lambda}\lambda^{n\bar{x}}$. The result follows since this density is of the gamma form with parameters $n\bar{x} + 1$ and $\frac{1}{n}$.

11.20. Another improper prior density for Poisson mean λ is $\pi(\lambda) = 1/\lambda, \lambda > 0$. Show that the posterior density in this case is gamma.

We have $\pi(\lambda | x) \propto e^{-n\lambda}\lambda^{n\bar{x}} \cdot \frac{1}{\lambda} \propto e^{-n\lambda}\lambda^{n\bar{x}-1}$. The posterior is therefore gamma with parameters $n\bar{x}$ and $\frac{1}{n}$.

11.21. An improper prior density for the Poisson mean, known as *Jeffreys' prior* for the Poisson, is given by $\pi(\lambda) = 1/\sqrt{\lambda}, \lambda > 0$. Find the posterior density under this prior.

Given the observation vector $\boldsymbol{x}$, the posterior density of λ is $\pi(\lambda | \boldsymbol{x}) \propto e^{-n\lambda}\lambda^{n\bar{x}} \cdot \left(\frac{1}{\sqrt{\lambda}}\right) \propto e^{-n\lambda}\lambda^{n\bar{x}-\frac{1}{2}}, \lambda > 0$.

This is a gamma density with parameters $n\bar{x} + \frac{1}{2}$ and $\frac{1}{n}$.

11.22. X is binomial with known n and unknown success probability θ. An improper prior density for θ, known as *Haldane's prior*, is given by $\pi(\theta) = \frac{1}{\theta(1-\theta)}, 0 < \theta < 1$. Find the posterior density of θ based on the observation x.

$\pi(\theta | x) = \binom{n}{x}\theta^x(1-\theta)^{n-x} \cdot \frac{1}{\theta(1-\theta)} = \binom{n}{x}\theta^{x-1}(1-\theta)^{n-x-1}, 0 < \theta < 1$. This is a beta density with

$\alpha = x$ and $\beta = n - x$, so we get $\pi(\theta | x) = \frac{\theta^{x-1}(1-\theta)^{n-x-1}}{B(x, n-x)}, 0 < \theta < 1$.

11.23. Do Problem 11.22 assuming Jeffreys' prior for the binomial, given by $\pi(\theta) = \frac{1}{\sqrt{\theta(1-\theta)}}, 0 < \theta < 1$.

$\pi(\theta | x) = \binom{n}{x}\theta^x(1-\theta)^{n-x} \cdot \frac{1}{\theta^{1/2}(1-\theta)^{1/2}} = \binom{n}{x}\theta^{x-\frac{1}{2}}(1-\theta)^{n-x-\frac{1}{2}}, 0 < \theta < 1$. This is a beta density with

$\alpha = x + \frac{1}{2}$ and $\beta = n - x + \frac{1}{2}$.

11.24. Suppose we are sampling from an exponential distribution (page 118) with unknown parameter θ, which has the improper prior density $\pi(\theta) = 1/\theta, \theta > 0$. Find the posterior density $\pi(\theta | \boldsymbol{x})$.

$\pi(\theta | \boldsymbol{x}) \propto \theta^n e^{-\theta\sum_i x_i} \cdot \frac{1}{\theta} \propto \theta^{n-1}e^{-\theta\sum_i x_i}, \theta > 0$. The posterior density for θ is therefore gamma with parameters

$\alpha = n$ and $\beta = 1/\sum_i x_i$.

11.25. X is normal with unknown mean θ and known variance σ^2. The prior distribution of θ is improper and is given by $\pi(\theta) = 1, -\infty < \theta < \infty$. Determine the posterior density $\pi(\theta | \boldsymbol{x})$.

$\pi(\theta | \boldsymbol{x}) \propto e^{-\frac{1}{2\sigma^2}\sum_i (x_i-\theta)^2} \cdot 1 \propto e^{-\frac{n}{2\sigma^2}(\theta-\bar{x})^2}$. The posterior distribution is thus normal with mean $\bar{x}$ and variance σ^2/n.

11.26. X is normal with mean 0 and unknown variance θ. The variance has the improper prior density $\pi(\theta) = 1/\sqrt{\theta}, \theta > 0$. Find the posterior distribution of θ.

$$\pi(\theta | x) \propto \frac{1}{\theta^{n/2}}e^{-\sum x^2/2\theta} \cdot \frac{1}{\sqrt{\theta}}, \theta > 0$$

$$\propto \theta^{-\left(\frac{n+1}{2}\right)}e^{-\sum x^2/2\theta}, \theta > 0$$

$$\propto \theta^{-\left(\frac{n-1}{2}\right)-1}e^{-\sum x^2/2\theta}, \theta > 0$$

This is an inverse gamma density (see Problem 11.99) with $\alpha = \frac{n-1}{2}$ and $\beta = \frac{\sum x^2}{2}$.

Conjugate prior distributions

11.27. A poll to predict the fate of a forthcoming referendum found that 1010 out of 2000 people surveyed were in favor of the referendum. Assuming a prior uniform density for the unknown population proportion θ, find the chances that the referendum would lose. Comment on your result with reference to Problems 11.11 and 11.12.

The posterior distribution of θ, given the poll result, is beta with parameters 1011, 991. We need the probability that $\theta < 0.5$. This comes out to be 0.33, so we can be 33% sure that the referendum would lose. This is the same result that we obtained in Problem 11.12 using for prior the posterior beta distribution derived in Problem 11.11. Since the beta family is conjugate with respect to the binomial distribution, we are able to update the posterior sequentially in Problem 11.12.

11.28. A random sample of size 10 drawn from a geometric distribution with success probability θ (see page 117) yields a mean of 4.2. The prior density of θ is uniform in the interval [0, 1]. Determine the posterior distribution of θ.

The prior distribution is beta with parameters $\alpha = \beta = 1$. We know from Example 11.12 that the posterior distribution is also beta. The parameters are given by $\alpha + n = 11$ and $\beta + n\bar{x} - n = 33$.

11.29. A random sample of size n is drawn from a negative binomial distribution with parameter θ (see page 117): $f(x; \theta) = \binom{x-1}{r-1}\theta^r(1-\theta)^{x-r}, x = r, r+1, \ldots$. Suppose that the prior density of θ is beta with parameters α and β. Show that the posterior density of θ is also a beta, with parameters $\alpha + nr$ and $\beta + n\bar{x} - nr$, where $\bar{x}$ is the sample mean. In other words, show that the beta family is conjugate with respect to the negative binomial distribution.

$\pi(\theta \mid \boldsymbol{x}) \propto \theta^{nr}(1-\theta)^{n(\bar{x}-r)} \cdot \theta^{\alpha-1}(1-\theta)^{\beta-1} \propto \theta^{nr+\alpha-1}(1-\theta)^{n(\bar{x}-r)+\beta-1}$, which is a beta density with parameters $\alpha + nr$ and $\beta + n\bar{x} - nr$.

11.30. The interarrival time of customers at a bank is exponentially distributed with mean $1/\theta$, where θ has a gamma distribution with parameters $\alpha = 1$ and $\beta = 0.2$. Twelve customers were observed over a period of time and were found to have an average interarrival time of 6 minutes. Find the posterior distribution of θ.

Applying Theorem 11-4 with $\alpha = 1$ and $\beta = 0.2$, $n = 11$ (12 customers $\Rightarrow$ 11 interarrival times), $\bar{x} = 6$, we see that the posterior density is gamma with parameters 12 and 0.014.

11.31. In the previous problem, suppose that a second, independent sample of 10 customers was observed and was found to have an average interarrival time of 6.5 minutes. Find the posterior distribution of θ.

Since the gamma family is conjugate for the exponential distribution, this problem can be done in two ways: (i) by starting with the prior gamma distribution with parameters 1 and 0.2 and applying Theorem 11-4 with $n = 11 + 9 = 20$ and $\bar{x} = ((11 \times 6) + (9 \times 6.5))/20 \approx 6.225$ or (ii) by starting with the prior gamma distribution with parameters 12 and 0.014 and applying Theorem 11-4 with $n = 9$. Both ways lead to the result that the posterior density is gamma with parameters 21 and 0.0077.

11.32. The following density is known as a *Rayleigh density:* $f(x) = \theta x e^{-(\theta x^2/2)}, x > 0$. It is a special case of the Weibull density (see page 118), with $b = 2$ and $a = \theta$. Show that the gamma family is conjugate with respect to the Rayleigh distribution. Specifically, show that if X has a Rayleigh density and θ has a gamma prior density with parameters α and β, then the posterior density of θ given a random sample $\boldsymbol{x} = (x_1, x_2, \ldots, x_n)$ of observations from X is also a gamma.

$\pi(\theta \mid x) \propto f(x \mid \theta) \cdot \pi(\theta) \propto \theta^n e^{-\theta \sum_i x_i^2/2} \cdot \theta^{\alpha-1} e^{-\theta/\beta} \propto \theta^{(\alpha+n)-1} e^{-\theta\left(\frac{1}{\beta}+\sum_i x_i^2/2\right)}, \theta > 0$. This is a gamma density with parameters $\alpha + n$ and $\dfrac{2\beta}{2 + \beta \sum x_i^2}$.

11.33. Show that the inverse gamma family (see Problem 11.99) is conjugate with respect to the normal distribution with known mean but unknown variance θ.

Assume that the mean of the normal density is 0. We have

$$f(x\,|\,\theta) = \frac{1}{(2\pi)^{n/2}\theta^{n/2}}\exp\left\{-\frac{1}{2\theta}\sum_{i=1}^{n}x_i^2\right\} \quad \text{and}$$

$$\pi(\theta) = \frac{\beta^{\alpha}\theta^{-\alpha-1}e^{-\beta/\theta}}{\Gamma(\alpha)}, \theta > 0$$

The posterior density is given by

$$\pi(\theta\,|\,x) = f(x\,|\,\theta)\cdot\pi(\theta) \propto \theta^{-n/2}e^{-\frac{1}{2\theta}\sum_i x_i^2}\cdot\theta^{-\alpha-1}e^{-\beta/\theta} \propto \theta^{(-\frac{n}{2}-\alpha)-1}e^{-\frac{1}{\theta}\left(\beta+\frac{\sum x_i^2}{2}\right)}, \theta > 0$$

This is also an inverse gamma, with parameters $\frac{n}{2} + \alpha$ and $\beta + \frac{\sum_i x_i^2}{2}$.

11.34. A random sample of n observations is taken from the exponential density with mean θ: $f(x\,|\,\theta) = (1/\theta)\exp\{-x/\theta\}, x > 0$. Assume that θ has an inverse gamma prior distribution (see Problem 11.99) and show that its posterior distribution is also in the inverse gamma family.

$$f(x\,|\,\theta) = (1/\theta)^n\exp\left\{-\sum_i x_i/\theta\right\}, x > 0$$

$$\pi(\theta) = \frac{\beta^{\alpha}\theta^{-\alpha-1}e^{-\beta/\theta}}{\Gamma(\alpha)}, \theta > 0$$

The posterior density, given by $\pi(\theta\,|\,x) \propto f(x\,|\,\theta)\cdot\pi(\theta) \propto \theta^{-n}e^{-\frac{\sum_i x_i}{\theta}}\cdot\theta^{-\alpha-1}e^{-\beta/\theta} \propto \theta^{-(n+\alpha)-1}e^{-\frac{1}{\theta}(\beta+\sum x_i)}, \theta > 0$, is inverse gamma with parameters $n + \alpha$ and $\beta + \sum_i x_i$.

11.35. In the previous problem, suppose that a second sample of m observations from the same population yields the observations $y_1, y_2, \ldots, y_m$. Find the posterior density incorporating the result from both samples.

Since the inverse gamma family is conjugate with respect to the exponential distribution, we can update the posterior parameters obtained in Problem 11.34 to $m + (n + \alpha)$ and $\left(\beta + \sum_i x_i\right) + \sum_j y_j$. The posterior density is thus inverse gamma with parameters $m + n + \alpha$ and $\beta + \sum_i x_i + \sum_j y_j$.

11.36. A random sample of n observations is taken from the gamma density:

$f(x\,|\,\theta) = \dfrac{x^{\alpha-1}e^{-x/\theta}}{\theta^{\alpha}\Gamma(\alpha)}, x > 0$. Assume that θ has an inverse gamma prior distribution with parameters γ and β and show that its posterior distribution is also in the inverse gamma family.

$$f(x\,|\,\theta) \propto (1/\theta^{\alpha})^n\exp\left\{-\sum_i x_i/\theta\right\}, x > 0$$

$$\pi(\theta) = \frac{\beta^{\gamma}\theta^{-\gamma-1}e^{-\beta/\theta}}{\Gamma(\gamma)}, \theta > 0$$

The posterior density, given by $\pi(\theta\,|\,x) \propto f(x\,|\,\theta)\cdot\pi(\theta) \propto \theta^{-n\alpha}e^{\frac{-\sum_i x_i}{\theta}}\cdot\theta^{-\gamma-1}e^{-\beta/\theta} \propto \theta^{-(n\alpha+\gamma)-1}e^{-\frac{1}{\theta}(\beta+\sum x_i)}, \theta > 0$ is inverse gamma with parameters $n\alpha + \gamma$ and $\beta + \sum_i x_i$.

Bayesian point estimation

11.37. In Problem 11.5, find the Bayes estimate of ρ with squared error loss function.

(a) The Bayes estimate is the mean of the posterior distribution, which is

$$\sum_{\rho=1}^{n}\rho\cdot\binom{n-1}{\rho-1}p^{\rho-1}(1-p)^{n-\rho} = 1 + \sum_{\rho=1}^{n}(\rho-1)\cdot\binom{n-1}{\rho-1}p^{\rho-1}(1-p)^{n-\rho} = 1 + (n-1)p$$

(b) The posterior mean is $\sum_{\rho=0}^{n-1}\rho\cdot\binom{n-1}{\rho}p^{\rho}(1-p)^{n-1-\rho} = (n-1)p$.

11.38. Show that the Bayes estimate μ_{post} for θ obtained in Theorem 11-5 is a convex combination of the maximum likelihood estimate of θ and the prior mean of θ.

$$\mu_{\text{post}} = \frac{x+\alpha}{n+\alpha+\beta} = \left(\frac{n}{n+\alpha+\beta}\right)\left(\frac{x}{n}\right) + \left(\frac{\alpha+\beta}{n+\alpha+\beta}\right)\left(\frac{\alpha}{\alpha+\beta}\right)$$

11.39. In Problem 11.10, find the Bayes estimate with squared error loss function for (a) θ (b) $1/\theta$.

(a) The posterior distribution is beta with parameters 25 and 17. The Bayes estimate, which equals the posterior mean, is $25/52 \approx 0.48$.

(b) The Bayes estimate of $1/\theta$ is the posterior mean of $1/\theta$, given by

$$\frac{1}{B(25,17)}\int_0^1 \frac{1}{\theta}\cdot\theta^{24}(1-\theta)^{16}d\theta = \frac{B(24,17)}{B(25,17)} = \frac{41}{24} \approx 1.71$$

11.40. In Problem 11.15, find the Bayes estimate with squared error loss function for θ.

The Bayes estimate is the posterior mean, which we know from Problem 11.15 to be 4.38.

11.41. In Problem 11.33, assume that $\alpha = \beta = 1$ and find the Bayes estimate for the variance with squared error loss.

The posterior distribution is inverse gamma (see Problem 11.99) with parameters $\frac{n}{2}+1$ and $1+\frac{\sum_i x_i^2}{2}$. The Bayes estimate is the posterior mean, given by $\frac{2+\sum_i x_i^2}{n}$.

11.42. Find the Bayes estimate of θ with squared error loss function in Problem 11.24 and compare it to the maximum likelihood estimate.

The parameters of the posterior are n and $1\Big/\sum_i x_i$. Therefore, the Bayes estimate, which is the posterior mean, is $1/\bar{x}$. This is the same as the maximum likelihood estimate for θ (see Problem 11.98).

11.43. In Example 11.10, determine the Bayes estimate for θ under the squared error loss function.

The posterior distribution of θ is normal with mean $\bar{x}$ and variance σ^2/n. The Bayes estimate of θ under the squared error loss, which is the posterior mean, is given by $\bar{x}$.

11.44. In Problem 11.30, find the Bayes estimate for θ with squared error loss function. Find the squared error loss of the estimate for each $\boldsymbol{x} = (x_1, x_2, \ldots, x_n)$ and compare it to the loss of the maximum likelihood estimate.

The Bayes estimate under squared error loss is the posterior mean $= \beta(\alpha+n)/(1+n\beta\bar{x})$. The squared error loss for each $\boldsymbol{x}$ is the posterior variance $\beta^2(\alpha+n)/(1+n\beta\bar{x})^2$. With $\alpha = 1$ and $\beta = 0.2$, $n = 11$ and $\bar{x} = 6$, this comes to 0.00238. The maximum likelihood estimate for θ is $1/\bar{x}$ and its squared error loss is $E\left[\left(\frac{1}{\bar{x}}-\theta\right)^2\middle|\,\boldsymbol{x}\right] = \left(\frac{1}{\bar{x}}\right)^2 - \frac{2}{\bar{x}}E(\theta\,|\,x) + E(\theta^2\,|\,x)$. With $\alpha = 1$ and $\beta = 0.2$, $n = 11$ and $\bar{x} = 6$, this comes to 0.00239.

11.45. If X is a Poisson random variable with parameter λ and the prior density of λ is gamma with parameters α and β, then show that the Bayes estimate for λ is a weighted average of its maximum likelihood estimate and the prior mean.

By Theorem 11-2, the posterior distribution is gamma with parameters $n\bar{x}+\alpha$ and $\beta/(1+n\beta)$. The posterior mean is $\frac{\beta(n\bar{x}+\alpha)}{(1+n\beta)} = \frac{n\beta}{1+n\beta}\cdot\bar{x} + \frac{1}{1+n\beta}\cdot\alpha\beta$.

11.46. In Problem 11.16, find the Bayes estimate of θ with (a) squared error loss and (b) absolute error loss.

(a) The Bayes estimate with squared error loss is the posterior mean of θ, which is 1.17.

(b) The Bayes estimate with absolute error loss is the posterior median, which is the same as the posterior mean in this case since the posterior distribution is normal.

11.47. In Problem 11.32, find the Bayes estimate for θ with squared error loss function.

The posterior distribution of θ is gamma with parameters $\alpha + n$ and $\frac{2\beta}{2 + \beta\sum x^2}$. Therefore, the posterior mean is $\frac{2\beta(\alpha + n)}{\left(2 + \beta\sum x_i^2\right)}$.

11.48. The time (in minutes) that a bank customer has to wait in line to be served is exponentially distributed with mean $1/\theta$. The prior distribution of θ is gamma with mean 0.4 and standard deviation 1. The following waiting times were recorded for a random sample of 10 customers: 2, 3.5, 1, 5, 4.5, 3, 2.5, 1, 1.5, 1. Find the Bayes estimate for θ with (a) squared error and (b) absolute error loss function.

The gamma distribution with parameters α and β has mean $\alpha\beta$ and variance $\alpha\beta^2$. Therefore, the parameters for our gamma prior must be $\alpha = 0.16$ and $\beta = 2.5$. The posterior distribution is (see Theorem 11-4) gamma with parameters $\alpha + n = 10.16$ and $\beta/(1 + n\beta\bar{x} = 0.04)$.

(a) The posterior mean is $10.16 \times 0.04 = 0.41$.

(b) The median of the posterior density, obtained using computer software, is 0.393.

11.49. In Problem 11.6, find the Bayes estimate with squared error loss for θ in each case and evaluate it assuming $n = 500$ and $x = 200$.

(a) From Theorem 11-6 we know that the Bayes estimate here is the posterior mean. The mean of the beta density with parameters $x + 2$ and $n - x + 1$ is $(x + 2)/(n + 3) = 0.4016$.

(b) Similar to the preceding. The Bayes estimate is $(x + 3)/(n + 4) = 0.4028$.

(c) The Bayes estimate is $(x + 4)/(n + 5) = 0.4040$.

11.50. In Problem 11.6, part (a), find the Bayes estimate with squared error loss for the population standard deviation, $\sqrt{n\theta(1 - \theta)}$.

The required estimate is the posterior expectation of $\sqrt{n\theta(1 - \theta)}$, which equals

$$\frac{\sqrt{n}}{B(x + 2, n - x + 1)}\int_0^1 \theta^{\frac{1}{2}}(1 - \theta)^{\frac{1}{2}} \cdot \theta^{x+1}(1 - \theta)^{n-x}\,d\theta = \sqrt{n} \times \frac{B\left(x + \frac{5}{2}, n - x + \frac{3}{2}\right)}{B(x + 2, n - x + 1)}$$

11.51. In Problem 11.6, find the Bayes estimate with absolute error loss for θ in each case assuming $n = 500$ and $x = 200$.

By Theorem 11-6, the Bayes estimate of θ with absolute error loss is the median of the posterior distribution of θ. Since there is no closed form expression for the median of a beta density, we have obtained the following median values using computer software: (a) 0.4015; (b) 0.4027; (c) 0.4038.

11.52. In Problem 11.14, estimate λ using a Bayes estimate (a) with squared error loss and (b) with absolute error loss.

The posterior density was obtained in Problem 11.14 as a gamma with parameters 25 and 0.091.

(a) The Bayes estimate with squared error loss is the posterior mean, which in this case is $\alpha\beta = 2.275$.

(b) The Bayes estimate with absolute error loss is the posterior median. Using computer software to calculate the median of the gamma posterior distribution, we get the estimate 2.245.

11.53. A random sample $x = (x_1, x_2, \ldots, x_n)$ of size n is taken from a population with density function $f(x|\theta) = 3\theta x^2 e^{-\theta x^3}, 0 < x < \infty$, where θ has a prior gamma density with parameters α and β. Find the Bayes estimate for θ with squared error loss.

$\pi(\theta|x) \propto \theta^n e^{-\theta\sum x^3} \cdot \theta^{\alpha-1}e^{-\theta/\beta} \propto \theta^{(n+\alpha)-1}e^{-\theta\left(\sum x^3 + \frac{1}{\beta}\right)}$, which is a gamma density with parameters $\alpha + n$ and $\frac{\beta}{1 + \beta\sum x^3}$.

The posterior mean estimate of θ is therefore $\frac{\beta(\alpha + n)}{1 + \beta\sum x^3}$.

11.54. In Problem 11.24, find the Bayes estimate of $e^{-t\theta}$ with respect to the squared error loss function.

The Bayes estimate is $E(e^{-t\theta}|\mathbf{x}) = \int_0^\infty e^{-t\theta}\pi(\theta|\mathbf{x})\,d\theta = \int_0^\infty e^{-t\theta}\theta^{n-1}e^{-\theta\sum_i x_i}\,d\theta = \int_0^\infty \theta^{n-1}e^{-\theta(t+\sum_i x_i)}\,d\theta = \dfrac{\Gamma(n)}{(t+n\bar{x})^n}$

11.55. The random variable X is normally distributed with mean θ and variance σ^2. The prior distribution of θ is standard normal. (a) Find the Bayes estimator of θ with squared error loss function based on a random sample of size n. (b) Is the resulting estimator unbiased (see page 195)? (c) Compare the Bayes estimate to the maximum likelihood estimate in terms of the squared error loss.

(a) By Theorem 11-3, the Bayes estimate is $\left(\dfrac{n\bar{x}}{n+\sigma^2}\right)$. With $c = \left(\dfrac{n}{n+\sigma^2}\right)$, the squared error loss for this estimate is $E[(c\bar{x}-\theta)^2|\mathbf{x}] = c^2\bar{x}^2 - 2c\bar{x}\cdot 0 + 1 = c^2\bar{x}^2 + 1$.

(b) Since $E\left(\dfrac{n\bar{X}}{n+\sigma^2}\right) = \dfrac{n\theta}{n+\sigma^2}$, the estimator is biased. It is, however, asymptotically unbiased.

(c) The maximum likelihood estimate of θ is $\bar{x}$. The squared error loss for this estimate is $\bar{x}^2 + 1$. Clearly, since $c < 1$, the loss is less for the Bayes estimate. For large values of n, the losses are approximately equal.

11.56. In Problem 11.22, show that the Bayes estimate of θ is the same as the maximum likelihood estimate.

The Bayes estimate is the posterior mean of θ, given by $\dfrac{\alpha}{\alpha+\beta} = \dfrac{x}{n}$. The maximum likelihood estimate is found by maximizing the likelihood $L \propto \theta^x(1-\theta)^{n-x}$ with respect to θ (see page 198). Solving the equation $\dfrac{dL}{d\theta} = x\theta^{x-1}(1-\theta)^{n-x} + (n-x)\theta^x(1-\theta)^{n-x-1} = 0$ for θ, we get the maximum likelihood estimate x/n.

11.57. In Problem 11.48, find the Bayes estimate for $1/\theta$ with squared error loss function.

The Bayes estimate is the expectation of $1/\theta$ with respect to the posterior distribution of θ:

$$E\left(\frac{1}{\theta}\Big|\mathbf{x}\right) = \frac{1}{(0.04)^{10.16}\Gamma(10.16)}\int_0^\infty \frac{1}{\theta}\cdot\theta^{9.16}e^{-\theta/0.04}\,d\theta = \frac{(0.04)^{9.16}\Gamma(9.16)}{(0.04)^{10.16}\Gamma(10.16)} \approx 2.73$$

Bayesian interval estimation

11.58. Measurements of the diameters of a random sample of 200 ball bearings made by a certain machine during one week showed a mean of 0.824 inch and a standard deviation of 0.042 inch. The diameters are normally distributed. Find a (a) 90%, (b) 95%, and (c) 98% Bayesian HPD credibility interval for the mean diameter θ of all ball bearings made by the machine. Assume that the prior distribution of θ is normal with mean 0.8 inch and standard deviation 0.05.

The posterior mean and standard deviation are respectively 0.824 and 0.0030.

(a) The 90% HPD interval is given by $0.824 \pm (1.645 \times 0.003)$ or [0.819, 0.829].

(b) The 95% HPD interval is given by $0.824 \pm (1.96 \times 0.003)$ or [0.818, 0.830].

(c) The 98% HPD interval is given by $0.824 \pm (2.33 \times 0.003)$ or [0.817, 0.831].

11.59. A sample poll of 100 voters chosen at random from all voters in a given district indicated that 55% of them were in favor of a particular candidate. Suppose that, prior to the poll, we believe that the true proportion θ of voters in that district favoring that candidate has Jeffreys' prior (see Problem 11.23) given by $\pi(\theta) = \dfrac{1}{\sqrt{\theta(1-\theta)}}$, $0 < \theta < 1$. Find 95% and 99% equal tail area Bayesian credibility intervals for the proportion θ of all voters in favor of this candidate.

We have $n = 100$ and $x = 55$. From Problem 11.23, the posterior density of θ is beta with parameters $\alpha = 55.5$ and $\beta = 45.5$. This density has the following percentiles: $x_{0.005} = 0.423$, $x_{0.025} = 0.452$, $x_{0.975} = 0.645$, $x_{0.995} = 0.673$. This gives us the 95% Bayesian equal tail credibility interval [0.452, 0.645] and the 99% Bayesian equal tail credibility interval [0.423, 0.673]. (It is instructive to compare these with the traditional intervals we obtained in Problem 6.13.)

11.60. In the previous problem, assume that θ has a uniform prior distribution on [0, 1] and find (a) 95% and (b) 99% equal tail area credibility intervals for θ.

The posterior distribution of θ is beta with parameters 56 and 46 (see Theorem 11-1).

(a) We need the percentiles $x_{0.025}$ and $x_{0.975}$ of the preceding beta distribution. These are respectively 0.452 and 0.644. The 95% interval is [0.452, 0.644].

(b) We need the percentiles $x_{0.005}$ and $x_{0.995}$ of the preceding beta distribution. These are respectively 0.422 and 0.644. The 99% interval is [0.422, 0.672].

11.60. In 40 tosses of a coin, 24 heads were obtained. Find a 90% and 99.73% credibility interval for the proportion of heads θ that would be obtained in an unlimited number of tosses of the coin. Use a uniform prior for θ.

By Theorem 11-1, the posterior density of θ is beta with $\alpha = 25$ and $\beta = 17$. This density has the following percentiles: $x_{0.00135} = 0.367$, $x_{0.05} = 0.469$, $x_{0.95} = 0.716$, $x_{0.99865} = 0.800$. The 90% and 99.73% Bayesian equal tail area credibility intervals are, respectively, [0.469, 0.716] and [0.367, 0.800]. (The traditional confidence intervals are given in Problem 6.15.)

11.62. A sample of 100 measurements of the diameter of a sphere gave a mean $\bar{x} = 4.38$ inch. Based on prior experience, we know that the diameter is normally distributed with unknown mean θ and variance 0.36. (a) Find 95% and 99% equal tail area credibility intervals for the actual diameter θ assuming a normal prior density with mean 4.5 inches and variance 0.4. (b) With what degree of credibility could we say that the true diameter is 4.38 ± 0.01?

(a) From Theorem 11-3, we see that the posterior mean and variance for θ are 4.381 and 0.004. The 95% credibility interval is $[4.381 - (1.96 \times 0.063), 4.381 + (1.96 \times 0.063)] = [4.26, 4.50]$. Similarly, the 90% credibility interval is $[4.381 - (1.645 \times 0.063), 4.381 + (1.645 \times 0.063)] = [4.28, 4.48]$.

(b) We need the area under the posterior density from 4.37 to 4.39. This equals the area under the standard normal density between $(4.37 - 4.381)/0.063 = -0.17$ and $(4.39 - 4.381)/0.063 = 0.14$. This equals 0.1232, so the required degree of credibility is roughly 12%.

11.63. In Problem 11.16, construct a 95% credibility interval for θ.

From Problem 11.16, we see that the posterior mean and variance for θ are 1.17 and 0.006. The 95% credibility interval is $[1.17 - (1.96 \times 0.077), 1.17 + (1.96 \times 0.077)] = [1.02, 1.32]$.

11.64. In Problem 11.25, what can you say about the *HPD* Bayesian credibility interval for θ compared to the conventional interval shown in (1), Chapter 6?

The posterior distribution of θ is normal with mean $\bar{x}$ and variance σ^2/n. The *HPD* credibility intervals we obtain would be identical to the conventional confidence intervals centered at $\bar{x}$.

11.65. The number of individuals in a year who will suffer a bad reaction from injection of a given serum has a Poisson distribution with unknown mean λ. Assume that λ has Jeffreys' improper prior density $\pi(\lambda) = 1/\sqrt{\lambda}, \lambda > 0$ (see Problem 11.21). Table 11-8 gives the number of such cases that occurred in each of the past 10 years.

(a) Derive a 98% equal tail credibility interval for λ. (b) With what degree of credibility can you assert that λ does not exceed 3?

Table 11-8

Year	1	2	3	4	5	6	7	8	9	10
Number	2	4	1	2	2	1	2	3	3	0

(a) We know from Problem 11.21 that the posterior distribution for λ is gamma with parameters $n\bar{x} + \frac{1}{2}$ and $\frac{1}{n}$, which in our case are 20.5 and 0.1. We thus need the 1st and 99th percentiles of the gamma distribution with these parameters. Using computer software, we get $x_{0.01} = 1.146$ and $x_{0.99} = 3.248$. The 98% credibility interval is [1.146, 3.248].

(b) We need the posterior probability that λ does not exceed 3. This is the area to the left of 3 under the gamma density with parameters 20.5 and 0.1. Since this area is 0.972, we can be about 97% certain that λ does not exceed 3.

11.66. In Problem 11.14, obtain the 95% Bayesian equal tail area credibility interval for λ.

The posterior density was obtained in Problem 11.14 as a gamma with parameters 25 and 0.091. The percentiles of this density relevant for our credibility interval are $x_{0.975} = 3.25$ and $x_{0.025} = 1.47$. The 95% Bayesian credibility interval is [1.47, 3.25].

11.67. Obtain an equal tail 95% credibility interval for θ in Problem 11.22 assuming $n = 10, x = 3$.

The posterior is beta with parameters 3 and 7. The percentiles are $x_{0.025} = 0.075$ and $x_{0.975} = 0.600$. The interval is [0.075, 0.600].

11.68. Obtain an equal tail area 95% credibility interval for θ in Problem 11.23 assuming $n = 10, x = 3$.

The posterior is beta with parameters 3.5 and 7.5. The percentiles are $x_{0.025} = 0.093$ and $x_{0.975} = 0.606$. The interval is [0.093, 0.606].

11.69. In Problem 11.48, obtain a 99% equal tail area credibility interval for (a) θ and (b) $1/\theta$.

(a) The posterior distribution of θ is gamma with parameters 10.16 and 0.04. We obtain the following percentiles of this distribution using computer software: $x_{0.005} = 0.15$ and $x_{0.995} = 0.81$. The credibility interval is [0.15, 0.81].

(b) Since $\theta < 0.15 \Leftrightarrow 1/\theta > 1/0.15$ and $\theta > 0.81 \Leftrightarrow 1/\theta < 1/0.81$, the equal tail area interval for $1/\theta$ is $[1/0.81, 1/0.15] = [1.23, 6.67]$.

Bayesian hypothesis tests

11.70. The mean lifetime (in hours) of fluorescent light bulbs produced by a company is known to be normally distributed with an unknown mean θ but a known standard deviation of 120 hours. The prior density of θ is normal with $\mu = 1580$ hours and $v^2 = 16900$. A mean lifetime of a sample of 100 light bulbs is computed to be 1630 hours. Test the null hypothesis $H_0{:}\theta \le 1600$ against the alternative hypothesis $H_1{:}\theta > 1600$ using a Bayes (a) 0.05 test and (b) 0.01 test.

(a) By Theorem 11.3, the posterior density is normal with mean 1629.58 and standard deviation 11.95. The posterior probability of H_0 is $P(\theta \le 1600|\mathbf{x}) = P\left(Z \le \frac{1600 - 1629.58}{11.95}\right) \approx 0.007$. Since this probability is less than 0.05, we can reject H_0.

(b) Since the posterior probability of the null hypothesis, obtained in (a), is less than 0.01, we can reject H_0.

11.71. Suppose that in Example 11.18 a second sample of 100 observations yielded a mean reaction time of 0.25 sec. Test the null hypothesis $H_0{:}\theta \le 0.3$ against the alternative $H_1{:}\theta > 0.3$ using a Bayes 0.05 test.

We take the prior distribution of θ to be the posterior distribution obtained in Example 11.18: Normal with mean 0.352 and variance 0.004. Applying Theorem 11-3 with this prior and the new data, we get a posterior mean 0.269 and variance 0.0007. The posterior probability of the null hypothesis is 0.88. Since this is not less than 0.05, we cannot reject the null hypothesis.

11.72. In Problem 11.21, suppose a sample of size 10 yielded the values 2, 0, 1, 1, 3, 0, 2, 4, 2, 2. Test $H_0{:}\lambda \le 1$ against $H_1{:}\lambda > 1$ using a Bayes 0.05 test.

We need the posterior probability of H_0. From Problem 11.21, it is the area from 0 to 1 under a gamma density with parameters $n\bar{x} + \frac{1}{2} = 17.5$ and $\frac{1}{n} = 0.1$. Using computer software, we see this probability is 0.02. Since this is less than the specified threshold of 0.05, we reject the null hypothesis.

11.73. In Problem 11.65, test the null hypothesis $H_0{:}\lambda \le 1$ against the alternative hypothesis $H_1{:}\lambda > 1$ using a Bayes 0.05 test.

The Bayes 0.05 test would reject the null hypothesis if the posterior probability of the hypothesis $\lambda \le 1$ is less than 0.05. In our case, this probability is given by the area to the left of 1 under a gamma distribution with parameters 20.5 and 0.1 and is 0.002. Since this is less than 0.05, we reject the null hypothesis.

11.74. In Problem 11.6, assume that $n = 40$ and $x = 10$ and test the null hypothesis $H_0{:}\theta \le 0.2$ against the alternative $H_1{:}\theta > 0.2$ using a Bayes 0.05 test.

(a) The posterior probability of the null hypothesis is given by the area from 0 to 0.2 under a beta density with parameters 12 and 31, which is determined to be 0.12 using computer software. Since this is not less than 0.05, we cannot reject the null hypothesis.

(b) The posterior probability is the area from 0 to 0.2 under beta density with parameters 13 and 31, which is 0.07. Since this is not less than 0.05, we cannot reject the null hypothesis.

(c) The posterior probability is the area from 0 to 0.2 under a beta density with parameters 14 and 31, which is 0.04. Since this is less than 0.05, we reject the null hypothesis.

11.75. In Problem 11.48, test the null hypothesis $H_0{:}\theta \ge 0.7$ against $H_1{:}\theta < 0.7$ using a Bayes 0.025 test.

The posterior distribution of θ is gamma with parameters 10.16 and 0.04. Therefore, the posterior probability of the null hypothesis is 0.022. Since this is less than 0.025, we reject the null hypothesis.

11.76. The life-length X of a computer component has the exponential density given by (see page 118) $f(x|\theta) = \theta e^{-\theta x}, x > 0$ with unknown mean $1/\theta$. Suppose that the prior density of θ is gamma with parameters $\alpha = 0.2$ and $\beta = 0.15$. If a random sample of 10 observations on X yielded an average life-length of 7 years, use a Bayes 0.05 test to test the null hypothesis that the expected life-length is at least 12 years against the alternative hypothesis that it is under 12 years.

The null and alternative hypothesis are respectively equivalent to $H_0: \theta \le 1/12 = 0.083$ and $H_1: \theta > 0.083$. From Theorem 11-4, the posterior distribution of θ is gamma with parameters 10.2 and 0.013. The posterior probability of the null hypothesis is 0.10. Since this is larger than 0.05, we cannot reject the null hypothesis.

Bayes factor

11.77. In Example 11.4, find the Bayes factor of $H_0: \lambda = 1$ relative to $H_1: \lambda \ne 1$.

$$BF = \{P(H_0|x)/[1 - P(H_0|x)]\} \div \{P(H_0)/[1 - P(H_0)]\} = (0.49/0.51) \div ((1/3)/(2/3)) \approx 1.92$$

11.78. It is desired to test the null hypothesis $\theta \le 0.6$ against the alternative $\theta > 0.6$, where θ is the probability of success for a Bernoulli trial. Assume that θ has a uniform prior distribution on [0, 1] and that in 40 trials there were 24 successes. What is your conclusion if you decide to reject the null hypothesis if $BF < 1$?

The posterior density of θ is beta with $\alpha = 25$ and $\beta = 17$. The posterior probability of the null hypothesis is 0.52. Posterior odds ratio is $0.52/0.48 = 1.0833$ and prior odds ratio is $6/4 = 1.5$. $BF = 0.72$. We reject the null hypothesis.

11.79. Prove that the *ad hoc* rule (see the Remark following Theorem 11-10) to reject H_0 if $BF \le 1$ is equivalent to the Bayes α test with $\alpha = P(H_0)$.

$$BF \le 1 \Leftrightarrow \left(\frac{P(H_0|x)}{P(H_1|x)}\right) \Big/ \left(\frac{P(H_0)}{P(H_1)}\right) \le 1 \Leftrightarrow P(H_0|x)[1 - P(H_0)] \le [1 - P(H_0|x)]P(H_0) \Leftrightarrow P(H_0|x) \le P(H_0)$$

11.80. In the preceding problem, find c such that the Bayes factor criterion to reject the null hypothesis if $BF < c$ is equivalent to the Bayes 0.05 rule.

By Theorem 11-10, $c = \dfrac{\alpha[1 - P(H_0)]}{(1 - \alpha)P(H_0)} = \dfrac{(0.05)(1 - 0.6)}{(1 - 0.05)(0.6)} \approx 0.035.$

11.81. Work Problem 11.71 using the decision to reject the null hypothesis if the Bayes factor is less than 1. We know from Problem 11.79 that the rule to reject H_0 if $BF < 1$ is equivalent to rejecting the null hypothesis if $P(H_0|x) \le P(H_0)$. We know from Problem 11.71 that $P(H_0|x) = 0.88$. From Example 11.18, we know that the prior distribution of θ is normal with mean 0.4 and variance 0.13; therefore, $P(H_0) = P\left(Z \leq \frac{0.3 - 0.4}{0.361}\right) \approx 0.39$. We cannot reject the null hypothesis.

11.82. In Problem 11.74, perform the test in each case using the Bayes factor rule to reject the null hypothesis if $BF \leq 4$.

(a) $BF = \{P(H_0|x)/[1 - P(H_0|x)]\} \div \{P(H_0)/[1 - P(H_0)]\} = (0.12/0.88) \div (0.04/0.96) \approx 3.27$. We reject the null hypothesis.

(b) $BF = \{P(H_0|x)/[1 - P(H_0|x)]\} \div \{P(H_0)/[1 - P(H_0)]\} = (0.07/0.93) \div (0.008/0.992) \approx 9.33$. We cannot reject the null hypothesis.

(c) $BF = \{P(H_0|x)/[1 - P(H_0|x)]\} \div \{P(H_0)/[1 - P(H_0)]\} = (0.04/0.96) \div (0.002/0.998) \approx 20.79$. We cannot reject the null hypothesis.

11.83. In Problem 11.21, determine what can be concluded using the Bayes factor criterion: Reject H_0 if $BF < 1$.

Since the prior distribution in this problem is improper, the prior odds ratio is not defined. Therefore, the Bayes factor criterion cannot be employed here.

11.84. Suppose that in Example 11.18 a second sample of 100 observations yielded a mean reaction time of 0.25 sec. Test the null hypothesis $H_0: \theta \geq 0.3$ against the alternative$H_1: \theta < 0.3$ using the Bayes factor criterion to reject the null hypothesis if $BF < 0.05$.

We take the prior distribution of θ to be the posterior distribution obtained in Example 11.18: Normal with mean 0.352 and variance 0.004. Applying Theorem 11-3 with this prior and the new data, we get a posterior mean 0.269 and variance 0.0007. Using this, we obtain the posterior probability of the null hypothesis as 0.12.

Note that the prior probability of the null hypothesis that is needed for calculating the Bayes factor in this problem should be based on the prior distribution given in Example 11.18: Normal with mean 0.4 and variance 0.13. Using this, we get the prior probability of the null hypothesis as 0.61. The Bayes factor is 0.087. Since this is larger than 0.05, we cannot reject the null hypothesis.

11.85. In Problem 11.48, test the null hypothesis $H_0: \theta \geq 0.7$ against $H_1: \theta < 0.7$ using the Bayes factor rule to reject the null hypothesis if $BF < 1$.

The prior distribution of θ is gamma with parameters 0.16 and 2.5. The posterior distribution of θ is gamma with parameters 10.16 and 0.04. The prior and posterior probabilities of the null hypothesis are respectively 0.154 and 0.022. Since $P(H_0|\mathbf{x}) \leq P(H_0)$, we reject the null hypothesis (see Problem 11.79).

Bayesian predictive distributions

11.86. The random variable X has a Bernoulli distribution with success probability θ, which has a prior beta distribution with parameters $\alpha = \beta = 2$. A sample of n trials on X yielded n successes. A future sample of two trials is being contemplated. Find (a) the predictive distribution of the number of future successes and (b) the predictive mean.

(a) $P(Y = y|\theta) = \binom{2}{y}\theta^y(1 - \theta)^{2-y}$, $y = 0, 1, 2$ and by Theorem11-1, $\pi(\theta|\mathbf{x})$ is beta with parameters $\alpha = n + 2$ and $\beta = 2$.

$$f^*(y) = \int_0^1 \binom{2}{y}\theta^y(1 - \theta)^{2-y}\frac{\theta^{n+1}(1 - \theta)}{B(n + 2, 2)}d\theta = \binom{2}{y}\frac{B(y + n + 2, 4 - y)}{B(n + 2, 2)}, y = 0, 1, 2.$$ This is tabulated as follows:

Table 11-9

y	0	1	2
$f^*(y)$	$6/[(n + 4)(n + 5)]$	$4(n + 2)/[(n + 4)(n + 5)]$	$[(n + 2)(n + 3)]/[(n + 4)(n + 5)]$

(b) The predictive mean is $(2n^2 + 14n + 20)/[(n + 4)(n + 5)]$.

11.87. X is a Poisson random variable with parameter λ. An initial sample of size n gives λ a gamma posterior distribution with parameters $n\bar{x} + \alpha$ and $\beta/(1 + n\beta)$. It is planned to make one further observation on the original population. (a) Find the predictive distribution of this observation. (b) Show that the predictive mean is the same as the posterior mean.

(a) Let Y denote the future observation. The predictive density of Y is given by

$$f^*(y) \propto \int_0^\infty \frac{e^{-\lambda}\lambda^y}{y!} \cdot \frac{(1 + n\beta)^{n\bar{x}+\alpha}\lambda^{(n\bar{x}+\alpha)-1}e^{-\lambda(n\beta+1)/\beta}}{\beta^{n\bar{x}+\alpha}\Gamma(nx + \alpha)} d\lambda$$

$$\propto \int_0^\infty \frac{(1 + n\beta)^{n\bar{x}+\alpha}\lambda^{(n\bar{x}+y+\alpha)-1}e^{-\lambda(n\beta+\beta+1)/\beta}}{y!\beta^{n\bar{x}+\alpha}\Gamma(n\bar{x} + \alpha)} d\lambda$$

$$= \frac{(1 + n\beta)^{n\bar{x}+\alpha}}{y!\beta^{n\bar{x}+\alpha}\Gamma(n\bar{x} + \alpha)} \cdot \frac{\beta^{n\bar{x}+y+\alpha}\Gamma(n\bar{x} + y + \alpha)}{(n\beta + \beta + 1)^{n\bar{x}+y+\alpha}}$$

$$f^*(y) = \binom{n\bar{x} + y + \alpha - 1}{n\bar{x} + \alpha - 1}\left(\frac{n\beta + 1}{n\beta + \beta + 1}\right)^{n\bar{x}+\alpha}\left(\frac{\beta}{n\beta + \beta + 1}\right)^y, \quad y = 0, 1, \ldots$$

With $u = n\bar{x} + \alpha + y$ the right hand side above may be written as

$$\binom{u - 1}{(n\bar{x} + \alpha) - 1}\left(\frac{n\beta + 1}{n\beta + \beta + 1}\right)^{n\bar{x}+\alpha}\left(\frac{\beta}{n\beta + \beta + 1}\right)^{u-(n\bar{x}+\alpha)}, \quad u = n\bar{x} + \alpha, (n\bar{x} + \alpha) + 1, \ldots$$

which is a negative binomial probability function with parameters $r = n\bar{x} + \alpha$ and $p = \dfrac{n\beta + 1}{n\beta + \beta + 1}$.

(b) The mean of this distribution is $\dfrac{r}{p} = \dfrac{(n\bar{x} + \alpha)(n\beta + \beta + 1)}{(n\beta + 1)}$. The predictive mean of Y is therefore

$$\frac{(n\bar{x} + \alpha)(n\beta + \beta + 1)}{(n\beta + 1)} - (n\bar{x} + \alpha) = \frac{\beta(n\bar{x} + \alpha)}{(n\beta + 1)}$$

which is the same as the posterior mean of λ.

11.88. In Problem 11.21, find the distribution of the mean of a future sample of size m.

$f^*(y) = \int_0^\infty f(y|\lambda)\pi(\lambda|x)d\lambda \propto \dfrac{1}{\Pi y_i!}\int_0^\infty e^{-(n+m)\lambda}\lambda^{n\bar{x}+m\bar{y}-\frac{1}{2}}d\lambda, \lambda > 0$. Normalizing this gamma density, we get

$$f^*(\bar{y}) = \frac{n^{n\bar{x}+\frac{1}{2}}\Gamma\left(n\bar{x} + m\bar{y} + \frac{1}{2}\right)}{\Pi y_i!\Gamma\left(n\bar{x} + \frac{1}{2}\right)(n + 1)^{n\bar{x}+m\bar{y}+\frac{1}{2}}}, \quad \bar{y} = \frac{0}{m}, \frac{1}{m}, \frac{2}{m}, \ldots$$

11.89. The number of accidents per month on a particular stretch of a highway is known to follow the Poisson distribution with mean λ. A total of 24 accidents occurred on that stretch during the past 10 months. What are the chances that there would be more than 3 accidents there next month? Assume Jeffreys' prior for λ: $\pi(\lambda) = 1/\sqrt{\lambda}, \lambda > 0$.

The predictive distribution of the number of accidents Y during the next month may be obtained from Problem 11.88 with $n = 10, n\bar{x} = 24, m = 1$:

$$f^*(y) = \frac{10^{24+\frac{1}{2}}\Gamma\left(24 + y + \frac{1}{2}\right)}{y!\Gamma\left(24 + \frac{1}{2}\right)11^{24+y+\frac{1}{2}}}, \quad y = 0, 1, 2, \ldots.$$

The probability we need is $1 - [f^*(0) + f^*(1) + f^*(2) + f^*(3)] = 1 - [0.097 + 0.216 + 0.250 + 0.201] = 0.236$.

11.90. In Problem 11.65, what are the chances that the number of bad reactions next year would not exceed 1?

We need the predictive distribution for one future observation. We have the posterior in Problem 11.65 as gamma with parameters 20.5 and 0.1. Combining this with the probability function of Y, we get

$$f(y;\lambda) = f(y|\lambda)\pi(\lambda|\mathbf{x}) = \frac{e^{-\lambda}\lambda^y}{y!}\cdot\frac{10^{20.5}\lambda^{19.5}e^{-10\lambda}}{\Gamma(20.5)},\quad y = 0, 1, 2, \ldots; \lambda > 0.$$

The marginal probability function for Y, obtained by integrating out λ, is

$$f^*(y) = \int_0^\infty \frac{10^{20.5}\lambda^{y+19.5}e^{-11\lambda}}{y!\Gamma(20.5)}\,d\lambda = \frac{10^{20.5}\Gamma(y + 20.5)}{y!\Gamma(20.5)11^{y+20.5}}$$

The probabilities corresponding to y-values 0 through 7 are given in Table 11-10. The probability that the number of bad reactions would be 0 or 1 is 0.4058.

Table 11-10

y	0	1	2	3	4	5	6	7
$f^*(y)$	0.1417	0.2641	0.2581	0.1760	0.0940	0.0419	0.0162	0.0056

11.91. In Theorem 11-4, suppose that another, independent sample of size 1 is drawn from the exponential population. (a) Determine its predictive distribution. (b) Estimate the result of the future observation using the predictive mean.

(a) Denote the future observation by Y. We then have the following joint distribution of Y and the posterior density of θ.

$$f(y;\theta) = f(y|\theta)\pi(\theta|\mathbf{x}) = \frac{\theta e^{-\theta y}(1 + n\beta\bar{x})^{n+\alpha}\theta^{n+\alpha-1}e^{-\theta\left(\frac{1}{\beta}+n\bar{x}\right)}}{\beta^{n+\alpha}\Gamma(n+\alpha)} = \frac{(1 + n\beta\bar{x})^{n+\alpha}\theta^{n+\alpha}e^{-\theta\left(\frac{1}{\beta}+n\bar{x}+y\right)}}{\beta^{n+\alpha}\Gamma(n+\alpha)},\quad \theta > 0.$$

Integrating out θ,

$$f^*(y) = \int_0^\infty \frac{(1 + n\beta\bar{x})^{n+\alpha}\theta^{n+\alpha}e^{-\theta\left(\frac{1}{\beta}+n\bar{x}+y\right)}}{\beta^{n+\alpha}\Gamma(n+\alpha)}\,d\theta = \frac{(1 + n\beta\bar{x})^{n+\alpha}\beta^{n+\alpha+1}\Gamma(n+\alpha+1)}{(1 + n\beta\bar{x} + \beta y)^{n+\alpha+1}\beta^{n+\alpha}\Gamma(n+\alpha)}$$

$$= \frac{(1 + n\beta\bar{x})^{n+\alpha}\beta(n+\alpha)}{(1 + n\beta\bar{x} + \beta y)^{n+\alpha+1}},\ y > 0$$

(b) The mean of this predictive distribution is $\displaystyle\int_0^\infty \frac{y(1 + n\beta\bar{x})^{n+\alpha}\beta(n+\alpha)}{(1 + n\beta\bar{x} + \beta y)^{n+\alpha+1}}\,dy = \frac{1 + n\beta\bar{x}}{\beta(n+\alpha-1)}$.

11.92. In Problem 11.29, find the predictive density and predictive mean of a future observation.

$$f^*(y) = \binom{y-1}{r-1}\frac{1}{B(\alpha + nr, \beta + n\bar{x} - nr)}\int_0^1 \theta^{\alpha+nr+r-1}(1-\theta)^{\beta+n\bar{x}+y-nr-r-1}\,d\theta,\ y = r, r+1, \ldots$$

$$= \binom{y-1}{r-1}\frac{B(\alpha + nr + r, \beta + n\bar{x} + y - nr - r)}{B(\alpha + nr, \beta + n\bar{x} - nr)},\ y = r, r+1, \ldots$$

11.93. A couple has two children and they are both autistic. Find the probability that their next child will also be autistic assuming that the incidence of autism is independent from child to child and has the same probability θ. Assume that the prior distribution of θ is (a) uniform, (b) beta with parameters $\alpha = 2, \beta = 3$.

(a) Applying Theorem 11-11 with $n = 2, x = 2, m = 1$, and $\alpha = \beta = 1$, we see that the predictive distribution of Y is $f^*(y) = \dfrac{B(3+y, 2-y)}{B(3,1)} = \dfrac{(2+y)!(1-y)!}{8}$, $y = 0, 1$. The probability that the next child will be autistic is $3/4$.

(b) Applying Theorem 11-11 with $n = 2, x = 2, m = 1$, and $\alpha = 2, \beta = 3$, we see that the predictive distribution of Y is $f^*(y) = \dfrac{B(4 + y, 4 - y)}{B(4, 3)} = \dfrac{(3 + y)!(3 - y)!}{84}$, $y = 0, 1$. The probability that the next child will be autistic is 4/7.

11.94. A random sample of size 20 from a normal population with unknown mean θ and variance 4 yields a sample mean of 37.5. The prior distribution of θ is normal with mean 30 and variance 5. Suppose that an independent observation is subsequently made from the same population. Find (a) the predictive probability that this observation would not exceed 37.5 and (b) the equal tail area 95% predictive interval for the observation. From Theorem 11-12, the predictive density is normal with mean 37.21 and standard deviation 2.05.

(a) Equals the area to the left of 0.14 under the predictive density: 0.56

(b) $37.21 \pm (1.96 \times 2.05) = [33.19, 41.23]$

11.95. All 10 tosses of a coin resulted in heads. Assume that the prior density for the probability for heads is $\pi(\theta) = 6\theta^5, 0 < \theta < 1$ and find (a) the predictive distribution of the number of heads in four future tosses, (b) the predictive mean, and (c) the predictive mode.

(a) Note that the prior density is beta with parameters $\alpha = 6$ and $\beta = 1$. From (19), with $m = 10, n = 4$, $\alpha = 6, \beta = 1$, and $x = 10$, we get $f^*(y) = \dbinom{4}{y}\dfrac{B(16 + y, 5 - y)}{B(16, 1)}$, $y = 0, 1, 2, 3, 4$. The numerical values are shown in Table 11-11.

Table 11-11

y	0	1	2	3	4
$f^*(y)$	0.0002	0.0033	0.0281	0.1684	0.8000

(b) the predictive mean is 3.76

(c) the predictive mode is 4

11.96. Prove Theorem 11-12.

Since $\bar{Y}$ is normal with mean θ and variance σ^2/m, the proof is essentially the same as for the case $m = 1$ with σ^2 replaced with σ^2/m. This is shown as follows.

The predictive density $f^*(\bar{y})$ of $\bar{Y}$ is given by

$$f^*(\bar{y}) = f(\bar{y}|\mathbf{x}) = \int f(\bar{y}, \theta|\mathbf{x})\,d\theta = \int f(\bar{y}|\theta)\cdot\pi(\theta|\mathbf{x})\,d\theta \propto \int_{-\infty}^{\infty} e^{-\frac{m}{2\sigma^2}(\bar{y}-\theta)^2} e^{-\frac{1}{2v_{\text{post}}}(\theta-\mu_{\text{post}})^2}\,d\theta$$

After some simplification, we get

$$f^*(\bar{y}) = \int_{-\infty}^{\infty} e^{-\frac{1}{2(\sigma^2 v^2_{\text{post}})/(\sigma^2+mv^2_{\text{post}})}\left[\theta-\frac{(mv^2_{\text{post}}\bar{y}+\alpha^2\mu_{\text{post}})}{\sigma^2+mv^2_{\text{post}}}\right]^2} \times e^{-\frac{1}{2(\sigma^2 v^2_{\text{post}})/(\sigma^2+mv^2_{\text{post}})}\left[-\left(\frac{mv^2_{\text{post}}\bar{y}+\sigma^2\mu_{\text{post}}}{\sigma^2+mv^2_{\text{post}}}\right)^2+\frac{mv^2_{\text{post}}\bar{y}+\sigma^2\mu_{\text{post}}}{\sigma^2+mv^2_{\text{post}}}\right]}\,d\theta$$

The exponent in the second factor may be further simplified to yield

$$f^*(y) \propto \int_{-\infty}^{\infty} e^{-\frac{1}{2(\sigma^2 v^2_{\text{post}})/(\sigma^2+uv^2_{\text{post}})}\left[\theta-\frac{(uv^2_{\text{post}}\bar{y}+\sigma^2\mu_{\text{post}})}{\sigma^2+mv^2_{\text{post}}}\right]^2} \times e^{-\frac{m}{2(\sigma^2+mv^2_{\text{post}})}(\bar{y}-\mu_{\text{post}})^2}\,d\theta$$

The second factor here is free from θ. The first factor is a normal density in θ and integrates out to an expression free from θ and $\bar{y}$. Therefore, we have the following normal predictive density for $\bar{Y}$:

$$f^*(\bar{y}) \propto e^{-\frac{m}{2(\sigma^2+mv^2_{\text{post}})}(\bar{y}-\mu_{\text{post}})^2}$$

11.97. The random variable X has a binomial distribution with $n = 6$ and unknown success probability θ which has the Haldane prior $\pi(\theta) = \dfrac{1}{\theta(1-\theta)}$, $0 < \theta < 1$. An observation on X results in three successes. If another observation is made on X, how many successes could be expected?

The predictive distribution of the number of successes in the second observation may be obtained from Theorem 11-11 (with $m = n = 6, x = 3, \alpha = \beta = 0$) as

$$f^*(y) = \binom{6}{y}\frac{B(3+y, 9-y)}{B(3,3)} \qquad y = 0, 1, \ldots, 6$$

This is shown in Table 11-12.

Table 11-12

y	0	1	2	3	4	5	6
$f^*(y)$	0.0606	0.1364	0.1948	0.2165	0.1948	0.1364	0.0606

The expectation of this distribution is 3. We could therefore expect to see three successes in the six future trials.

Miscellaneous problems

11.98. Show that the maximum likelihood estimate of α in the exponential distribution (see page 124) is $1/\bar{x}$.

We have $L = \alpha^n e^{-\alpha \sum x_k}$. Therefore, $\ln L = n \ln \alpha - \alpha \sum x_k$. Differentiating with respect to α and setting it equal to 0 gives $\dfrac{n}{\alpha} - \sum x_k = 0$ or $\alpha = \dfrac{n}{\sum x_k} = \dfrac{1}{\bar{x}}$.

11.99. The random variable X has a gamma distribution with parameters α and β. Show that $Y = 1/X$ has the *inverse gamma density* with parameters α and β, defined by

$$g(y) = \begin{cases} \dfrac{\beta^\alpha y^{-\alpha-1} e^{-\beta/y}}{\Gamma(\alpha)} & y > 0 \\ 0, & y \le 0 \end{cases} \qquad (\alpha, \beta > 0)$$

From (33), Chapter 2, we have

$$g(y) = \frac{(1/y)^{\alpha-1} e^{-1/(\beta y)}}{\beta^\alpha \Gamma(\alpha)} \cdot \frac{1}{y^2} = \frac{y^{-\alpha-1} e^{-\frac{1}{\beta y}}}{\beta^\alpha \Gamma(\alpha)} \qquad y > 0$$

The mean, mode, and variance are:

$$\text{Mean} = \frac{\beta}{\alpha - 1} \quad \text{for } \alpha > 1, \quad \text{Mode} = \frac{\beta}{\alpha + 1}, \quad \text{Variance} = \frac{\beta^2}{(\alpha-1)^2(\alpha-2)} \quad \text{for } \alpha > 2.$$

11.100. Show that the Bayes estimate with absolute error loss function is the posterior median. Assume that the posterior distribution is continuous (see page 83).

We have to show that if m is the median of the posterior density $\pi(\theta \mid x)$, then

$$\int_{-\infty}^{\infty} |\theta - m|\, \pi(\theta \mid x)\, d\theta \le \int_{-\infty}^{\infty} |\theta - a|\, \pi(\theta \mid x)\, d\theta \text{ for all } a.$$

Assume $a \le m$.

$$\int_{-\infty}^{\infty} (|x - m| - |x - a|) f(x)\, dx = \int_{-\infty}^{a} (m - a)\pi(\theta \mid x)\, d\theta + \int_{a}^{m} (m + a - 2x)\pi(\theta \mid x)\, d\theta + \int_{m}^{\infty} (a - m)\pi(\theta \mid x)\, d\theta$$

$$\le \int_{-\infty}^{a} (m - a)\pi(\theta \mid x)\, d\theta + \int_{a}^{m} (m - a)\pi(\theta \mid x)\, d\theta + \int_{m}^{\infty} (a - m)\pi(\theta \mid x)\, d\theta$$

(since, in the middle integral, $m + a - 2x = (m - x) - (x - a) \le m - x \le m - a$)

$$\le (m - a)\left\{ \int_{-\infty}^{m} \pi(\theta \mid x)\, d\theta - \int_{m}^{\infty} \pi(\theta \mid x)\, d\theta \right\} = 0$$

The proof when $a > m$ is similar.

11.101. Generalize the results in Problem 11.91 to the sample mean of m future observations.

(a) Denote the mean of the future sample of size m by $\bar{Y}$. We then have the following joint distribution of $\bar{Y}$ and the posterior density of θ.

$$f(\bar{y};\theta) = f(\bar{y}|\theta)\pi(\theta|\mathbf{x}) = \frac{\theta^m e^{-\theta m\bar{y}}(1+n\beta\bar{x})^{n+\alpha}\theta^{n+\alpha-1}e^{-\theta\left(\frac{1}{\beta}+n\bar{x}\right)}}{\beta^{n+\alpha}\Gamma(n+\alpha)} = \frac{(1+n\beta\bar{x})^{n+\alpha}\theta^{m+n+\alpha}e^{-\theta\left(\frac{1}{\beta}+n\bar{x}+m\bar{y}\right)}}{\beta^{n+\alpha}\Gamma(n+\alpha)} \qquad \theta > 0$$

Integrating out θ, we have

$$f^*(\bar{y}) = \int_0^\infty \frac{(1+n\beta\bar{x})^{n+\alpha}\theta^{m+n+\alpha}e^{-\theta\left(\frac{1}{\beta}+n\bar{x}+m\bar{y}\right)}}{\beta^{n+\alpha}\Gamma(n+\alpha)}\,d\theta = \frac{(1+n\beta\bar{x})^{n+\alpha}\beta^{m+n+\alpha}\Gamma(m+n+\alpha)}{(1+n\beta\bar{x}+m\beta\bar{y})^{m+n+\alpha}\beta^{n+\alpha}\Gamma(n+\alpha)}$$

$$= \frac{(1+n\beta\bar{x})^{n+\alpha}\beta^{m}\Gamma(m+n+\alpha)}{(1+n\beta\bar{x}+m\beta\bar{y})^{m+n+\alpha}\Gamma(n+\alpha)} \qquad \bar{y} > 0$$

(b) The mean of this distribution is

$$\int_0^\infty \bar{y}\cdot\frac{(1+n\beta\bar{x})^{n+\alpha}\beta^{m}\Gamma(m+n+\alpha)}{(1+n\beta\bar{x}+m\beta\bar{y})^{m+n+\alpha}\Gamma(n+\alpha)}\,d\bar{y} = \frac{\Gamma(m+n+\alpha-2)}{m^2\Gamma(n+\alpha)}\left(\frac{\beta}{1+n\beta\bar{x}}\right)^{m-2}$$

SUPPLEMENTARY PROBLEMS

Subjective probability

11.102. Identify the type of probability used: (a) I have no idea whether I will or will not pass this exam, so I would say I am 50% sure of passing. (b) The chances are two in five that I will come up with a dime because I know the box has two dimes and three nickels. (c) Based on her record, there is an 80% chance that she will score over 40 baskets in tomorrow's game. (d) There is a 50-50 chance that you would run into an economist who thinks we are headed for a recession this year. (e) My investment banker believes the odds are five to three that this stock will double in price in the next two months.

Prior and posterior probabilities

11.103. A box contains a biased coin with $P(H) = 0.2$ and a fair coin. A coin is chosen at random from the box and tossed once. If it comes up heads, what is the probability of the event B that the chosen coin is biased?

11.104. The random variable X has a Poisson distribution with an unknown parameter λ. As shown in Table 11-13, the parameter λ has the subjective prior probability function, indicating prior ignorance. A random sample of size 2 yields the X-values 2 and 0. Find the posterior distribution of λ.

Table 11-13

λ	0.5	1.0	1.5
$\pi(\lambda)$	1/3	1/3	1/3

11.105. X is a binomial random variable with known n and unknown success probability θ. Find the posterior density of θ assuming a prior density $\pi(\theta) = 4\theta^3, 0 < \theta < 1$.

Sampling from a binomial distribution

11.106. The number of defective tools in each lot of 10 produced by a manufacturing process has a binomial distribution with parameter θ. Assume a vague prior density for θ (uniform on (0, 1)) and determine its posterior density based on the information that two defective tools were found in the last lot that was inspected.

11.107. In 50 tosses of a coin, 32 heads were obtained. Find the posterior distribution of the proportion of heads θ that would be obtained in an unlimited number of tosses of the coin. Use a noninformative prior (uniform on (0, 1)) for the unknown probability.

11.108. Continuing the previous problem, suppose an additional 50 tosses of the coin were made and 35 heads were obtained. Find the latest posterior density.

Sampling from a Poisson distribution

11.109. The number of accidents during a six-month period at an intersection has a Poisson distribution with mean λ. It is believed that λ has a gamma prior density with parameters $\alpha = 2$ and $\beta = 5$. If a total of 14 accidents were observed during the first six months of the year, find the (a) posterior density, (b) posterior mean, and (c) posterior variance.

11.110. The number of defects in a 2000-foot spool of yarn manufactured by a machine has a Poisson distribution with unknown mean λ. The prior distribution of λ is gamma with parameters $\alpha = 4$ and $\beta = 2$. A total of 42 defects were found in a sample of 10 spools that were examined. Determine the posterior density of λ.

Sampling from a normal distribution

11.111. A random sample of 16 observations is taken from a normal population with unknown mean θ and variance 9. The prior distribution of θ is standard normal. Find (a) the posterior mean, (b) its precision, and (c) the precision of the maximum likelihood estimator.

11.112. The reaction time of an individual to certain stimuli is known to be normally distributed with unknown mean θ but a known standard deviation of 0.30 sec. A sample of 20 observations yielded a mean reaction time of 2 sec. Assume that the prior density of θ is normal with mean 1.5 sec. and variance $v^2 = 0.10$. Find the posterior density of θ.

Improper prior distributions

11.113. The random variable X has the Poisson distribution with parameter λ. The prior distribution of λ is given $\pi(\lambda) = 1/\sqrt{\lambda}$, $\lambda > 0$. A random sample of 10 observations on X yielded a sample mean of 3.5. Find the posterior density of λ.

11.114. A population is known to be normal with mean 0 and unknown variance θ. The variance has the improper prior density $\pi(\theta) = 1/\sqrt{\theta}$, $\theta > 0$. If a random sample of size 5 from the population consists of 2.5, 3.2, 1.8, 2.1, 3.1, find the posterior distribution of θ.

Conjugate prior distributions

11.115. A random sample of size 20 drawn from a geometric distribution with parameter θ (see page 117) yields a mean of 5. The prior density of θ is uniform in the interval [0, 1]. Determine the posterior distribution of θ.

11.116. The interarrival time of customers at a bank is exponentially distributed with mean $1/\theta$, where θ has a gamma distribution with parameters $\alpha = 1$ and $\beta = 2$. Ten customers were observed over a period of time and were found to have an average interarrival time of 5 minutes. Find the posterior distribution of θ.

11.117. A population is known to be normal with mean 0 and unknown variance θ. The variance has the inverse gamma prior density with parameters $\alpha = 1$ and $\beta = 1$ (see Problem 11.99). Find the posterior distribution of θ based on the following random sample from the population: 2, 1.5, 2.5, 1.

Bayesian point estimation

11.118. The waiting time to be seated at a restaurant is exponentially distributed with mean $1/\theta$. The prior distribution of θ is gamma with mean 0.1 and variance 0.1. A random sample of six customers had an average waiting time of 9 minutes. Find the Bayes estimate for θ with (a) squared error (b) absolute error loss function.

11.119. The life-length X of a computer component has the exponential density given by (see page 118) $f(x|\theta) = \theta e^{-\theta x}$, $x > 0$ with unknown mean $1/\theta$. Suppose that the prior density of θ is gamma with parameters α and β. Based on a random sample of n observations on X, find the Bayes estimate of (a) θ and (b) $1/\theta$ with respect to the squared error loss function.

11.120. In Problem 11.29, find the Bayes estimate of (a) θ and (b) $\theta(1 - \theta)$ with squared error loss function.

11.121. In Problem 11.33, find the Bayes estimate of θ with squared error loss.

11.122. In Problem 11.26, find the Bayes estimate of θ with squared error loss.

11.123. In Problem 11.6, part (a), find the Bayes estimate with squared error loss for the variance of the population, $n\theta(1 - \theta)$.

Bayesian interval estimation

11.124. Ten Bernoulli trials with probability of success θ result in five successes. θ has the prior density given by $\pi(\theta) = \dfrac{1}{\theta(1-\theta)}$, $0 < \theta < 1$. Find the 90% Bayes equal tail area credibility interval for θ.

11.125. A random sample of size 10 drawn from a geometric distribution with success probability θ yields a mean of 5. The prior density of θ is uniform in the interval [0, 1]. Find the 88% equal tail area credibility interval for θ.

11.126. In Problem 11.30, find the 85% Bayesian equal tail area credibility interval for θ.

11.127. In Problem 11.119, suppose that the prior density of θ is gamma with parameters $\alpha = 0.2$ and $\beta = 0.15$. A random sample of 10 observations on X yielded an average life-length of seven years. Find the 85% equal tail area Bayes credibility interval for (a) θ and (b) $1/\theta$.

Bayesian tests of hypotheses

11.128. In Problem 11.21, suppose a sample of size 10 yielded the values 2, 0, 1, 1, 3, 0, 2, 4, 2, 2. Test $H_0: \lambda \leq 1$ against $H_1: \lambda > 1$ using a Bayes 0.05 test.

11.129. In Problem 11.6, assume that $n = 50$ and $x = 14$ and test the null hypothesis $H_0: \theta \leq 0.2$ against the alternative $H_1: \theta > 0.2$ using a Bayes 0.025 test.

11.130. Suppose that in Example 11.18 a second sample of 100 observations yielded a mean reaction time of 0.35 sec. Test the null hypothesis $H_0: \theta \leq 0.3$ against the alternative $H_1: \theta > 0.3$ using the Bayes 0.05 test.

Bayes factor

11.131. It is desired to test the null hypothesis $\theta \leq 0.6$ against the alternative $\theta > 0.6$, where θ is the probability of success for a Bernoulli trial. Assume that θ has a uniform prior distribution on [0, 1] and that in 30 trials there were 17 successes. What is your conclusion if you decide to reject the null hypothesis if $BF < 1$?

11.132. The time (in minutes) that a bank customer has to wait in line to be served is exponentially distributed with mean $1/\theta$. The prior distribution of θ is gamma with parameters $\alpha = 0.2$ and $\beta = 3$. A random sample of 10 customers waited an average of 3 minutes. Test the null hypothesis $H_0: \theta \geq 0.7$ against $H_1: \theta < 0.7$ using the Bayes factor rule to reject the null hypothesis if $BF < 1$.

Bayesian predictive distributions

11.133. In Problem 11.13, find the predictive distribution and predictive mean of the number of accidents during the last six months of the year.

11.134. Suppose that 4 successes were obtained in 10 Bernoulli trials with success probability θ. An independent set of 5 more Bernoulli trials with the same success probability is being contemplated. Find the predictive distribution of the number of future successes. Assume a prior uniform density for θ.

11.135. The number of accidents per month on a particular stretch of a highway is known to follow the Poisson distribution with parameter λ. A total of 24 accidents occurred on that stretch during the past 10 months. What are the chances that there would be fewer than four accidents there next month? Assume Jeffreys' prior for λ: $\pi(\lambda) = 1/\sqrt{\lambda}, \lambda > 0$.

11.136. Suppose that all 10 out of 10 Bernoulli trials were successes. What are the chances that all five out of five future Bernoulli trials would be successes? Assume a uniform prior density for the probability of success.

11.137. A sample of size 20 from a normal population with unknown mean θ and variance 4 yields a sample mean of 37.5. The prior distribution of θ is normal with mean 30 and variance 3. Suppose that an independent observation from the same population is subsequently made. Find the predictive probability that this observation would not exceed 37.5.

ANSWERS TO SUPPLEMENTARY PROBLEMS

11.102. (a) subjective; (b) classical; (c) frequency; (d) insufficient information: an equally convincing case could be made for this being a classical, frequency, or subjective probability; (e) subjective

11.103. 2/7 **11.104.** The posterior distribution is given in Table 11-14.

Table 11-14

λ	0.5	1.0	1.5
$\pi(\lambda \mid x)$	0.42	0.41	0.17

11.105. $\pi(\theta|x) = \dfrac{1}{B(x+4, n-x+1)}\theta^{x+3}(1-\theta)^{n-x}, 0 < \theta < 1$

11.106. The posterior density is beta with parameters 3 and 9.

11.107. The posterior density of θ is beta with $\alpha = 33$ and $\beta = 19$.

11.108. The posterior density of θ is beta with $\alpha = 68$ and $\beta = 34$.

11.109. (a) The posterior density is gamma with parameters $n\bar{x} + \alpha = 14 + 2 = 16$ and $\beta/(1 + n\beta) = 5/6 \approx 0.83$; (b) the posterior mean $= 80/6 \approx 13.33$; (c) the posterior variance $= 400/36 \approx 11.11$

11.110. The posterior density is gamma with parameters $n\bar{x} + \alpha = 42 + 4 = 46$ and $\beta/(1 + n\beta) = 2/21 \approx 0.10$.

11.111. (a) The posterior mean of θ is $\left(\dfrac{16}{16+9}\right)\bar{x} = \dfrac{16\bar{x}}{25}$; (b) the precision is roughly 4.34;
(c) the precision is about 1.78

11.112. The posterior density is normal with mean 1.98 and variance 0.0043.

11.113. The posterior density of λ is gamma with parameters 35.5 and 0.1 (see Problem 11.25).

11.114. The posterior density is inverse gamma with $\alpha = 2$ and $\beta = 16.875$ (see Problem 11.26).

11.115. The posterior density is beta with parameters 21 and 81 (see Example 11.12).

11.116. The posterior density is gamma with parameters 11 and 0.02.

11.117. The posterior density is inverse gamma with parameters 3 and 7.75.

11.118. (a) 0.11; (b) 0.11 **11.119.** (a) $E(\theta|x) = \dfrac{\beta(\alpha + n)}{1 + n\beta\bar{x}}$; (b) $E\left(\dfrac{1}{\theta}|x\right) = \dfrac{1 + n\beta\bar{x}}{\beta(n + \alpha - 1)}$

11.120. (a) $\dfrac{\alpha + nr}{\alpha + \beta + n\bar{x}}$; (b) $\dfrac{(\alpha + nr)(\beta + n\bar{x} - nr)}{(\alpha + \beta + n\bar{x} + 1)(\alpha + \beta + n\bar{x})}$ **11.121.** $\left(\dfrac{\sum x^2}{2} + \beta\right)\Big/\left(\dfrac{n}{2} + \alpha - 1\right)$

11.122. $\dfrac{\sum x^2}{n - 3}$ **11.123.** $n \times \dfrac{B(x + 3, n - x + 2)}{B(x + 2, n - x + 1)} = \dfrac{n(x + 2)(n - x + 1)}{(n + 4)(n + 3)}$

11.124. [0.25, 0.75] **11.125.** [0.13, 0.30] **11.126.** [0.10, 0.28] **11.127.** (a) [0.078, 0.196]; (b) [5.10, 12.82]

11.128. The posterior probability of the null hypothesis is 0.02. Since this is less than 0.05, we reject the null hypothesis.

11.129. (a) The posterior probability of the null hypothesis is 0.04. Since this is not less than 0.025, we cannot reject the null hypothesis.

(b) The posterior probability of the null hypothesis is 0.026. Since this is not less than 0.025, we cannot reject the null hypothesis.

(c) The posterior probability of the null hypothesis is 0.015. Since this is less than 0.025, we reject the null hypothesis.

11.130. The posterior probability of the null hypothesis is 0.03. Since this is less than 0.05, we reject the null hypothesis.

11.131. The posterior odds ratio is 0.66/0.34 = 1.94 and the prior odds ratio is 6/4 = 1.5. *BF* = 1.29. We cannot reject the null hypothesis.

11.132. Reject the null hypothesis since the posterior probability of the null hypothesis is 0.033 while the prior probability of the null hypothesis is 0.216.

11.133. $f^*(y) = \binom{y + 9}{9}\left(\dfrac{6}{11}\right)^{10}\left(\dfrac{5}{11}\right)^{y}$, $y = 0, 1, 2, \ldots$. Predictive mean = 50/6.

11.134.

Table 11-15

y	0	1	2	3	4	5
$f^*(y)$	0.106	0.240	0.288	0.224	0.112	0.029

11.135. 0.764 **11.136.** 11/16 **11.137.** 0.59

Mathematical Topics

Special Sums

The following are some sums of series that arise in practice. By definition, $0! = 1$. Where the series is infinite, the range of convergence is indicated.

1. $\sum_{j=1}^{m} j = 1 + 2 + 3 + \cdots + m = \dfrac{m(m+1)}{2}$

2. $\sum_{j=1}^{m} j^2 = 1^2 + 2^2 + 3^2 + \cdots + m^2 = \dfrac{m(m+1)(2m+1)}{6}$

3. $e^x = 1 + x + \dfrac{x^2}{2!} + \dfrac{x^3}{3!} + \cdots = \sum_{j=0}^{\infty} \dfrac{x^j}{j!}$ all x

4. $\sin x = x - \dfrac{x^3}{3!} + \dfrac{x^5}{5!} - \dfrac{x^7}{7!} + \cdots = \sum_{j=0}^{\infty} \dfrac{(-1)^j x^{2j+1}}{(2j+1)!}$ all x

5. $\cos x = 1 - \dfrac{x^2}{2!} + \dfrac{x^4}{4!} - \dfrac{x^6}{6!} + \cdots = \sum_{j=0}^{\infty} \dfrac{(-1)^j x^{2j}}{(2j)!}$ all x

6. $\dfrac{1}{1-x} = 1 + x + x^2 + x^3 + \cdots = \sum_{j=0}^{\infty} x^j$ $\quad |x| < 1$

7. $\ln(1-x) = -x - \dfrac{x^2}{2} - \dfrac{x^3}{3} - \dfrac{x^4}{4} - \cdots = -\sum_{j=1}^{\infty} \dfrac{x^j}{j}$ $\quad -1 \le x < 1$

Euler's Formulas

8. $e^{i\theta} = \cos\theta + i\sin\theta, \qquad e^{-i\theta} = \cos\theta - i\sin\theta$

9. $\cos\theta = \dfrac{e^{i\theta} + e^{-i\theta}}{2}, \qquad \sin\theta = \dfrac{e^{i\theta} - e^{-i\theta}}{2i}$

The Gamma Function

The *gamma function*, denoted by $\Gamma(n)$, is defined by

$$\Gamma(n) = \int_0^{\infty} t^{n-1} e^{-t} dt \qquad n > 0 \tag{1}$$

A *recurrence formula* is given by

$$\Gamma(n+1) = n\Gamma(n) \tag{2}$$

where $\Gamma(1) = 1$. An extension of the gamma function to $n < 0$ can be obtained by the use of (2).

If n is a positive integer, then

$$\Gamma(n+1) = n! \tag{3}$$

For this reason $\Gamma(n)$ is sometimes called the *factorial function*. An important property of the gamma function is that

$$\Gamma(p)\Gamma(1-p) = \frac{\pi}{\sin p\pi} \tag{4}$$

For $p = \frac{1}{2}$, (4) gives

$$\Gamma\left(\frac{1}{2}\right) = \sqrt{\pi} \tag{5}$$

For large values of n we have *Stirling's asymptotic formula*:

$$\Gamma(n + 1) \sim \sqrt{2\pi n}\, n^n e^{-n} \tag{6}$$

where the sign $\sim$ indicates that the ratio of the two sides approaches 1 as $n \to \infty$. In particular, if n is a large positive integer, a good approximation for $n!$ is given by

$$n! \sim \sqrt{2\pi n}\, n^n e^{-n} \tag{7}$$

The Beta Function

The *beta function*, denoted by $B(m, n)$, is defined as

$$B(m, n) = \int_0^1 u^{m-1}(1 - u)^{n-1}\,du \qquad m > 0, n > 0 \tag{8}$$

It is related to the gamma function by

$$B(m,n) = \frac{\Gamma(m)\Gamma(n)}{\Gamma(m + n)} \tag{9}$$

Special Integrals

The following are some integrals which arise in probability and statistics.

10. $\displaystyle\int_0^\infty e^{-ax^2}\,dx = \frac{1}{2}\sqrt{\frac{\pi}{a}} \qquad a > 0$

11. $\displaystyle\int_0^\infty x^m e^{-ax^2}\,dx = \frac{\Gamma\left(\frac{m+1}{2}\right)}{2a^{(m+1)/2}} \qquad a > 0, m > -1$

12. $\displaystyle\int_0^\infty e^{-ax^2}\cos bx\,dx = \frac{1}{2}\sqrt{\frac{\pi}{a}}\,e^{-b^2/4a} \qquad a > 0$

13. $\displaystyle\int_0^\infty e^{-ax}\cos bx\,dx = \frac{a}{a^2 + b^2} \qquad a > 0$

14. $\displaystyle\int_0^\infty e^{-ax}\sin bx\,dx = \frac{b}{a^2 + b^2} \qquad a > 0$

15. $\displaystyle\int_0^\infty x^{p-1}e^{-ax}\,dx = \frac{\Gamma(p)}{a^p} \qquad a > 0, p > 0$

16. $\displaystyle\int_{-\infty}^\infty e^{-(ax^2+bx+c)}\,dx = \sqrt{\frac{\pi}{a}}\,e^{(b^2-4ac)/4a} \qquad a > 0$

17. $\displaystyle\int_0^\infty e^{-(ax^2+bx+c)}\,dx = \frac{1}{2}\sqrt{\frac{\pi}{a}}\,e^{(b^2-4ac)/4a}\,\mathrm{erfc}\left(\frac{b}{2\sqrt{a}}\right) \qquad a > 0$

    where

    $$\mathrm{erfc}(u) = 1 - \mathrm{erf}(u) = 1 - \frac{2}{\sqrt{\pi}}\int_0^u e^{-x^2}\,dx = \frac{2}{\sqrt{\pi}}\int_u^\infty e^{-x^2}\,dx$$

    is called the *complementary error function.*

18. $\displaystyle\int_0^\infty \frac{\cos \omega x}{x^2 + a^2}\,dx = \frac{\pi}{2a}e^{-a\omega} \qquad a > 0, \omega > 0$

19. $\displaystyle\int_0^{\pi/2} \sin^{2m-1}\theta \cos^{2n-1}\theta\,d\theta = \frac{\Gamma(m)\Gamma(n)}{2\Gamma(m + n)} \qquad m > 0, n > 0$

APPENDIX B

Ordinates y of the Standard Normal Curve at z

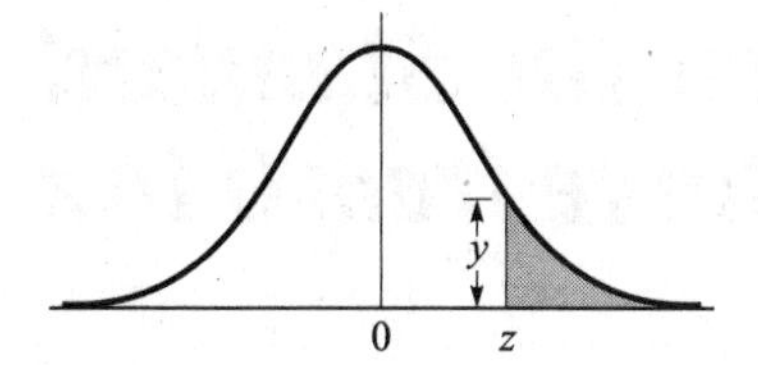

z	0	1	2	3	4	5	6	7	8	9
0.0	.3989	.3989	.3989	.3988	.3986	.3984	.3982	.3980	.3977	.3973
0.1	.3970	.3965	.3961	.3956	.3951	.3945	.3939	.3932	.3925	.3918
0.2	.3910	.3902	.3894	.3885	.3876	.3867	.3857	.3847	.3836	.3825
0.3	.3814	.3802	.3790	.3778	.3765	.3752	.3739	.3725	.3712	.3697
0.4	.3683	.3668	.3653	.3637	.3621	.3605	.3589	.3572	.3555	.3538
0.5	.3521	.3503	.3485	.3467	.3448	.3429	.3410	.3391	.3372	.3352
0.6	.3332	.3312	.3292	.3271	.3251	.3230	.3209	.3187	.3166	.3144
0.7	.3123	.3101	.3079	.3056	.3034	.3011	.2989	.2966	.2943	.2920
0.8	.2897	.2874	.2850	.2827	.2803	.2780	.2756	.2732	.2709	.2685
0.9	.2661	.2637	.2613	.2589	.2565	.2541	.2516	.2492	.2468	.2444
1.0	.2420	.2396	.2371	.2347	.2323	.2299	.2275	.2251	.2227	.2203
1.1	.2179	.2155	.2131	.2107	.2083	.2059	.2036	.2012	.1989	.1965
1.2	.1942	.1919	.1895	.1872	.1849	.1826	.1804	.1781	.1758	.1736
1.3	.1714	.1691	.1669	.1647	.1626	.1604	.1582	.1561	.1539	.1518
1.4	.1497	.1476	.1456	.1435	.1415	.1394	.1374	.1354	.1334	.1315
1.5	.1295	.1276	.1257	.1238	.1219	.1200	.1182	.1163	.1145	.1127
1.6	.1109	.1092	.1074	.1057	.1040	.1023	.1006	.0989	.0973	.0957
1.7	.0940	.0925	.0909	.0893	.0878	.0863	.0848	.0833	.0818	.0804
1.8	.0790	.0775	.0761	.0748	.0734	.0721	.0707	.0694	.0681	.0669
1.9	.0656	.0644	.0632	.0620	.0608	.0596	.0584	.0573	.0562	.0551
2.0	.0540	.0529	.0519	.0508	.0498	.0488	.0478	.0468	.0459	.0449
2.1	.0440	.0431	.0422	.0413	.0404	.0396	.0387	.0379	.0371	.0363
2.2	.0355	.0347	.0339	.0332	.0325	.0317	.0310	.0303	.0297	.0290
2.3	.0283	.0277	.0270	.0264	.0258	.0252	.0246	.0241	.0235	.0229
2.4	.0224	.0219	.0213	.0208	.0203	.0198	.0194	.0189	.0184	.0180
2.5	.0175	.0171	.0167	.0163	.0158	.0154	.0151	.0147	.0143	.0139
2.6	.0136	.0132	.0129	.0126	.0122	.0119	.0116	.0113	.0110	.0107
2.7	.0104	.0101	.0099	.0096	.0093	.0091	.0088	.0086	.0084	.0081
2.8	.0079	.0077	.0075	.0073	.0071	.0069	.0067	.0065	.0063	.0061
2.9	.0060	.0058	.0056	.0055	.0053	.0051	.0050	.0048	.0047	.0046
3.0	.0044	.0043	.0042	.0040	.0039	.0038	.0037	.0036	.0035	.0034
3.1	.0033	.0032	.0031	.0030	.0029	.0028	.0027	.0026	.0025	.0025
3.2	.0024	.0023	.0022	.0022	.0021	.0020	.0020	.0019	.0018	.0018
3.3	.0017	.0017	.0016	.0016	.0015	.0015	.0014	.0014	.0013	.0013
3.4	.0012	.0012	.0012	.0011	.0011	.0010	.0010	.0010	.0009	.0009
3.5	.0009	.0008	.0008	.0008	.0008	.0007	.0007	.0007	.0007	.0006
3.6	.0006	.0006	.0006	.0005	.0005	.0005	.0005	.0005	.0005	.0004
3.7	.0004	.0004	.0004	.0004	.0004	.0004	.0003	.0003	.0003	.0003
3.8	.0003	.0003	.0003	.0003	.0003	.0002	.0002	.0002	.0002	.0002
3.9	.0002	.0002	.0002	.0002	.0002	.0002	.0002	.0002	.0001	.0001

APPENDIX C

Areas under the Standard Normal Curve from 0 to z

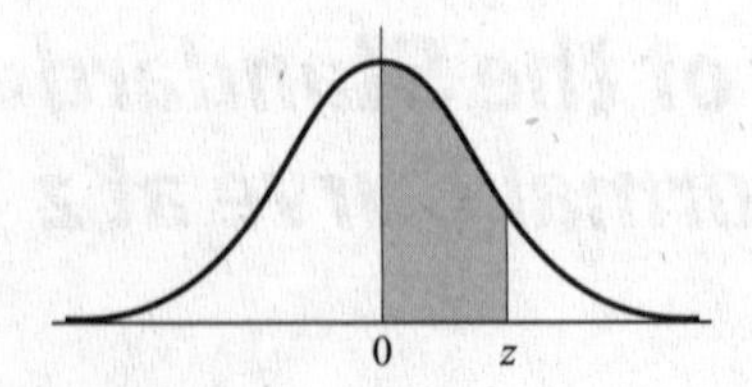

z	0	1	2	3	4	5	6	7	8	9
0.0	.0000	.0040	.0080	.0120	.0160	.0199	.0239	.0279	.0319	.0359
0.1	.0398	.0438	.0478	.0517	.0557	.0596	.0636	.0675	.0714	.0754
0.2	.0793	.0832	.0871	.0910	.0948	.0987	.1026	.1064	.1103	.1141
0.3	.1179	.1217	.1255	.1293	.1331	.1368	.1406	.1443	.1480	.1517
0.4	.1554	.1591	.1628	.1664	.1700	.1736	.1772	.1808	.1844	.1879
0.5	.1915	.1950	.1985	.2019	.2054	.2088	.2123	.2157	.2190	.2224
0.6	.2258	.2291	.2324	.2357	.2389	.2422	.2454	.2486	.2518	.2549
0.7	.2580	.2612	.2642	.2673	.2704	.2734	.2764	.2794	.2823	.2852
0.8	.2881	.2910	.2939	.2967	.2996	.3023	.3051	.3078	.3106	.3133
0.9	.3159	.3186	.3212	.3238	.3264	.3289	.3315	.3340	.3365	.3389
1.0	.3413	.3438	.3461	.3485	.3508	.3531	.3554	.3577	.3599	.3621
1.1	.3643	.3665	.3686	.3708	.3729	.3749	.3770	.3790	.3810	.3830
1.2	.3849	.3869	.3888	.3907	.3925	.3944	.3962	.3980	.3997	.4015
1.3	.4032	.4049	.4066	.4082	.4099	.4115	.4131	.4147	.4162	.4177
1.4	.4192	.4207	4222	.4236	.4251	.4265	.4279	.4292	.4306	.4319
1.5	.4332	.4345	.4357	.4370	.4382	.4394	.4406	.4418	.4429	.4441
1.6	.4452	.4463	.4474	.4484	.4495	.4505	.4515	.4525	.4535	.4545
1.7	.4554	.4564	.4573	.4582	.4591	.4599	.4608	.4616	.4625	.4633
1.8	.4641	.4649	.4656	.4664	.4671	.4678	.4686	.4693	.4699	.4706
1.9	.4713	.4719	.4726	.4732	.4738	.4744	.4750	.4756	.4761	.4767
2.0	.4772	.4778	.4783	.4788	.4793	.4798	.4803	.4808	.4812	.4817
2.1	.4821	.4826	.4830	.4834	.4838	.4842	.4846	.4850	.4854	.4857
2.2	.4861	.4864	.4868	.4871	.4875	.4878	.4881	.4884	.4887	.4890
2.3	.4893	.4896	.4898	.4901	.4904	.4906	.4909	.4911	.4913	.4916
2.4	.4918	.4920	.4922	.4925	.4927	.4929	.4931	.4932	.4934	.4936
2.5	.4938	.4940	.4941	.4943	.4945	.4946	.4948	.4949	.4951	.4952
2.6	.4953	.4955	.4956	.4957	.4959	.4960	.4961	.4962	.4963	.4964
2.7	.4965	.4966	.4967	.4968	.4969	.4970	.4971	.4972	.4973	.4974
2.8	.4974	.4975	.4976	.4977	.4977	.4978	.4979	.4979	.4980	.4981
2.9	.4981	.4982	.4982	.4983	.4984	.4984	.4985	.4985	.4986	.4986
3.0	.4987	.4987	.4987	.4988	.4988	.4989	.4989	.4989	.4990	.4990
3.1	.4990	.4991	.4991	.4991	.4992	.4992	.4992	.4992	.4993	.4993
3.2	.4993	.4993	.4994	.4994	.4994	.4994	.4994	.4995	.4995	.4995
3.3	.4995	.4995	.4995	.4996	.4996	.4996	.4996	.4996	.4996	.4997
3.4	.4997	.4997	.4997	.4997	.4997	.4997	.4997	.4997	.4997	.4998
3.5	.4998	.4998	.4998	.4998	.4998	.4998	.4998	.4998	.4998	.4998
3.6	.4998	.4998	.4999	.4999	.4999	.4999	.4999	.4999	.4999	.4999
3.7	.4999	.4999	.4999	.4999	.4999	.4999	.4999	.4999	.4999	.4999
3.8	.4999	.4999	.4999	.4999	.4999	.4999	.4999	.4999	.4999	.4999
3.9	.5000	.5000	.5000	.5000	.5000	.5000	.5000	.5000	.5000	.5000

APPENDIX D

Percentile Values t_p for Student's t Distribution with ν Degrees of Freedom

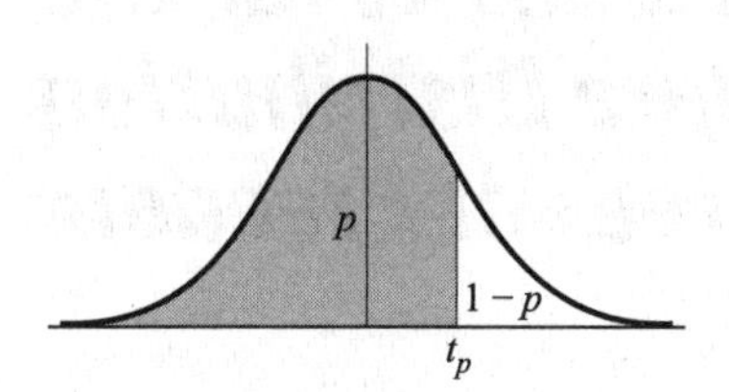

ν	$t_{.55}$	$t_{.60}$	$t_{.70}$	$t_{.75}$	$t_{.80}$	$t_{.90}$	$t_{.95}$	$t_{.975}$	$t_{.99}$	$t_{.995}$
1	.158	.325	.727	1,000	1.376	3.08	6.31	12.71	31.82	63.66
2	.142	.289	.617	.816	1.061	1.89	2.92	4.30	6.96	9.92
3	.137	.277	.584	.765	.978	1.64	2.35	3.18	4.54	5.84
4	.134	.271	.569	.741	.941	1.53	2.13	2.78	3.75	4.60
5	.132	.267	.559	.727	.920	1.48	2.02	2.57	3.36	4.03
6	.131	.265	.553	.718	.906	1.44	1.94	2.45	3.14	3.71
7	.130	.263	.549	.711	.896	1.42	1.90	2.36	3.00	3.50
8	.130	.262	.546	.706	.889	1.40	1.86	2.31	2.90	3.36
9	.129	.261	.543	.703	.883	1.38	1.83	2.26	2.82	3.25
10	.129	.260	.542	.700	.879	1.37	1.81	2.23	2.76	3.17
11	.129	.260	.540	.697	.876	1.36	1.80	2.20	2.72	3.11
12	.128	.259	.539	.695	.873	1.36	1.78	2.18	2.68	3.06
13	.128	.259	.538	.694	.870	1.35	1.77	2.16	2.65	3.01
14	.128	.258	.537	.692	.868	1.34	1.76	2.14	2.62	2.98
15	.128	.258	.536	.691	.866	1.34	1.75	2.13	2.60	2.95
16	.128	.258	.535	.690	.865	1.34	1.75	2.12	2.58	2.92
17	.128	.257	.534	.689	.863	1.33	1.74	2.11	2.57	2.90
18	.127	.257	.534	.688	.862	1.33	1.73	2.10	2.55	2.88
19	.127	.257	.533	.688	.861	1.33	1.73	2.09	2.54	2.86
20	.127	.257	.533	.687	.860	1.32	1.72	2.09	2.53	2.84
21	.127	.257	.532	.686	.859	1.32	1.72	2.08	2.52	2.83
22	.127	.256	.532	.686	.858	1.32	1.72	2.07	2.51	2.82
23	.127	.256	.532	.685	.858	1.32	1.71	2.07	2.50	2.81
24	.127	.256	.531	.685	.857	1.32	1.71	2.06	2.49	2.80
25	.127	.256	.531	.684	.856	1.32	1.71	2.06	2.48	2.79
26	.127	.256	.531	.684	.856	1.32	1.71	2.06	2.48	2.78
27	.127	.256	.531	.684	.855	1.31	1.70	2.05	2.47	2.77
28	.127	.256	.530	.683	.855	1.31	1.70	2.05	2.47	2.76
29	.127	.256	.530	.683	.854	1.31	1.70	2.04	2.46	2.76
30	.127	.256	.530	.683	.854	1.31	1.70	2.04	2.46	2.75
40	.126	.255	.529	.681	.851	1.30	1.68	2.02	2.42	2.70
60	.126	.254	.527	.679	.848	1.30	1.67	2.00	2.39	2.66
120	.126	.254	.526	.677	.845	1.29	1.66	1.98	2.36	2.62
∞	.126	.253	.524	.674	.842	1.28	1.645	1.96	2.33	2.58

Source: R. A. Fisher and F. Yates, *Statistical Tables for Biological, Agricultural and Medical Research*, published by Longman Group Ltd., London (previously published by Oliver and Boyd, Edinburgh), and by permission of the authors and publishers.

APPENDIX E

Percentile Values χ_p^2 for the Chi-Square Distribution with ν Degrees of Freedom

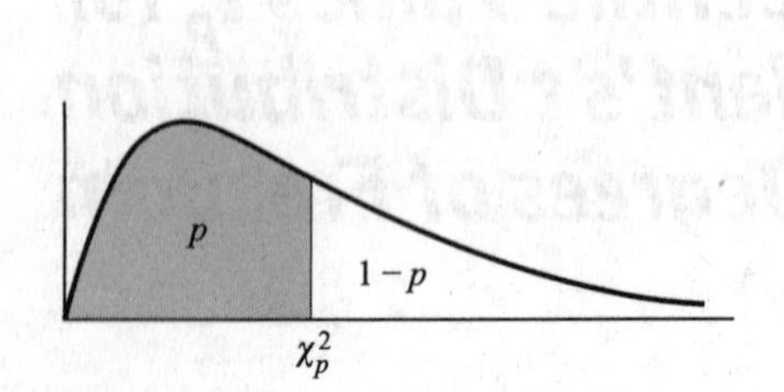

ν	$\chi^2_{.005}$	$\chi^2_{.01}$	$\chi^2_{.025}$	$\chi^2_{.05}$	$\chi^2_{.10}$	$\chi^2_{.25}$	$\chi^2_{.50}$	$\chi^2_{.75}$	$\chi^2_{.90}$	$\chi^2_{.95}$	$\chi^2_{.975}$	$\chi^2_{.99}$	$\chi^2_{.995}$	$\chi^2_{.999}$
1	.0000	.0002	.0010	.0039	.0158	.102	.455	1.32	2.71	3.84	5.02	6.63	7.88	10.8
2	.0100	.0201	.0506	.103	.211	.575	1.39	2.77	4.61	5.99	7.38	9.21	10.6	13.8
3	.0717	.115	.216	.352	.584	1.21	2.37	4.11	6.25	7.81	9.35	11.3	12.8	16.3
4	.207	.297	.484	.711	1.06	1.92	3.36	5.39	7.78	9.49	11.1	13.3	14.9	18.5
5	.412	.554	.831	1.15	1.61	2.67	4.35	6.63	9.24	11.1	12.8	15.1	16.7	20.5
6	.676	.872	1.24	1.64	2.20	3.45	5.35	7.84	10.6	12.6	14.4	16.8	18.5	22.5
7	.989	1.24	1.69	2.17	2.83	4.25	6.35	9.04	12.0	14.1	16.0	18.5	20.3	24.3
8	1.34	1.65	2.18	2.73	3.49	5.07	7.34	10.2	13.4	15.5	17.5	20.1	22.0	26.1
9	1.73	2.09	2.70	3.33	4.17	5.90	8.34	11.4	14.7	16.9	19.0	21.7	23.6	27.9
10	2.16	2.56	3.25	3.94	4.87	6.74	9.34	12.5	16.0	18.3	20.5	23.2	25.2	29.6
11	2.60	3.05	3.82	4.57	5.58	7.58	10.3	13.7	17.3	19.7	21.9	24.7	26.8	31.3
12	3.07	3.57	4.40	5.23	6.30	8.44	11.3	14.8	18.5	21.0	23.3	26.2	28.3	32.9
13	3.57	4.11	5.01	5.89	7.04	9.30	12.3	16.0	19.8	22.4	24.7	27.7	29.8	34.5
14	4.07	4.66	5.63	6.57	7.79	10.2	13.3	17.1	21.1	23.7	26.1	29.1	31.3	36.1
15	4.60	5.23	6.26	7.26	8.55	11.0	14.3	18.2	22.3	25.0	27.5	30.6	32.8	37.7
16	5.14	5.81	6.91	7.96	9.31	11.9	15.3	19.4	23.5	26.3	28.8	32.0	34.3	39.3
17	5.70	6.41	7.56	8.67	10.1	12.8	16.3	20.5	24.8	27.6	30.2	33.4	35.7	40.8
18	6.26	7.01	8.23	9.39	10.9	13.7	17.3	21.6	26.0	28.9	31.5	34.8	37.2	42.3
19	6.84	7.63	8.91	10.1	11.7	14.6	18.3	22.7	27.2	30.1	32.9	36.2	38.6	43.8
20	7.43	8.26	9.59	10.9	12.4	15.5	19.3	23.8	28.4	31.4	34.2	37.6	40.0	45.3
21	8.03	8.90	10.3	11.6	13.2	16.3	20.3	24.9	29.6	32.7	35.5	38.9	41.4	46.8
22	8.64	9.54	11.0	12.3	14.0	17.2	21.3	26.0	30.8	33.9	36.8	40.3	42.8	48.3
23	9.26	10.2	11.7	13.1	14.8	18.1	22.3	27.1	32.0	35.2	38.1	41.6	44.2	49.7
24	9.89	10.9	12.4	13.8	15.7	19.0	23.3	28.2	33.2	36.4	39.4	43.0	45.6	51.2
25	10.5	11.5	13.1	14.6	16.5	19.9	24.3	29.3	34.4	37.7	40.6	44.3	46.9	52.6
26	11.2	12.2	13.8	15.4	17.3	20.8	25.3	30.4	35.6	38.9	41.9	45.6	48.3	54.1
27	11.8	12.9	14.6	16.2	18.1	21.7	26.3	31.5	36.7	40.1	43.2	47.0	49.6	55.5
28	12.5	13.6	15.3	16.9	18.9	22.7	27.3	32.6	37.9	41.3	44.5	48.3	51.0	56.9
29	13.1	14.3	16.0	17.7	19.8	23.6	28.3	33.7	39.1	42.6	45.7	49.6	52.3	58.3
30	13.8	15.0	16.8	18.5	20.6	24.5	29.3	34.8	40.3	43.8	47.0	50.9	53.7	59.7
40	20.7	22.2	24.4	26.5	29.1	33.7	39.3	45.6	51.8	55.8	59.3	63.7	66.8	73.4
50	28.0	29.7	32.4	34.8	37.7	42.9	49.3	56.3	63.2	67.5	71.4	76.2	79.5	86.7
60	35.5	37.5	40.5	43.2	46.5	52.3	59.3	67.0	74.4	79.1	83.3	88.4	92.0	99.6
70	43.3	45.4	48.8	51.7	55.3	61.7	69.3	77.6	85.5	90.5	95.0	100	104	112
80	51.2	53.5	57.2	60.4	64.3	71.1	79.3	88.1	96.6	102	107	112	116	125
90	59.2	61.8	65.6	69.1	73.3	80.6	89.3	98.6	108	113	118	124	128	137
100	67.3	70.1	74.2	77.9	82.4	90.1	99.3	109	118	124	130	136	140	149

Source: E. S. Pearson and H. O. Hartley, *Biometrika Tables for Statisticians*, Vol. 1 (1966), Table 8, pages 137 and 138, by permission.

95th Percentile Values (0.05 Levels), $F_{0.95}$, for the F Distribution

0.95

0.05

$F_{0.95}$

ν_1 degrees of freedom in numerator
ν_2 degrees of freedom in denominator

ν_2 \ ν_1	1	2	3	4	5	6	7	8	9	10	12	15	20	24	30	40	60	120	∞
1	161	200	216	225	230	234	237	239	241	242	244	246	248	249	250	251	252	253	254
2	18.5	19.0	19.2	19.2	19.3	19.3	19.4	19.4	19.4	19.4	19.4	19.4	19.4	19.5	19.5	19.5	19.5	19.5	19.5
3	10.1	9.55	9.28	9.12	9.01	8.94	8.89	8.85	8.81	8.79	8.74	8.70	8.66	8.64	8.62	8.59	8.57	8.55	8.53
4	7.71	6.94	6.59	6.39	6.26	6.16	6.09	6.04	6.00	5.96	5.91	5.86	5.80	5.77	5.75	5.72	5.69	5.66	5.63
5	6.61	5.79	5.41	5.19	5.05	4.95	4.88	4.82	4.77	4.74	4.68	4.62	4.56	4.53	4.50	4.46	4.43	4.40	4.37
6	5.99	5.14	4.76	4.53	4.39	4.28	4.21	4.15	4.10	4.06	4.00	3.94	3.87	3.84	3.81	3.77	3.74	3.70	3.67
7	5.59	4.74	4.35	4.12	3.97	3.87	3.79	3.73	3.68	3.64	3.57	3.51	3.44	3.41	3.38	3.34	3.30	3.27	3.23
8	5.32	4.46	4.07	3.84	3.69	3.58	3.50	3.44	3.39	3.35	3.28	3.22	3.15	3.12	3.08	3.04	3.01	2.97	2.93
9	5.12	4.26	3.86	3.63	3.48	3.37	3.29	3.23	3.18	3.14	3.07	3.01	2.94	2.90	2.86	2.83	2.79	2.75	2.71
10	4.96	4.10	3.71	3.48	3.33	3.22	3.14	3.07	3.02	2.98	2.91	2.85	2.77	2.74	2.70	2.66	2.62	2.58	2.54
11	4.84	3.98	3.59	3.36	3.20	3.09	3.01	2.95	2.90	2.85	2.79	2.72	2.65	2.61	2.57	2.53	2.49	2.45	2.40
12	4.75	3.89	3.49	3.26	3.11	3.00	2.91	2.85	2.80	2.75	2.69	2.62	2.54	2.51	2.47	2.43	2.38	2.34	2.30
13	4.67	3.81	3.41	3.18	3.03	2.92	2.83	2.77	2.71	2.67	2.60	2.53	2.46	2.42	2.38	2.34	2.30	2.25	2.21
14	4.60	3.74	3.34	3.11	2.96	2.85	2.76	2.70	2.65	2.60	2.53	2.46	2.39	2.35	2.31	2.27	2.22	2.18	2.13
15	4.54	3.68	3.29	3.06	2.90	2.79	2.71	2.64	2.59	2.54	2.48	2.40	2.33	2.29	2.25	2.20	2.16	2.11	2.07
16	4.49	3.63	3.24	3.01	2.85	2.74	2.66	2.59	2.54	2.49	2.42	2.35	2.28	2.24	2.19	2.15	2.11	2.06	2.01
17	4.45	3.59	3.20	2.96	2.81	2.70	2.61	2.55	2.49	2.45	2.38	2.31	2.23	2.19	2.15	2.10	2.06	2.01	1.96
18	4.41	3.55	3.16	2.93	2.77	2.66	2.58	2.51	2.46	2.41	2.34	2.27	2.19	2.15	2.11	2.06	2.02	1.97	1.92
19	4.38	3.52	3.13	2.90	2.74	2.63	2.54	2.48	2.42	2.38	2.31	2.23	2.16	2.11	2.07	2.03	1.98	1.93	1.88
20	4.35	3.49	3.10	2.87	2.71	2.60	2.51	2.45	2.39	2.35	2.28	2.20	2.12	2.08	2.04	1.99	1.95	1.90	1.84
21	4.32	3.47	3.07	2.84	2.68	2.57	2.49	2.42	2.37	2.32	2.25	2.18	2.10	2.05	2.01	1.96	1.92	1.87	1.81
22	4.30	3.44	3.05	2.82	2.66	2.55	2.46	2.40	2.34	2.30	2.23	2.15	2.07	2.03	1.98	1.94	1.89	1.84	1.78
23	4.28	3.42	3.03	2.80	2.64	2.53	2.44	2.37	2.32	2.27	2.20	2.13	2.05	2.01	1.96	1.91	1.86	1.81	1.76
24	4.26	3.40	3.01	2.78	2.62	2.51	2.42	2.36	2.30	2.25	2.18	2.11	2.03	1.98	1.94	1.89	1.84	1.79	1.73
25	4.24	3.39	2.99	2.76	2.60	2.49	2.40	2.34	2.28	2.24	2.16	2.09	2.01	1.96	1.92	1.87	1.82	1.77	1.71
26	4.23	3.37	2.98	2.74	2.59	2.47	2.39	2.32	2.27	2.22	2.15	2.07	1.99	1.95	1.90	1.85	1.80	1.75	1.69
27	4.21	3.35	2.96	2.73	2.57	2.46	2.37	2.31	2.25	2.20	2.13	2.06	1.97	1.93	1.88	1.84	1.79	1.73	1.67
28	4.20	3.34	2.95	2.71	2.56	2.45	2.36	2.29	2.24	2.19	2.12	2.04	1.96	1.91	1.87	1.82	1.77	1.71	1.65
29	4.18	3.33	2.93	2.70	2.55	2.43	2.35	2.28	2.22	2.18	2.10	2.03	1.94	1.90	1.85	1.81	1.75	1.70	1.64
30	4.17	3.32	2.92	2.69	2.53	2.42	2.33	2.27	2.21	2.16	2.09	2.01	1.93	1.89	1.84	1.79	1.74	1.68	1.62
40	4.08	3.23	2.84	2.61	2.45	2.34	2.25	2.18	2.12	2.08	2.00	1.92	1.84	1.79	1.74	1.69	1.64	1.58	1.51
60	4.00	3.15	2.76	2.53	2.37	2.25	2.17	2.10	2.04	1.99	1.92	1.84	1.75	1.70	1.65	1.59	1.53	1.47	1.39
120	3.92	3.07	2.68	2.45	2.29	2.18	2.09	2.02	1.96	1.91	1.83	1.75	1.66	1.61	1.55	1.50	1.43	1.35	1.25
∞	3.84	3.00	2.60	2.37	2.21	2.10	2.01	1.94	1.88	1.83	1.75	1.67	1.57	1.52	1.46	1.39	1.32	1.22	1.00

Source: E. S. Pearson and H. O. Hartley, *Biometrika Tables for Statisticians*, Vol. 2 (1972), Table 5, page 178, by permission.

99th Percentile Values (0.01 Levels), $F_{0.99}$, for the F Distribution

0.99 0.01 $F_{0.99}$

ν_1 degrees of freedom in numerator
ν_2 degrees of freedom in denominator

ν_2 \ ν_1	1	2	3	4	5	6	7	8	9	10	12	15	20	24	30	40	60	120	∞
1	4052	5000	5403	5625	5764	5859	5928	5981	6023	6056	6106	6157	6209	6235	6261	6287	6313	6339	6366
2	98.5	99.0	99.2	99.2	99.3	99.3	99.4	99.4	99.4	99.4	99.4	99.4	99.4	99.5	99.5	99.5	99.5	99.5	99.5
3	34.1	30.8	29.5	28.7	28.2	27.9	27.7	27.5	27.3	27.2	27.1	26.9	26.7	26.6	26.5	26.4	26.3	26.2	26.1
4	21.2	18.0	16.7	16.0	15.5	15.2	15.0	14.8	14.7	14.5	14.4	14.2	14.0	13.9	13.8	13.7	13.7	13.6	13.5
5	16.3	13.3	12.1	11.4	11.0	10.7	10.5	10.3	10.2	10.1	9.89	9.72	9.55	9.47	9.38	9.29	9.20	9.11	9.02
6	13.7	10.9	9.78	9.15	8.75	8.47	8.26	8.10	7.98	7.87	7.72	7.56	7.40	7.31	7.23	7.14	7.06	6.97	6.88
7	12.2	9.55	8.45	7.85	7.46	7.19	6.99	6.84	6.72	6.62	6.47	6.31	6.16	6.07	5.99	5.91	5.82	5.74	5.65
8	11.3	8.65	7.59	7.01	6.63	6.37	6.18	6.03	5.91	5.81	5.67	5.52	5.36	5.28	5.20	5.12	5.03	4.95	4.86
9	10.6	8.02	6.99	6.42	6.06	5.80	5.61	5.47	5.35	5.26	5.11	4.96	4.81	4.73	4.65	4.57	4.48	4.40	4.31
10	10.0	7.56	6.55	5.99	5.64	5.39	5.20	5.06	4.94	4.85	4.71	4.56	4.41	4.33	4.25	4.17	4.08	4.00	3.91
11	9.65	7.21	6.22	5.67	5.32	5.07	4.89	4.74	4.63	4.54	4.40	4.25	4.10	4.02	3.94	3.86	3.78	3.69	3.60
12	9.33	6.93	5.95	5.41	5.06	4.82	4.64	4.50	4.39	4.30	4.16	4.01	3.86	3.78	3.70	3.62	3.54	3.45	3.36
13	9.07	6.70	5.74	5.21	4.86	4.62	4.44	4.30	4.19	4.10	3.96	3.82	3.66	3.59	3.51	3.43	3.34	3.25	3.17
14	8.86	6.51	5.56	5.04	4.70	4.46	4.28	4.14	4.03	3.94	3.80	3.66	3.51	3.43	3.35	3.27	3.18	3.09	3.00
15	8.68	6.36	5.42	4.89	4.56	4.32	4.14	4.00	3.89	3.80	3.67	3.52	3.37	3.29	3.21	3.13	3.05	2.96	2.87
16	8.53	6.23	5.29	4.77	4.44	4.20	4.03	3.89	3.78	3.69	3.55	3.41	3.26	3.18	3.10	3.02	2.93	2.84	2.75
17	8.40	6.11	5.19	4.67	4.34	4.10	3.93	3.79	3.68	3.59	3.46	3.31	3.16	3.08	3.00	2.92	2.83	2.75	2.65
18	8.29	6.01	5.09	4.58	4.25	4.01	3.84	3.71	3.60	3.51	3.37	3.23	3.08	3.00	2.92	2.84	2.75	2.66	2.57
19	8.18	5.93	5.01	4.50	4.17	3.94	3.77	3.63	3.52	3.43	3.30	3.15	3.00	2.92	2.84	2.76	2.67	2.58	2.49
20	8.10	5.85	4.94	4.43	4.10	3.87	3.70	3.56	3.46	3.37	3.23	3.09	2.94	2.86	2.78	2.69	2.61	2.52	2.42
21	8.02	5.78	4.87	4.37	4.04	3.81	3.64	3.51	3.40	3.31	3.17	3.03	2.88	2.80	2.72	2.64	2.55	2.46	2.36
22	7.95	5.72	4.82	4.31	3.99	3.76	3.59	3.45	3.35	3.26	3.12	2.98	2.83	2.75	2.67	2.58	2.50	2.40	2.31
23	7.88	5.66	4.76	4.26	3.94	3.71	3.54	3.41	3.30	3.21	3.07	2.93	2.78	2.70	2.62	2.54	2.45	2.35	2.26
24	7.82	5.61	4.72	4.22	3.90	3.67	3.50	3.36	3.26	3.17	3.03	2.89	2.74	2.66	2.58	2.49	2.40	2.31	2.21
25	7.77	5.57	4.68	4.18	3.86	3.63	3.46	3.32	3.22	3.13	2.99	2.85	2.70	2.62	2.54	2.45	2.36	2.27	2.17
26	7.72	5.53	4.64	4.14	3.82	3.59	3.42	3.29	3.18	3.09	2.96	2.82	2.66	2.58	2.50	2.42	2.33	2.23	2.13
27	7.68	5.49	4.60	4.11	3.78	3.56	3.39	3.26	3.15	3.06	2.93	2.78	2.63	2.55	2.47	2.38	2.29	2.20	2.10
28	7.64	5.45	4.57	4.07	3.75	3.53	3.36	3.23	3.12	3.03	2.90	2.75	2.60	2.52	2.44	2.35	2.26	2.17	2.06
29	7.60	5.42	4.54	4.04	3.73	3.50	3.33	3.20	3.09	3.00	2.87	2.73	2.57	2.49	2.41	2.33	2.23	2.14	2.03
30	7.56	5.39	4.51	4.02	3.70	3.47	3.30	3.17	3.07	2.98	2.84	2.70	2.55	2.47	2.39	2.30	2.21	2.11	2.01
40	7.31	5.18	4.31	3.83	3.51	3.29	3.12	2.99	2.89	2.80	2.66	2.52	2.37	2.29	2.20	2.11	2.02	1.92	1.80
60	7.08	4.98	4.13	3.65	3.34	3.12	2.95	2.82	2.72	2.63	2.50	2.35	2.20	2.12	2.03	1.94	1.84	1.73	1.60
120	6.85	4.79	3.95	3.48	3.17	2.96	2.79	2.66	2.56	2.47	2.34	2.19	2.03	1.95	1.86	1.76	1.66	1.53	1.38
∞	6.63	4.61	3.78	3.32	3.02	2.80	2.64	2.51	2.41	2.32	2.18	2.04	1.88	1.79	1.70	1.59	1.47	1.32	1.00

Source: E. S. Pearson and H. O. Hartley, *Biometrika Tables for Statisticians*, Vol. 2 (1972), Table 5, page 180, by permission.

APPENDIX G AND H

Values of $e^{-\lambda}$

$(0 < \lambda < 1)$

λ	0	1	2	3	4	5	6	7	8	9
0.0	1.0000	.9900	.9802	.9704	.9608	.9512	.9418	.9324	.9231	.9139
0.1	.9048	.8958	.8869	.8781	.8694	.8607	.8521	.8437	.8353	.8270
0.2	.8187	.8106	.8025	.7945	.7866	.7788	.7711	.7634	.7558	.7483
0.3	.7408	.7334	.7261	.7189	.7118	.7047	.6977	.6907	.6839	.6771
0.4	.6703	.6636	.6570	.6505	.6440	.6376	.6313	.6250	.6188	.6126
0.5	.6065	.6005	.5945	.5886	.5827	.5770	.5712	.5655	.5599	.5543
0.6	.5488	.5434	.5379	.5326	.5273	.5220	.5169	.5117	.5066	.5016
0.7	.4966	.4916	.4868	.4819	.4771	.4724	.4677	.4630	.4584	.4538
0.8	.4493	.4449	.4404	.4360	.4317	.4274	.4232	.4190	.4148	.4107
0.9	.4066	.4025	.3985	.3946	.3906	.3867	.3829	.3791	.3753	.3716

$(\lambda = 1, 2, 3, \ldots, 10)$

λ	1	2	3	4	5	6	7	8	9	10
$e^{-\lambda}$	.36788	.13534	.04979	.01832	.006738	.002479	.000912	.000335	.000123	.000045

NOTE: To obtain values of $e^{-\lambda}$ for other values of λ, use the laws of exponents.

Example: $e^{-3.48} = (e^{-3.00})(e^{-0.48}) = (.04979)(.6188) = .03081.$

Random Numbers

51772	74640	42331	29044	46621	62898	93582	04186	19640	87056
24033	23491	83587	06568	21960	21387	76105	10863	97453	90581
45939	60173	52078	25424	11645	55870	56974	37428	93507	94271
30586	02133	75797	45406	31041	86707	12973	17169	88116	42187
03585	79353	81938	82322	96799	85659	36081	50884	14070	74950
64937	03355	95863	20790	65304	55189	00745	65253	11822	15804
15630	64759	51135	98527	62586	41889	25439	88036	24034	67283
09448	56301	57683	30277	94623	85418	68829	06652	41982	49159
21631	91157	77331	60710	52290	16835	48653	71590	16159	14676
91097	17480	29414	06829	87843	28195	27279	47152	35683	47280

Subject Index

Index for Solved Problems